Lecture Notes in Mechanical Engineering

Lecture Notes in Mechanical Engineering (LNME) publishes the latest developments in Mechanical Engineering—quickly, informally and with high quality. Original research reported in proceedings and post-proceedings represents the core of LNME. Volumes published in LNME embrace all aspects, subfields and new challenges of mechanical engineering. Topics in the series include:

- Engineering Design
- Machinery and Machine Elements
- Mechanical Structures and Stress Analysis
- Automotive Engineering
- Engine Technology
- Aerospace Technology and Astronautics
- Nanotechnology and Microengineering
- Control, Robotics, Mechatronics
- MEMS
- Theoretical and Applied Mechanics
- Dynamical Systems, Control
- Fluid Mechanics
- Engineering Thermodynamics, Heat and Mass Transfer
- Manufacturing
- Precision Engineering, Instrumentation, Measurement
- Materials Engineering
- Tribology and Surface Technology

To submit a proposal or request further information, please contact the Springer Editor of your location:

China: Ms. Ella Zhang at ella.zhang@springer.com
India: Priya Vyas at priya.vyas@springer.com
Rest of Asia, Australia, New Zealand: Swati Meherishi at swati.meherishi@springer.com
All other countries: Dr. Leontina Di Cecco at Leontina.dicecco@springer.com

To submit a proposal for a monograph, please check our Springer Tracts in Mechanical Engineering at https://link.springer.com/bookseries/11693 or contact Leontina.dicecco@springer.com

Indexed by SCOPUS. All books published in the series are submitted for consideration in Web of Science.

More information about this series at https://link.springer.com/bookseries/11236

Ramin Karim · Alireza Ahmadi ·
Iman Soleimanmeigouni ·
Ravdeep Kour · Raj Rao
Editors

International Congress and Workshop on Industrial AI 2021

Springer

Editors
Ramin Karim
Division of Operation
and Maintenance Engineering
Luleå University of Technology
Luleå, Sweden

Alireza Ahmadi
Division of Operation
and Maintenance Engineering
Luleå University of Technology
Luleå, Sweden

Iman Soleimanmeigouni
Division of Operation
and Maintenance Engineering
Luleå University of Technology
Luleå, Sweden

Ravdeep Kour
Division of Operation
and Maintenance Engineering
Luleå University of Technology
Luleå, Sweden

Raj Rao
COMADEM International
Birmingham, UK

ISSN 2195-4356 ISSN 2195-4364 (electronic)
Lecture Notes in Mechanical Engineering
ISBN 978-3-030-93638-9 ISBN 978-3-030-93639-6 (eBook)
https://doi.org/10.1007/978-3-030-93639-6

This Springer imprint is published by the registered company Springer Nature Switzerland AG
The registered company address is: Gewerbestrasse 11, 6330 Cham, Switzerland

Preface

Industrial artificial intelligence (IAI) is an umbrella term for a set of technologies adapted to industrial contexts, such as business, operation, maintenance, and asset management. IAI materialized through the concept of eMaintenance that aims to accelerate transformation of industry. IAI-empowered eMaintenance enables fact-based decision-making through an enhanced analytics approach based on a combination of data-driven and model-driven approaches, also called hybrid analytic approach. The benefits of IAI-empowered eMaintenance are reflected in organizations' overall performance efficiency and effectiveness.

After successful conduct of the five acts of the International Workshop and Congress on Industrial AI and eMaintenance since 2010, time has arrived to conduct the sixth act. Originally, this act was planned for May 2020, but due to the global pandemic (associated with COVID-19), the event was postponed to 6–7 October 2021, in an online conference. The ambition of the current act has been to bring insight into the potential opportunities and challenges through implementation of AI in industries apart from the future developments with special reference to operation and maintenance of industrial assets.

We are pleased to announce that the sixth act of the congress has been organized in close collaboration with International Journal of Condition Monitoring and Diagnostic Engineering Management (COMADEM).

We have received excellent support from both industry and academia in terms of number of technical papers and number of participants as well. The eMaintenance Workshop and the Congress is planned to provide a regular platform every year to initiate discussion amongst various partners to provide directions for effective utilization of industrial AI and eMaintenance, besides the new and emerging technologies. IAI and eMaintenance solutions are the fusion and integration of various emerging technologies and methodologies. Thus, a number of challenges and issues related to wide domain of disciplines related to eMaintenance are included and considered during the congress.

The purpose and theme of the congress is to provide a timely review of research efforts on the topic covering both theoretical and applied researches that will contribute towards understanding the strategic role of industrial AI and eMaintenance in asset management and performance of operation and maintenance of complex systems. The presentations and papers included in this proceeding cover most of the areas related and relevant to the main themes of the Congress, as mentioned below:

- AI for Fault Diagnostic
- Asset Management
- Maintenance Planning and Optimization
- Reliability, Safety and Security
- Performance Measurement
- AI and Maintenance Analytics
- Cyber Security
- Vibration and Acoustic
- Diagnostic and Prognostic.

Some of the selected papers from the proceedings will be published in International Journals, namely International Journal of Condition Monitoring and Diagnostic Engineering Management (COMADEM), Journal of Quality in Maintenance Engineering (JQME), and International Journal of System Assurance Engineering and Management (IJSAEM).

We thank all the authors for their papers and the reviewers for their reviewing support. We would also like to thank all the members of the International Advisory Committee, the Programme and Organizing committees for their active support.

Ramin Karim
Alireza Ahmadi
Raj Rao
Uday Kumar

Conference Organization

Honorary Chair

Uday Kumar	Luleå University of Technology, Luleå, Sweden

Conference Chair

Ramin Karim	Luleå University of Technology, Luleå, Sweden

Conference Co-chair

Raj Rao	COMADEM International, UK
Alireza Ahmadi	Luleå University of Technology, Luleå, Sweden

Program Committee

Diego Galar	Luleå University of Technology, Luleå, Sweden
PhillipTretten	Luleå University of Technology, Luleå, Sweden
Aditya Parida	Luleå University of Technology, Luleå, Sweden
Ian Sherrington	University of Central Lancashire
Antonio Ramos Andrade	Universidade de Lisboa
Abhinav Saxena	General Electric
Suprakash Gupta	Banaras Hindu University
Len Gelman	The University of Huddersfield
Peter Soderholm	Trafikverket
Mahmood Shafiee	Cranfield University
Tore Markeset	University of Stavanger
Javad Barabady	UIT The Arctic University of Tromsø
Mohamed Ben-Daya	American University of Sharjah
Lihui Wang	KTH Royal Institute of Technology
Parmod Kumar Kapur	AIBS, Amity University Noida

Rakesh Mishra	University of Huddersfield
Mohammad Ali Farsi	Aerospace Research Institute, Iran
Mirka Kans	Linnaeus University
Kouroush Jenab	Morehead State University, USA
Benoit Iung	Lorraine University
Berna Ulutas	Eskisehir Osmangazi University
Gopinath Chattopadhyay	Federation University
Xun Xiao	Massey University
John Andrews	University of Nottingham
Pierre Dehombreux	Université de Mons - UMONS
Anita Mirijamdotter	Linnaeus University
Md. Rezaul Karim	University of Rajshahi
Neil Eklund	Schlumberger, USA
Fernando Abrahao	ITA
David Baglee	University of Sunderland
Jyoti Sinha	University of Manchester, UK

Organizing Committee

Miguel Castano	Luleå University of Technology, Luleå, Sweden
Veronica Jägare	Luleå University of Technology, Luleå, Sweden
Adithya Thaduri	Luleå University of Technology, Luleå, Sweden
Iman Soleimanmeigouni	Luleå University of Technology, Luleå, Sweden
Ravdeep Kour	Luleå University of Technology, Luleå, Sweden
Jaya Kumari	Luleå University of Technology, Luleå, Sweden
Amit Patwardhan	Luleå University of Technology, Luleå, Sweden

Editorial Committee

Ramin Karim	Luleå University of Technology, Luleå, Sweden
Alireza Ahmadi	Luleå University of Technology, Luleå, Sweden
Raj Rao	COMADEM International, UK
Uday Kumar	Luleå University of Technology, Luleå, Sweden
Iman Soleimanmeigouni	Luleå University of Technology, Luleå, Sweden
Ravdeep Kour	Luleå University of Technology, Luleå, Sweden

Contents

Towards Reinforcement Learning Control of an Electromechanical Pinball Machine

Mirco Alpen(✉), Sven Herzig, and Joachim Horn

Helmut Schmidt University, Holstenhofweg 85, 22043 Hamburg, Germany
{mirco.alpen,sven.herzig,joachim.horn}@hsu-hh.de

Abstract. In recent years, the application of artificial intelligence in production processes is becoming increasingly important. Initially focused on the quality management process, artificial intelligence has experienced more usage in production control processes and other complex dynamic systems. In addition to learning speed, the robustness of the resulting algorithm is an important quality criterion.

This paper describes the steps towards a reinforcement learning trained Q-network that operates an electromechanical pinball machine. The aim is that the algorithm generates inputs for the system such that the resulting playing time and the associated point gain is maximized and shows a superior performance compared to a human player. This paper presents the planned learning concept and contains exemplary measurements to determine transition probabilities. Based on these results a Monte Carlo simulation should be used to train the Q-network and thus be able to get an optimal initialization for the practical realization.

Keywords: AI · Reinforcement learning · Q-learning · Discrete systems

1 Introduction

Since the beginning of the computer age, there exists a vision that machines and robots learn to gain all the skills and knowledge that humans have and that one day outdo us with their skills [4]. In the 1970s, attempts to implement artificial intelligence essentially consisted of interviewing specialists through engineers about their activities and then feeding a computer with the resulting if-then instructions [5]. The big difficulty, however, was to ask the right questions on the one hand and on the other hand to implement the answers appropriately. In the mid-1980s, the topic got a new impetus from a research volume, which led to the establishment of the topic in science. Today, several robotic and industrial applications are based on using artificial intelligence [3] and at many universities it has become an important element in studying engineering.

In 2015, Mnih and others published a paper that uses artificial intelligence to play old computer games from the 1990s [7]. The AI was trained based on the visual information on the screen using reinforcement learning. One of the games considered is a pinball simulator. The evaluations presented in the paper show that, considering a certain learning phase, the AI performs significantly better than human operators. The difference to the content presented in this paper is the implementation of a real pinball machine, which

R. Karim et al. (Eds.): IAI 2021, LNME, pp. 1–11, 2022.
https://doi.org/10.1007/978-3-030-93639-6_1

Fig. 1. Pinball machine with motion tracking cameras

has significantly higher uncertainties than a simulation. At the moment, work is also being done on playing current 3D multiplayer games based on reinforcement learning at a level comparable to human operators [2].

Playing a real digital pinball machine by a computer was tried in 2004. Lichtenberg and others [6] have evaluated the pinballs state machine to find the optimal trigger point for flipping the ball. In this case no optical information was used and the current position of the ball was unknown. This approach leads to no improvement in contrast to a human operator.

This paper represents a combination of the publications [6] and [7]. The selected application example of the electromechanical pinball machine chosen for this paper is characterized by high uncertainties and changing dynamics caused by temperature, voltage fluctuations and general wear of the ball. If a self-learning algorithm is able to mastering such a system, the results can be transferred to various other applications such as traffic control or highly variable chemical processes or in the financial sector.

The rest of the paper is structured as follows: Sect. 2 gives a system description of the used electromechanical pinball machine. The used algorithm and its practical realization is part of Sect. 3. Section 4 gives information about the references used for validation followed by the results in Sect. 5. This paper closes with a conclusion and an outlook to further work.

2 The Pinball Machine

The algorithm presented in this paper is validated on an electromechanical pinball machine. It is the '300' from Gottlieb (see Fig. 1). It was produced in 1975 and has two flippers in the lower part of the playfield. Up to four players can play in the same round. There is only one ball in the game at a time and each player has three balls available. The four displays for the points achieved can be seen on the lightbox as well as the magazine for the bonus-points which is located in the lower left corner. If a bonus is triggered in the game, this is saved by inserting a red ball into this magazine. By hitting a certain target or running out of the ball, this magazine is emptied and the corresponding amount of bonus points is counted. Further information on the pinball machine can be found in the Internet database [1].

Some modifications were necessary to use the pinball machine as a demonstrator for machine learning. These are briefly introduced in the following paragraphs.

First, three cameras were installed above the playfield. These belong to a motion capture system using the Vicon Tracker software. With the use of ultraviolet light emitted by the cameras this system is able to track a large number of specially directed reflecting marks with a frequency of 100 Hz [10]. The ball was covered with a reflective film to make it visible to this system. Thus, the position of the ball is detected 100 times per second with an accuracy of 10^{-3} m.

The second modification concerns the control devices of the pinball machine. Here, relays controlled by a microcontroller are used to actuate the right and the left flipper as well as the plunger, which has been replaced by a solenoid. Both functions can still be triggered manually to maintain the use by a human user.

The last change or extension relevant for this paper is the use of another microcontroller to count the points. This is implemented in parallel to the electromechanical counting and provides both the current score and the number of bonus events triggered.

With these changes and extensions described above and the use of a commercially available PC, it is possible to play the pinball machine using an algorithm implemented in MATLAB®.

3 Learning Algorithm

The majority of machine learning algorithms can be divided into supervised and unsupervised learning. Reinforcement learning belongs to the second group and is based on the natural learning behavior of humans. Human learning, especially in the early stages of learning, often takes place through simple exploration of the environment. Human activities in the context of the learning problem are defined by a certain scope of action. In response to its actions, the humans receive feedback from their surroundings, represented abstractly in the form of a reward or punishment. Thus, this kind of learning in combination with a Q-network is used for the application presented in this paper. The Following subsections show the used approach in detail.

3.1 Reinforcement Learning

Basically, reinforcement learning consists of five important components, the agent and the environment as well as the state s, the action a and the reward r at the current sample k [8, 9]. Figure 2 illustrates the interaction of the components. Basically, the process can be described as follows: Assuming a deterministic agent with n states $s_1, \ldots, s_n$ and m actions $a_1, \ldots, a_m$ is acting in a defined environment. At a certain sampling time k, the agent is in the state $s_{i,k}$ and selects an action $a_{j,k}$. This leads to a reaction of the environment in the form of a reward r_{k+1} and the agent ends up in another state $s_{k+1}(s_{j,k}, a_{j,k})$. The reaction of the environment to the action $a_{j,k}$ of last turn now influences the choice of the action $a_{j,k+1}$. Over numerous iterations, the agent is able to approximate a connection between its actions and the expected future rewards in each status and thus to behave accordingly optimally [9].

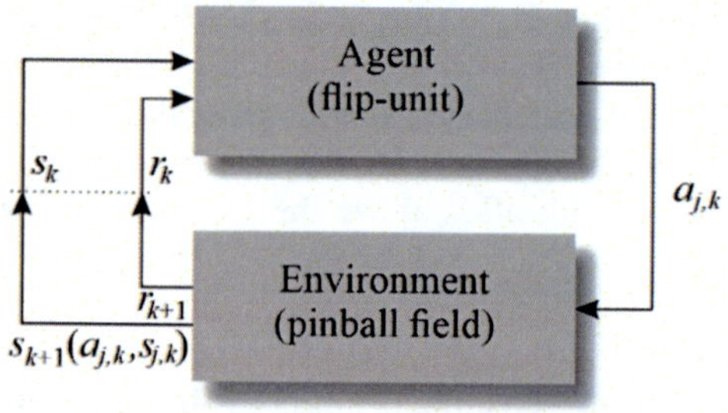

Fig. 2. Reinforcement learning concept adapted from [9]

Based on this basic function of reinforcement learning, it is necessary to define a finite number of states (initial situations) and actions. In addition, a learning objective must be defined, which is the basis for the reward system. Rewards can be both positive and negative. In combination with a learning rate α these rewards lead to certain quantities Q by moving from one state to another. These are stored in a network and will be updated if the system makes the same change of state again. Because the Q-value itself represents the long term expectation return depending on the action $a_{j,k}$ at the current state $s_{i,k}$, it is written as $Q(s_{i,k}, a_{j,k})$. Based on these quantities the algorithm choses a certain action at a certain state. The corresponding update equation for this quantities regarding the presented application will be discussed in Sect. 3.3.

3.2 Definition of State and Actions

The only way to influence the ball of a pinball machine is to vary the timing of the flipping and thus to influence the firing angle of the ball. The goal is to bring the ball up getting the highest possible score on the one hand and that it comes down again such that one has the chance to hit it again on the other. Basically, there are significantly more points and bonuses to be collected in the upper area of the playing field. In order to maximize the score per shot, the ball should enter the area above the spinning target (named as *SPIN* in Fig. 3) if possible. The two bumpers B_1 and B_2 in particular drive up the current score quickly. Due to the uncertain rebound angle and the additional acceleration of the ball, it is difficult to predict in which way it will leave the upper area and from which direction it will return to the flipper. This is the area of tension in which the algorithm has to produce good results.

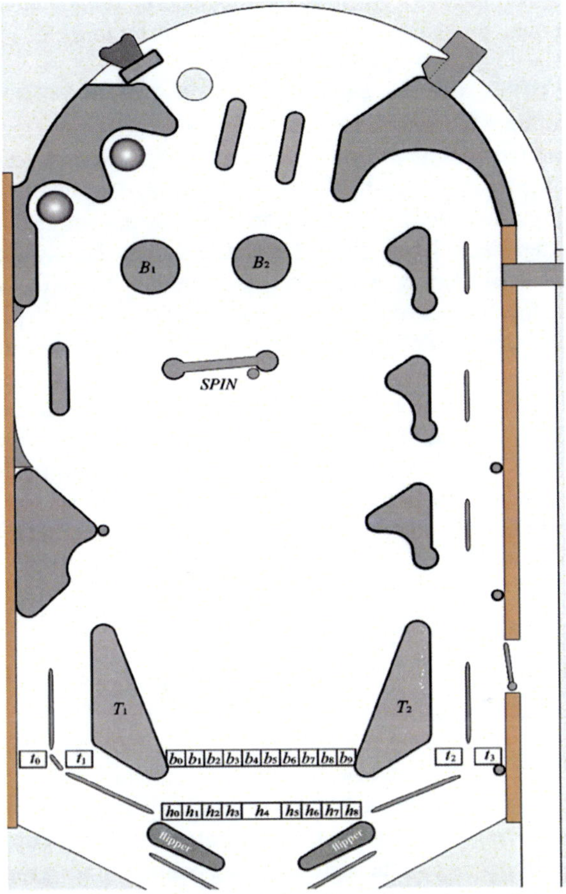

Fig. 3. Playfield with discretization for state detection

The combination of direction and speed of the ball to be flipped provides the respective states $s_{i,k}$. Figure 3 shows a schematic representation of the playing field. The directional component of the state results from the combination of the sections b0 to b9 and h0 to h8 which can be seen in the lower area. If the ball crosses the line given by b_0 to b_9, based on the current velocity and direction a prediction for the balls position at the line given by h_0 to h_8 is calculated. This prediction is necessary to avoid measurement errors due to ricochets occurring in the area of the lower targets T_1 and T_2 and to compensate the dead time of the flippers. It takes about 100 ms from triggering the flipper by software to its maximum deflection. In addition to these 90 possible directions, there are four further entry options t_0 to t_3 for the ball, which results from the two paths or traces in the left and right edge area of the lower playing field. This results in a total of 94 directions of incidence for the ball.

Various measurements have shown that the speed of the ball varies greatly in the respective direction of incidence. Therefore, a quantization based on the predicted time the ball needs between the lines given by b_0 to b_9 and h_0 to h_8 is done. Several measurements have shown, that a quantization in 12 steps is reasonable. Thus, there are all in all, $n = 94 \cdot 12 = 1128$ states are obtained.

As mentioned before, the only way to influence the ball of a pinball machine is to vary the moment of time to flip it. Therefore, the possible triggering time of the flippers results in the permitted actions a_j in the context of reinforcement learning. Based on measurements of the dynamics of the flippers and the system-related dead time, nine permissible tripping times from $f_1 = 20\,\text{ms}$ to $f_9 = 100\,\text{ms}$ with steps of 10 ms were determined. This step size is based on the fact that due to the clock speed of the motion capture system, the self-learning algorithm gets an update of the current ball position every 10 ms. Thus, the number of possible actions a_j is $m = 9$.

Based on these definitions of states and actions and the assumption that each action leads to another state, the resulting Q-network will consist of 10152 quantities Q. A dynamical model of the pinball machine is not available. Thus, all these quantities as well as the transition probabilities from one state to another have to be determined by playing. In order to reach a significant sample for each state-action pair a high number of experiments is needed. The first estimates showed a playing time of several months. This is the reason for focusing on the transition probabilities first and trying to use them for a simulation later on.

3.3 Realization of the Learning Concept

Figure 4 shows the flow chart of the designated learning concept used for the presented application. All starts with shooting up the ball. Based on its movement across the playfield, points are collected by hitting targets. If the ball then moves towards the flippers, the state s_{k+1} is identified based on the definition in Sect. 3.2.

The possible states can be divided into two groups. If the direction of incidence is such that the ball can be hit by flippers, the state belongs to the group '*in*', if it cannot be hit, it runs out of the field and the state therefore belongs to the group '*out*'.

If the state s_{k+1} belongs to the group '*in*', a corresponding update of the Q-network will be determined according to the left area of Fig. 4. A corresponding Q-value is entered at the node from the previous state $s_{i,k}$, the used action $a_{j,k}$ and the current state $s_{k+1}(s_{j,k}, a_{j,k})$. In general, the update of the Q-value is given by the assignment of Eq. (1).

$$Q(a_{j,k}, s_{i,k}) \leftarrow Q(a_{j,k}, s_{i,k}) + \alpha\left[r_{k+1} + \gamma \cdot ds - Q(a_{j,k}, s_{i,k})\right] \tag{1}$$

The learning rate $\alpha \in [0\,1]$ specifies the influence with which the outcome of the step currently being carried out has in relation to the knowledge already achieved. With

$$ds = p(s_{i+j,k}) \cdot \max_a\left(Q\left(a_{k+1}, s_{k+1}(s_{j,k}, a_{j,k})\right)\right) \tag{2}$$

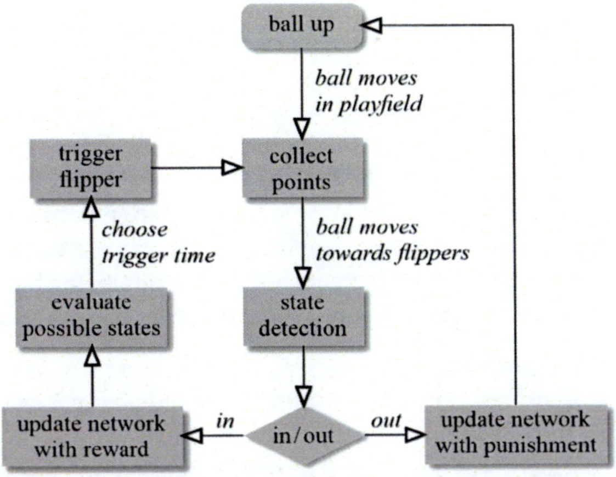

Fig. 4. Flow chart of the used learning algorithm

giving the utility of the desired state based on the Q-value of the best action one can choose. Due to the high uncertainty of the system, the associated transition probability $p(s_{i+j,k})$ has to be taken into account as well. The value calculated in Eq. (2) is taken into account in the weighting of the discount factor $\gamma \in [0\ 1]$ when updating the Q-value based on Eq. (1). The discount factor γ has influence how much one cares about getting a reward sooner rather than later. Taking a look at the square brackets in Eq. (1), it becomes clear that it will be zero as soon as the quantity already obtained matches the quantity to be expected. If this is the case, the algorithm has converged. The decision which action to choose in the current state is based on Eq. (1) as well. If the decision is made, the corresponding tripping time f_i is chosen and the ball is flipped.

If the current state $s_{i,k}$ belongs to the group '*out*' the Q-network is determined according to the right area of Fig. 4. The highest possible punishment '-1' is entered at the node from the previous state $s_{i,k}$, the used action $a_{j,k}$ and the current state $s_{k+1}(s_{j,k}, a_{j,k})$. The punishment is chosen as the maximum, because the overall goal is to keep the ball in the game. After that, the ball is shot up again by the ball launcher.

As already mentioned at the end of Sect. 3.2, it would take a long time to acquire all the quantities and thus also put a lot of strain on the old pinball table. Therefore, the learning phase just described is preceded by a step in which the focus is initially on learning the transition probabilities $p(s_{i+j,k})$. For this purpose, every possible action a_j is carried out 100 times for each possible state s_i. During this phase, the reward based only on the evaluation of the score achieved by the last shot is determined as well. Both information should then be used in order to obtain associated Q-values using a Monte Carlo simulation, which represents an initialization for the practical implementation of the learning phase later on.

4 Reference Data

This section describes two scenarios with which the performance of the learning algorithm has to be compared. On the one hand, the operation by a human and, on the other hand, automated operation, which essentially follows the rule: 'As soon as the ball can be hit, the flippers trigger'.

Human Operation
During an event at our university, about 10 different people played on the pinball machine. Among other things, the trajectory of the ball throughout the game, the triggering times of the pinball fingers and the score achieved per shot were recorded. The subsequent analysis of the data showed that in total approx. 600 flips were made in this test. On average, around 1900 points were scored per shot. Furthermore, the ball has run out 136 times. This means that the human test players lost the ball on average after 5 flips.

Automated Untrained Operation
Before the learning algorithm presented in Sect. 3, we defined two catch areas, one for the right and one for the left flipper. Whenever the ball was within one of these capture areas, the associated flipper was triggered. The only exception was when the ball entered the area indicated by t_2 or t_3 in Fig. 3. In this case, the flip time was optimized to hit the spinning target.

With this comparatively simple approach, an average of around 1500 points per flip were scored. Furthermore, the ball was flipped an average of 5.5 times before it was lost. These values are based on a data set consisting of more than 1000 flips.

The score achieved per flip are slightly lower compared to the human operator, but the ball remains in play longer. Overall, the results are therefore comparable. Because of this comparability, the points achieved on average are used as a basis for evaluating the reward required for the learning algorithm. So positive rewards are only achieved when more than 1700 points are scored in one shot. If it is less, a negative reward, i.e. a punishment, will be saved in proportion to the difference.

5 Results

As described in Sect. 3 the implementation of the learning algorithm is essentially based on the data analysis of the recorded test games by human operators and that of the automated non-learning operation. The experience of one of the authors, who owns the pinball machine for many years, has also been incorporated.

Since the generation of data for the learning algorithm is very time-consuming, the focus of this section is on the transition probabilities $p(s_{i+j,k})$. This analysis was started with the comparatively easy to handle states that result when the ball rolls onto the flippers along the paths marked t_1 to t_2 in Fig. 3. For all results shown in this section, the ball was released manually over the paths mentioned so that the resulting speed is comparable to that of a real game situation and within the chosen range of quantization.

Figure 5 and Fig. 6 show two distributions for exemplary investigated states. In this two cases the ball was defined to roll on the flipper along the path, which is labeled by t_1 in Fig. 3. As described in Sect. 3.2, the speed component for determining the state is divided into 12 increments. Releasing the ball by hand, essentially two speeds represented by the increments 6 and 7 occurred. Figure 5 is based on 100 measurements caused by the same action and therefore the same triggering time, Fig. 6 is based on 120 measurements.

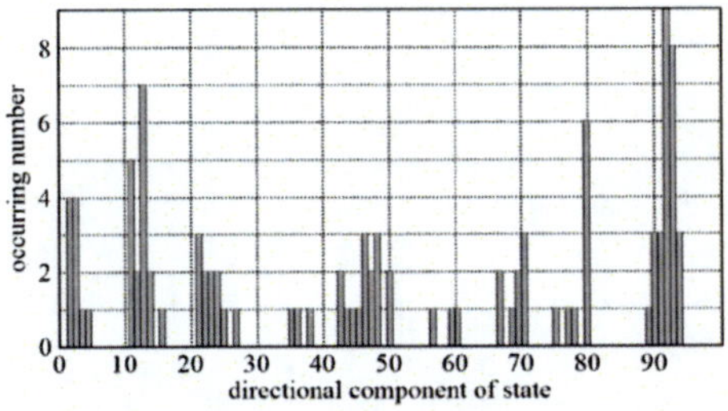

Fig. 5. Distribution of states reached after one flip with initial direction corresponding with t_1 and speed of 6

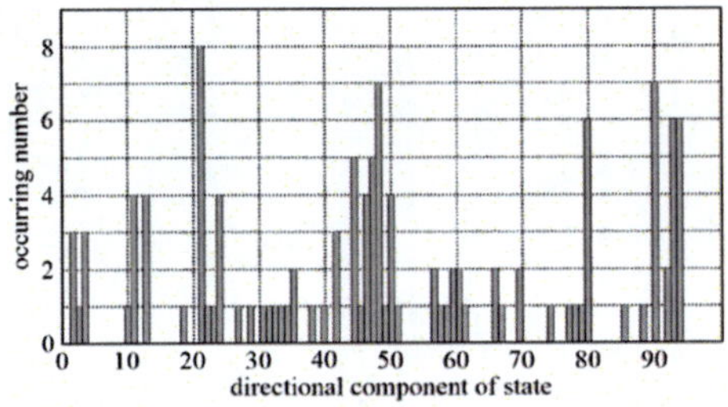

Fig. 6. Distribution of states reached after one flip with initial direction corresponding with t_1 and speed of 7

For clarity, only the 94 state's directions components of the coming down ball based on the combination of b_i and h_i (see Fig. 3) is shown in both figures. The associated speed component was recorded in the measurement, but is not shown in the figures due to the low variation. Thus, there are 94 possible states on the x-axis. The small variation in the speeds suggests that the speeds associated with the direction of incidence strongly depend on the initial state.

Figure 5 shows as well as Fig. 6, that most of the directions occur only once or twice and a few occur significantly more often. States 0 to 39 represent the directions of incidence between the lower targets T_1 and T_2 on the left flipper, states 50 to 89 those on the right. States 90 and 91 describe the path via t_0 and t_1, 92 and 93 correspond to the path of the ball via t_2 or t_3. The area between the two flippers is coded with the numbers 41 to 50. Here, the ball coming down cannot be reached by the flippers and it is therefore a condition that has to be avoided.

As can be seen in Fig. 5 overall, this situation has occurred 15 times. In Fig. 6 the ball runs out 30 times. This means a risk for losing of 25% in contrast to 15% of Fig. 5. Both figures have in common that most of the states occurring more often are in the marginal areas of the playing field.

Figure 7 and Fig. 8 show two distributions where the ball was rolled in towards the flipper along the path, which is labeled by t_2 in Fig. 3. Similar to the two cases discussed before, most of the directions occur only once or twice and a few occur significantly more often and these are located in the marginal areas as well. In Fig. 7 which is based on 86 samples, the ball runs out 11 times. Figure 8 is based on 100 samples. One can

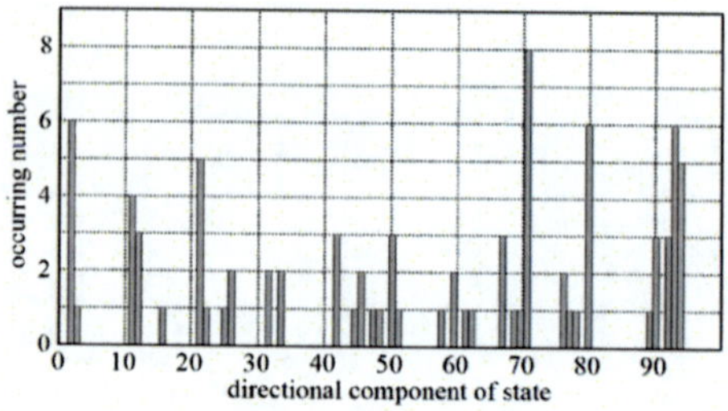

Fig. 7. Distribution of states reached after one flip with initial direction corresponding with t_2 and speed of 7

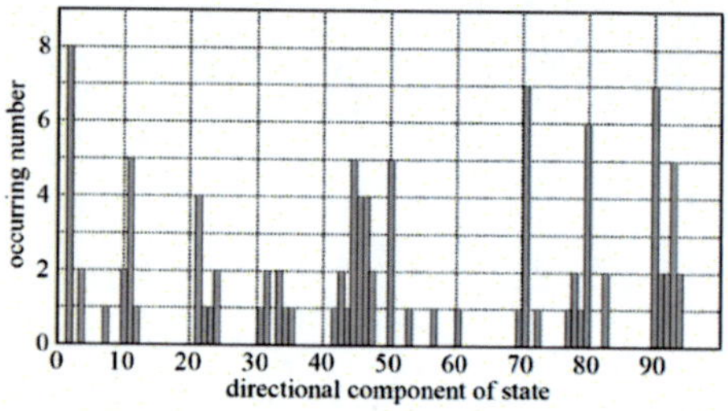

Fig. 8. Distribution of states reached after one flip with initial direction corresponding with t_2 and speed of 8

see, that the flippers are not able to hit the ball for 23 times. Thus, here the risk of losing the ball is 13% in contrast to 23%.

The interesting point in this case is, that for the initial condition which leads to Fig. 8, the fired ball most of the time runs through the spinning target directly into the bumpers B_1 and B_2. Thus the possible score of this shot is quite high. But the risk to lose the ball is comparatively high as well. This result underpins the initially outlined field of tension between maximizing the playing time on the one hand and maximizing the score on the other. Lucrative targets are associated with a significantly higher risk of losing the ball.

Overall, the Figs. 5, 6, 7 and 8 show that there are certain clusters of the states reached by a certain action $a_{j,k}$ at a certain state $s_{i,k}$, but that the amounts do not differ so much. Furthermore, the figures underline the high uncertainty of the system. Finally, it should be noted that the number of approx. 100 measurements per action and state represents a promising database for a Monte Carlo simulation that is still to be done.

6 Conclusion

In this paper an approach for a learning algorithm is presented, which should enable the AI-based operation of an electromechanical pinball machine. The demonstrator used is characterized by a high degree of uncertainty. Therefore, with the reinforcement learning or more precisely Q-learning, an approach was chosen that allows the consideration of uncertainties. Furthermore, it is a model free approach. This is quite important for the presented application because no dynamical model is available at present.

One main aspect of this paper is the description of the designated learning concept and the definition of states and actions regarding the pinball machine. A Q-network consisting of 10152 quantities has been introduced. Getting all these quantities converged by playing real game would be very time consuming. Thus, a combination of learning by real experiments and simulation was chosen. An approximation of the transition probabilities is done by experiments and based on this a Monte Carlo simulation should be done.

The presented results show that, based on the selected states and actions, there are indications of accumulations in the transitions from one state to the other, but these are not particularly pronounced. The expected number of games that are necessary to achieve a visible learning success therefore is very high. It will take weeks and months to obtain representative results for the entire Q-network. Unfortunately, these are not available at this time.

A simulation without any initial measurements is not possible because there is no digital model of the pinball machine. The results presented therefore focus on the exemplary evaluation of measurements from the initialization phase.

In the near future all needed transition probabilities should be evaluated with a sufficiently high number of samples. Afterwards the Monte Carlo simulation will be implemented. Based on its results, the Q-network will be initialized an tested on the pinball machine.

Acknowledgments. The authors thank the volunteer test persons who made themselves available for generating a database as a comparison for the results presented in this paper.

Further thanks are going to sponsors of IAI2020 Conference for their intellectual and financial support.

References

1. Internet Pinball Database. https://www.ipdb.org/machine.cgi?id=2539. Accessed January 2020
2. Jaderberg, M., et al.: Human-level performance in 3D multiplayer games with population-based reinforcement learning. Science **364**(6443), 859–865 (2019)
3. Kober, J., Bagnell, A., Peters, J.: Reinforcement learning in robotics: a survey. Int. J. Robot. Res. **32**(11), 1238–1274 (2013)
4. Kodratoff, Y.: Introduction to Machine Learning. Morgan Kaufmann Publishers, Inc., London (1988)
5. Kubat, M.: An Introduction to Machine Learning. Springer, Heidelberg (2015). https://doi.org/10.1007/978-3-319-20010-1
6. Lichtenberg, G., Neidig, J.: Discrete event control of a pinball machine. In: Proceedings of 7th IFAC Workshop on Discrete Event Systems, Reims, France, pp 133–138 (2004)
7. Mnih, V., et al.: Human-level control through deep reinforcement learning. Nature **518**, 529–533 (2015)
8. Sugiyama, M.: Statistical Reinforcement Learning - Modern Machine Learning Approaches. Taylor & Francis Group, LLC, Boca Raton (2015)
9. Sutton, R., Barto, A.: Reinforcement Learning: An Introduction, 2nd edn. MIT Press, Cambridge (2018)
10. Vicon Motion Systems: Vicon Tracker User Guide (2018). https://docs.vicon.com/display/Tracker37. Accessed January 2020

The Research of Civil Aero-Engine Remaining Useful Life Estimation Based on Gaussian Process

Rui Wu, Chao Liu, and Dongxiang Jiang(✉)

The Department of Energy and Power Engineering, Tsinghua University, Beijing, China
wurui18@mails.tsinghua.edu.cn, cliu5@tsinghua.edu.cn, jiangdx@mail.tsinghua.edu.cn

Abstract. Nowadays, with the improvement of deep learning theory, more and more nonlinear system problems are well solved. The capability of deep learning algorithms for seeking the relationships of the data enhances a lot. This paper proposes a data driven approach using the Gaussian Process (GP) and deep Gaussian Process (DGP) for civil aero-engine remaining useful life (RUL) estimation. The DGP is the extension of the Gaussian Process (GP), which is described as the distribution of functions. The raw data with selection and normalization are directly used as input of the GP and DGP network, which forms a mapping between the input and the RUL. In order to demonstrate the superiority of the proposed approach, some experiments are carried out based on the C-MAPSS dataset. The result suggests that this approach can get an expected performance in RUL estimation.

Keywords: Gaussian process · Deep learning · Remaining useful life · Civil aero-engine

1 Introduction

As the development of the economy, people put forward higher requirements for the convenience, safety, and comfort of transportation. In the next 20 years, the total number of civil aircrafts in the world will reach 35,300, with a total value of 4.8 trillion dollar. The aircraft demand of China will increase to 5,500, accounting for 16% of the global market [1]. The maintenance cost accounts for 10%–20% of the civil aviation operating cost. And the aero-engine maintenance cost accounts for about 30% [3] of the maintenance cost. The research on aero-engine condition monitoring maintenance technology is of great significance for improving the safety and reliability, optimizing the flight plan and reducing the operating cost.

The maintenance of aero-engines goes through three stages: fault repair, scheduled maintenance, and condition-based maintenance. Fault repair usually refers to the maintenance strategy that starts after an aero-engine failure occurs. The cost of this unplanned maintenance is the highest. Scheduled maintenance has a prior plan for aero-engine maintenance with a fixed period. This strategy has advantages over unplanned maintenance, but there is also an excessive maintenance state, which may cause unnecessary

R. Karim et al. (Eds.): IAI 2021, LNME, pp. 12–23, 2022.
https://doi.org/10.1007/978-3-030-93639-6_2

maintenance cost. Condition-based maintenance focus on the relationship between the engine state and the historical data, which can be used to evaluates and predicts the state of the engine, provide the guidance of maintenance, and reduce the cost and operational risk effectively. It is widely used by aero-engine manufacturers such as GE, Rolls-Royce, and Pratt & Whitney.

Aero-engine degradation and remaining useful life estimation is the key technology in aero-engine condition-based maintenance. By using the mathematical models and historical data, the current state and the degradation process of aero-engine can be described, which can be used to estimate the remaining useful life. Pecht divides the prediction methods of RUL into three types: model-based, data-driven, and hybrid approaches [5]. Theoretically, the more precise a model-based approach is, the more accurate prediction of the RUL will be. The model-based approaches need to simulate the system itself and the uncertain operating environment, which is difficult to be described by a certain formula. When approximate estimation is used, the error of RUL produces. Besides, the model-based approaches cannot give the density function (PDF) and reliability of prediction. The data-driven approaches mainly use the historical operation data and artificial intelligence or statistical probability methods to obtain the estimation. Artificial intelligence cannot give the reliability of the prediction, too. The statistical methods can naturally give confidence intervals at a certain probability. However, features are passed inside both the artificial intelligence network and the statistical model, which makes it difficult to obtain the knowledge behind the phenomena. The hybrid approaches combine the model-based approaches and the data-driven approaches to control the parameters. Due to the challenges of model-based prediction, there are numerous researches and applications based on data-driven methods, such as Support Vector Machine (SVM), Convolutional Neural Networks (CNN), Long short-Term memory neural network (LSTM), and Weibull process, Hidden Markov Model (HMM), etc. Recently, with the development of deep learning research, methods based on deep learning, such as DGP and Deep Convolutional Neural Networks (DCNN), have better performance on the remaining useful life estimation.

As for C-MAPSS dataset, more than 70 papers use it for engine degradation and remaining useful life estimation research. The methods can be divided into three forms. One is to use the state data and the remaining useful life data to form a mapping, mainly using a neural network model, for training and predicting. Such as Heimes [6] uses a recurrent neural network (RNN) model, Riad [7] uses multi-layer perceptron (MLP), Malhotra P [8] uses LSTM, Li [9] uses DCNN for the RUL prediction. Further research has shown that feature extraction methods such as Extended Kalman Filter Model (EKF), Gene Algorithms (GA) can improve the performance of the model. Some health parameters are extracted from the train data, which can describe the process of degradation. Some methods use the prior knowledge of the data for predicting, such as SVM, HMM and least squares support vector regression (LS-SVR). Hu [10] compare the SVM, CVM, and RNN result. Li Qian [11] uses the fusion method LS-SVR to predict the remaining useful life of the engine. The third methods use the similarity matching methods to establish databases including historical data and known failure time, which can help to seek the similarity between a certain engine and the database samples.

Ramasso [12] builds a health indicator library, using polygon coverage to predict the remaining useful life of the engine, which is proved to achieve excellent results.

This paper is divided as four parts. The second part of this paper will briefly introduce the Gaussian process and deep Gaussian process. The third part will show the application of Gaussian process and deep Gaussian process for the remaining useful life estimation based on C-MAPSS dataset. The fourth part is the conclusion.

2 Study Method

2.1 Gaussian Process

The Gaussian process is a supervised machine learning method based on Bayesian theory, which can solve specific regression and classification problems. Gaussian process regression is more suitable for nonlinear regression and has applied in robot motion control and finance. It can be proved from the weight space and function space [13].

From the perspective of weight space, it assumed that the training dataset $D = \{(x_i, y_i)|i = 1, 2, \ldots, n\}$, where x_i is the input vector of d dimension, y_i is the corresponding target value. The size of D is $(d + 1) \times n$. For the linear regression model, the target y can be represented by the input vector x and the weight ω. Usually the data has the noise ε, so the relationship between x and y can be described as Eq. (1).

$$f(x) = x^T \omega, y = f(x) + \varepsilon \tag{1}$$

Assume that ε satisfies the Gaussian distribution,

$$\varepsilon \sim N(0, \sigma_n^2) \tag{2}$$

the likelihood function satisfies the Gaussian distribution,

$$p(y|x, \omega) = \frac{1}{(2\pi\sigma_n^2)^{\frac{n}{2}}} \exp(-\frac{(y - X^T\omega)^2}{2\sigma_n^2}) \tag{3}$$

The Gaussian process performs a priori estimation of ω. It assumes that ω also satisfies the Gaussian distribution, whose mean and the covariance is zero vector and Σ. According to Bayes theorem, the posterior probability of ω is described as

$$p(\omega|X, y) = \frac{p(y|X, \omega)p(\omega)}{p(y|X)} \tag{4}$$

The marginal probability is

$$p(y|X) = \int p(y|X, \omega)p(\omega)d\omega \tag{5}$$

The posterior probability can be proved to be the Gaussian distribution, which means the posterior probability is uniquely determined by the mean and the covariance. The following formulas represent the posterior distribution of ω.

$$p(y|X) = \int p(y|X, \omega)p(\omega)d\omega \tag{6}$$

Where

$$A = \sigma_n^2 XX^{\mathrm{T}} + \Sigma^{-1} \tag{7}$$

The new sample point x_* and its corresponding output value f_* satisfies the Gaussian distribution, too. The f_* is determined by

$$\begin{aligned} p(f_*|x_*, \omega) &= \int p(f_*|x_*, \omega)p(\omega|X, y)d\omega \\ &= \int x_*^T \omega p(\omega|X, y)d\omega \\ &= N\left(\frac{1}{\sigma_n^2} x_*^T A^{-1} Xy, x_*^T A^{-1} x_*\right) \end{aligned} \tag{8}$$

In this formula, $\mu = \frac{1}{\sigma_n^2} x_*^T A^{-1} Xy$ is the best estimate of f_*. However, the linear model may not get good performance for some nonlinear system. A series of the nonlinear functions, which can map the input values in the low-dimensional space to the high-dimensional space, are used to solve nonlinear problems in the high-dimensional feature space. Assuming that the mapping function $\varphi(\cdot)$ can map the d-dimensional space to the N-dimensional space ($N > d$), then the weight ω in this space is also N-dimensional.

$$f(x) = \phi(x)^T \omega \tag{9}$$

$$f_*|x_*, \omega \sim N\left(\frac{1}{\sigma_n^2}\phi(x_*)^T A^{-1} \phi y, \phi(x_*)^T A^{-1} \phi(x_*)\right) \tag{10}$$

In the equation,

$$\phi = \phi(x), A = \sigma_n^2 XX^{\mathrm{T}} + \Sigma^{-1} \tag{11}$$

So,

$$\phi_* = \phi(x_*), K = \phi^T \Sigma \phi \tag{12}$$

We can prove that

$$f_*|x_*, \omega \sim (\mu, \sigma^2) \tag{13}$$

Where

$$\mu = N\phi_*^T \Sigma \phi \left(K + \sigma_n^2 I\right)^{-1} y \tag{14}$$

$$\sigma^2 = \phi_*^T \Sigma \phi_* - \phi_*^T \Sigma \phi \left(K + \sigma_n^2 I\right)^{-1} \phi^T \Sigma \phi_* \tag{15}$$

It is obvious that the value f_* is determined by $\phi^T \Sigma \phi$, $\phi_*^T \Sigma \phi$, and $\phi_*^T \Sigma \phi_*$ in the feature space. Assuming that a kernel function k(x, x′) is

$$k(x, x') = \phi(x)^T \Sigma \phi(x') \tag{16}$$

It shows that different basic functions and kernel functions have significant influence on the predicted values. The kernel function is a measure of the degree of similarity between two points in the feature space. Depending on the data structure, there may be different kernel functions. Some common kernel functions are shown in Table 1 below. the Gaussian process method can also customize the form of the kernel function. It is based on an understanding of the data.

2.2 Deep Gaussian Process

The deep Gaussian process is a deep confidence model [14], which is a combination of multivariate Gaussian processes. It has the same structure as the deep neural network, which is composed of multi-layers and multiple hidden nodes. The input of each Gaussian layer is the output of previous Gaussian layer. The output is the input of the next Gaussian layer, too. The deep Gaussian process is a model with better generalization performance and uncertainty estimation. More than one hidden layer can break through the nonlinear fitting limitation using a single-layer Gaussian process. Furthermore, for a practical problem, it is necessary to define a new kernel function to describe the structure of the data, which is usually complicated and has no rules to follow. However, the hidden layers of the deep Gaussian process can learn the data with relatively simple kernel functions. By setting different numbers of layers and kernel functions, the deep gaussian model has a good fitting result of nonlinear mapping, which makes the model more flexible and general [15] (Fig. 1).

Table 1. Some kernel function for Gaussian process

名称	函数形式
Linear	$k(x, x') = x^T x' + c$
Polynomial	$k(x, x') = (\alpha x^T x' + c)^d$
RBF	$k(x, x') = \sigma^2 \exp\left(-\frac{(x-x')^2}{2l^2}\right)$
RQ	$k(x, x') = \sigma^2 \exp\left(-\frac{(x-x')^2}{2\alpha l^2}\right)^{-\alpha}$
Matern 3/2	$k(x, x') = \sigma^2\left(1 + \frac{\sqrt{3}\lvert x-x'\rvert}{\sigma_l}\right)\exp\left(-\frac{\sqrt{3}\lvert x-x'\rvert}{\sigma_l}\right)$
ARD SE	$k(x, x') = \sigma^2 \exp\left(-\sum_{m=1}^{d}\frac{(x-x')^2}{2l_m^2}\right)$

Assume that $Y \in R^{N\times D}$ is the output, the original input $Z \in R^{N\times Q_z}$, and the hidden node is X. The output of the previous Gaussian process f^X satisfies

$$f^X \sim gp(0, k^X(Z, Z)) \tag{17}$$

and the input of the later Gaussian process f^Y satisfies

$$f^Y \sim gp(0, k^Y(X, X)) \tag{18}$$

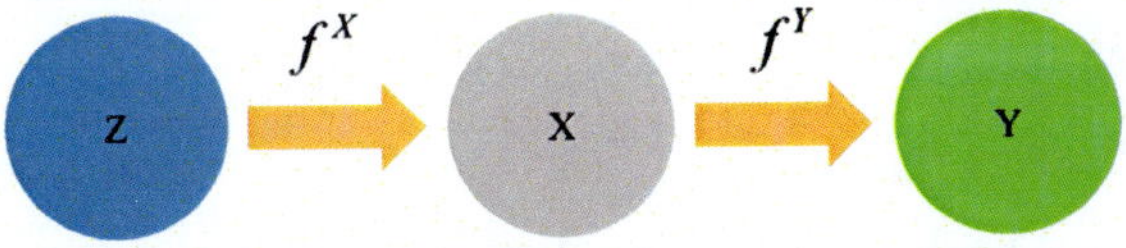

Fig. 1. Two layers deep Gaussian process structure

In details,

$$x_{nq} = f^X(z_n) + \varepsilon_{nq}^X, q = 1, 2, \ldots, Q, z_n \in R^{Q_z} \tag{19}$$

$$y_{nd} = f^Y(x_n) + \varepsilon_{nd}^Y, d = 1, 2, \ldots, D, x_n \in R^Q \tag{20}$$

When the entire structure has L Gaussian layers, its basic frame is the same as the frame of one Gaussian layer.

$$p(f_l|\theta_l) = gp(f_l; 0, K_l), l = 1, 2, \ldots, L \tag{21}$$

$$p\left(h_l|f_l, h_{l-1}, \sigma_l^2\right) = \prod_n gp\left(h_{l,n}; f_l\left(h_{l-1,n}\right), \sigma_l^2\right), \tag{22}$$

$$h_{1,n} = x_n \tag{23}$$

$$p\left(y|f_L, h_{L-1}, \sigma_L^2\right) = \prod_n gp\left(y_n; f_L\left(h_{L-1,n}\right), \sigma_L^2\right) \tag{24}$$

It is difficult to train a deep Gaussian process model and obtain the posterior distribution and the marginal likelihood function under completely independent training conditions. The transfer algorithm EP and the fully independent training condition approximation are used to save computing resources for parameters training [15] (Fig. 2).

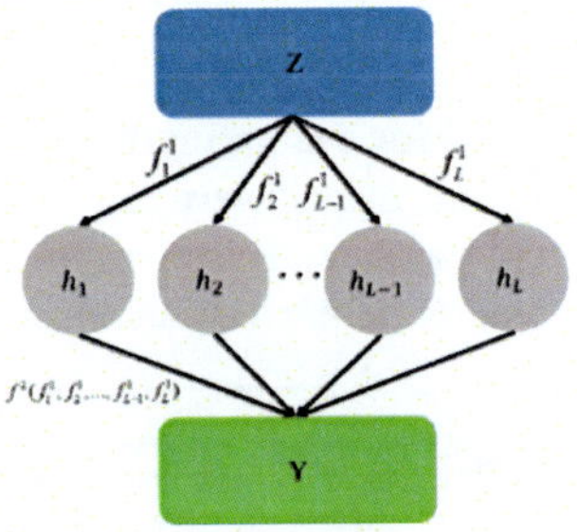

Fig. 2. The general structure of deep Gaussian process

3 Experiment

3.1 C-MAPSS Dataset

The remaining useful life prediction technology research in this paper is based on the competition data set (C-MAPSS) released by the NASA in 2008 Prognosis and Health Management Conference. The dataset has four sub-datasets, each of which includes a training set, a test set and a target set of remaining useful life. The training set includes a full-cycle parameter monitoring data of a certain number of engines, and the test set has the parameter monitoring data of the engine from the beginning to a specific cycle. The target set is the real remaining useful life value of the corresponding engine running to a specific cycle in the test set. Different data sets are distinguished by the different operating environments and the type of the fault. The data set is shown in Table 2 below. Each dataset has 26 parameters, including engine number, cycle number, 3 operating parameters, and 21 sensor parameters. The sensor data represent the low-pressure compressor outlet temperature, shaft speed and pressure ratio, etc. The data are artificially added with unknown Gaussian white noise, which is closer to the real operating situation.

Table 2. The C-MAPSS dataset

Dataset	C-MAPSS			
	FD001	FD002	FD003	FD004
Engines for training	100	260	100	249
Engines for testing	100	259	100	248
Operation modes	1	6	1	6
Fault modes	1	1	2	2

3.2 Performance Metrics

In this paper, three metrics are used for evaluating the performance, i.e. the root mean square error (RMSE), accuracy, and mean average percentage error (MAPE). These metrics is widely used by many researchers. Based on the definitions, lower RMSE and MAPE values mean the better prediction performance.

$$RMSE = \sqrt{\frac{\sum_{i=1}^{n} (\bar{y} - y)^2}{n}} \quad (24)$$

$$|\bar{y} - y| \in [-10, 13] \quad (25)$$

$$MAPE = \frac{1}{n} \sum \left| \frac{\bar{y} - y}{y} \right| \times 100\% \quad (26)$$

3.3 The Result of the Gaussian Process

Gaussian process regression is used to predict the remaining useful life of aero-engine in this research. 20631 sample points, which is composed of 24 characteristic variables from FD001 data set are used as input values, and the corresponding remaining useful life is used as the target. In consideration of different basic functions and kernel functions making big difference on prediction, some combination experiments are carried out during the model training. The experiments are shown in Table 3 below.

Table 3. The experiment with different Gaussian basic and kernel functions

	exp	Sq-exp	ardexp	ardSq-exp	matern32
None	√	√	√	√	√
Const	√	√	√	√	–
Linear	√	√	√	√	√
pureQu adratic	√	√	√	√	–

According to the calculation, the prediction result under the condition of none-exponential is the best, the RMSE is 31.24, the Accuracy is 45%, and the MAPE is 37.88%. The training performance of first ten aero-engine data is shown in Fig. 3. It shows that the model fits well when the engine fault is obvious, but at the early stage of the engine state, the fitting result is poor, which has a significant influence on the RMSE and MAPE of the model.

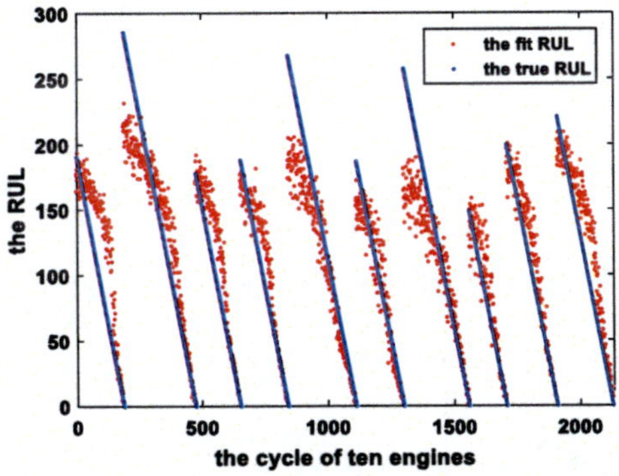

Fig. 3. The RUL prediction result on the training data

The prediction result of test data shows in the Fig. 4. The position of the point is closer to $y = x$, the prediction is more accurate. In the picture, when the remaining useful life is near to the end, the prediction is close to the true value, and when the engine is at very beginning of its working life, the prediction is far from the true value. It is because that the degradation of the engine is becoming more and more obvious as the working time goes on, which is easier for the model to extract degradation feature.

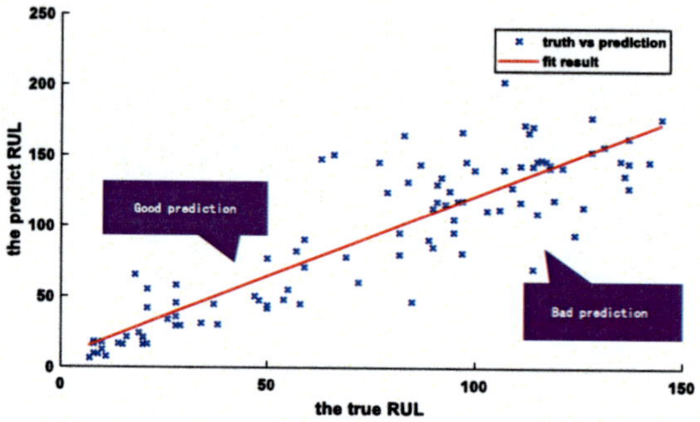

Fig. 4. Comparison of the real and predicted remaining useful life on test data

Researches on the engine degradation process show that the degradation is not apparent at the early stage of the engine working life. And the health parameters used to characterize the engine state don't change a lot, which leads to bad prediction of the model. In the middle and later stages, the degradation speed gradually increases, which reflect on the changes of the health parameters. The prediction error is more and more small as the engine life closing to the end. Therefore, the linear remaining useful life assumption is not suitable for the practical situation. The multi-segment linear remaining useful life response can be used to obtain a better prediction result. This paper assumes that the degradation is not obvious during the initial period of the engine, and the remaining useful life does not change during that period. When the health of the first 30% of the cycle is assumed to be unchanged, the predicted result is the best. Figure 5 is a comparison of the real and predicted remaining useful life of the test set. The figure shows improvements in the early remaining useful life prediction. The RMSE under multi-segment linear conditions is 18.64, the accuracy rate is 54%, and the MAPE is 25.01%, which is greatly improved compared with the simple linear result.

3.4 The Result of Deep Gaussian Process

The multi-segment linear remaining useful life response is used for deep Gaussian process model training, too. The data is prepared in the same way as in [9], and the engine life is set at 125 cycles. This model uses the remaining useful life as the output. Fourteen sensor variables were combined, and 30 cycle parameter values were concatenated into a set of samples as input to the model. The deep Gaussian process structure is set with three hidden layers. The cross-validation and early stopping methods are used to avoid overfitting of the training data. The number of nodes in each layer and the number of

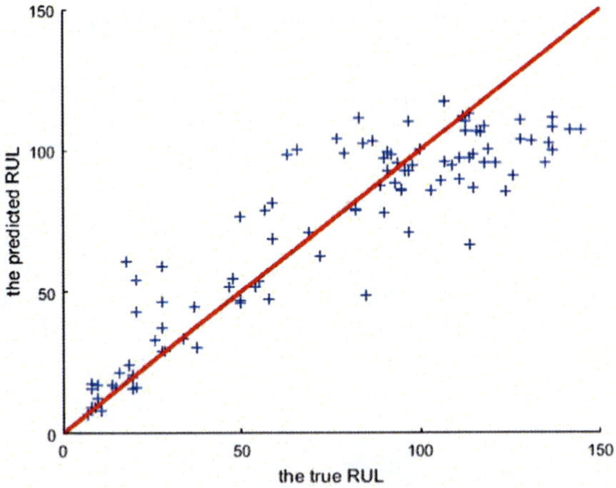

Fig. 5. Comparison of the real and predicted remaining useful life on test data by using the multi-segment linear method

inducing points, learning rate, and mini-batch size parameters are used as hyperparameters of the model, which are determined by searching the optimal value. The model training effect of the first ten engines is shown in Fig. 6. It shows that the training effect is better than that of the single-layer Gaussian model. For the engine of the test set, the root mean square error RMSE is 13.61, the accuracy is 64%, and the MAPE is 14.19%. Compared with the multi-segment linear Gaussian process model, it is further improved.

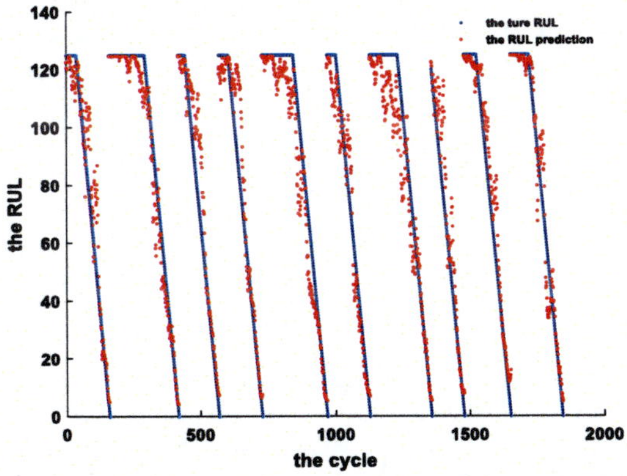

Fig. 6. The RUL estimation on the training data by using multi-segment linear and deep Gaussian process method

By comparing with the literature results, it shows that multi-segment linear deep Gaussian process method improves the accuracy of the remaining useful life prediction of the C-MAPSS dataset. The specific comparison results are shown in Table 4.

Table 4. The result comparison between different kinds of literature for RUL prediction on C-MAPSS dataset

Model	RMSE	Accuracy	MAPE
GP (linear RUL)	**31.24**	**45%**	**37.88%**
GP (polyline RUL)	**18.64**	**54%**	**25.01%**
DGP (polyline RUL)	**13.61**	**64%**	**14.19%**
Echo state network-KF [16]	63.46	–	–
SVM-classifier [17]	29.82	–	–
First attempt DCNN [18]	18.45	–	–
Random Forest [19]	17.91	–	–
DCNN [9]	*12.61*	–	–
Belief Function [20]	–	53%	–
FS – LSSVR [11]	–	–	38%
ANN [11]	–	–	43%

4 Conclusion

In this paper, the Gaussian process and deep Gaussian process is used to estimate remaining useful life of the aero-engine. Gaussian process regression, as a regression model based on Bayesian theory, has higher adaptability to nonlinear regression problems and has its unique advantages in small sample size data. It is meaningful for future research to study the structure and hyperparameter optimization, which can make it easier and more effective to extract feature and format the mapping on nonlinear system data.

Acknowledgments. This work was partly supported by National Key R\&D Program of China (Grant No. 2019YFF0216104 and 2019YFF0216101).

References

1. Xu, L.: Research on trend prediction of gas path parameters of aero-engine. Civil Aviation Flight University of China, p. 93 (2016)
2. Kumar, U.D., Knezevic, J., Crocker, J.: Maintenance free operating period – an alternative measure to MTBF and failure rate for specifying reliability? Int. J. Emerg. Electr. Power Syst. **64**(3), 127–131 (2015)
3. Dixon, M.: Maintenance costs of aging aircraft. Costs of Aging Aircraft Insights from Commercial Aviation (2006)
4. Huang, L.: The research of health management of aviation engine and engine fleet health assessment methods. Civil Aviation Flight University of China (2014)
5. Pecht, M.: Prognostics and Health Management of Electronics. Wiley, Hoboken (2008)
6. Heimes, F.: Recurrent neural networks for remaining useful life estimation. In: IEEE International Conference on Prognostics and Health Management (2008)

7. Riad, A., Elminir, H., Elattar, H.: Evaluation of neural networks in the subject of prognostics as compared to linear regression model. Int. J. Eng. Technol. **10**, 52–58 (2010)
8. Malhotra, P., Vishnu, T.V., Ramakrishnan, A., et al.: Multi-sensor prognostics using an unsupervised health index based on LSTM encoder-decoder (2016)
9. Li, X., Ding, Q., Sun, J.Q.: Remaining useful life estimation in prognostics using deep convolution neural networks. Reliab. Eng. Syst. Saf. **172**, 1–11 (2018)
10. Hu, C., Youn, B., Wang, P., Yoon, J.: Ensemble of data-driven prognostic algorithms for robust prediction of remaining useful life. Reliab. Eng. Syst. Saf. **103**, 120–135 (2012)
11. Li, X., Qian, J., Wang, G.: Fault prognostic based on hybrid method of state judgment and regression. Adv. Mech. Eng. **2013**(149562), 1–10 (2013)
12. Ramasso, E.: Investigating computational geometry for failure prognostics in presence of imprecise health indicator: results and comparisons on C-MAPSS datasets. In: European Conference on Prognostics and Health Management
13. Rasmussen, C.E., Williams, C.K.I.: Gaussian Processes for Machine Learning (Adaptive Computation and Machine Learning). The MIT Press, Cambridge (2005)
14. Damianou, A.C., Lawrence, N.D.: Deep Gaussian processes. Comput. Sci. 207–215 (2012)
15. Bui, T.D., Hernández-Lobato, D., Li, Y., et al.: Deep Gaussian processes for regression using approximate expectation propagation (2016)
16. Peng, Y., Wang, H., Wang, J., et al.: A modified echo state network based remaining useful life estimation approach. In: Prognostics and Health Management, pp. 1–7. IEEE (2012)
17. Louen, C., Ding, S.X., Kandler, C.: A new framework for remaining useful life estimation using support vector machine classifier. In: Proceedings of IEEE Conference on Control and Fault-Tolerant Systems, pp. 228–233 (2013)
18. Sateesh Babu, G., Zhao, P., Li, X.-L.: Deep convolutional neural network based regression approach for estimation of remaining useful life. In: Navathe, S.B., Wu, W., Shekhar, S., Du, X., Wang, X.S., Xiong, H. (eds.) DASFAA 2016. LNCS, vol. 9642, pp. 214–228. Springer, Cham (2016). https://doi.org/10.1007/978-3-319-32025-0_14
19. Zhang, C., Lim, P., Qin, A.K., Tan, K.C.: Multiobjective deep belief networks ensemble for remaining useful life estimation in prognostics. IEEE Trans. Neural Netw. Learn. Syst. **PP**(99), 1–13 (2016)
20. Ramasso, E., Rombaut, M., Zerhouni, N.: Joint prediction of continuous and discrete states in time-series based on belief functions. IEEE Trans. Syst. Man Cybern. Part B Cybern. Publ. IEEE Syst. Man Cybern. Soc. **43**(1), 37–50 (2012)

Spare Part Management Considering Risk Factors

Reza Barabadi[1], Mohamad Ataei[1], Reza Khalokakaie[1], Abbas Barabadi[2(✉)], and Ali Nouri Qarahasanlou[3]

[1] Shahrood University of Technology, Shahrood, Iran
{Ataei,R_kakaie}@shahroodut.ac.ir
[2] UiT The Arctic University of Norway, Tromsø, Norway
abbas.b.abadi@uit.no
[3] Faculty of Technical and Engineering, Imam Khomeini International University, Qazvin, Iran
Ali_Nouri@eng.ikiu.ac.ir

Abstract. The spare parts provision is a complex process, which needs a precise model to analyze all factors with their possible effects on the required number of spare parts. The required number of spare parts for an item can be calculated based on its reliability performance. Various factors can influence the reliability characteristics of an item, including operational environment, maintenance policy, operator skill, etc. Thus, the statistical approach of choice for reliability performance analysis should assess the effects of these factors. In this study, Reliability Regression Models (RRM) with risk factors have been used to estimate the required number of crane shovels in the Jajarm bauxite mine. For this, at the first stage, all risk factors and failure data have been collected. The required data were extracted from a database of 15 months, which were collected from different sources, such as daily reports, workshop reports, weather reports, meetings, and direct observations in the format of time to failures and risk factors. After that, the potential distribution has been nominated to model the reliability of the crane shovels bucket teeth. The Akaike information criterion and Bayesian information criterion have been used to identify the best fit distribution. The candidate distribution with the smallest AIC and BIC value is the best distribution that fits the data. After that, the required number of spare parts is calculated. The results show 18% differences between the forecasted number of required spare parts when considering and non-considering the risk factors.

Keywords: Spare part · Reliability · Risk factors · Jajarm bauxite mine

1 Introduction

A system without failure can never be designed due to technological and economic issues. Thus, to provide support and spare parts to guarantee a proper level of availability throughout the system's lifecycle, appropriate and well-scheduled activities need to be performed [1]. However, the provision of spare parts appears to be a complex process,

R. Karim et al. (Eds.): IAI 2021, LNME, pp. 24–39, 2022.
https://doi.org/10.1007/978-3-030-93639-6_3

which needs an accurate analysis of all factors affecting the required number of spare parts.

The spare parts availability is surely a highly important factor, increasing the system's performance and effectiveness. If all needed spare parts for the repair can be immediately provided in a system failure, we can significantly reduce downtime. Otherwise, the increased waiting time can cause dramatic production losses. Yet, the overstocking of unnecessary spare parts or the many outdated stored units may also lead to huge losses due to the investment costs. Thus, we need to provide an accurate prediction rate of spare parts in the design and operation phases as a major factor in the product support activity [2, 3]. However, spare parts prediction and optimization are also complex problems, which need to identify all effective factors and choose a proper model to quantify their effects on the required number of spare parts. Some major effective factors are operational conditions, climatic conditions (temperature, wind, snow, dust, ice, etc.), operators and maintenance crew skills, the history of repair activities done on the machine, etc. [4, 5].

The first step in the reliability-based spare parts provision is to identify the item's reliability performance and failure rate. We can then estimate the number of the required spare parts and the availability rate of spare parts [6]. However, we need to consider all factors influencing the reliability performance of the item for effective forecasting. The factors with a possible effect on the reliability performance of an item are called risk factors, ignoring which may lead to inaccurate results in the reliability performance analysis and the provision of spare parts [7–10].

In the recent two decades, Kumar, Ghodrati, Barabadi, and Nouri Qarahasanlou introduced Proportional Hazard Model (PHM) in the spare parts provision process [11–16]. For example, in 2015–2018, Nouri Qarahasanlou demonstrated the Cox regression method for mining fleet, spare tire analysis of dump truck in Sungun mine, Iran.

The required number of spare parts is calculated in the reliability-based statistical approaches according to the item reliability. Hence, we should measure their effects on the item reliability performance to assess the impact of operational conditions on the required number of spare parts. However, operating time is the only variable evaluated in most available studies, and operational conditions have not been considered a variable [17]. Thus, the RRM has been rarely used and implemented as a proportional hazard model for spare parts predictions.

A review of the relevant literature showed the occasional use of reliability models with risk factors for spare part predictions. This paper was designed to examine the use of RRM in the provision of spare parts for bucket teeth in the Jajarm bauxite mine, Iran. Bucket teeth are considered important parts of the crane shovels, and the shortage of such items can stop production in the mine. The operational conditions in a mine are more difficult than in most other industries. The reliability characteristics of the bucket teeth are believed to be influenced by operational conditions in the Jajarm bauxite mine. Hence, investigating the issue seems important to accurately estimate the number of spare parts needed while considering operational conditions to reduce downtime.

Moreover, as different types of bucket teeth can be used for the loading process, we need to find the most cost-effective one to minimize the loading process costs. The reliability performance of the bucket teeth by considering operational conditions can provide essential information for such a cost analysis. One of the most important issues in using a regression model such as the PHM or family model is the baseline function. Because the risk factors shift it up or down. Most industrial machines or systems deprived of baseline function and researchers are forced to find it from Time between Failures (TBFs). In this way, most of them did not discuss "how" or "why" selected special baseline function. This paper used the goodness of fit test for choosing the best parametric regression model and best analysis method. Model selection plays a fundamental role in choosing the best model from a series of candidate models for data-driven modeling and system identification problems. In general, system identification and data-driven modeling consist of several important steps, including data collection, data processing, selection of representation functions, model structure selection, model validation, and model refinement [17]. Akaike Information Criterion (AIC) and Bayesian Information Criterion (BIC) are two most popular measures among various model selection methods. The rest of the paper is organized as follows: Description of the basic concept and methodology for spare parts prediction by RRM in Sect. 2. Section 3 demonstrated using this methodology through a case study followed by showing how an appropriate RRM can be found for specific data sets. Finally, Sect. 4 provided relevant conclusions.

2 Reliability Analysis Considering Risk Factors Effects

The RRM can be categorized into two main groups: Parametric and non-parametric models. In the parametric method, like the family of accelerated failure time models, the lifetime of a system is assumed to have a specific distribution, such as lognormal distribution. However, parametric methods may be misleading if the historical data does not follow the selected distribution pattern associated with incorrect assumptions about the parametric method. On the other hand, in non-parametric methods such as the proportional hazard models family, no specified distribution is assumed for a system lifetime [17–19].

A major contribution to the concept of non-parametric models for modeling the effects of risk factors is the PHM suggested by Cox [18]. In general, the basic theory of these non-parametric models is to build the baseline hazard function using historical failure data and the risk factor function by using the risk factor data. The baseline hazard function is the hazard rate experienced by an item when the effect of the risk factors is equal to zero. The risk factor function shows how the baseline hazard model will be changed due to risk factors. To see the most available RRM see references [19–21]. After identifying the distribution of failure data using the appropriate model, we can calculate the number of item failures in a specific period. Finally, the required spare parts can be calculated using an existing model like birth and death or Palm's theorem and considering other factors such as expected preventive maintenance frequency and repair rates for the repairable items [21]. This is a continuous procedure, which should be updated by upcoming historical data.

The PHM model is based on the proportionality of the hazard ratio (PH). The risk factors are time-independent variables, suggesting that the ratio of any two hazard rates is constant concerning time [22]. In PHM, the hazard rate Proportional Hazard Model (PHM) for an item is a product of the baseline hazard function, λ_0 (t) of the item and a function $\psi(z, \alpha)$ incorporating the effect of risk factors. The generalized form of PHM that is most commonly used is written as [23]:

$$\lambda(t, z) = \lambda_0(t)\psi(z, \alpha) \tag{1}$$

Where λ_0 (t) is the baseline hazard function, and $\psi(z, \alpha)$ is a function incorporating risk factors. And z is a row vector consisting of the covariates, and α is a column vector consisting of the regression parameters. If the $\psi(z, \alpha)$ function be a log-linear, therefore, the common form of PHM is expressed as Eq. (2) [13]:

$$\lambda(t, z) = \lambda_0(t)\psi(z\alpha) = \lambda_0(t)exp\left(\sum_{i=1}^{n} z_i\alpha_i\right) \tag{2}$$

Where Z_i, i = 1, 2 … n, are the covariates associated with the system and α_i I = 1, 2 … n, are the model's unknown parameters, defining the effects of each one of the n covariates. For example, the "t" multiplicative factor, $exp(z\alpha)$ may be termed the relative risk of failure due to the presence of the covariate z.

The reliability influenced by the risk factors is given as [11]:

$$R(t, z) = (R_0(t))^{exp\left(\sum_{i=1}^{n} z_i\alpha_i\right)} \tag{3}$$

Where $\lambda(t, z)$ and $R(t, z)$ are the hazard and reliability functions, respectively; and α (column vector) is the unknown parameter of the model or regression coefficient of the corresponding n risk factors, and z is row vector consisting of the risk factor parameters indicating the degree of influence of each risk factor on the hazard function; and $\lambda_0(t)$ and $R_0(t)$ are the baseline failure rate and baseline reliability, respectively.

As mentioned earlier, the PH assumption suggests that the risk factors are time-independent variables; thus, the ratio of any two hazard rates is constant concerning time. Different approaches have been used to determine whether PH assumption fits a given data set. The graphical procedure, a goodness-of-fit testing procedure, and a procedure that involves the use of time-dependent variables have been used most widely in PH assumption evaluations [23]. There are two general approaches to check the time-dependency of risk factors: i) graphical procedure, ii) goodness-of-fit testing procedure [18]. The developed graphical procedure can generally be categorized into three main groups as i) cumulative hazards plots, ii) average hazards plots, and iii) residual plots [24]. For example, in the cumulative hazard plots, the data will be categorized based on different risk factors to be checked for time dependency. Hence, if the assumption of PH is justified, then the logarithm plots of the estimated cumulative baseline hazard rates versus time for the defined categories should simply be shifted by an additive constant coefficient of risk factors. In other words, they should be approximately paralleled and separated, corresponding to the different values of the risk factors. Departure from parallelism of the above plots for different categories may suggest that z_r is a time-dependent risk factor.

For the review of other graphical approaches, see [25]. Like the cumulative baseline hazard rate, a Log-log Kaplan-Meier curve over different (combinations of) categories of variables can be used to examine the PH assumption. A log-log reliability curve is simply a transformation of an estimated reliability curve, which results from taking the natural log of an estimated survival probability twice. If we use a PHM model and plot the estimated log-log reliability curves for the defined categories on the same graph, the two plots would be approximately parallel [40]. In the residuals plots at the first step, the residual should be estimated by using the estimated values of the cumulative hazard rate, $H_0(t_i)$, and the regression vector η as:

$$e_i = -H_0(t_i)\exp(\eta_r z_r) \tag{4}$$

Where $H_0(t_i)$ is cumulative hazard rate, η regression vector; and z_r is r^{th} row vector consisting of the risk factor parameter.

If the PH assumption is justified, then the logarithm of the estimated reliability function of e_i against the residuals should lie approximately on a straight line with slope -1.

When the risk factor is time-dependent, the component will have different failure rates based on different values of the time-dependent risk factors. In this case, the Stratified Cox Regression Method (SCRM) can analyse the data [25]. The "stratified Cox model" is an extension of the PHM, allowing the control process by "stratification" of a predictor not to satisfy the PH assumption. Each level is defined as a stratum in this model when there are n levels for the time-dependent risk factors. Under these circumstances, the historical data will be classified into various strata. Then, separate baseline reliability functions are computed for each stratum, while the regression coefficients for all strata are equal. We can write the hazard rate using the stratification approach in the sth stratum as follows [22]:

$$\lambda_s(t, z) = \lambda_{0s}(t)exp\left(\sum_{i=1}^{n} z_i\alpha_i\right) \quad s = 1, 2, \ldots, r \tag{5}$$

The component reliability influenced by risk factors in the sth stratum Eq. 6:

$$R_s(t, z) = (R_{0s}(t))^{exp\left(\sum_{i=1}^{n} z_i\alpha_i\right)} \quad s = 1, 2, \ldots, r \tag{6}$$

Where $\lambda_s(t, z)$ and $R_s(t, z)$: are the hazard and reliability functions in the sth stratum; and $z\alpha = \sum_{i=1}^{n} z_i\alpha_i$, and α (column vector) is the unknown parameter of the model or regression coefficient of the corresponding n risk factors; and z row vector consisting of the risk factor parameters, indicating the degree of influence which each risk factor has on the hazard function; and $\lambda_{0s}(t)$ and $R_{0s}(t)$ are the baseline failure rate and baseline reliability in the sth stratum.

3 Reliability-Based Spare Part Provision Considering Risk Factors

The Reliability Spare Part Provision (RSPP) was used to provide spare parts based on the renewal theory, which is one of the popular mathematical models. The renewal process model provides an approach to describe the rate of events occurring, in our case, the number of failures, over time. It seems reasonable to assume that the number of spare parts required is equal to the number of failures since the non-repairable components are thrown away. The renewal process can be employed whenever the failure rate is not constant. However, at constant failure rates, we use the homogeneous Poisson process as a special case of a renewal process to forecast the demands for spare parts. It is important to note that the above statement is valid only for non-repairable spares [12]. In cases of quite a long operation time (and planning horizon) of the machine, during which the parts are installed, and when we should make several replacements during this period, the average number of failures in time t, E[N(t)] = M(t) will be stabilized to the asymptotic value of the function as follows [12]:

$$M(t) = E[N(t)] = \frac{t}{\overline{\overline{T}}} + \frac{\zeta^2 - 1}{2} \tag{7}$$

where ζ denotes the coefficient of variation of the time to failure and defined as [12]:

$$\zeta = \frac{\sigma(T)}{\overline{\overline{T}}} \tag{8}$$

where $\overline{T}$ is the average time to failure for replacements of a part and $\sigma(T)$ is the standard deviation of time to failure [19–21]. The approximated number of spares (N_t) needed during period of planning horizon with a probability of shortage $= 1 - p$ is given by [12]:

$$N_t = \frac{t}{\overline{\overline{T}}} + \frac{\zeta^2 - 1}{2} + \zeta\sqrt{\frac{t}{\overline{\overline{T}}}}\Phi^{-1}(p) \tag{9}$$

Where $\Phi^{-1}(p)$ is the inverse normal distribution function Thus, estimation of N_t need to calculate ζ in different distribution, specified t, and p. As mentioned before, PHM or SCRM used for the model time dataset in incorporating the effects of risk factors. The problem originates here that determining $\overline{T}$ and $\sigma(T)$ for PHM. Hence, we had to change the parameter of the best fit classic distributions (e.g., Exponential, Weibull, Lognormal, etc.) in the reliability baseline function to consider the risk factors' effects. Unfortunately, as stated above, most of the studies conducted on RSPP (almost all of them) had used just two exponential and Weibull distributions instead of the best-fit one. Thus, we tried to fix it in our study.

4 Spare Parts Inventory Management

Every inventory management system's major goal is to obtain a good spare part rate with a minimum inventory investment and the lowest managerial costs. This may be fulfilled directly by reducing the ordering cost by ordering spare parts more than required. Insufficient provision level increases unacceptably long downtime, while unreasonably high levels can also trap the capital sources in the inventory section [19–21]. We can use the Economic Order Quantity (EOQ) to balance the inventory management, which minimizes the total inventory costs in holding and ordering phases by eliminating the shortages that can be calculated as follows [11]:

$$EOQ = \sqrt{\frac{2DS}{H}} \tag{10}$$

Where: "D" is the annual demand (units/year) [equals N_t in one year], "S" is the cost of ordering or setting up one lot ($/lot), and "H" is the cost of holding one unit in the inventory for a year (often calculated as a proportion of the item's value). We had to calculate the "Reorder Point (ReP)" for obtaining the "continuous review system" as inventory position controlling and management. The ReP is [11]:

$$ReP = d \times L + \sigma_D \times \sqrt{L\Phi_{(P/2)}} \tag{11}$$

where d: is the average demand, L is the lead time, $\Phi_{(P/2)}$: is the confidence level of the cycle service, and σ_D: is the number of standard deviations from mean and calculated as [11]:

$$\sigma_D = \sqrt{\frac{t}{\overline{T}}} \tag{12}$$

5 Case Study

Spare parts for maintenance tasks, except for preventive maintenance activities, are usually required at random intervals. Hence, due to the uncertainty about the time of failure, the spare parts can be modeled using the probability distribution illustrated in previous sections. As Fig. 1 presents, the methodology is based on five main tasks:

- Establishing the context.
- Data collection, identification, and formulation of risk factors.
- Identification of the model of failure data considering risk factor effects.
- Calculation of the required number of spare parts.
- Inventory management.

5.1 Establishing the Context

The case study refers to the crane shovels bucket teeth ($m = 5$) from the Kaj-Mahya company put into service in the Jajarm bauxite mine. Jajarm bauxite mine in Iran has 19 main open mines in the city of Jajarm. The longitudinal expanse of the mine from west to east (namely: Golbini 1–8, Zou 1–4, Tagouei 1–6, and Sangtarash) is 16 km. The length of these sections is as follows: Golbini: totally 4.7 km, Zou mines: totally 3.3 km, Tagouei mines, totally 5 km, and Sangtarash mine is about 3 km in length. The Jajarm bauxite falls in the lens-like layer category. The expanse of bauxite is mostly in the form of layers. The mineral lies on the karstic-dolomites that make up the Elika formation, which lies under the shales and sandstones of Shemshak formation. The bauxite layer is not of even thickness and consistent quality. In general, the bauxite layer ranges from less than 1 m to about 40 m in thickness. The main design characteristics (weight, size, maximum load capacity, etc.) of the crane shovels are nearly identical.

5.2 Data Collection

Using the developed framework in Fig. 1, the failure data and associated observed risk factors should be collected at the first stage. For this aim, the observed risk factors should be identified. Table 1 shows the selected observed risk factors. As this table shows, 6 risk factors are identified which may affect the reliability of the crane shovels bucket teeth. The number in the branches in Table 1 is used to nominate the risk factors. For example, crane shovels work in three different shifts named morning, afternoon, and night shifts; here, zero, 1, and 2 are used to represent these shifts, respectively. Table 2 shows a sample of data.

Table 1. The identified observed risk factors for the crane shovels.

Risk factor	Risk factor level	Risk factor	Risk factor level
Working shift (z_{wf})	Morning shift [0]	Rock kind (z_{rk})	H. Bauxite [1]
	Afternoon shift [1]		
	Night shift [2]		LG. Bauxite [2]
Humidity (z_p)	Continuous		Kaolin Bauxite [3]
Temperature (z_t)	Continuous		Chile Bauxite [4]
System ID (Crane shovels number)	DT1 (1) to DT4 (4)		Tailings [5]
			Dolomite [6]

To formulate the risk factors, we used observation, repair shop cards and reports, and the experience of managers, operators, and maintenance crews, especially with the field data. The part of data with risk factors is shown in Table 2.

Table 2. A sample of failure data and their associated risk factors.

No.	TBF (Hours)	z_{id}	z_{wf}	z_{bt}	z_{rh}	z_t oC	z_{rk}
1	408	1	2	5	53	5	2
2	422	1	1	1	28	5	5
3	447	2	3	2	57	1	3

5.3 Reliability Model Identification

We present the test of Harrell and Lee (1986), a variation of a test originally proposed by Shenfield (1982) and based on the residuals defined by Shenfield, now called the Shenfield residuals. This study used the goodness-of-fit (GOF) test to check the PH assumption. The GOF testing approach is attractive because it provides a test statistic and p-value (P (PH)) for checking the PH assumption for a given predictor of interest. Thus, a more objective decision provides by a statistical test than a graphical approach. The P (PH) is used for evaluating the PH assumption for that variable. An insignificant (i.e., large) P(PH), say greater than 0.10, suggests that the PH assumption is reasonable. In contrast, a small P(PH), say less than 0.05, suggests that the variable being tested does not satisfy this assumption [26]. Table 3 is illustrated the mean value and the statistical GOF test outcomes of influence risk factors for data.

Table 3. Statistical test approach results for PH assumption

		Rock type	Temperature	Humidity	System ID	Shift
TBF	Pearson correlation (P-PH)	.[a]	.[a]	.[a]	−.087	.279
	Sig. (2-tailed)				.681	.176
	N	0	0	0	25	25

a. Cannot be computed because at least one of the variables is constant

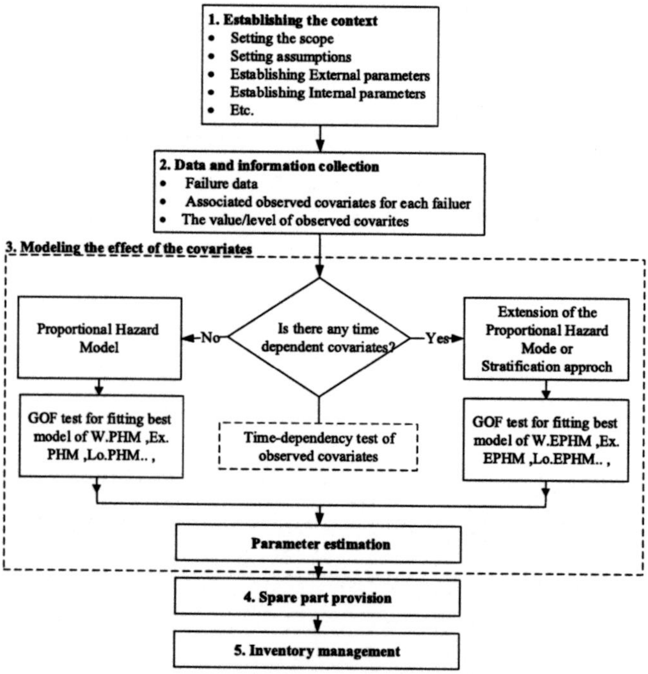

Fig. 1. A methodology for calculating required numbers of spare parts considering the effect of risk factors.

The P(PH) values given in this table provide GOF tests for each variable in the fitted model adjusted for the other variables in the model. The P (PH) values are quite high for all variables satisfying the PH assumption. Also, the log minus log survival plot was used as a graphical test for PH assumption. In this test, if the risk factors are time-independent, Log Minus Log (LML) survival. To check the time-dependency of risk factor effect on equipment performance, collected data of mine equipment were stratified based on rock types and system ID. Plot or log cumulative failure plot versus time graphs for the different selected risk factors yield parallel curves. The results show that the plotted curves are parallel for five types using LML and log cumulative failure plots. For example, Fig. 2 shows the results of such analysis for teeth in both rock types and system ID. Thus, according to Fig. 1, the PHM can assess the risk factors of the teeth.

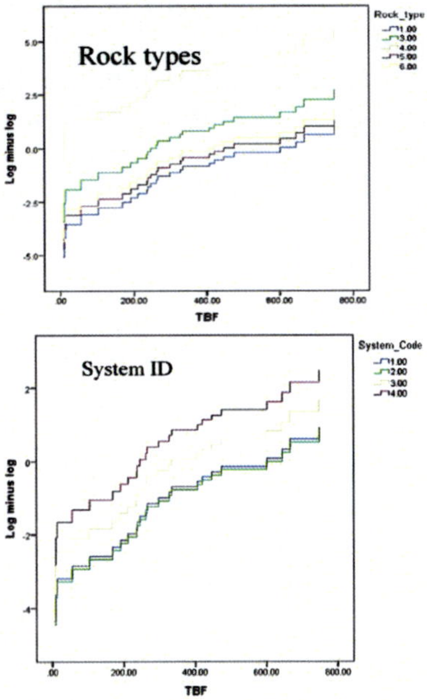

Fig. 2. The Log minus log graph for the time between crane shovels based on rock kind and system ID.

According to methodology steps in Fig. 1 on the left side of the algorithm, the GOF test needs to fit the best baseline function for data. The AIC and BIC can be used to find the best fit distribution for the baseline hazard rate [27]. The candidate distribution with the smallest AIC and BIC value is the best fit distribution to model the baseline hazard rate [28]. Many variations of AIC have been developed for model selection. The AIC was designed to estimate the Kullback–Leiber information of models in 1998; also, the delta AIC and the Akaike weights were introduced to measure how much better the best model is when compared with the other models. The AIC, delta AIC and AIC weights are calculated for each candidate model in the model selection process. Usually, the 'best' model is chosen to be the model with the smallest AIC. The BIC model selection criterion proposed by Schwarz in 1978. It referred to as the Schwarz information criterion, or the

Schwarz BIC. Similar to AIC, BIC is also calculated for each candidate model and the model with the smallest BIC is chosen to be the best mode. The only difference between AIC and BIC is that BIC uses a larger penalty on the increment of the model terms. In recent years, BIC has also been increasingly used as model selection criterion [30, 31]. It can be noted that both AIC and BIC have their own advantages and limitations. It cannot be guaranteed that one is better than another regardless of application scenarios. The reason is that the data, model type and other aspects of the modelling problems can be significantly important in determining which of the criteria is more suitable.

As mentioned, the AIC and BIC are applied to select the best fit distribution for the baseline hazard rate under two different techniques for model estimation (complete and backward stepwise) with for different distribution (Weibull, Exponential, Lognormal and Log-Logistic). Table 4 shows the values of the AIC and BIC for the different nominated distributions for the baseline hazard rate with the same risk factors. As a result in Table 4: shows, the Weibull PHM is the most suitable model for the data, as it has the smallest AIC or BIC among all the models. Therefore, the model with unobserved heterogeneity can better estimate the reliability of the teeth data. In stepwise methods, the score statistic is used to select variables for the model. In this study, corresponding estimates are obtained by a backward stepwise method and tested for their significance based on the Wald statistic (P-value).

Table 4. Goodness of fit of different reliability models.

Model	AIC	BIC
Weibull Model - Estimation stepwise	349.77	354.26
Weibull Model - Estimation complete	359.47	369.94
Exponential Model - Estimation complete	357.74	368.21
Lognormal Model - Estimation stepwise	425.31	425.31
Lognormal Model - Estimation complete	403.84	414.31
Log-Logistic Model - Estimation stepwise	356.13	362.12
Log-Logistic Model - Estimation complete	360.92	371.40

SYSTAT software is used to estimate the value of the regression vector. The asymptotic distribution of the Z statistic is chi-square with degrees of freedom equal to the number of parameters estimated. In the backward stepwise procedure, the effects of one risk factor, "Temperature" ($\mathbf{z_t}$) is found significant at the 10% level. The estimates of α (coefficient of the risk factor) and parameters of two parameters, Weibull baseline distribution (Shape and Scale), are listed in Table 5.

The operational reliability considering the environmental conditions are represented respectively as:

$$R(t, z) = \left(exp\left(-\frac{t}{238.766}\right)^{1.344}\right)^{exp(0.031z_t)} \tag{13}$$

Table 5. Estimation of reliability baseline parameters and risk factor coefficient.

Parameter	Estimate	Standard error	Z	p-Value
Shape	1.344	0.228	5.904	0
Scale	238.766	123.304	1.936	0.053
Temperature	0.031	0.032	0.975	0.329

The reliability and hazard rate of the teeth of crane shovels is now calculated and plotted for the mean value (150C), low value (−70C), and high value (200C) as normal, cold, and hot weather of zt, as shown in Fig. 3. The results show the teeth in hot weather are less reliable than the teeth in other weather conditions. As can be seen, their reliability reaches about 58% after about 100 h of operation. Furthermore, there is a 93% and a 95% chance that teeth will work without failure for 24 h in normal and cold weather, respectively. The results can help engineers and managers decide operation planning, maintenance strategy, sales contract negotiations, spare parts management, etc.

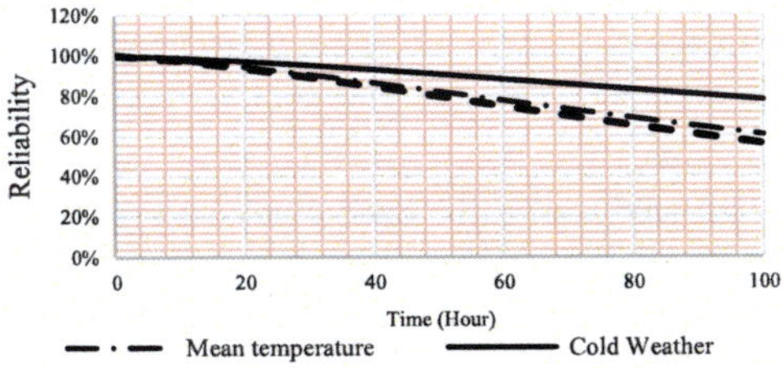

Fig. 3. Comparison of reliability performance of teeth in normal, cold, and hot weather.

5.4 Spare Part-Provision

According to the existing literature, if the distribution of baseline hazard rate of an item is Weibull, the effect of risk factors only changes the scale parameter of the distribution, and the shape parameter remains unchanged. Therefore, shape parameter (β) and scale parameter (η), of Weibull distribution considering the effect of risk factors are defined by [12]:

$$\begin{cases} \beta = \beta_0 \\ \eta = \eta_0 \left[exp\left(\sum_{i=1}^{n} z_i \alpha_i \right) \right]^{-\frac{1}{\beta_0}} \end{cases} \tag{14}$$

The $\overline{T}_s$ and $\sigma_s(T)$ of the Weibull distribution and the Power Law Process (PLP) can be calculated based on the shape and scale parameter, expressed as Eq. 15:

$$\overline{T}_s = \eta_s \Gamma\left(1 + \frac{1}{\beta_s}\right) \tag{15}$$

$$\sigma_s(T) = \eta_s \sqrt{\Gamma\left(1 + \frac{2}{\beta_s}\right) - \Gamma^2\left(1 + \frac{1}{\beta_s}\right)} \tag{16}$$

The number of required spare parts for teeth is calculated using Eqs. (13) and (14) for considering the effect of risk factors and without them. The operation with the probability of storage is equal to 95%. The results of the analysis for five years are shown in Table 6.

Table 6. Spare part-provision based on WPHM and without risk factor effect.

Year	Spare part	
	With risk factors	Without risk factors
1	21.49	15.67
2	40.24	29.09
3	58.51	42.09
4	76.52	54.89
5	94.37	67.55

The result of the data analysis of the case study shows that the required number of spare parts according to the WPHM approach is more than ignoring the effect of risk factors. In addition, Table 7 provides the number of spare parts required in 5 years considering the influence factor. As Table 7 shows, there is a big difference between cold and hot weather spare part required that is about hot weather two times bigger than a cold one.

Table 7. Required number of spare parts for different weather conditions over 5 years.

Year	Spare part	
	Cold weather	Hot weather
1	13.60	23.60
2	25.14	44.31
3	36.31	64.50
4	47.28	84.43
5	58.12	104.19

5.5 Spare Part Inventory Management

We start with the following assumptions:

- The cost of one tooth equals 20 USD$
- The cost of ordering one lot equals 2 USD$
- The annual holding cost equals 2 USD$ of the part cost
- The average lead-time is 5 days
- Cycle service confidence level is 95%.

The EQO and ReP concerning annual demand rates in different scenarios are calculated based on Eqs. (10) and (11) and tabulated in Table 8 for considering and ignoring the condition. Table 8 shows that for 1 year with considering risk factors' effect whenever the inventory position reaches 6.56 units/teeth, we should order 3.04. However, ignoring the risk factor, the EOQ and ReP of teeth for one year are equal to 5.6 and 2.51, respectively. In comparison, the EOQ and ReP in both conditions, with or without considering the operating environment's effect, illustrate the significance of these factors and their role in the actual life of the parts. In other words, the operating environment parameters should be considered in the process management of machines, in this case, the crane shovels.

Table 8. Economic order quantity.

Year	With risk factors		With risk factors	
	EOQ	Reorder point (ReP)	EOQ	Reorder point (ReP)
1	6.56	3.04	5.60	2.51
2	8.97	4.44	7.63	3.65
3	10.82	5.56	9.18	4.56
4	12.37	6.54	10.48	5.35
5	13.74	7.44	11.62	6.07

6 Conclusion

The operational environment may have a significant influence on the required number of spare parts. Hence, any method, which is used, for spare parts provision must be able to quantify such effects. The reliability-based spare part provision considering the effect of risk factors can quantify the effect of the operational environment. In these methods, the operational environment can be considered a risk factor. Their effects on the reliability characteristic and consequently on the required number of spare parts can be analyzed. Available regression methods such as PHM can be used by defining the risk factors for spare parts provision properly to quantify influence factors. However, it is necessary to examine the historical data to find an appropriate model, which fits the data more appropriately. For example, the reliability analysis of the bucket teeth in Jajarm mine using Weibull PHM shows that the reliability of a part in cold weather is higher than in other conditions. Moreover, the temperature has a significant effect on the reliability characteristics of the bucket teeth and, consequently, on the required number of spare parts. The noticeable difference in spare parts estimation is caused by considering and neglecting the temperature effect. The economic order quantity and reorder point calculation show about an 18% difference between the two cases.

References

1. Lanza, G., Niggeschmidt, S., Werner, P.: Optimization of preventive maintenance and spare part provision for machine tools based on variable operational conditions. CIRP Ann. Manuf. Technol. **58**(1), 429–443 (2009)
2. Jaarsveld, W.V., Dekker, R.: Spare parts stock control for redundant systems using reliability centered maintenance data. Reliab. Eng. Syst. Saf. **96**(11), 1576–1586 (2011)
3. Marseguerra, M., Zio, E., Podofillini, L.: Multiobjective spare part allocation by means of genetic algorithms and Monte Carlo simulation. Reliab. Eng. Syst. Saf. **87**(3), 325–335 (2005)
4. Barabadi, A.: Reliability and spare part provision considering operational environment. A case study. Int. J. Performab. Eng. **8**(5), 497–506 (2012)
5. Ghodrati, B.: Weibull and exponential renewal models in spare parts estimation: a comparison. Int. J. Performab. Eng. **2**(2), 135–147 (2006)
6. Kumar, U.D., Crocker, J., Knezevic, J., El-Haram, M.: Reliability, Maintenance and Logistic Support – A Life Cycle Approach. Kluwer Academic Publishers, New York (2000)
7. Xie, L.Y., Zhou, J.Y., Wang, Y.Y., Wang, X.M.: Load-strength order statistics interference models for system reliability evaluation. Int. J. Performab. Eng. **1**(1), 23–36 (2005)
8. Kumar, D., Klefsjo, B.: Proportional hazards model: a review. Reliab. Eng. Syst. Saf. **44**(2), 177–188 (1994)
9. Barabadi, A., Barabady, J., Markeset, T.: Application of accelerated failure model for the oil and gas industry in Arctic Region. In: The Proceeding of the IEEE International Conference on Industrial Engineering and Management (IEEM2010), Macao, China, pp. 2244–2248 (2010)
10. Louit, D., Pascual, R., Banjevic, D., Jardine, A.K.S.: Condition-based spares ordering for critical components. Mech. Syst. Signal Process. **25**(5), 1837–1848 (2011)
11. Ghodrati, B., Kumar, U., Kumar, D.: Product support logistics based on product design characteristics and operating environment, p. 21. Society of Logistics Engineers, Huntsville (2003)
12. Ghodrati, B., Banjevic, D., Jardine, A.: Developing effective spare parts estimations results in improved system availability, pp. 1–6. IEEE (2010)
13. Barabadi, A., Barabady, J., Markeset, T.: Application of reliability models with risk factors in spare part prediction and optimization–a case study. Reliab. Eng. Syst. Saf. **123**, 1–7 (2014)
14. Nouri Qarahasanlou, A.: Production assurance of mining fleet based on dependability and risk factor (case study: Sungun Copper mine). Ph.D. thesis in Mineral Exploita, Shahrood University of Technology Faculty of Mining, Petroleum & Geophysics, Iran, Shahrood (2017)
15. Nouri Qarahasanlou, A., et al.: Spare part requirement prediction under different maintenance strategies. Int. J. Min. Reclamat. Environ. **33**(3), 169–182 (2019)
16. Barabadi, A., Barabady, J., Markeset, T.: A methodology for throughput capacity analysis of a production facility considering environment condition. Reliab. Eng. Syst. Saf. **96**(12), 1637–1646 (2011)
17. Preacher, K.J., Merkle, E.C.: The problem of modelselection uncertainty in structural equation modeling. PsychologicalMethods **17**(1), 1–14 (2012). https://doi.org/10.1037/a0026804

18. Cox, D.R.: Regression models and life-tables. J. Roy. Stat. Soc. **34**(2), 187–220 (1972)
19. Kumar, D., Westberg, U.: Some reliability models for analysing the effects of operating conditions. Int. J. Reliab. Qual. Saf. Eng. **4**(2), 133–148 (1997)
20. Gorjian, N., Ma, L., Mittinty, M., Yarlagadda, P., Sun, Y.: A review on reliability models with risk factors. In: proceedings of the Fourth World Congress on Engineering Asset Management. Marriott Athens Ledra Hotel, Athens, 28–30 September 2009
21. Kumar, U.D., Crocker, J., Knezevic, J., El-Haram, M.: Reliability, Maintenance and Logistic Support: A Life Cycle Approach. Kluwer Academic Publishers, USA (2000)
22. Barabadi, A., Barabady, J., Markeset, T.: Maintainability analysis considering time-dependent and time-independent risk factors. Reliab. Eng. Syst. Saf. **96**, 210–217 (2011)
23. Kleinbaum, D.G.: Survival Analysis. Springer, Heidelberg (2011)
24. Kumar, D., Klefsjö, B.: Proportional hazards model: a review. Reliab. Eng. Syst. Saf. **44**(2), 177–188 (1994)
25. Barabadi, A.: Reliability and spare parts provision considering operational environment: a case study. Int. J. Performab. Eng. **8**, 497 (2012)
26. Kleinbaum, D.G., Klein, M.: Survival Analysis: A Self-Learning Text (2012)
27. Lee, E.T., Wang, J.: Statistical Methods for Survival Data Analysis. Wiley, Hoboken (2003)
28. Garmabaki, A., Ahmadi, A., Block, J., Pham, H., Kumar, U.: A reliability decision framework for multiple repairable units. Reliab. Eng. Syst. Saf. **150**, 78–88 (2016). https://doi.org/10.1016/j.ress.2016.01.020
29. Akaike, H.: Information theory and an extension of the maximum likelihood principle BT - selected papers of Hirotugu Akaike. In: Second International Symposium on Information Theory, pp. 199–213 (1998). https://doi.org/10.1007/978-1-4612-1694-0_15
30. Cobos, C., Munoz- Collazos, H., Urbano-Munoz, R., Mendoza, M., Leon, E., Herrera-Viedma, E.: Clustering of web search results based on the cuckoo search algorithm and balanced Bayesian information criterion. Inf. Sci. **281**, 248–264 (2014). https://doi.org/10.1016/j.ins.2014.05.047
31. Hooten, M.B., Hobbs, N.T.: A guide to Bayesian model selection for ecologists. Ecol. Monogr. **85**(1), 3–28 (2015). https://doi.org/10.1890/14-0661.1

Analysis of Breakdown Reports Using Natural Language Processing and Machine Learning

Mobyen Uddin Ahmed[1], Marcus Bengtsson[1,2], Antti Salonen[1](✉), and Peter Funk[1]

[1] Mälardalen University, Högskoleplan 1, 721 23 Västerås, Sweden
{mobyen.ahmed,marcus.bengtsson,antti.salonen,peter.funk}@mdh.se
[2] Volvo Construction Equipment, Västerås, Sweden

Abstract. Proactive maintenance management of world-class standard is close to impossible without the support of a computerized management system. In order to reduce failures, and failure recurrence, the key information to log are failure causes. However, Computerized Maintenance Management System (CMMS) seems to be scarcely used for analysis for improvement initiatives. One part of this is due to the fact that many CMMS utilizes free-text fields which may be difficult to analyze statistically. The aim of this study is to apply Natural Language Processing (NPL), Ontology and Machine Learning (ML) as a means to analyze free-textual information from a CMMS.

Through the initial steps of the study, it was concluded though that none of these methods were able to find any suitable hidden patterns with high-performance accuracy that could be related to recurring failures and their root causes. The main reason behind that was that the free-textual information was too unstructured, in terms of for instance: spelling- and grammar mistakes and use of slang. That is the quality of the data are not suitable for the analysis. However, several improvement potentials in reporting and to develop the CMMS further could be provided to the company so that they in the future more easily will be able to analyze its maintenance data.

Keywords: Natural language processing · Machine learning · Computerized maintenance management system · Recurring breakdowns · Root cause failure analysis

1 Introduction

Upholding a high machine utilization and availability through, for instance, maintenance has been one, out of several, important prerequisites for companies to focus on in order to be competitive. Lean production, with for instance, reduction of buffers, has made this even more evident.

In order to stay competitive, it is necessary for companies to continuously improve their efficiency and productivity. Improvements should not only be directed towards digitalization, but also basic maintenance must be improved [1, 2]. However, basic maintenance can in many cases be improved using new technologies found within digitalization, in many cases as a means of improving decision-making. Many maintenance

R. Karim et al. (Eds.): IAI 2021, LNME, pp. 40–52, 2022.
https://doi.org/10.1007/978-3-030-93639-6_4

decisions are though seldom fact-based [3, 4] point out data-driven decision-making as one out of four dimensions of smart maintenance. Improvements in both maintenance efficiency and effectiveness is important [5]. In order to prioritize these improvements, it is essential for companies to analyze its current state, as well as the historic behavior of the equipment [6]. As a tool to use in this process a Computerized Maintenance Management System (CMMS) is essential [7, 8]. Also, in order to improve the program for preventive maintenance it is vital to analyze the failures that occur in the equipment [9, 10]. As a means, [9] states that ideally, a CMMS may offer a platform for decision analysis, thus providing management support. Further [2] state: *"Through logging in computerized maintenance management systems much data on, e.g., failure reports, disturbances, scheduled preventive maintenance, etc., has laid the foundation to possible maintenance efficiency and effectiveness actions."* (p. 5).

In order to reduce machine failures and recurring failure, it is of key importance to log the failure causes in the CMMS [11]. However, it seems that CMMS are scarcely used for analysis of maintenance records [9]. In addition, on a general level, [12, p. 112] state that: *"Cross-industry studies show that on average, less than half of an organization's structured data is actively used in making decisions—and less than 1% of its unstructured data is analyzed or used at all. More than 70% of employees have access to data they should not, and 80% of analysts' time is spent simply discovering and preparing data"*. One major issue with the data in a CMMS is that it is often of free-text character and this type of data may be difficult to analyze using automated systems. Therefore, it is quite common that this data is analyzed manually which not only consume resources but also takes a long time to get done [13], often with a limited result regarding time-period and in cross-case analysis. That is, only shorter periods of times are analyzed and the analysis is often only performed on one or a few machine equipment's as opposed to all machines of a selection and all its years in operations. See for instance [13, 14] in where cases with the aim to cluster and classify maintenance logs using text data mining and Natural Language Processing has been presented.

There are two ways forward, one is to minimize or avoid free text in failure logs and the other is to use more sophisticated approaches to process the free text. Both approaches have weaknesses and strengths. Free text is valuable since it enables technicians to describe rare or complicated faults in a rich way transferring experience to the next person reading it. It also gives the opening for insufficient and incomplete information and content and quality may by nature have a wide variation. Minimizing or eliminating free text and instead requiring highly structured text would make the failure reports and action highly accessible for automated processing and analysis, but at the same time heavily restricts the way how faults and actions can be described with the risk that in complex domain valuable experience and knowledge is lost.

In this work we have adopted the approach of NLP processing [15] using computational linguistic and artificial intelligence to understand the meaning of the free text and extract the important information from it. There is also promising work in the field of Natural Language processing of failure reports enabling improved maintenance e.g. [16].

The aim of this paper is to apply Natural Language Processing (NPL), Ontology and Machine Learning (ML) to increase the efficiency in analyzing free-textual information from a CMMS in order to find improvement potential for maintenance departments.

2 Approach and Method

The approach taken is to take the available free text information and process it in order to extract the most valuable information from it so this information can be used in statistical processing and for decisions in predictive and preventive maintenance. The overall approach taken divides the proves in 4 phases as described in Fig. 1.

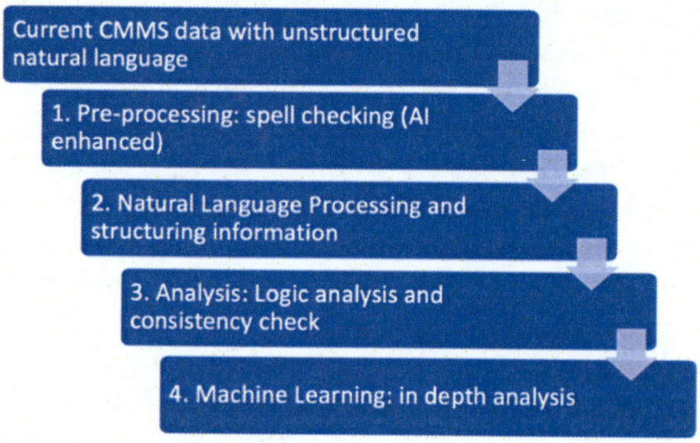

Fig. 1. Overview over the proposed process to extract relevant information from the maintenance reports.

By using a large amount of real maintenance reports (1700) we immediately see the difficulties and potential of the different techniques from NLP, AI and ML applied. In the project we took a minimal approach where we did implement basic versions of the different techniques in order to see how far we can reach with standard approaches of NLP.

The current work is the initial step to explore such an approach and how much effort is needed to implement a framework that extracts all relevant data from work orders. The initial work and implementation was made by six students at masters' level with AI background who got access to the data and received extended supervision and guidance. The total amount of time spent on the project was six person months i.e. 1 month for each of the six persons (for more details on the project see [17]).

2.1 Case Company

The project was performed within an industrial company. The company is a discrete item manufacturing plant supplying internal customers with components for the automotive industry.

2.1.1 Case Company Context

The plant has roughly 700 employees. Its manufacturing processes include machining, curing, assembling, testing and painting of components such as axles and transmissions. The plant has roughly 300 manufacturing machines; a heat treatment facility; various assembly equipment's such as presses, torch wrenches, and manual guided vehicles; test benches; and a paint shop process.

In the reference company, roughly 70 employees are working with maintenance. Repairmen perform most of the hands-on work with tasks including, e.g.: corrective and preventive maintenance including condition monitoring and improvement work. Other functions include maintenance engineers and developers, procurers, and storage personnel. The reference company has worked with the same Computerized Maintenance Management System (CMMS) since 1999. All maintenance work orders are logged in the CMMS with information containing, e.g.: work order request (free text, most often performed by operators), work order type, timestamps, spare parts used, maintenance cost (automatically based on working hours, spare part cost, external service cost etc.), and work order report back (free text, performed by repairmen).

2.1.2 Improvement Potential in Case Company

Self-perceived and to some extent analyzed [6, 18], the reference company is suffering from recurring failures. In order to find these recurring failures maintenance engineers manually analyze the work orders in the CMMS, machine for machine, and specifically the free text fields in the work order request as well as the work order report back. The process is according to the maintenance engineers at the reference company rather time consuming and it is difficult to cross-reference machines and their recurring failures with other machines and their recurring failures. Therefore, the case company came up with the idea of the student project.

2.1.3 Case Study Data

Data was downloaded for a complete manufacturing cell containing five machining centers, as well as one washing machine, one storage crane, and four overhead cranes. Data was obtained with all maintenance work orders performed since the equipment were installed in 2007–2008 until the first quarter of 2019. The work orders of breakdowns, amounted to a total of almost 1700 during all years. All information in the work order request and work order report back was downloaded and shared with the student team. This data thusly includes, i.e., timestamps, information on spare parts, free text fields. The student team also visited the reference company and got to meet with production and maintenance engineers that work with this manufacturing cell.

3 Implementation

Overall implementation of the selected methods identified by the literature study have been conducted through MATLAB, a programming language developed by Math Works. The framework contains both the unsupervised approach (clustering) and also supervised approch (classification). Both the approches are implemented and evaluated completely in different experiments.

3.1 Pre-processing

The textual data is pre-processed in order to convert the raw data sets into an understandable format by the computer. Various techniques are applied from the NLP method, such as tokenization, stemming, stop-word removal, and lemmatization. A Bag of Word (BoW) was generated that contain the word frequency in the documents. In the *tokenization,* all the text sentences are broken down into individual tokens. A *stemming* process is conducted to remove text extension e.g. x-, z-, and y- while referring to the same kind of machine part in the free text. *Lemmatization* is conducted to remove unnecessary symbols and punctuation. Here, an ontology was applied that consists of two columns, referring to the term and its meaning to overcome the misspellings. Again, the ontology also splits some complex terms into smaller terms, e.g. "electrictablecontroller" -> "electric", "table" and "controller", to include the relations and meaning in the tokens. A standard *stop words* list was made and used along with the domain-specific words to remove unnecessary terms from the tokenized documents by comparing the stop words list. This term list is than sent through a *stemming function* that reduces the word to its stem. Here, the stemming function is used to handle the Swedish language i.e. removes suffixes from the tokens using the snowball algorithm. Finally, a BoW is generated which determine the word frequency in the tokenized documents as tf–idf or TFIDF, short for term frequency–inverse document frequency. As one limitation of BoW is that the most frequent words have more power, so some of the common frequently used words been removed by manual inspection before feeding to the ML algorithm. The research work re-uses some existing libraries for the pre-processing but as support for Swedish spelling correction is not present in MATLAB, this task had to be performed manually, with corrected spelling being added as part of the ontology.

3.2 Clustering

In order to identify both common and hidden patterns based on the unlabeled data set several clustering approaches are applied. Here, the aim of the clustering approaches is to divide the data into several groups and to achieve the divided clusters that can be associated with the root cause of the breakdown. The clustering approaches are unsupervised machine learning algorithms, such as K-means clustering, Agglomerative Hierarchical clustering, and DBSCAN. In *K-means,* Euclidean distance is used to separate data points into groups that are similar to one another [19]. Here, a random centroid data point is selected as the starting point of every cluster, and an iterative process is conducted to optimize the cluster. The iterative process stops when the optimized value of centroids is stabilized, and the number of K is determined with the Elbow method [20]. *Agglomerative Hierarchical* clustering algorithm start building cluster based on a single data point in each cluster, then based on the similarity value it merges the clusters into larger clusters and process continues until all of the objects are finally lying in a single cluster [21]. *DBSCAN* separates high-density clusters from low-density clusters with two critical parameters i.e. epsilon ('eps' is the radius of the neighborhood), and minimum points ('MinPts' is the minimum number of neighbors within the" eps" radius). For each core point, if it has not been assigned to a cluster, it creates a new cluster and finds all its connected density points and assigns them to the same cluster as the core point and iterates itself [22].

3.3 Classification

The data sets are manually labelled into six classes based on the representing cause of the shutdown to apply the supervised machine learning and observed classification accuracy. The six classes are: 1) Leakages, 2) Mechanical problems, 3) Electrical faults, 4) Hydraulic faults, 5) Computer problems and 6) Uncleanness. Here, three supervised machine learning classification algorithms are employed, 1) Naïve Bayes, 2) Random forest, and 3) Logistic Regression. Naïve Bayes is a probabilistic machine learning algorithm designed to accomplish classification tasks. Given the data set, *Naïve Bayes* converts into a frequency table with generated likelihood by finding the probabilities and the class with the highest probability is the outcome of the prediction [23]. The algorithm *Random forest* generate decision trees during the training and provide an output of the individual decision tree. These decision trees vote on the most proficient cases to group a given case of the information [24], with a purpose to avoid over-fitting. The algorithm *Logistic Regression* measures the relationship between the categorical dependent variable and one or more independent variables by estimating probabilities using a logistic function. Here, a multinomial logistic regression is a form of logistic regression used to predict a target variable have more than 2 classes. It is a modification of logistic regression using the softmax function instead of the sigmoid function the cross entropy loss function It uses a black box function to understand the relation between the categorical dependent variable and the independent variables [25]. Since both the Logistic Regression and Naïve Bayes does not work efficiently with a large number of dimensions, a Term Frequency-Inverse Document Frequency (TF-IDF) is applied to reduce the number of dimensions.

4 Evaluation

An evaluation to observe the importance of structured data i.e. if the data is separable or not has been conducted based on three tests using t-distributed Stochastic Neighbor Embedding (t-SNE) [26]. The tests are: 1) with raw data, 2) with processed data and 3) with processed using ontology.

The test results are presented in the following Figs. 2, 3 and 4, here, x- and y-axis represents the numerical weights, colors are different class (Black: IT, Blue: Electrical, Green: Mechanical, Red: Uncleanness, Yellow: Hydraulic, and Cyan: Leakage), detailed of the figures from [17]. As can be observed from Fig. 2, the raw data set (≈7000) is very compact and agglomerated without any filtration. Figure 3 present, the data (≈4500) is more spread out after the pre-processing with stop word removal, stemming, and filtering. Data set (≈3900) is more spread out while ontology has been performed together with the pre-processing as presented in Fig. 4, however, the groups are still bundled together without presenting any distinct divisions.

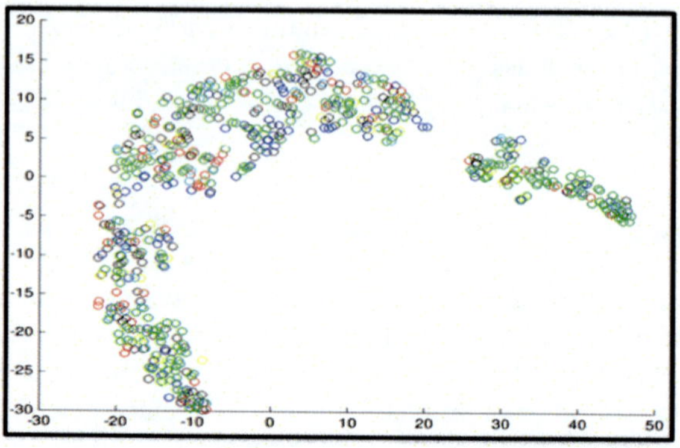

Fig. 2. T-SNE graph based on the raw data sets.

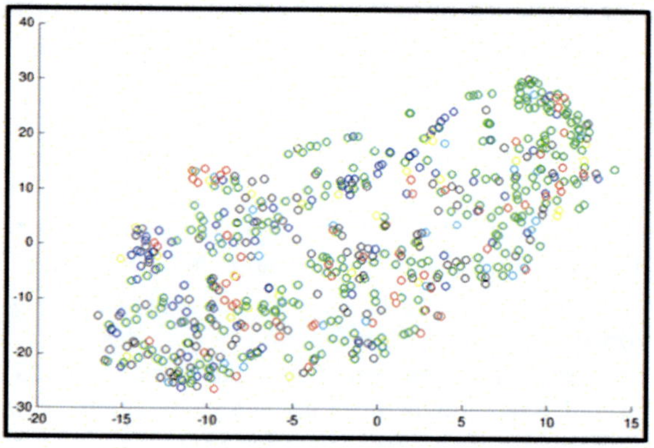

Fig. 3. T-SNE graph based on the processed data sets

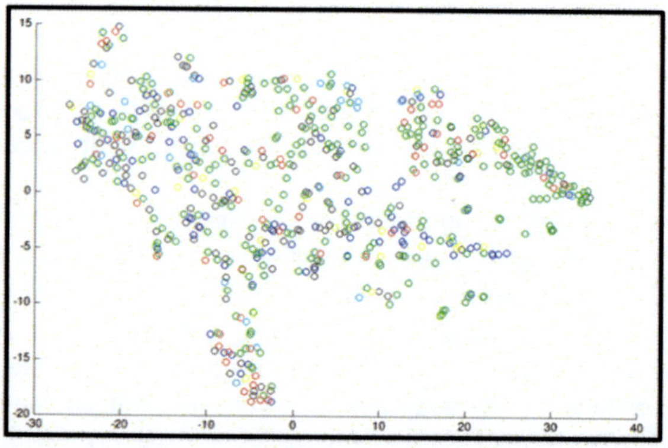

Fig. 4. T-SNE graph based on the processed data sets using ontology.

4.1 Clustering

The evaluation results for three clusters i.e. K-means clustering, Agglomerative Hierarchical clustering, and DBSCAN are presented in the Fig. 5, 6 and 7 respectively. As can be seen form the figures, the work orders assigned to the clusters produced from the clustering algorithms do not have any strong correlation between the clusters and the root cause of machine breakdowns.

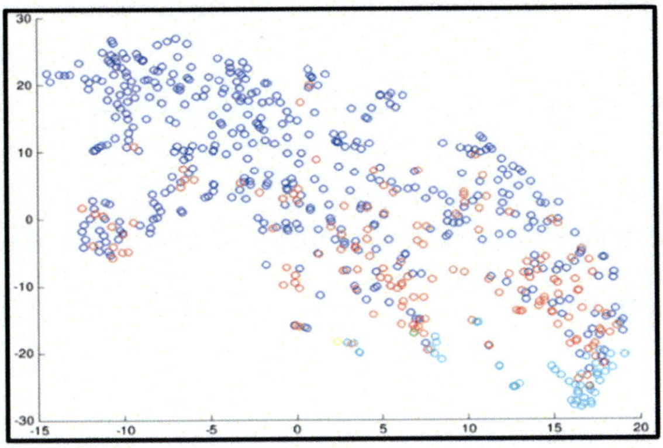

Fig. 5. Clustering results using K-means with 6 classes

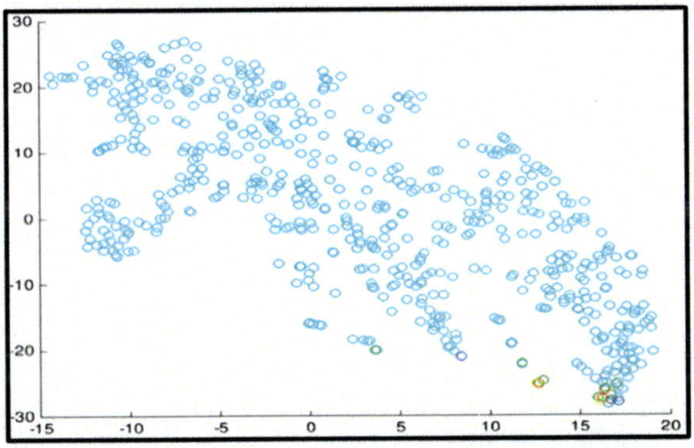

Fig. 6. Clustering results using agglomerative hierarchical clustering with 6 dimensions.

4.2 Classification

Regarding the evaluation of the performance of supervised learning algorithms, that is the accuracy of the classification is observed based on 604 entries. In order to protect against overfitting, k-fold cross validation was used when training the classification models, here the configured K was 10 folds. Furthermore, the measurement was also sampled 10 times using Naïve-Bayes, Random Forest, and Logistic Regression. For this

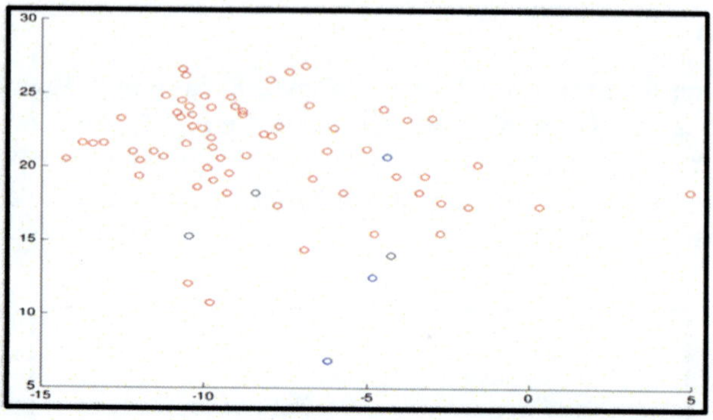

Fig. 7. Clustering results using DBSCAN with 3 minimum neighbors.

evaluation, Random Forest was configured with a depth of 30 decisions. The average result can be seen in Table 1. As can be seen from the table, Random forest using BoW achieved the highest amount of accuracy with 53%. Out of the three classification algorithms that used TF-IDF, Logistic Regression proved to have the highest amount of accuracy.

Table 1. Average results of 10-fold cross validation in classification accuracy.

Classification methods	Total correctly classification data	Total samples	Calculated accuracy in percentage
Naive-Bayes	149	604	25%
Logistic Regression	267		45%
Random Forest with TF-IDF	244		40%
Random Forest with Bow	318		53%

4.3 Estimated Time Analysis

A time estimation on the replacement of spare parts has been observed and analyzed i.e. the total amount of replaced spare parts, and their time of replacement between and during of spare parts changes is compared with labelled data. A *Mean Time Between Work orders* is calculated that presents the average time for between two labelled classes. Also, a *Mean Down Time* is calculated that present the average time for between all labelled classes. The number of spare parts used per class is also analyzed and the results are presented in Table 2, here, the total time with Mean Down Time per class are presented in 'HH:MM' and Mean Time Between Work orders in 'DD:HH:MM' format. As can be seen, the most common type of failure is mechanical, which is typical for industrial machines.

Table 2. Result from the time estimate analysis

Classes	No. of occurrences	Total time	Mean Down Time	Mean time between work orders
IT	128	1808:25	14:14	33:14:06
Electrical	89	1535:45	17:27	48:06:42
Mechanical	269	6098:22	22:40	15:19:17
Uncleanness	61	895:12	14:40	70:14:13
Hydraulic	31	418:21	13:29	141:08:08
Leakage	26	1139:23	43:49	163:17:17
Total	604	11869:31	21:03	78:21:22

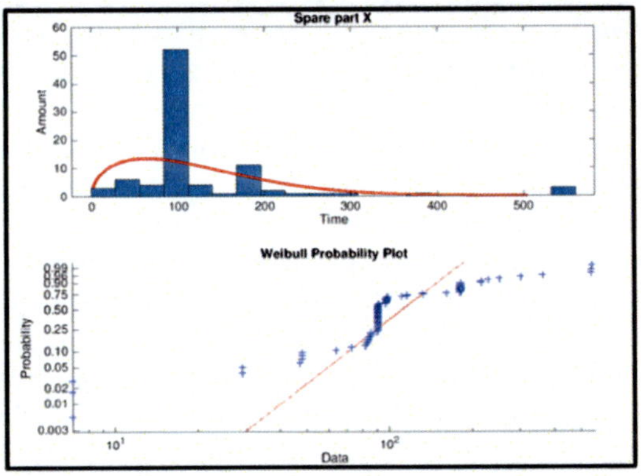

Fig. 8. Representation of average time in machine 1 to 5 between changes of one spare part.

A Weibull Probability plot of the replacement of one spare part for a machine group (Fig. 8). Every spare part has been grouped and then analyzed in a similar fashion to show the average life length of the different parts.

The *Mean Down Time* for the classes of work orders in percentage where spare parts were used is presented in Table 3. As can be observed, the largest percentage per class of spare parts are used in "Mechanical" and "Leakage" classes. Here, the percentage of the total work orders with spare parts are 23.3% and 19.2% respectively and more time has been taken to change parts in these classes compared to others.

5 Discussion and Future Work

According to the result, the characteristic and the content of the work orders diverse a lot, which was anticipated, and this presents to put more effort in the structure to the free text without preventing the technicians to use free text. This could be done by improving the work order interface and providing different required and optional text fields. This study

Table 3. Time estimation analysis for spare parts where total time and MDT in HH:MM format

Classes	Percentage	Total time	Mean down time
IT	2.3% (3/128)	36:45	12:15
Electrical	5.6% (5/89)	54:35	10:55
Mechanical	23.3% (60/269)	2100:18	35:00
Uncleanness	6.6% (4/61)	53:02	13:15
Hydraulic	3.2% (1/31)	4:29	4:29
Leakage	19.2% (5/26)	882:13	176:26
Total	12.9% (78/604)	3131:25	42:03

considers a very basic approach to extract relevant information from the textual work orders, which is not enough according to the results observation. A more advanced NLP approach and improved ontology needs to be applied to extract all available information at the same level or close to the performance of a technician reading the report. Also, the performance of the implemented system should be evaluated against the performance of a technician reading the work order, as they also missed to extract relevant information for the un-structured data samples.

As the work orders from the CMMS database are unstructured, it would be recommended to structure the free text and other columns in a more standardized way. Where the description would be a clear indicator of what has gone wrong and the free text field for the repairmen would be centered on what was done to fix the problem. Also, to create a spellchecker, language checker, the improvement of the domain specific ontology for the free text inputs into the database would increase the accuracy of the algorithms applied in the project.

To continue to group words depending on their domain-specific nature until as few un-grouped words remain as possible. However, there is a need to implement a prediction model based on the analysis of the results from the clustering and classification. Since the data is not sufficient enough to make a complete statistical model, the previously mentioned steps need to be done before this could be implemented. Lastly, the labeled data was unbalanced, thus, the current training process does not take this into account, which likely is affecting the classification accuracy. The training process could be improved if some form of balancing was performed, for example, random under-sampling.

Acknowledgments. The study was conducted in the context of the XPRES framework at Mälardalen University. The authors would like to acknowledge the students Albert Bergström, Subaharan Kailayanathan, Saji Kamdod, Dmitrii Shabunin, Simon Monié, Martin Norrbom; for their work on the data and effort in the course project held in Mälardalen University, Sweden, many thanks to them.

References

1. Bengtsson, M., Lundström, G.: On the importance of combining "the new" with "the old" – one important prerequisite for maintenance in industry 4.0. Proc. Manuf. **25**, 118–125 (2018)
2. Kans, M., Campos, J., Salonen, A., Bengtsson, M.: The thinking industry – an approach for gaining highest advantage of digitalisation within maintenance. J. Maint. Eng. **2**, 147–158 (2017)
3. Gopalakrishnan, M.: Data-driven decision support for maintenance prioritisation – connecting maintenance to productivity. Ph.D. diss., Chalmers University of Technology Sweden: Department of Industrial and Materials Science (2018)
4. Bokrantz, J., Skoogh, A., Berlin, C., Wuest, T., Stahre, J.: Smart maintenance: an empirically grounded conceptualization. International Journal of Production Economics **223**, 107534 (2019)
5. Bengtsson, M., Salonen, A.: Requirements and needs – a foundation to reducing maintenance-related waste. In: Koskinen, K., et al. (eds.) Proceedings of the 10th World Congress on Engineering Asset Management, pp. 105–112. Lecture Notes in Mechanical Engineering. Springer, Cham (2015). https://doi.org/10.1007/978-3-319-27064-7_10
6. Salonen, A., Bengtsson, M., Fridholm, V.: The possibilities of improving maintenance through CMMS data analysis. Submitted to SPS2020 (2020)
7. Duffuaa, S.O., Raouf, A.: Planning and Control of Maintenance Systems. Springer, Cham (2015). https://doi.org/10.1007/978-3-319-19803-3
8. Labib, A.: A decision analysis model for maintenance policy selection using a CMMS. J. Qual. Maint. Eng. **10**(3), 191–202 (2004)
9. Labib, A.: World class maintenance using a computerized maintenance management system. J. Qual. Maint. Eng. **4**(1), 66–75 (1998)
10. Garg, A., Deshmukh, S.G.: Maintenance management: literature review and directions. J. Qual. Maint. Eng. **12**(3), 205–238 (2006)
11. Rausand, M., Øien, K.: The basic concepts of failure analysis. Reliab. Eng. Syst. Saf. **53**(1), 73–83 (1996)
12. DalleMule, L., Davenport, T.H.: What's your data strategy. Harv. Bus. Rev. **95**(3), 112–121 (2017)
13. Stenström, C., Al-Jumaili, M., Parida, A.: Natural language processing of maintenance records data. Int. J. COMADEM **18**(2), 33–37 (2015)
14. Edwards, B., Zatorsky, M., Nayak, R.: Clustering and classification of maintenance logs using text data mining. In: Proceedings of the 7th Australasian Data Mining Conference (2008)
15. Wagner, S.: Natural language processing is no free lunch. In: Perspectives on Data Science for Software Engineering, pp. 175–179. Morgan Kaufmann, Boston (2016)
16. Carchiolo, V., Longheu, A., di Martino, V., Consoli, N.: Power plants failure reports analysis for predictive maintenance. In: Proceedings of the 15th International Conference on Web Information Systems and Technologies (WEBIST 2019) (2019)
17. Bergström, A., et al.: Detection of breakdowns using historical work orders data analysis for preventive maintenance. Technical report at Mälardalen University, January 2020
18. Fridholm, V.: Improve maintenance effectiveness and efficiency by using historical breakdown data from a CMMS. Master thesis, Mälardalen University, Sweden (2018)
19. Kanungo, T., Mount, D.M., Netanyahu, N.S., Piatko, C.D., Silverman, R., Wu, A.Y.: An efficient k-means clustering algorithm: analysis and implementation. IEEE Trans. Pattern Anal. Mach. Intell. **24**(7), 881–892 (2002)
20. Kodinariya, T., Makwana, P.: Review on determining number of cluster in K-means clustering. Int. J. Adv. Res. Comput. Sci. Manag. Stud. **1**(6), 90–95 (2013)

21. Bouguettaya, A., Yu, Q., Liu, X., Zhou, X., Song, A.: Efficient agglomerative hierarchical clustering. Expert Syst. Appl. **42**(5), 2785–2797 (2015)
22. Schubert, E., Sander, J., Ester, M., Kriegel, H.P., Xu, X.: DBSCAN revisited, revisited: why and how you should (still), use DBSCAN. ACM Trans. Database Syst. **42**(3), 19:1-19:21 (2017)
23. Tang, B., Kay, S., He, H.: Toward optimal feature selection in naive bayes for text categorization. IEEE Trans. Knowl. Data Eng. **28**(9), 2508–2521 (2016)
24. Srivastava, A., Han, E.-H., Kumar, V., Singh, V.: Parallel formulations of decision-tree classification algorithms. In: Guo, Y., Grossman, R. (eds.) High Performance Data Mining: Scaling Algorithms, Applications and Systems, pp. 237–261. Springer, Boston (2002). https://doi.org/10.1007/0-306-47011-X_2
25. Dreiseitl, S., Ohno-Machado, L.: Logistic regression and artificial neural network classification models: a methodology review. J. Biomed. Inform. **35**(5), 352–359 (2002)
26. Gisbrecht, A., Schulz, A., Hammer, B.: Parametric nonlinear dimensionality reduction using kernel t-SNE. Neurocomputing **147**(1), 71–82 (2015)

An ICT System for Gravel Road Maintenance – Information and Functionality Requirements

Mirka Kans(✉), Jaime Campos, and Lars Håkansson

Linnaeus University, 351 95 Växjö, Sweden
{mirka.kans,jaime.campos,lars.hakansson}@lnu.se

Abstract. The gravel road network is an important function for rural residents and entrepreneurs. Traditional maintenance of gravel roads is well-functioning but provides a relatively high maintenance cost per unit length of the road, and every maintenance action as well as extraction and transport of new gravel contributes to increased climate impact and resource depletion. Today, maintenance planning is carried out periodically based on the maintenance history, which also is reflected in the economic models and procurement methods. Current maintenance plans may be enhanced and will not be a reliable basis in the future, e.g. due to climate change. Instead, real needs and conditions must be given greater consideration. Today, appropriate maintenance management systems are lacking, e.g. in order to be able to evaluate maintenance deficiencies, prioritize objects and choose the appropriate maintenance action. Moreover, the knowledge available at specific stakeholders is not shared with other actors. In this paper, an Information and Communication Technology (ICT) system for gravel road maintenance is proposed in the form of a cloud-based system covering the information needs of stakeholders in the gravel road maintenance ecosystem. Requirement specifications are given for the sub-systems intended for the maintenance executioner and the maintenance planner. The specifications are based on workshops and interviews conducted with stakeholders, where requirements were acquired e.g. in the form of User stories.

Keywords: Gravel road maintenance · ICT system · Requirements specification · Maintenance planning

1 Introduction

Gravel roads constitute a substantial part of the Swedish road network, as they make up about 35% of the public roads. In total, there are about 350,000 km of gravel road in Sweden, of which about 75 000 km are private-public roads with governmental subsidiaries and open for public use, and approximately 30 000 km are state/municipality owned roads [1]. These gravel roads constitute the last branch of the road network and have an important social as well as traffic function for sparsely populated areas; without them, it is difficult to conduct business and live in the countryside, and it is therefore

R. Karim et al. (Eds.): IAI 2021, LNME, pp. 53–64, 2022.
https://doi.org/10.1007/978-3-030-93639-6_5

important that they are kept in good condition. For best access, gravel roads must be maintained regularly. The interval between the maintenance actions depends on factors such as traffic frequency, environmental conditions as well as road condition [2, 3].

Planing, graveling and edge trimming is usually performed every three to five years in the southern part of Sweden. Traditional maintenance of gravel roads is well-functioning, but provides a relatively high maintenance cost per unit length of the road, and each maintenance action, such as the extraction and transport of new gravel, contributes to an increased climate impact and depletion of natural resources. Approximately 60,000 road owners are responsible for the management of the national private road network, of which about 40% receive state subsidies [4]. In order for a road to receive subsidies, the road holding must be organized in a road association, or the similar. However, it is difficult to find people to take the chair, and the average age is often high, while the general understanding of road maintenance and its execution is decreasing in the road associations [5].

In addition to limited resources and difficulty in engaging road owners, there are a number of other problems that need to be addressed, including urbanization that affects the rural communities' economic opportunities, seasonal variations in traffic density (especially the tourism industry can contribute to large variation), reduced number of contractors and reduced competence regarding gravel roads, as well as unclear and disseminating responsibility and financing models [5]. In Sweden, the government supports a vibrant rural area, with an emphasis on digitalisation, innovation power and infrastructure. However, the Government's investments in upgrading and maintenance of infrastructure have mainly been directed at the rail and the major roads, while gravel road maintenance has been a low priority [5]. In the spring budget 2019, the Government has promised to allocate about 60 million extra for maintenance of a private-public road, which is needed as a large stifled maintenance need of e.g. bridges exist in the road network. However, this investment is temporary and does not solve the underlying systematic problems. This is a recognized problem not only in Sweden, but in large parts of the world where gravel roads form an essential part of the road network [6–8]. Furthermore, climate change has a major impact on the gravel road maintenance, for example through more damage to roads due to more frequent and heavy rainfall, or difficulty to execute maintenance actions due to prolonged heat waves [9]. This means that current maintenance plans will not be a reliable basis for planning in the future, but real needs and conditions must be taken into account to a higher degree. Through a data-driven maintenance approach, the planning and decision-making become fact-based and based on real conditions, regardless of the maintenance strategy applied. Data-driven maintenance is especially suitable for predictive and prescriptive maintenance strategies [10]. An appropriate ICT system is needed for supporting this approach.

The purpose of this paper is to achieve a deepened understanding of the information and functionality requirements of the main stakeholders in the gravel road maintenance ecosystem. The main goal is consequently to specify information and functionality requirements for a cloud-based system supporting the stakeholders' needs.

2 Theoretical Framework

The research base for gravel road maintenance is relatively small compared to other infrastructure [6], but national as well as international studies exist, for example on the construction and maintenance of gravel roads, which affects maintenance needs, condition assessment, planning and optimizing of maintenance, and collaboration models regarding maintenance. One of the most comprehensive research initiatives is the ROADEX project [11, 12], which was conducted by countries in Northern Europe 1998–2012, addressing many of the above-mentioned areas. Gravel roads are found in sparsely populated parts of the world, but construction and maintenance differ depending on geographical, climatic and economic conditions. Large amounts of water affect road structure and maintenance in e.g. Northern Europe [3] and Australia [13]. In the northern hemisphere, snow and frost also affect maintenance needs [14, 15]. Resource shortage is a recurring problem: in addition to financial problems [8, 9], a lack of natural resources such as water [6] and gravel [15] is reported. Lack of competent personnel and equipment [6] are examples of organizational resource problems that occur. The strong link between the ability to assess the condition of fixed assets and to make effective maintenance-related decisions is known for the gravel road and other infrastructure [11, 12, 16]. The digital development today allows the use of sensors and real-time data management for robust and reliable collection of condition data [11, 12]. The research on predictive maintenance for road transport infrastructure is focused, with very few exceptions, on paved roads, and cannot be applied directly for gravel roads, as damage development differs from paved roads [17]. Current measurement systems for assessing the condition of gravel roads are not adapted for predictive maintenance; they are either subjective, and therefore uncertain [2], or time-consuming when they are to be performed manually, see e.g. [18]. Automated measurement systems are often costly, non-robust or provide too low a degree of detail to provide a reliable overall picture of the condition of the road [10, 19, 20]. Lack of relevant data leads to routine or ad hoc maintenance of gravel road maintenance [21].

Some decision support for gravel road has been developed, e.g. for the assessment of gravel layer conditions, prioritization of maintenance activities and more efficient use of gravel pit resources [19, 21–23], but these are usually local or specific applications and do not cover the total need. The need to develop information systems for gravel road maintenance is therefore recognized [11, 17, 18, 24]. The use of gravel road maintenance systems varies; from being well-developed in South Africa to deficient in Australia, the United States and Europe [19]. A national audit showed that the Swedish Road Administration, now the Swedish Transport Administration, lacked appropriate maintenance management systems, e.g. in order to be able to evaluate maintenance deficiencies, prioritize objects and choose the appropriate maintenance action. Moreover, the knowledge available at the Swedish Transport Administration was not shared with other actors [25]. In a study conducted by the Swedish Road Administration, it was concluded that one cannot significantly affect customer satisfaction with maintenance efforts alone [26]. Instead, the development of methods for information sharing, dialogue and maintenance management was recommended. In other words, there is a need to develop systems that not only handle maintenance-related information, but gravel road information in general between actors, see also [27].

3 ICT Solution for Gravel Road Maintenance

3.1 General Approach and Study Design

This study forms a part of the three year project Sustainable maintenance of gravel road, in which new technical solutions are evaluated and proposed in the area of gravel road maintenance, such as a device for gravel reuse, a method for gravel road condition monitoring, and an ICT system for gravel road maintenance management. Our approach for the development of the latter is to thoroughly map the needs and conditions of all relevant actors in the ecosystem "Maintenance of gravel road" and then propose appropriate forms of collaboration, planning tools and digital solutions to support these actors. The project will cover the individual actors' needs as well as the overall needs that exist in the ecosystem (see Fig. 1), from a systems perspective.

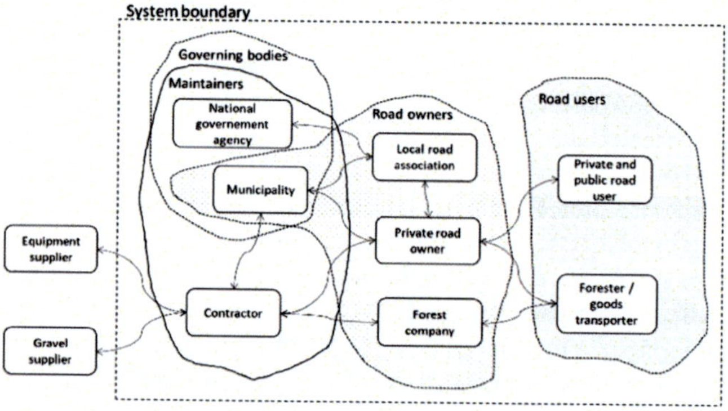

Fig. 1. System boundaries and main modules.

The project involves a great deal of interaction between researchers and other actors, where results are achieved in collaboration. An action-based approach is therefore used in the project, and the research method is based on collaborative design methodology [28, 29]. The requirements elicitation process included four main steps: 1) Stakeholder analysis, 2) User context analysis, 3) Use case scenario development, and 4) Requirements specification and validation [27]. These steps, especially steps 2–4, were iterated in several rounds, sometimes focusing on the subsystem, and other times on the overall system. The main purpose of the stakeholder analysis was to identify the main stakeholders in the ecosystem. Unstructured interviews, governmental reports, statistics and organizational descriptions, such as web sites and annual reports, were used as a basis. In total 17 interviews were carried out with actors representing all stakeholders except private road owners and foresters according to Fig. 1. The user context analysis included interviews with different stakeholders, as well as observations onsite for gaining an understanding of the environments and working conditions in which the system should be implemented. Two workshops were also arranged where information and functionality requirements were gathered using the User stories method. User stories are short and simple descriptions of what to be achieved with an ICT system told from the perspective of the user or the customer [30]. The format is: As a <type of user>, I want

<some goal> so that <some reason>. The first workshop was limited to the project participants, and mainly focused on the maintainer's perspective. The second workshop included participants representing ICT developers as well as maintainers.

3.2 Information and Functionality Requirements in the Ecosystem

Figure 1 describes the main stakeholders in the ecosystem. To the right, the users of the infrastructure, i.e. the gravel roads, are found (Road users). These comprise people living in the area, but also companies using the roads for transport of raw material such as timber or food supplies. In addition, people using the roads to access recreation areas, emergency vehicles, and social functions such as garbage collection frequently uses the roads. The roads could be owned by private persons, companies, municipalities, or the government (Road owners). Private roads are managed in the form of road associations. The national government agency, The Swedish Transport Administration, is regulating the gravel road maintenance, and manages subsidiaries for private road owners. Some municipalities give additional funding for road maintenance. These two stakeholders are both Governing bodies. The roads are often maintained by third-party contractors, but also municipalities and The Swedish Transport Administration have their own maintenance organisations (Maintainers). Outside the system boundary, suppliers are found. Gravel is the main raw material resource, while heavy vehicles suppliers delivers equipment such as graders.

Four generic stakeholders were identified as candidate system users: Road users, Road owners, Governing bodies, and Maintainers. These stakeholders have similar needs and requirements. The main need of the Road user is accessibility and driving comfort. Information required to achieve this is the current road condition and road shutdowns, in order to view road status for a specific road. Road owners strive for an efficient and adequate level of maintenance, and therefore the management of contracts is essential. Information required for this includes current road conditions, planned and conducted maintenance, as well as maintenance costs and budgets. The main need of the Governing bodies is to assure accessibility, safety, environment and health for public infrastructure, but also efficient handling of subsidies. Information required to achieve this is the current road condition, in order to manage and control road quality, and in addition road information such as ownership and length in order to manage subsidies. Maintainers naturally require quite a lot of information for achieving cost-efficiency and high resource utilisation, such as current road condition, weather forecasts and ground information, planned and conducted maintenance, as well as economic information. Two major functions are "planning, follow up and optimisation of maintenance", and "financing and contract management". The maintainer also require information for the daily work. The maintainer was therefore divided into two main user roles: maintenance executioner and maintenance planner.

Technology wise the aim is to develop a system based on the Cloud computing approach. The Cloud computing is a network-based computing over the Internet, which appeared during the late 2007's. Its main objective is to deliver on demand high quality ICTs services containing high reliability, scalability and availability in a distributed environment. In Cloud computing, the entire network-based computing over the Internet is seen as a service. The Clouds can be categorized as Infrastructure as a Service (Iaas), Platform as a Service (PaaS) and Software as a Service (SaaS). The important aspects of

Cloud computing in connection to the resource management of IaaS, such as scalability, customization and reusability as well as the performance metrics, namely delay, bandwidth reliability and security are discussed in comprehensive survey paper of Manvi and Shyam [31]. Thus, in the case of the ICT solution for gravel road maintenance the services provided are Iaas, PaaS and SaaS, more specifically for the different users, such as, road user, maintenance executioner, maintenance planner etc. Figure 2 below highlight the data that the system gathers and store for later use in the different software, which supports the various uses in their work tasks. When using cloud architecture, there is a need to understand how the security issues will be taken into account [32]. For example, where should the data for diagnostics and analysis be done etc. because depending on where this is handled are the security aspects that should be taken into account.

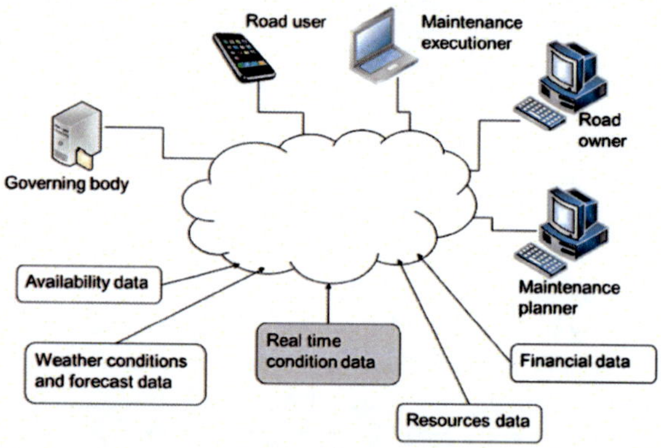

Fig. 2. Conceptual ICT model.

The full ICT system will support several different stakeholders (identified as the main stakeholders), and as a part of the solution, a prototype will be developed covering the Maintainers, representing the users' denoted Maintenance executioner and Maintenance planner in Fig. 2. The further descriptions are thus delimited to the two subsystems included in the prototype.

4 Prototype Development

The prototype system consists of two main subsystems: subsystem Execution, which mainly supports the role of the maintenance provider (the one conducting maintenance), and subsystem Planning, which mainly supports the role of the planner. The subsystems interact with each other as well as with peripheral devices such as mobile phones for information sharing. The prototype system will be implemented partly on the grader in the form of a robust industry computer containing subsystem software for maintenance execution located in the operator cabin, and partly as a maintenance management subsystem accessed by a stationary PC in the office. The robust industry computer will provide an on-board storage of road condition data, etc., and an interface for the operator. The data from the on-board computer will be transferred to a stationary PC at certain transfer stops that have good internet coverage, or where Wi-Fi is available (alternatively the data may be transferred via cable). The two subsystems uses road condition information, which is obtained from a prototype system for measuring road conditions, conceptuallty described in [10]. In the condition monitoring prototype system, a three meter long aluminium beam is equipped with a number of radar distance sensors mounted longitudinally on the beam. The number of radar sensors is selected so as to obtain adequate road condition information. To determine the position of the beam, two accelerometers are mounted at each end measuring the vertical movement. In addition, a gyroscope measures the slope of the beam.

In the following, the functionality and information requirements for the subsystems are accounted for, as well as the main User stories that represent the main functionality needs as expressed by the stakeholders.

4.1 Subsystem "Execution of Maintenance"

Table 1 accounts for the User stories connected with the subsystem *Execution of maintenance*. As can be seen, the person using the system could be a contractor, a road owner, or someone else conducting maintenance actions. The role of this actor is called *Maintainer* in the requirements documentation.

Table 1. User stories subsystem Execution of maintenance

#	As a[a]...	I want...	So that...
1	Maintainer (or road owner)	To report maintenance needs after maintenance actions	I can recommend road associations in which maintenance should be performed
2	Contractor	To get easy-to-understand data/information about the condition of the road and how the maintenance is done	Correct/optimal road maintenance could be performed, and I can document how the maintenance has been done during invoicing
3	Maintainer	To know what material is added	It is possible to supplement with more material to achieve a certain standard

[a] The actual term used by the workshop participants

With the system, the *Maintainer* receives support for estimating the road's condition regarding road profile and roughness in real-time before planning is performed, as well as for reporting performed maintenance actions and current classification after the maintenance action (Connected to User story 1 and 2). If necessary, the Maintainer should also be able to produce classification information based on history, and not on real-time data, as well as report failures (Connected to User story 1). Failure reporting can only be done by the Maintainer in the prototype version, but is a function that should be available for other users, such as road owners, in the next version. There is also a desirable functionality for the future system to view sieve curve for gravel that is returned on the road after planing, as well as the opportunity to send the sieve curve information to the person performing the maintenance action of adding gravel to the planned road (Connected to User story 3). This requires the development of a system for measuring the quality of the gravel that is reused, i.e. returned to the road after planing. Such systems does not exist today, and therefore the ICT system will not include this feature in the prototype version.

The functionality and information specifications are found in Table 2 (desirable in italics).

Table 2. Specifications subsystem Execution of maintenance

Functions	Information
1. See road condition 1.1 See slope, 1.2 See crossfall, 1.3 See roughness, 1.4 See classification,	Detailed road condition information (slope, crossfall, roughness, GPS position), Road quality class
2. Report maintenance action 2.1 Select work order, 2.2 See work order information, 2.3 See road information, 2.4 Enter report details	Work order ID, Time stamp, Planned maintenance, Material, Planned start and end date, Road area, GPS positioning Information created: Time stamp, Responsible, Road quality class, Maintenance action description
3. Report failure 3.1 Select road section, 3.2 Enter report details	Road area, GPS positioning Information created: Time stamp, Responsible, Failure information (Description, Classification, Severity)
4. See sieve curve	*Sieve curve information (Particle size, passing amount in weight %)*

4.2 Subsystem "Maintenance Planning"

Table 3 accounts for the User stories connected with the subsystem *Maintenance planning*. As can be seen, the person using the system could be a contractor, a maintainer, or someone else planning maintenance actions. The role of this actor is called *Planner* in the requirements documentation. With the system, the *Planner* receives support for planning (Connected to User story 1 and 2) and scheduling (preparing) maintenance (Connected to User story 8), handling quotation requests (Connected to User story 1–4) as well as creating documents for the customer (Connected to User story 5 and 6). Planning can be based on previous plans, based on real condition, or a combination of both. Planning is done on a medium to long term basis, i.e. weeks to years.

Scheduling involves detailed planning on a short term basis (day/week) and includes the planning of road sections, action, resource and materials. The prototype system assumes that materials are available at scheduling time, and does not need to be ordered. The purchasing of materials is thus not included. Also, weather forecasting is not included in the prototype system, but is assumed to be done manually if necessary, e.g. using a regular smartphone. However, it is interesting to be able to include weather forecasts and weather information in the future. The planner should also be able to handle a quotation request, i.e. calculate maintenance needs and costs to be able to produce prices. In the prototype, each procurement situation is handled as a separate event. In a future version, the opportunity to handle several contracts/customers in order to coordinate maintenance measures in the immediate area (between two or more customers) is a desirable feature (Connected to User story 7). Planners should be able to compile information about a certain section of the road in order to be able to create documentation for a customer.

Table 3. User stories subsystem Maintenance planning

#	As a[a]...	I want...	So that...
1	Contractor	To know the condition of the road	I can specify class on the road in procurement
2	Contractor	To know the condition of a certain road	I can adapt procurement so that maintenance is premeditated or postponed
3	Contractor	A holistic understanding of the condition of the roads	I can coordinate procurements with other municipalities (or actors)
4	Contractor	To enter the prerequisites for a certain stretch of the road such as length, condition, or humidity	I can calculate the price of maintenance
5	Contractor	To produce documentation (sieve curve) on gravel that is recycled	I can attach this to clients
6	Maintainer	To get easy-to-understand data/information about the condition of the road and how the maintenance is done	I can document how the maintenance has been performed during invoicing
7	Road maintainer	Coordination between different road holders	reduce the distance between the roads (the distance travelled by the maintenance person when doing maintenance work) to be maintained
8	Planner	To find available resources	I can optimize planning/scheduling

[a] The actual term used by the workshop participants

The functionality and information specifications are found in Table 4 (desirable in italics).

Table 4. Specifications subsystem Maintenance planning

Functions	Information
1. Request management 1.1 See maintenance history, condition and class for a specific road section, 1.2 Calculate material consumption, 1.3 Calculate price, *1.4 Coordination of road sections/quotation request*	Road information (area, quality class, owner), Detailed road condition information, Contract period, Maintenance history [for the previous period], Material quantity and quality, Total cost
2. Customer documentation 2.1 Retrieve maintenance history and class for a specific road section, *2.2 Create sieve curve for a specific road section*	Date, Road information, Maintenance history, *Sieve curve information*
3. Maintenance planning 3.1 See road condition for a specific road section, 3.2 See maintenance history for a specific road section, 3.3 Create new maintenance plan, 3.4 Update existing maintenance plan	Road information, Maintenance history Information created: Maintenance plan (Time stamp, Road area, Responsible, Maintenance action type, Planned start/end date, Materials required)
4. Maintenance scheduling 4.1 List resources, List materials, 4.2 See maintenance plan, 4.3 See road condition, 4.4 Create a schedule for a specific road section, 4.5 Update schedule, *4.6 See weather forecast*	Recourses information (Personnel, equipment, vehicles), Materials information (Quantity, quality, supplier), Maintenance plan, Detailed road condition information, *Weather forecast* Information created: Maintenance schedule (Time stamp, Road area, Responsible, Maintenance action type, Start date, End date, Materials required, Resources required)

5 Conclusions

The use of a co-design approach, i.e., where the users' involvement is emphasized during the development process, was shown to be beneficial to gather the different stakeholders' requirements and needs of the system, which is crucial, as well, for its later acceptance. In this case, the stakeholders' needs of the final product were gathered through workshops and interviews where the different users' aspects were emphasized, such as its functionalities and information specifications. The different stakeholders' interactions with the system will later be illustrated with the support of use case scenarios based on the Unified Modelling Language (UML). Cloud computing is proposed for the integration of the different modules of the system, because it gives advantages connected with, for instance, a wide variety of services that it offers. However, it is important to consider the security aspects when a cloud solution is applied. The security aspects are not part of this paper; however, they are crucial for the systems successful use.

Acknowledgments. The research has been conducted as part of the project named Sustainable maintenance of gravel roads funded by The Kamprad Family Foundation. The project develops new methods and technologies for gravel road maintenance.

References

1. Swedish Transport Agency. Sveriges vägnät (2019). https://www.trafikverket.se/resa-och-trafik/vag/Sveriges-vagnat/. Accessed 02 May 2019
2. Alzubaidi, H.: Operations and maintenance of gravel roads. A literature study. VTI meddelande 852a. VTI, Linköping (1999)
3. Kuttah, D.: The performance of a trial gravel road under accelerated pavement testing. Transp. Geotech. **9**, 161–174 (2016)
4. Sveriges Kommuner och Landsting: Mer grus under Maskineriet. Handbok för tillståndsbedömning och underhåll av grusvägar. LTAB, Stockholm (2015)
5. Sveriges Kommuner och Landsting: Vägen till glesbygdens framtid. SKL, Stockholm (2014)
6. Tarimo, M., Wondimu, P., Odeck, J., Lohne, J., Lædre, O.: Sustainable roads in Serengeti national park - gravel roads construction and maintenance. Proc. Comput. Sci. **121**, 329–336 (2017)
7. Smadi, A., Hough, J., Schulz, L., Birst, S.: North Dakota gravel road management alternative strategies. Transp. Res. Rec. J. Transp. Res. Board **1652**(1), 16–22 (1999)
8. Henderson, M., Van Zyl, G.: Management of unpaved roads: developing a strategy and refining models. In: 36th Southern African Transport Conference, 10–13 July 2017, Pretoria, South Africa (2017)
9. Kalantari, Z.: Road structures under climate and land use change. Bridging the gap between science and application. KTH, Stockholm (2014)
10. Kans, M., Campos, J., Håkansson, L.: Condition monitoring of gravel roads-current methods and future directions. In: Ball, A., Gelman, L., Rao, B. (eds.) Advances in Asset Management and Condition Monitoring. Smart Innovation, Systems and Technologies, vol. 166, pp. 451–461. Springer, Cham (2020). https://doi.org/10.1007/978-3-030-57745-2_38
11. Saarenketo, T.: Monitoring, communication and information systems & tools for focusing actions. Roadex II report (2005)
12. Saarenketo, T.: Monitoring low volume roads. Roadex III report (2006). www.roadex.com
13. Forsyth, A.R., Bubb, K.A., Cox, M.E.: Runoff, sediment loss and water quality from forest roads in a southeast Queensland coastal plain Pinus plantation. For. Ecol. Manag. **221**, 194–206 (2006)
14. Aho, S., Saarenketo, T.: Utformning och utförande av åtgärder på vägar som lider av försvagning vid tjällossning. ROADEX II report (2006). www.roadex.com
15. Pilli-Sihvola, E., Aapaoja, A., Leviäkangas, P., Kinnunen, T., Hautala, R., Takahashi, N.: Evolving winter road maintenance ecosystems in Finland and Hokkaido, Japan. Intell. Transp. Syst. **9**(6), 633–638 (2015)
16. Edvardsson, K.: Lågtrafikerade vägar En litteraturstudie utifrån nytta, standard, tillstånd, drift och underhåll. VTI rapport 775. VTI, Linköping (2013)
17. Alferor, R.M., McNiel, S.: Method for determining optimal blading frequency of unpaved roads. Transp. Res. Rec. **1252**, 21–32 (2017)
18. Trafikverket: Bedömning av grusväglag. TDOK 2014:0135 (2014)
19. van Wijk, I., Williams, D., Serati, M.: Roughness deterioration models for unsealed road pavements and their use in pavement management. Int. J. Pavement Eng. (2018). https://doi.org/10.1080/10298436.2018.1511991

20. Steyn, W.J.: Optimization of gravel road blading. J. Test. Eval. (2018). https://doi.org/10.1520/JTE20180022
21. Radeshi, R., Maher, M., Barakzai, K.: Defining needs for optimized management of gravel road networks. In: The 2018 Conference of the Transportation Association of Canada, Saskatoon, SK (2018)
22. Oladele, A.S., Vokolkova, V., Egwurube, J.A.: Pavement performance modeling using artificial intelligence approach: a case of botswana district gravel road networks. J. Eng. Appl. Sci. **5**(2), 23–31 (2014)
23. Ross, D., Townshend, M.: An economics-based road classification system for South Africa. In: 37th Annual Southern African Transport Conference, 9–12 July, Pretoria, South Africa, pp. 11–21 (2018)
24. Enkell, K., Svensson, J.: Grusvägsstyrsystem Förstudie. VTI notat 44-2000. VTI, Linköping (1999)
25. Riksrevisionen: Underhåll av belagda vägar. RiR 2009:16. Riksdagstryckeriet, Stockholm (2009)
26. Alzubaidi, H.: Vi vill kundanpassa grusvägsunderhåll. Publikation 2006:70. Vägverket (2007)
27. Campos, J., Kans, M., Håkansson, L.: Information system requirements elicitation for gravel road maintenance – a stakeholder mapping approach. In: Ball, A., Gelman, L., Rao, B.K.N. (eds.) Advances in Asset Management and Condition Monitoring. SIST, vol. 166, pp. 377–387. Springer, Cham (2020). https://doi.org/10.1007/978-3-030-57745-2_32
28. Petersson, A.M.: Collaborative conceptual design methods in the context of the Swedish deregulated railway market. Luleå universitet, Luleå (2017)
29. Sanders, E.B.N., Stappers, P.J.: Co-creation and the new landscapes of design. Co-Design **4**(1), 5–18 (2008)
30. Cohn, M.: User Stories Applied: For Agile Software Development. Addison-Wesley Professional (2004)
31. Manvi, S.S., Shyam, G.K.: Resource management for Infrastructure as a service (IaaS) in cloud computing: a survey. J. Netw. Comput. Appl. **41**, 424–440 (2014)
32. Campos, J., Sharma, P., Jantunen, E., Baglee, D., Fumagalli, L.: The challenges of cybersecurity frameworks to protect data required for the development of advanced maintenance. Proc. CIRP **47**, 222–227 (2016)

Autonomous Anomaly Detection and Handling of Spatiotemporal Railway Data

Murat Kulahci[1], Bjarne Bergquist[1](✉), and Peter Söderholm[2]

[1] Quality Technology and Logistics, Luleå University of Technology, Luleå, Sweden
{murat.kulahci,bjarne.bergquist}@ltu.se
[2] The Swedish Transport Administration, and Quality Technology and Logistics, Luleå University of Technology, Luleå, Sweden
peter.soderholm@trafikverket.se

Abstract. Prognostics is a vital application for Industrial AI. However, data quality sometimes does not suit regular prediction models. For instance, data sampling procedures occasionally include irregular sampling, and many prognostic methods require evenly sampled time series. One example is measurement trains that travel along the railway line, measuring the geometry of the railway track.

In the railway case, the time series often display discontinuities due to maintenance events and outliers. Such discontinuities must be handled differently for a prognostic model to work sufficiently. We have studied different approaches such as moving range statistics and Cook's distance statistics for outlier detection and removal and detection and model adjustments for maintenance.

We also provide a practical solution that manages the autonomous anomaly detection and predictions of future track geometry condition through a developed R Shiny App to support users.

Keywords: Time series analysis · Data cleansing · Data filtering · Outlier detection · Prognostics

1 Introduction

Degradation of physical systems due to wear and tear is inevitable following the second law of thermodynamics. Hence, systems are maintained to retain and restore their required functions. The maintenance can be preventive or corrective, i.e., before or after a fault has occurred. If preventive maintenance is a set of time-based activities that does not account for the actual system condition, we define it as time-based maintenance. On the other hand, if the preventive maintenance is scheduled based on the condition (state) of the system at the time of inspection, it is called condition-based maintenance. See, e.g., [1]. If a prediction of the future state is instead what controls the preventive maintenance, the maintenance is called predictive [1]. Predictive maintenance has gained prominence lately, particularly in systems where unnecessary stoppages to perform unplanned maintenance contribute significantly to production costs. This study focuses on the railway track maintenance problems and the analysis of data generated through track geometry measurements.

R. Karim et al. (Eds.): IAI 2021, LNME, pp. 65–72, 2022.
https://doi.org/10.1007/978-3-030-93639-6_6

Railway track geometry is one of the primary indicators of the railway tracks' condition [2]. Measurement trains, also called track geometry cars, usually collect these data. These so-called measurement runs are performed at varying intervals depending on the inspection class of that track and the availability of the measurement trains. The inspection classes are, in turn, set by the combined load and speed of the passenger and freight trains using the track [3].

Maintenance personnel analyse the track geometry data from these measurement runs, particularly the produced alarms. If any measurement falls outside tolerable geometrical limits, those deviations will trigger alarms to initiate corresponding maintenance actions. The required maintenance actions could be either corrective or preventive, depending on the size of the deviation and the criticality of the geometrical limit.

Tamping is a common corrective action involving lifting the rails and sleepers while adjusting the trackbed level. The adjustment includes pressing down forks into the trackbed top layer, the ballast and then pack it to adjust for inaccuracies in the track geometry. As opposed to corrective tamping, scheduled preventive tampings are in northern Sweden planned for the summer months where the weather conditions are more favourable with minimal risk of ground frost. The short summers make the scheduling of preventive maintenance activities of all types crucial.

In this presentation, our focus will be on pre-processing the track geometry data collected from this region to provide effective predictive models of the track geometry.

2 Data

Measurement trains collected the data for this study, and the data consists of different track geometry properties measured at 25 cm intervals. Trafikverket (the Swedish transport administration) provided the data. Unfortunately, these data lack the positioning accuracy required by many analysis methods. The measurement cars that provided the measurement data rely on two systems for positioning: GPS and dead reckoning. The positioning accuracy of GPS is currently accurate within 15 m [4]. Such precision is unsatisfactory for maintenance workers out in the field, and it is unsatisfactory for some analyses. To monitor the development of a point defect, for instance, requires better positioning accuracy. GPS demand open skies to obtain satellite positioning information which is impossible in tunnels.

The measurement cars complement the GPS positioning with a system based on dead reckoning to improve the measurement situation in tunnels and elsewhere. The dead reckoning calculations involve counting wheel revolutions and multiplying the revolutions with the wheel circumference [5]. However, the dead reckoning approach has other measurement issues. Wheel slips and slides, as well as wheel wear, induce dead reckoning errors. Another feature of the dead reckoning positioning calculations is that they may undergo automated calibrations through a network of unique reference points such as radio frequency identification tags placed at stations or trackside along the track. The system requires that the measurement series are continuous and that every new measurement run connects to the neighbouring run. Automated calibration routines, therefore, compress or stretch the measurement series to fit these criteria. The segment recordings may be stretched out in one measurement train passage and shortened in the next.

There are two common approaches to overcome positioning uncertainties. One is to synchronise measurements, and the other is to look at longer track segments so that positioning error effects are minor compared to the segment length. In addition, researchers have sought methods to align data using pattern matching in the former approach, see, e.g., [5]. However, this approach needs to overcome problems besides a global positioning fault, including local faults due to the dead reckoning errors and automated calibration schemes. Another issue is that patterns may change, e.g., due to frost heave or maintenance actions (e.g., tamping).

Nonetheless, it is necessary to monitor a profile correctly to monitor the evolution of a point defect. For such purposes, the former solution is necessary. Another drawback besides that the profiles may change, which often hinders pattern matching from aligning profiles, is the computational complexity and needs for ample data storage during alignment.

The method to study distributional properties of a longer track interval will suffice if the maintenance planners wish to study how a property measured over a longer track length develops over time. Such properties could include the track alignment or the longitudinal level of the rail. In the studied case, the longitudinal level is the principal measure that base scheduled maintenance decisions. We, therefore, chose to study aggregate measures over a 200 m segment.

A first step for the analysis was thus to group the data into 200 m lengths. The grouping involved calculating various statistics such as maximum deviation from the nominal level of the property or the standard deviation for the 800 observations from each 200 m segment. Many statistics have tolerances, maintenance limits, where observations outside these limits trigger corrective maintenance actions. The size of the deviation from the tolerance limits and the criticality attributed to the variable both determine the corrective action's acuteness. This combination also determines if the track needs to shut down for traffic or if it needs immediate speed restrictions before it has been maintained. These calculations were performed for each segment and property. See, also e.g., Bergquist & Söderholm [6, 7].

Based on these statistics, we meant to construct predictive models for future planning of the maintenance activities. As the data accumulates over time, the dataset grows. As in many large datasets, the analyst's first crucial step is to "pre-process" the data before performing the actual analysis. Of course, the analyst can pre-process manually (which is often the case). However, this study aims to demonstrate how the current availability of computing power and software can automate this pre-processing and how to add algorithms to strengthen the analysis.

3 Pre-processing

If we study a particular geometrical track variable of a trafficked track, we can expect that wear and tear degrade conditions over time. We can also expect that maintenance in most cases will improve the situation and that such maintenance should manifest as step-change improvements in the data. We do not expect to see radical improvements or geometrical degradations from one measurement to the next without external interference. If we study an extended time series and observe one deviating observation in the middle of an otherwise slowly and constantly degrading track, it is likely an outlier, regardless of if there are log-book incidents that can explain why. However, if the conditions improve through a step change, perhaps indicated by a change of the level of the variable and its degradation rate, we could assume that the track was maintained.

We had two main concerns in the data pre-processing step. The first was related to records of past maintenance activities. A joint display of the maintenance information and the track geometry will usually reveal the maintenance impact on the maintained geometrical property. The maintenance log leaves evidence in the measured track geometry. However, when studied as a time series, the data often exhibit "improvements" that one can only attribute to unrecorded maintenance activities ("ghost tampings"), which means that the maintenance logs are incomplete.

If the analyst must deal with such missing log-book information, the degradation models, therefore, must reset after each maintenance activity, recorded or otherwise. It was thus crucial to identify all past maintenance activities. We defined those as observations with a significant improvement in the track geometry compared to earlier observation series. The second issue was the usual outlier detection, as outliers tend to gravely affect the models if the analyst fails to identify them. Below we describe the strategies we followed in both counts.

4 Empirical Identification of Maintenance Activities

This study primarily entertains two methods for identifying improvements empirically (i.e., potentially unknown tamping). Furthermore, these methods can be tuned to maximise the relationship between true tamping and "ghost tampings".

Method 1: This method is based on the successive difference in the data. Consider a time series of n observations $y_{t-n}, y_{t-n+1}, \ldots, y_{t-1}, y_t$. One-lag moving ranges are obtained using $y_{t-2} - y_{t-1}$, and we subsequently obtain the standard deviation based on these one-lag moving ranges. We then use

$$y_{t-2} - y_{t-1} > k_1 \tag{1}$$

where y_{t-2} and y_{t-1} are two consecutive measurements with at least one observation before (y_{t-3}) and one later (y_t), with a lower value indicating a better condition. If the y_{t-2} $-y_{t-1}$ difference is a test for degradation or improvement. A positive difference indicates improved conditions, which could be due to chance or due to a maintenance event. k_1 is a constant (in the test, we used $k_1 = 2 \times$ standard deviations). If (1) is true, the variation reduction is classified as a maintenance activity *if* the variation reduction between two successive observations is considerable *and* the variation differences between the preceding and following observations are small. If there is only one improved observation in a series of poor observations, we let the model classify it as a measurement outlier rather than a maintenance action. Hence, the maintenance is declared if

$$\{|y_{t-3} - y_{t-2}| < k_2 \ \& \ |y_{t-1} - y_t| < k_2 \ \& \ y_{t-2} - y_{t-1} > k_1\} \tag{2}$$

where y_{t-3} and y_t are the fourth to last and the last observations respectively, and k_2 is a constant (in the test $k_2 = 1 \times$ standard deviation). If (1) is true but (2) is not, the conclusion is that y_{t-1} is an outlier. If (1) and (2) are both true, we conclude that the variation reduction is due to maintenance between $t-2$ and $t-1$ rather than the result of an outlier.

Method 2: This method was suggested by [8]: If we observe a decrease of, e.g., 85%, in the variable of interest, we assume that an improvement from an unknown/unrecorded maintenance activity has occurred on the track segment. We can view the percentage decrease in the variable of interest as a tuning parameter that the user can change.

5 Outlier Detection

We also followed another approach to detect outliers. The second approach relates to the model we ultimately used for degradation. We found that the simple linear regression model based on log-transformed data performed best. The suitability of a linear model applied to log-transformed data was expected as the raw observation's degradation often follows an exponential decay pattern. The linear regression model in the log-transformed scale thus has the following model assumption:

$$y_t = \beta_0 + \beta_1 \, t + \varepsilon \tag{3}$$

where y_t is the value of the track geometry variable at time t. β_0 and β_1 are the regression coefficients and ε is the error term, which is assumed to be independent and normally distributed with mean 0 and variance σ^2. Montgomery et al. [8] provide further details in simple linear regression.

Cook's distance is a measure of how influential each observation is when fitting a linear regression model. Cook's distance measures the effect of removing a point from regression (as in leave-one-out cross-validation). D_i, the Cook's distance, measures the squared distance between the least-squares estimate $\hat{\boldsymbol{\beta}}$, including all observations and the estimate obtained by deleting the ith observation $\hat{\boldsymbol{\beta}}_{(i)}$ [9]. We deem observations with large D_i values to have considerable influence on the least-squares estimates $\hat{\boldsymbol{\beta}}$, and we consider such observations as potential outliers in our approach. We express Cook's distance as:

$$D_i = \frac{\left(\hat{\mathbf{y}}_{(i)} - \hat{\mathbf{y}}\right)^T \left(\hat{\mathbf{y}}_{(i)} - \hat{\mathbf{y}}\right)}{pMS_{\text{Res}}} \tag{4}$$

where $\hat{\mathbf{y}}_{(i)}$ is a vector with the estimated observation values with observation i removed during estimation, $\hat{\mathbf{y}}$ is a vector with the estimated observation values including all data, p is the number of estimated parameters (e.g. β_0, β_1) in the model and MS_{Res} is the mean square for residuals. For further details, see, e.g. [9]. We chose the threshold for Cook's distance to 2, but the user may change it to something else. That is, if $D_i > 2$ then we treat the observation as an outlier.

6 Automated Improvement and Outlier Detection

Figure 1 shows a particular track geometry characteristic for one of the studied 200m segments. The blue vertical lines show the empirically found improvements, which the model considers a maintenance activity. The green vertical lines represent the recorded tampings.

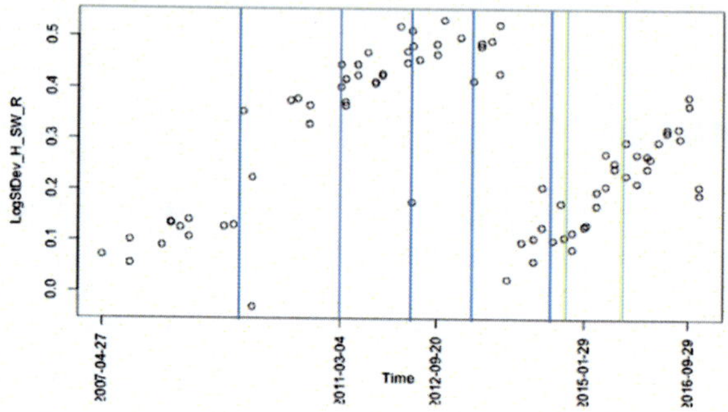

Fig. 1. Maintenance activities where blue lines are for the empirically found improvements and the green lines are for the recorded tamping.

Figure 2 shows the linear model on the logarithm of the track geometry characteristic after considering all empirically found and recorded maintenance activities.

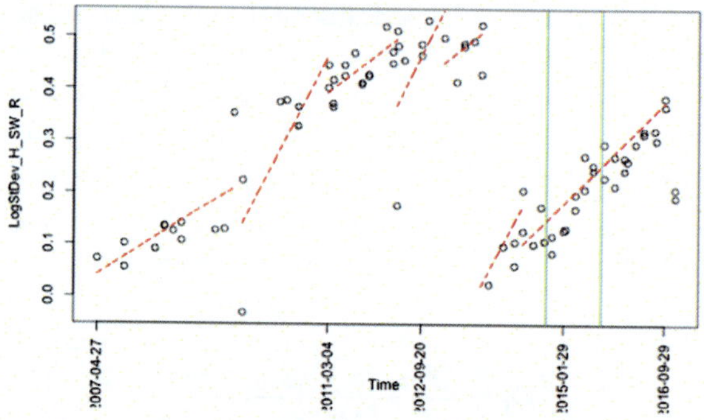

Fig. 2. Linear degradation models on the log of the track geometry characteristic

These models lend themselves to rapid calculations easily performed in computers or even on smartphones. Figure 3 illustrates how the linear model in Eq. (3) works in R Shiny [10] for one track segment using Method II to identify track geometry improvements due to potentially unknown maintenance activities. As the figure shows, the linear prediction model restarts at these empirically identified improvement points.

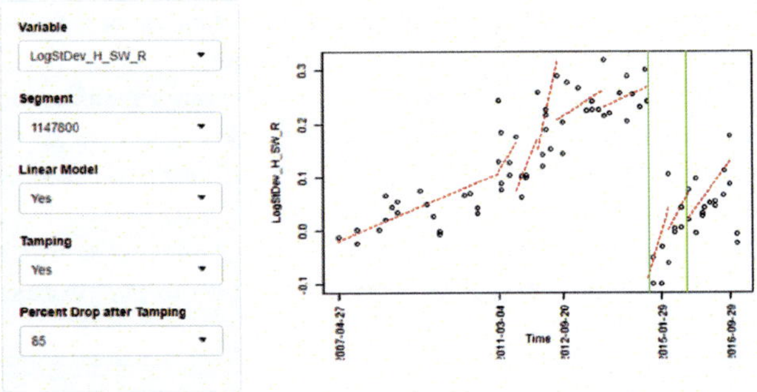

Fig. 3. Track geometry variable: Log (base 10) of the shortwave standard deviation for the right rail. The dashed red lines represent the linear models. Green vertical lines represent known tamping occasions.

7 Discussion

The paper presented two approaches for detecting maintenance actions. The fast method is to state that an improvement (decrease) of 85% of, e.g., the standard deviation of a property, leads us to declare that the track is maintained. The more comprehensive method includes surrounding measurements in the analyses, i.e., before and after the observed improvement. The motivation to use a simple criterion is simplicity and the possibility to study and judge the latest measurement in a time series. The motivation for the more comprehensive method is that we found that some segments were inherently unstable, displaying considerable variation between measurements. The added extra tests prevented maintenance declarations that were effects of the considerable variation rather than maintenance actions.

Furthermore, an inherent property of the data is that the measurements are made by different cars of unknown calibration frequencies. By studying the measurements, we can detect that the cars produce differences in measurement, which adds to the background noise when several cars provide measurements. In such cases, the extra protection of the added checks was useful. However, besides increasing model complexity and increased computation needs, the extra checks cannot be used directly when a new dataset is entered into the database since it analyses the relation between the second-to-last observation to its neighbours.

8 Conclusion

Pre-processing of data is typically a crucial preliminary step in any data analysis effort. Large datasets require further care in this step as they are more prone to missing values, outliers, or unrecorded changes in the data, as we saw here. Traditionally, this step calls for manual interventions and actions taken at the analyst's discretion. However, the sheer volume of current production data often requires an automated or at the very least semi-automated approach for this preliminary step. We developed such a scheme in a predictive maintenance study using railway track geometry measurements. The paper suggests two methods for finding and adjusting time series for performed maintenance actions. We recommend that the analyst tries the simple approach first. The bases for this recommendation are the drawbacks of the more comprehensive version and the usefulness of performing checks when the analyst adds new data to the database. If needed, the involved approach is suitable for off-line analyses and cases when the simple model fails by producing false maintenance events.

The scheme is implemented through an R shiny App, and it offers the user the necessary flexibility and control for a customised option to pre-process the available data.

Acknowledgement. We want to thank Arne Nissen and Trafikverket (the Swedish transport administration) for their support in this project.

References

1. SS-EN 13306:2017. Maintenance – Maintenance Terminology. Swedish Standards Institute
2. SS-EN 13848-1:2019 - Railway applications – Track – Track geometry quality – Part 1: Characterisation of track geometry. Swedish Standards Institute
3. KRAV - Säkerhetsbesiktning av fasta järnvägsanläggningar TDOK 2014:0240 Version 8.0. Trafikverket
4. Ross, R., Hoque, R.: Augmenting GPS with geolocated fiducials to improve accuracy for mobile robot applications. Appl. Sci. **10**(1), 146 (2020)
5. Khosravi, M., Soleimanmeigouni, I., Ahmadi, A., Nissen, A.: Reducing the positional errors of railway track geometry measurements using alignment methods: a comparative case study. Measurement **178**, 109383 (2021)
6. Bergquist, B., Söderholm, P.: Data analysis for condition-based railway infrastructure maintenance. Qual. Reliab. Eng. Int. **31**(5), 773–781 (2015)
7. Bergquist, B., Söderholm, P.: Predictive modelling for estimation of railway track degradation. In: Kumar, U., Ahmadi, A., Verma, A.K., Varde, P. (eds.) Current Trends in Reliability, Availability, Maintainability and Safety. LNME, pp. 331–347. Springer, Cham (2016). https://doi.org/10.1007/978-3-319-23597-4_24
8. Khajehei, H., Ahmadi, A., Soleimanmeigouni, I., Nissen, A.: Allocation of effective maintenance limit for railway track geometry. Struct. Infrastruct. Eng. **15**(12), 1597 (2019)
9. Montgomery, D.C., Peck, E.A., Vining, G.G.: Introduction to Linear Regression Analysis, 5th edn. Wiley, New York (2012)
10. Shiny. Shiny from R Studio (2019). https://shiny.rstudio.com/2021-01-01

An ILP Approach for the Maintenance Crew Scheduling Problem Considering Skillsets

Tiago Alves(✉) and António R. Andrade

IDMEC, Instituto Superior Técnico,
University of Lisbon, Av. Rovisco Pais, 1, 1049-001 Lisbon, Portugal
{tiago.p.alves,antonio.ramos.andrade}@tecnico.ulisboa.pt

Abstract. Public transportation services are essential to any metropolitan area. To meet the rising passenger demand, emergent and disruptive technologies have been used for the railway sector. This paper assesses the maintenance crew scheduling problem and presents a mathematical programming model which optimizes the daily scheduling of technicians in a railway depot. The aim consists in minimizing the labour costs, while assigning maintenance workers to maintenance tasks, considering the skillset required for each one of them. In fact, this is the key aspect added to the literature by the present maintenance crew scheduling model: the skills of technicians. Using data collected from a Portuguese train operating company, an integer linear programming model is formulated and applied to the case study while monitoring the rolling stock schedule and the maintenance tactical plan. The optimized results indicate that the maintenance team could be reduced, suggesting that maintenance crew scheduling and associated labour conditions may need more flexible approaches. This model is also part a broader framework, as it is connected to two different models: a tactical maintenance planning model, and an operational maintenance scheduling model. All together, these models provide a decision framework that can support maintenance planning and scheduling decisions.

Keywords: Railway management · Operations research · Maintenance · Crew scheduling · Integer linear programming

1 Introduction

In an era where technology evolves at a frenetic pace, new researches and approaches arise to solve more complex problems in different engineering areas, and railway systems are not an exception. The *"rising traffic demand, congestion, security of energy supply and climate changes"* [1] are some of the challenges that the European railway systems face, and new technologies can play a major role influencing the way future rail automation and maintenance are organised. An European rail initiative named Shift2Rail (S2R) was created in 2009, *"when key European rail sector players, under the coordination of the Association of the European Rail Industry (UNIFE), began investigating a policy instrument that could facilitate a step change for the European rail system"* [2]. Following these ideas, the current work aims to use such techniques to solve a maintenance

R. Karim et al. (Eds.): IAI 2021, LNME, pp. 73–85, 2022.
https://doi.org/10.1007/978-3-030-93639-6_7

crew scheduling problem for the case study of a Portuguese train operating company. Since the maintenance of an equipment is a significant part of the total operating costs in most industry sectors but its real impact is often under-estimated [3], it is important to efficiently plan and schedule maintenance while assigning the necessary resources, in particular the maintenance crew. Railway maintenance and associated asset management techniques have been the focus of research in the last decade, namely in railway track maintenance. Optimization models to support railway maintenance have been proposed to support railway maintenance decisions [4–6]. For instance, Vale et al. [4] put forward an integer programming model to optimize tamping decisions in railway tracks. Caetano and Teixeira [5] presented a strategic model to optimize railway track renewal operations at a network level, which assesses opportunistic renewal of railway track components from a network perspective, while potentially reusing some track components on different lines. Dao et al. [6] also discussed maintenance scheduling for railway tracks under limited possession time, showing that clustering maintenance operations in a single track possession is beneficial in general. Fewer research has focused on planning the necessary human resources to achieve railway maintenance.

The emergence of new and impactful technologies may have a big impact on how the transportation systems are designed and optimized. Therefore, in order to remain a viable and competitive solution, the railway sector must keep up with all these developments. One way to accomplish this is through carefully planned schedules, which must satisfy primarily the customers, but they should also comply with workers and company's requirements (e.g. rolling-stock timetables, maintenance and crew scheduling). This paper focuses in scheduling maintenance crew and presents a mathematical model that minimizes the costs associated with its scheduling for a Portuguese train operating company (Fertagus). To achieve this goal, a company's rolling-stock timetable and previously scheduled maintenance activities must logically be considered beforehand, so that the problem is contextualized. Hence, it is proposed to apply the decision model to the case study of Fertagus' operating rolling-stock schedule, adapting a previously obtained maintenance scheduling plan from [7], which considers the maintenance actions that have to be performed for each day of the week. The ultimate objective of this paper is to present a model that minimizes the costs associated with crew scheduling for a railway company and that creates its daily planning for a given week, while considering workers' skillsets as well as the required skills to perform a given maintenance task. In fact, taking these skillsets into account constitutes one of the key contributions of this work, as normally, this matter is not given much relevance on common crew scheduling problems.

The structure of the present paper is the following: Sect. 1 introduced the necessity of optimizing not only maintenance but also maintenance crew scheduling. Section 2 briefly reviews past researches on this subject. Section 3 contextualizes the maintenance crew scheduling problem and presents the mathematical model, by defining all its information. The model is then applied to the case study and the results discussed in Sect. 4. Finally, Sect. 5, draws the main conclusions and some guidelines for further research.

2 Related Literature

The topic of maintenance crew scheduling in railways has received some contribution in the past, usually explored within maintenance scheduling, though with more contributions in other transports (e.g., buses or aircrafts). For instance, Haghani and Shafahi [8] dealt with the problem of bus maintenance scheduling, aiming to design a daily schedule that minimized the number of unavailability hours for each vehicle, carrying out as many inspections as possible when buses are out of service, and thus, maximizing the usage of maintenance resources. Using an integer programming approach, their model output a maintenance schedule for each bus and the minimum number of maintenance lines that should be assigned for each type of action. Another research carried out by Fuentes *et al.* [9] tackled the crew scheduling problem in rapid transit networks, where distances are small, but the service frequency is extremely high. Different methods were studied and applied to a portion of RENFE's train operating company's rapid transit network, with the fix-and-relax algorithm proving to be the better option due to its performance in terms of computational time, a major concern when dealing with this type of network. To solve a signalling maintenance crew scheduling problem for a section of the Danish railway system, Pour *et al.* [10] proposed an hybrid Constraint Programming/Mixed Integer Programming framework. This framework is divided in two parts: i) the construction phase, where initial feasible solutions are obtained through a Constraint Programming (CP) model; and ii) the improvement phase, where a Mixed Integer Programming (MIP) solver was used to improve the quality of the initial solutions. By using this method, only feasible solutions previously obtained by CP are later improved, reducing the time required by the improvement phase to produce an improved solution. Recently, Martins *et al.* [11] presented an integer linear programming model to simultaneously schedule the maintenance crew and the maintenance tasks in a bus operating company. The proposed model uses a constructive heuristic approach, based on solving the maintenance scheduling problem for each bus separately. The authors could verify that the heuristic finds better solutions than exact methods (based on branch-and-bound techniques) in a much lower computational time, highlighting the relevance of such heuristic approaches for maintenance scheduling in practice.

There were also another couple of impactful research studies analysed, as both of them concerned the same train operating company as the current paper does (Fertagus). First, Mecháin *et al.* [12] focused on the maintenance planning problem, by using a mixed integer linear programming (MILP) model, aligned with a research carried out by Doganay and Bohlin [13] and followed by Bohlin and Wärja [14] on the technical planning problem, which takes into account the many technical and depot constraints. The aim of the study was to develop a model that outputs a technical maintenance plan for a time horizon of 52 weeks, while minimizing the cost related with preventive maintenance. The optimization model defined which maintenance actions needed to be carried out each week, the maintenance line where the maintenance took place, and even the number of spare parts necessary to fulfil the technical plan. It is worth mentioning that, even if lately improved maintenances approaches such as predictive maintenance have gained attention, among the existent maintenance approaches, namely corrective, predictive, risk or condition-based and preventive maintenance, it is the latter the most applied in the maintenance of mechanical systems. Following this study, more recently,

Mira *et al.* [7] also developed a mixed integer linear programming decision model by extending an existent model on the problem of robust rolling-stock planning [15]. Their model provided a weekly, optimal and robust rolling-stock schedule capable of including maintenance actions, considering a previously scheduled preventive maintenance plan for each of the different weeks of the year from a prior model [12]. The results indicated that by rearranging the operating rolling-stock schedule, it was possible to meaningfully reduce the deadheading distance covered by train units. Moreover, the obtained schedule, indicated which units and when they should carry out maintenance, as well as the maintenance type to be performed.

From the reviewed literature both, Mira *et al.* [7] and Pour *et al.* [10] models were chosen to serve as starting points of the decision model to be created. Even though the literature review was helpful to comprehend the area developments and some approaches to the maintenance crew scheduling problem, some modifications were necessary, as well as some additions, namely, the inclusion of maintenance crew competences.

3 An Integer Linear Programming Model to Schedule Maintenance Crew

This third section presents the mathematical model, including constants, sets, parameters, variables, objective function and associated constraints. All the data required to build this model had to be either defined or imported, some of it from the models presented in [7] and in [12]. An initial context of the present model is provided in Subsect. 3.1.

3.1 Context

The present formulation for the maintenance crew scheduling model was inspired by a recent model on the preventive signalling maintenance crew scheduling proposed by Pour *et al.* [10]. Their approach was adapted to meet the train operating company requirements and to fit the outputs of the operational scheduling model [7], so that an integrated decision framework could be proposed. The present model considers some simplifications (compared to [10]), namely there is a single depot and each maintenance task may require more than one skill.

The present model is a natural extension of the previous models proposed in [7] and [12], as suggested in Fig. 1. The first [12], is a tactical maintenance planning model that outputs a preventive maintenance plan for all the 52 weeks for a given year. The outputs from this first model (e.g. which units benefit from which maintenance tasks in a certain week) are defined as inputs in the operational scheduling model [7], which then finds an operational schedule, including the rolling-stock plan integrated with the maintenance scheduling. The second model [7] integrates the weekly maintenance plan derived in the first model within the rolling-stock schedule. In the present/third model, the maintenance crew must be assigned to the scheduled maintenance actions computed in the operational scheduling model. This third model, entitled maintenance crew scheduling model, considers the workers' competences/skills and ensure that a maintenance worker has the right competence/skill to perform a given maintenance action. All the constraints associated with infrastructure, crew, rolling stock and other temporal and logical constraints are considered and respected while defining this model. Finally, from a previous research [10], some constraints associated with the competences/skills served as starting points for the definition and implementation of these type of constraints in the present model given week.

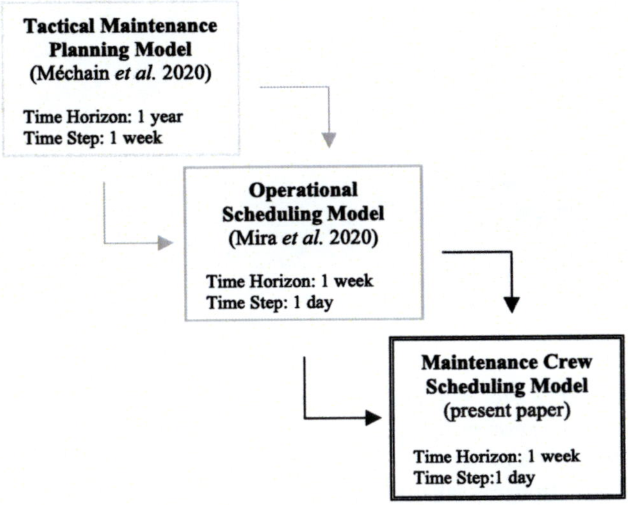

Fig. 1. Present work context within previous researches.

3.2 Model Implementation

Optimization is key to find the best solution possible for any problem, whether it is a daily question or work-related choices. It consists in a continuous search for a better solution, combining a real world instance with an algorithm that replicates it and outputs a solution to be then interpreted under it. Actually, it plays a major role in railway systems. Operational costs related to maintenance crew have a big impact on their expenses and optimization solvers can provide better solutions and thus, lead to reduced costs. The current optimization model can be divided into several inputs so that an easier understanding algorithm is achieved, as it follows on the next subsections.

3.2.1 Constants

NK Number of units
NS Number stations
NW Number of maintenance workers
NC Number of competences
NM Number of different types of maintenance actions
NT Number of tasks

ND Number of days

t_{min} Gap required by maintenance workers when changing unit in successive maintenance actions

t_{man} Gap required by units to set up for maintenance after arriving and before departing the depot

LN Large number.

3.2.2 Sets

K Set of units, k

S Set of stations, s

T Set of tasks, i

M Set of maintenance actions, m

C Set of competences, c

D Set of days, d

W Set of maintenance workers, w.

3.2.3 Parameters

cw	daily cost of each maintenance worker w
Sd_i	departure station of task i
Sa_i	arrival station of task i
Dd_i	departure time of task i
Da_i	arrival time of task i
MT_m	duration of maintenance action m
AWt_m	total amount of work required for each maintenance m
$AW_{m,c}$	amount of work per competence c, required for each maintenance m
$X_{k,i}$	tasks i carried out on unit k
$Y_{k,i,j}$	pair of tasks (i, j) linked by unit k
$YM_{k,i,j,m}$	maintenance actions m performed on unit k, between pair of tasks (i, j)
$KM_{k,m}$	maintenance actions m that need to be performed on each unit k
$ZM_{k,d}$	units k that cover a maintenance action, on a given day d
$MWC_{w,c}$	competences c, mastered by each maintenance worker w.

3.2.4 Variables

$w_{w,d}$ binary variable set to 1 if maintenance worker w works on day d, and set 0 otherwise

$z_{w,k,m,d}$ binary variable set to 1 if maintenance worker w, performs maintenance action m on unit k, on day d and set 0 otherwise

$t1_{w,k,m,d}$ maintenance worker w starting time, performing maintenance action m, on unit k, on day d

$t2_{w,k,m,d}$ maintenance worker w ending time, performing maintenance action m, on unit k, on day d.

3.2.5 Objective Function

$$\text{Minimize: } \sum\nolimits_{w \in W} \sum\nolimits_{d \in D} c_w \times w_{w,d}$$

3.2.6 Constraints

In order to implement the necessary specifications and requirements of the problem, the objective function must respect several constraints, as it follows on Table 1.

Constraints (1) and (2) guarantee that if a unit *k* does not go to depot to perform maintenance actions on day *d*, then no maintenance worker *w*, will be performing any maintenance on that unit *k*. Furthermore, if a maintenance action *m*, is not previously scheduled to be performed on a unit *k*, then no maintenance worker *w*, will be assigned to perform it. Constraint (3) establishes that if a maintenance worker performs any maintenance action on a day d, then he/she is assigned to work on that day. Constraint (4) secures that if a maintenance worker *w* is assigned, then its starting/finishing time must be greater than zero, meanwhile constraint (5) assures that the finishing time of a maintenance action, logically must be greater than its starting time. Constraint (6.1) and (6.2) assure that maintenance tasks are performed within the right time gap. More precisely, constraint (6.1) guarantees that maintenance actions performed between pair of tasks *(i, j)* can only start after the arrival time of the unit Da_i to the depot, plus a gap required by units to set up for maintenance t_{man}. On the other hand, constraint (6.2) assures that maintenance actions end before the departure time of the unit Dd_j from the depot, less t_{man}. Constraint (7) guarantees that the finishing hour of a maintenance action *m*, $t2_{w,k,m,d}$, must be greater than the starting hour $t1_{w,k,m,d}$, plus the duration necessary to carry out the maintenance action, MT_m. Additionally, constraint (8) ensures that the previously defined duration of a maintenance action MT_m, cannot be exceeded. Constraints (9), (10.1) and (10.2) assure that if a maintenance worker is assigned to two different maintenance actions m_1 and m_2, he/she can only start another maintenance action m_2, after finishing the one that was started first. While constraint (9) guarantees that this happens for two maintenance actions performed on the same unit *k*, constraints (10.1) and (10.2) impose that this temporal coherence is established for two different units k_1 and k_2, whether the maintenance actions to be performed are the same or not. Finally, constraint (11) states that the amount of work per competence required by each one of the maintenance actions, $AW_{m,c}$, must be satisfied by the maintenance workers assigned to the respective maintenance action. Logically a maintenance worker *w*, can only be assigned to a maintenance action *m*, in case he/she possesses at least one of the required competences to carry it out, i.e. $MWC_{w,c} = 1$.

Table 1. List of used constraints.

$z_{w,k,m,d} = 0$ $\forall w \in W, k \in K, m \in M, d \in D \mid KM_{k,m} = 0$	(1)
$z_{w,k,m,d} = 0$ $\forall w \in W, k \in K, m \in M, d \in D \mid ZM_{k,m} = 0$	(2)
$z_{w,k,m,d} \leq w_{w,d}$ $\forall w \in W, k \in K, m \in M, d \in D$	(3)
$t1_{w,k,m,d} + t2_{w,k,m,d} \leq LN \times z_{w,k,m,d}$ $\forall w \in W, k \in K, m \in M, d \in D$	(4)
$t2_{w,k,m,d} \geq t1_{w,k,m,d}$ $\forall w \in W, k \in K, m \in M, d \in D$	(5)
$t1_{w,k,m,d} \geq (Da_i + t_{man}) \times z_{w,k,m,d}$ $\forall i \in T, j \in T, w \in W, k \in K, m \in M, d \in D$ $\mid YM_{k,i,j,m} = 1$	(6.1)
$t2_{w,k,m,d} \leq (Dd_j - t_{man}) \times z_{w,k,m,d}$ $\forall i \in T, j \in T, w \in W, k \in K, m \in M, d \in D$ $\mid YM_{k,i,j,m} = 1$	(6.2)
$t2_{w,k,m,d} \geq t1_{w,k,m,d} + MT_m - (1 - z_{w,k,m,d}) \times LN$ $\forall i \in T, j \in T, w \in W, k \in K, m \in M, d \in D$ $\mid YM_{k,i,j,m} = 1$	(7)
$t2_{w,k,m,d} - t1_{w,k,m,d} \leq MT_m \times z_{w,k,m,d}$ $\forall i \in T, j \in T, w \in W, k \in K, m \in M, d \in D$ $\mid YM_{k,i,j,m} = 1$	(8)
$t1_{w,k,m2,d} \geq t2_{w,k,m1,d} - LN \times (1 - z_{w,k,m2,d})$ $\forall i \in T, j \in T, w \in W, k \in K, m \in M, d \in D$ $\mid YM_{k,i,j,m1} = 1 \wedge YM_{k,i,j,m2} = 1 \wedge m_1 \neq m_2$	(9)
$t1_{w,k2,m,d} \geq t2_{w,k1,m,d} + t_{min} - LN \times (1 - z_{w,k2,m,d})$ $\forall w \in W, k_1 \in K, k_2 \in K, m \in M, d \in D$ $\mid KM_{k1,m} = 1 \wedge KM_{k2,m} = 1 \wedge k_1 \neq k_2$	(10.1)
$t1_{w,k2,m2,d} \geq t2_{w,k1,m1,d} + t_{min} - LN \times (1 - z_{w,k2,m2,d})$ $\forall w \in W, k_1 \in K, k_2 \in K, m_1 \in M, m_2 \in M, d \in D$ $\mid KM_{k1,m1} = 1 \wedge KM_{k2,m2} = 1 \wedge k_1 \neq k_2 \wedge m_1 \neq m_2$	(10.2)
$AW_{m,c} \leq \sum_{w \in W \mid MWC_{w,c}=1} (t2_{w,k,m,d} - t1_{w,k,m,d})$ $\forall i \in T, j \in T, k \in K, m \in M, c \in C, d \in D$ $\mid YM_{k,i,j,m} = 1$	(11)
$w_{w,d} \in \{0,1\} \quad \forall w \in W, d \in D$	(12)
$z_{w,k,m,d} \in \{0,1\} \quad \forall w \in W, k \in K, m \in M, d \in D$	(13)
$t1_{w,k,m,d} \geq 0 \quad \forall w \in W, k \in K, m \in M, d \in D$	(14)
$t2_{w,k,m,d} \geq 0 \quad \forall w \in W, k \in K, m \in M, d \in D$	(15)

4 Case Study

Fertagus is a Portuguese private train operating company, a branch of the group Barraqueiro, which links Setúbal, in the south bank of Lisbon, to Roma-Areeiro, up north the Tejo river. It should be noted that Fertagus' trains are not the only ones using the rail track between Roma-Areeiro and Setúbal, which may constitute a problem since not every train unit has the same requirements. Besides operating the railway line, the company is also accountable for the maintenance of the rolling-stock units as well as the maintenance of some railway stations. In this case study, 17 train units are supposed to cover 196 daily tasks, in which some of them have to perform maintenance actions previously scheduled. There are 14 different types of maintenance activities and each one requires a certain set of competences/skills, so the amount of work required to perform it depends not only on the maintenance action but also on the competences needed, $AW_{m,c}$. There are 10 distinct competences which a worker can master, and each worker may use several competences at once. Finally, the maintenance crew team is formed by 16 workers. The ultimate aim is to obtain the best maintenance crew scheduling possible for one day of the week under a short computational time.

The tables displayed next, present parameters and the respective values used in this case study. For the sake of simplicity, only the most relevant data is displayed here. The constants used on this model are presented in Table 2. Table 3 establishes which tasks are assigned to each unit, whereas Table 4 specifies between which pair of tasks, each train unit, performs the scheduled maintenance actions. Table 5 displays information regarding the daily cost of employing a maintenance worker. Due to its large extension, additional supporting data may be provided by the authors upon reasonable request.

Table 2. Constants used

Constants	Unit	Value
NK	-	17
NS	-	15
NW	-	16
NC	-	10
NM	-	14
NT	-	196
ND	day	1
TMS	min	540
Δ_m	min	15
t_{min}	min	5
t_{man}	min	5
LN	-	10000

The maintenance crew scheduling model was executed for a specific day of a given week, adapting the scheduled maintenance actions from [7] model and using the actual Fertagus rolling-stock schedule. First, it was possible to observe that the computational time required to run this model is considerably small, 12.6 s more precisely. This is

Table 3. Tasks carried out by each unit

$X_{k,i}$		i											
		1	2	3	4	5	6		191	192	193	194	195	196
k	1	0	1	1	1	1	1		0	0	0	0	0	0
	2	0	0	1	1	1	1		0	0	0	0	0	0
	3	0	0	0	0	0	0		0	0	0	0	0	0
	4	0	0	0	0	0	0		0	0	0	0	0	0
	5	0	0	0	0	0	0		0	0	0	0	0	0
	6	0	0	0	0	0	0		0	0	0	0	0	0
	7	0	0	0	0	0	0		0	0	0	0	0	0
	8	0	0	0	0	0	0		0	0	0	0	0	0
	9	0	0	0	0	0	0	(...)	0	0	0	0	0	0
	10	0	0	0	0	0	0		0	0	0	0	0	0
	11	0	0	0	0	0	0		0	0	0	0	0	0
	12	0	0	0	0	0	0		0	0	0	0	0	0
	13	0	0	0	0	0	0		0	0	0	0	0	0
	14	0	0	0	0	0	0		0	0	0	0	0	0
	15	0	0	0	0	0	0		0	0	0	0	0	0
	16	0	0	0	0	0	0		0	0	0	0	0	0
	17	0	0	0	0	0	0		1	1	1	1	1	1

Table 4. Information on the unit and pair of tasks between each maintenance action is completed

$YM_{k,i,j,m}$		m													
		1	2	3	4	5	6	7	8	9	10	11	12	13	14
$Y_{k,i,j} = 1$	...								(...)						
	$Y_{1,7,8}$	0	0	1	0	0	0	0	0	0	0	0	0	1	0
	...								(...)						
	$Y_{3,23,24}$	0	0	0	0	0	1	0	0	0	0	0	0	0	0
	...								(...)						
	$Y_{9,99,100}$	0	0	0	0	0	0	0	0	1	0	0	0	0	0

Table 5. Cost of employing a maintenance worker

w	1	2	3	4	5	6	7	8	9	10	11	12	13	14	15	16
c_w	43	43	43	46	40	40	43	43	43	43	43	36	66	60	66	60

due to certain restrictions, such as the fact that in a single day, only three different train units can go to the depot to perform maintenance, which reduces the size of the problem significantly. The results obtained suggest that from the whole crew of 16 maintenance workers, only 10 are required to successfully carry out all maintenance actions and so, 6 of them are not assigned to work on this day. As the objective function focuses on minimizing the cost of employing workers, logically, the ones with an associated lower cost will be assigned if they have the required competences. For a faster visualization of the solution obtained, Fig. 2 was created, where *'x'* indicates which activities each worker carries out. Integrated with the starting and finishing times presented in the results' data file provided by the model, it allows a broader overview of the solution, through a clearer schematic planning sectioned by unit.

Since it is previously known which train units go to the depot for maintenance on this day, for the sake of understanding, only those units are presented. From the results obtained a couple of aspects are noteworthy of mention. First, it was possible to observe

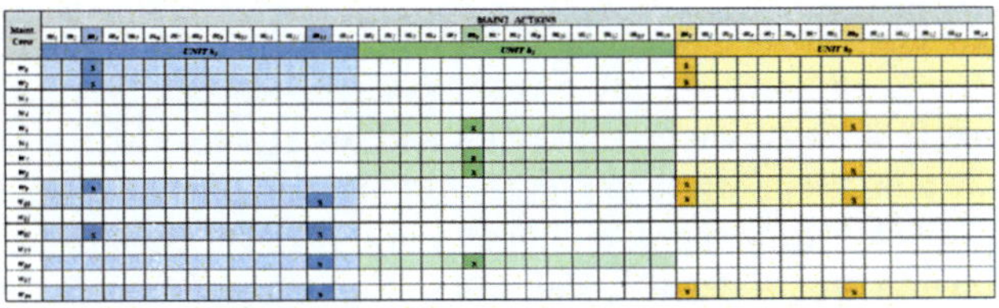

Fig. 2. Condensed results for the daily maintenance crew scheduling problem.

that while the same maintenance worker cannot perform two different maintenance activities simultaneously, there are cases where, for the same unit, two different activities are carried out at the same time. Moreover, it is also important to note that different maintenance workers can start the same maintenance activity, with some minutes of interval, and so, that action may have several starting and finishing times, as it happens for example for the maintenance action m_3 executed on train unit k_1.

5 Conclusions and Further Research

Following the reviewed work, with a higher focus on [7] study, it was decided to follow some thoughts exposed on his future research section, namely the crew scheduling that takes into account the different skills of maintenance technicians. This idea was in fact the first main objective for the present dissertation: a maintenance crew scheduling model that considers different skillsets for each worker and that could be applied to a train operating company, Fertagus. Information related to maintenance had to be gathered from [7] and [12] models and researches, as the current work is in fact a continuation of both of these works, as explained in Fig. 1. Moreover, some ideas implemented regarding the crew competences were inspired on [10], and so it is also noteworthy to mention. Actually, to the best of our knowledge, it was not found a mathematical model that executes the maintenance crew scheduling with a skillset associated to each worker, across all the reviewed and published work. This constitutes one of the main contributions of the present work, an innovative maintenance crew scheduling that takes into account the workers' different set of competences.

Additionally, the computational time required to run the models is significantly small, as expected, since the output consists on a daily schedule. Hereby, if desired, it is possible to obtain a weekly schedule in a practical time by running the model for each day of the respective week. No sensitivity analysis on the weight of the workers' employment costs was made, as these values are established by Fertagus company and are not meant to be modified. Finally, an optimality gap analysis was also not carried out since the computational time necessary to run the models is not large enough to make it relevant. Moreover, it is important to note that the model is flexible enough to be adapted to other instances or means of transport, just by modifying some inputs and constraints.

Since this scheduling problem is a continuation of researches carried out by [7] and [12], obviously some limitations were already established beforehand, as it happens for the number of units that can go to depot on a single day. However, for larger train companies, this number may be much larger and so the application of this model to a larger maintenance instance is suggested. Additionally, even though in this study maintenance actions do not have any type of relation between them, it may happen that maintenance activities have dependencies and for example, one action cannot be carried out while a different one is not finished. This kind of specific relations may as well be implemented. Furthermore, as future work, it might be interesting to extend such formulation into a multi-depot model, whose maintenance crew can travel between depots and be assigned to work in different days of the week to a certain depot. This crew flexibility might contribute to a more efficient scheduling.

Acknowledgments. The authors thank the support of the Fertagus train operating company, namely, individuals Eng. João Grossinho and Eng. João Duarte. The authors also gratefully acknowledge the financial support from the EU Horizon 2020 under the Grant Agreement No. 777627 (SMaRTE). This work was also supported by FCT, through IDMEC, under LAETA, project UID/EMS/50022/2020.

References

1. EC: European Commission Rail Research and Shift2Rail (2019). https://ec.europa.eu/transport/modes/rail/shift2rail_en. Accessed 12 Oct 2018
2. S2R: Shift2Rail History of the Initiative (2019). https://shift2rail.org/about-shift2rail/history-of-the-initiative/. Accessed 12 Oct 2018
3. Wienker, M., Henderson, K., Volkerts, J.: The computerized maintenance management system an essential tool for world class maintenance. Procedia Eng. **138**, 413–420 (2016)
4. Vale, C., Ribeiro, I.M., Calçada, R.: Integer programming to optimize tamping in railway tracks as preventive maintenance. J. Transp. Eng. **138**(1), 123–131 (2012)
5. Caetano, L.F., Teixeira, P.F.: Strategic model to optimize railway-track renewal operations at a network level. J. Infrastruct. Syst. **22**(2), 04016002 (2016)
6. Dao, C., Basten, R., Hartmann, A.: Maintenance scheduling for railway tracks under limited possession time. J. Transp. Eng. Part A: Syst. **144**(8), 04018039 (2018)
7. Mira, L., Andrade, A.R., Gomes, M.C.: Maintenance scheduling within rolling stock planning in railway operations under uncertain maintenance durations. J. Rail Transp. Plan. Manag. **14**, 100177 (2020)
8. Haghani, A., Shafahi, Y.: Bus maintenance systems and maintenance scheduling: model formulations and solutions. Transp. Res. Part A: Policy Pract. **36**(5), 453–482 (2002)
9. Fuentes, M., Cadarso, L., Marín, Á.: A fix & relax matheuristic for the crew scheduling problem. Transp. Res. Procedia **33**, 307–314 (2018)
10. Pour, S.M., et al.: A hybrid constraint programming/mixed integer programming framework for the preventive signaling maintenance crew scheduling problem. Eur. J. Oper. Res. **269**(1), 341–352 (2018)
11. Martins, R., Fernandes, F., Infante, V., Andrade, A.R.: Simultaneous scheduling of maintenance crew and maintenance tasks in bus operating companies: a case study. J. Qual. Maintenance Eng. (2021). https://doi.org/10.1108/JQME-09-2020-0099
12. Méchain, M., Andrade, A.R., Gomes, M.C.: Planning maintenance actions in train operating companies—a Portuguese case study. In: Ball, A., Gelman, L., Rao, B.K.N. (eds.) Advances in Asset Management and Condition Monitoring. SIST, vol. 166, pp. 1163–1181. Springer, Cham (2020). https://doi.org/10.1007/978-3-030-57745-2_96
13. Doganay, K., Bohlin, M.: Maintenance plan optimization for a train fleet. WIT Trans. Built Environ. **114**(12), 349–358 (2010)

14. Bohlin, M., Wärja, M.: Maintenance optimization with duration-dependent costs. Ann. Oper. Res. **224**(1), 1–23 (2012). https://doi.org/10.1007/s10479-012-1179-1
15. Tréfond, S., et al.: Optimization and simulation for robust railway rolling-stock planning. J. Rail Transp. Plan. Manag. **7**(1–2), 33–49 (2017)

Availability Importance Measure for Various Operation Condition

Abbas Barabadi[1](✉), Ali Nouri Qarahasanlou[2], Ali Hazrati[3], Ali Zamani[3], and Mehdi Mokhberdoran[4]

[1] Department of Engineering and Safety, UiT, The Arctic University of Norway, Tromsø, Norway
abbas.b.abadi@uit.no
[2] Faculty of Technical and Engineering, Imam Khomeini International University, Qazvin, Iran
Ali_Nouri@eng.ikiu.ac.ir
[3] School of Mining, College of Engineering, University of Tehran, Tehran, Iran
{Hazrati.ali,Ali.Arabshah}@ut.ac.ir
[4] SGS, Tabriz, Iran
Mahdi.Mokhberdoran@sgs.co

Abstract. The concept of availability importance measures can be used to identifying critical components from the availability performance point of view. The availability of an item depends on the combined aspects of its reliability and maintainability performance indices. These indices are considerably affected by operational and environmental conditions such as; ambient temperature, precipitation, wind, etc. Thus, different subsystems or components' availability in various conditions changes the performance priority of the system. In this way, the paper used the availability importance measure considering the operating environment for a mining fleet consisting of one shovel and six trucks. The reliability and maintainability characteristic of machines considering all influence factors (covariates) is analysed using by Cox regression model. The availability importance measure in two scenarios demonstrated that subsystem criticality changes in various conditions and the appropriate decisions should be made on different operational conditions.

Keywords: Importance measure · Operation condition · Reliability · Maintainability

1 Introduction

The increasing demand for material is forcing mining companies to increase their output. To meet production targets, large-scale equipment with high performance is needed.

Availability is a comprehensive metric for repairable systems performance management, which combines reliability and maintainability [1]. Since reliability and maintainability [2] characteristics of equipment are considerably affected by the operating environment (temperature, precipitation, etc.) thus, availability can change through the different conditions.

R. Karim et al. (Eds.): IAI 2021, LNME, pp. 86–98, 2022.
https://doi.org/10.1007/978-3-030-93639-6_8

As previously noted, availability is an important indicator of a repairable system. When the availability of a system is low, efforts are needed to improve it. In a system whose performance depends on the performance of its components, some of these components may play a more important role than others. Therefore, identifying the crucial components is an optimal way to improve system availability [3]. Importance measure is an effective approach that provides a guideline for performance improvement. Availability importance measure is an index that shows the contribution of components available on the system availability. The index provides a numerical ranking of components that highly ranked components have the greatest effect on system availability [4].

Up to this time, many studies have investigated component contribution on system performance. At first, Birnbaum proposed a quantitative definition that measures the contribution of component reliability to system reliability. Then, Gao *et al.* [5] have proposed a maintainability importance measure to identify critical components from a maintainability perspective. Gao *et al.* [6] and Wu and Coolen [7] have proposed a cost-effective manner to improve system reliability using the concept of Birnbaum importance measure. Availability importance measure that identifies the importance of each component on system availability investigated by some authors. Thus, the best strategy selection through decreasing failure rate or increasing repair rate is proposed [4]. Finally, Qarahasanlou *et al.* have presented availability importance measure considering operation environmental factors on RAM characteristics. The study shows different operation conditions can change the availability of components [8].

The previous research shows that the reliability, maintainability, and availability importance of components have been conducted using time series data such as Time Between Failures (TBF's) and Time To Repairs (TTR's). However, RAM characteristics of components can be affected by operation environmental factors (covariates); thus, accurate estimation of system performance requires both time series data and covariates. Furthermore, since various operation conditions may have different influences on component availability performance, thus availability importance of components may change through different operation conditions, which has not been addressed in the previous studies. In this paper, the availability of the system is calculated using both time and covariate data. After that, the concept of availability importance measure is used to identify critical components from the availability point of view in various conditions.

The paper is structured as follows: Sect. 2 introduces a methodology for reliability and maintainability analysis considering operation conditions after introducing availability and availability importance measures. Section 3 presents a case study describing component importance analysis in various conditions in Iran's Golgohar iron mine. Section 4, finally, concludes the paper.

2 Theoretical Background and Definitions

2.1 Reliability and Maintainability Analysis

The traditional reliability approach is based on the lifetime distribution of event records of a population of identical items. In this approach, the population characteristics [e.g., mean time to failure (MTTF) and probability of reliable operation] are estimated using historical failure time data. This popular technique was proposed as a standard tool for the planning and operation of automatic and complex mining system reliability/maintainability in the mid-1980s [9]. The approach does not require condition and operating environment data, a limitation in dynamic operating and environmental conditions. Therefore, covariate-based hazard models (regression models) approaches were developed [10]. A pioneering covariate-based hazard model is the proportional hazard model (PHM), introduced by Cox (1972). The common form of PHM is log-linear and is expressed as Eq. (1) [11].

$$\lambda(t, z) = \lambda_0(t)\psi(\alpha z) = \lambda_0(t)\exp\left(\sum_{i=1}^{n} \alpha_i z_i\right) \tag{1}$$

The component reliability influenced by covariates is expressed as Eq. (2).

$$R(t, z) = (R_0(t))^{\exp\left(\sum_{i=1}^{n} \alpha_i z_i\right)} \tag{2}$$

The mean time to failure (MTTF) can be given by Eq. (3) [6]. In this Equation, $\lambda(t, z)$ and $R(t, z)$ are the failure and reliability functions, respectively, z is row vector consisting of the covariate parameters (indicating the degree of influence which each covariate has on the failure function), α (column vector) is the unknown parameter of the model or regression coefficient of the corresponding n covariates (z), $\lambda_0(t)$ and $R_0(t)$ are baseline hazard rate and baseline reliability, respectively, dependent on time only and $exp \sum_{i=1}^{n}(\alpha_i z_i)$, exponential function commonly used for covariates term [12].

$$\begin{aligned} MTTF &= \int_0^{\infty} R(t, z)dt = \int_0^{\infty} (R_0(t))^{\exp\left(\sum_{i=1}^{n} \alpha_i z_i\right)} dt \\ &= \int_0^{\infty} \left(\exp\left(-\int_0^{t} \lambda_0(x)dx\right)\right)^{\exp\left(\sum_{i=1}^{n} \alpha_i z_i\right)} dt \end{aligned} \tag{3}$$

The proportional repair model (PRM) is proposed to predict repair rate considering the operating environment. PRM can be expressed as Eq. (4) [2].

$$\mu(t, z) = \mu_0(t)\phi(\beta w) = \mu_0(t)\exp\left(\sum_{i=1}^{m} \beta_i w_i\right) \tag{4}$$

The component maintainability influenced by covariates is expressed as [2].

$$M(t, z) = 1 - (1 - M_0(t))^{\exp\left(\sum_{i=1}^{m} \beta_i w_i\right)} \tag{5}$$

The mean time to repair (MTTR) can be given by [13].

$$MTTR = \int_0^{\infty} (1 - M(t, z))dt = \int_0^{\infty} (1 - M_0(t))^{\exp\left(\sum_{i=1}^{m} \beta_i w_i\right)} dt$$

$$= \int_0^{\infty} \left(\exp\left(-\int_0^t \mu_0(x)dx \right) \right)^{\exp\left(\sum_{i=1}^{m} \beta_i w_i\right)} dt \quad (6)$$

Where $\mu(t, w)$ and $M(t, w)$ are the repair and maintainability function, respectively; β is the regression coefficient of the corresponding m covariates (w); and $\mu_0(t, w)$ and $M_0(t, w)$ are the baseline repair rate and baseline maintainability (cumulative distribution function of TTRs), respectively. The main assumption in the PHM/PRM is that the covariates are time-independent variables (PH assumption [2].

There are several approaches to evaluating the proportional hazards (PH) assumption of the PHM: a graphical procedure, a goodness-of-fit testing procedure, and a procedure involving the use of time-dependent variables [14]. However, the PH assumption may not be valid in some cases. This means the effect of the environment on reliability performance is time-dependent. In this case, we can use the stratified Cox regression model (SCRM). The stratified Cox model modifies the PHM, allowing for control by "stratification" of a predictor that does not satisfy the PH assumption. Predictors assumed to satisfy the PH assumption are included in the model, but the stratified predictor is not. For example, suppose the population can be divided into r strata for failure (/g strata for failure), based on the discrete values of a single covariate or a combination of discrete values of a set of covariates. Then, the hazard rate of an asset in the sth stratum can be expressed by Eq. (7) [14].

$$\lambda_s(t, z) = \lambda_{0s}(t)\exp\left(\sum_{i=1}^{n} \alpha_i z_i\right), \quad s = 0, 1, \ldots .r \quad (7)$$

As with the original SCRM, there are two unknown components in the hazard model: the failure regression parameter α_i and the baseline failure function λ_{0s} (t) for each stratum. The baseline failure functions for "r" strata could be arbitrary and assumed completely unrelated. The repair rate of an asset in the g'th stratum can be expressed as Eq. (8) [11].

$$\mu_g(t, z) = \mu_{0g}(t)\exp\left(\sum_{j=1}^{m} \beta_j w_j\right), \quad g = 0, 1, \ldots .u \quad (8)$$

Where β_j is the regression parameter and $\mu_{0g}(t)$ the baseline repair function for each stratum. The baseline repair functions for "u" strata could be arbitrary and assumed completely unrelated.

2.2 Availability Performance

Availability of a system depending on its component uptime (reliability performance), downtime (maintainability performance), and the system structure (i.e., configuration) [15]. Suppose that 1 and 0 denote system up and downstate, respectively. Therefore, availability is the probability that the system is operational at time t. This measure can be represented mathematically by [16]:

$$A(t) = \Pr(X(t) = 1) \quad (9)$$

A(t) is a point or instantaneous availability. However, a common measure of availability is steady-state availability. Steady-state availability is defined as the long-term fraction of time that an item is available. The system's steady-state availability (As.s) is the limit of the point availability as time tends to infinity [17].

$$A_{s.s} = \lim_{t\to\infty} A(t) \tag{10}$$

Typical system structures are series, parallel and series-parallel. In the present paper, the series-parallel structure is discussed. For other structures, see references [4] and [18]. Steady-state availability of a series-parallel system that consists of n independent component in series and m independent component in parallel can be found in Eq. (11).

$$\begin{aligned} A_s(t) &= \prod_{k=1}^{n}\left(1-\prod_{l=1}^{m}(1-A_{kl}(t))\right) \\ &= \prod_{k=1}^{n}\left(1-\prod_{l=1}^{m}\left(1-\frac{MTBF_{kl}}{MTBF_{kl}+MTTR_{kl}}\right)\right) \end{aligned} \tag{11}$$

2.3 Availability Importance Measure

Importance analysis, as one of such tools, can be used to prioritize components in a system by mathematically measuring the importance level of each component on the system performance [4]. Availability importance measure can identify the weakest area of the system from an availability point of view. Availability importance measure is the partial derivative of the system's availability for the component availability, which is mathematically expressed by Eq. (12):

$$I_A^i = \frac{\partial A_S}{\partial A_i} \tag{12}$$

I_A^i is availability importance of component i, A_s and A_i is the system and ith component availability, respectively. In a system, components with high I_i^A have the greatest effect on the system availability [4]. From Eqs. (11) and (12), the availability importance measure of ijth component in a series-parallel system can be found in Eq. (13) [4].

$$\begin{aligned} I_A^{ij} = \frac{\partial A_s}{\partial A_{ij}} &= \frac{\partial\left(\prod_{k=1}^{n}\left(1-\prod_{l=1}^{m}(1-A_{kl}(t))\right)\right)}{\partial A_{ij}} \\ &= \prod_{\substack{k=1\\k\neq i}}^{n}\left(1-\prod_{l=1}^{m}(1-A_{kl})\right)\times\prod_{\substack{l=1\\l\neq j}}^{m}(1-A_{kl}) \end{aligned} \tag{13}$$

I_A^{ij} To determine the relative ranking of components, this index () should be normalized. represent the absolute value of importance measure, which may not be as significant as component relative ranking [19]. Therefore, normalized availability importance measure for component ij of a series-parallel system can be defined as [18]:

$$NI_{A}^{ij} = \frac{I_{A}^{ij}}{\sum_{k=1}^{n}\sum_{l=1}^{m} I_{A}^{kl}} \tag{14}$$

After detecting critical components, the best strategy for availability improvement through increasing TBF or decreasing repair time should be identified. For this purpose, reliability and maintainability-based availability importance measures are appropriate metrics. Indeed, reliability and maintainability-based availability importance measure shows the influence of reliability and maintainability of component i on the availability of the whole system and can be represented by Eqs. (15) and (16), respectively [4].

$$I_{A,MTBF_{ij}}^{ij} = I_{A}^{ij} \times A_{ij} \times \frac{MTTR_{ij}}{MTBF_{ij}(MTBF_{ij} + MTTR_{ij})} \tag{15}$$

$$I_{A,MTTR_{ij}}^{ij} = I_{A}^{ij} \times A_{ij} \times \frac{1}{(MTBF_{ij} + MTTR_{ij})} \tag{16}$$

Normalized reliability and maintainability-based availability importance measure are defined by Eqs. (17) and (18), respectively [18].

$$NI_{A,MTBF}^{ij} = \frac{I_{A,MTBF}^{ij}}{\sum_{k=1}^{n}\sum_{l=1}^{m} I_{A,MTBF}^{kl}} \tag{17}$$

$$NI_{A,MTTR}^{ij} = \frac{I_{A,MTTR}^{ij}}{\sum_{k=1}^{n}\sum_{l=1}^{m} I_{A,MTTR}^{kl}} \tag{18}$$

3 Case Study

Mining companies are using large-scale equipment with high investment. To meet the production target, high performance of this equipment is needed. Here, we present a case study to illustrate the proposed methodology. This mine is an open pit, and the shovel-truck fleet is used for material handling. Therefore, a shovel-truck system illustrated in Fig. 1 is selected as a case study. This fleet is considered a system, loading and hauling are considered a subsystem, and each machine is considered a component. Table 1 represents the model of the machines and corresponding codes.

After identifying the system, subsystem, and component, required data were collected over 18 months. For each machine, TBF, TTR, and corresponding covariate were sorted, classified, and quantified. Identification and quantification of all influence covariates are curtailed tasks. For instance, the classification and quantification of covariates for haulage subsystem failure are shown in Table 2. In addition, identified covariates for shovel and dump truck subsystem failure and repair are appeared in Table 3.

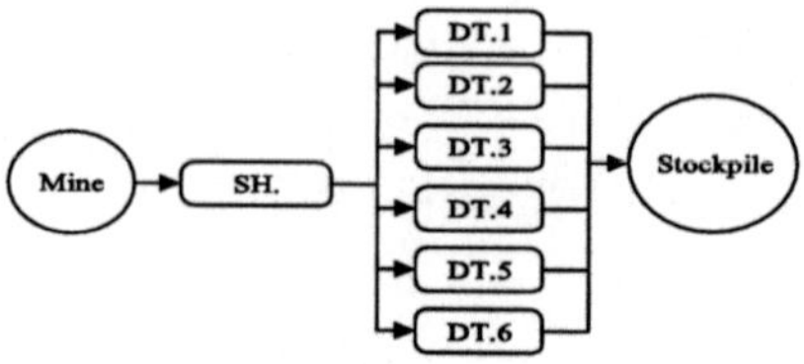

Fig. 1. Block diagram for shovel-truck fleet

Table 1. Components of the mining system and their code

Equipment	Model	Code
Shovel	Liebherr R9350	SH
Dump trucks	Terex-TR100	DT.1,2,3 and 4
Dump trucks	Caterpillar-777D	DT.5 and 6

Table 2. Classification and quantification of failure covariates for haulage subsystem

Covariate	Classification	Quantification
Shift (Zsh)	Morning	1
	Midday	2
	Night	3
Working Place (Z_{WP})	Bench 10–12	1
	Bench 13–15	2
	Bench 16–18	3
Math with loader (Z_{ML})	Excavator	1
	Shovel	2
Number of times service (Z_{NS})	Suitable	1
	Moderate suitable	2
	Unsuitable	3
Rock kind (Z_{RK})	Ore	2
	Waste	1
Team (Z_{Team})	Team A	1
	Team B	2
	Team C	3
	Team D	4

Table 3. Failure and repair covariates for subsystems

Failure covariates		Repair covariates
Dump trucks	Shovel	
Shift	Shift	Shift
Rock kind	Rock kind	Precipitation
Number of services	Number of services	Temperature
Match with loader	Match with loader	Wind
Team	Team	
Working place	Precipitation	
Precipitation	Temperature	
Temperature		

3.1 Reliability and Maintainability Performance Analysis

In this case, graphical and theoretical models are used for checking PH assumption. In Cox PHM or PRM, Weibull distribution widely used for baseline hazard rate λ_0 and baseline repair rate μ_0. So, for ensuring the Akaike Information Criterion (AIC) is used for goodness to fit baseline hazard rate and baseline repair rate for all subsystems. In the present study, test z for eliminated covariates was found to have no significant value from the subsequent calculations. The corresponding estimates of a regression coefficient were obtained and tested for their significance based on test z and (/or) p-value (obtained from the table of normal unit distribution). Used the p-value of 5% as the upper limit to check the significance of covariates. To avoid any bias in the results of PRM and PHM, the assumption of the proportional repair model and proportional failure model must be checked; Stata accommodates a statistical test of the PH assumption using the Schoenfeld residuals. The results of the PH assumption for DT.1 are shown in Table 4. In the theoretical model, if the p-value is bigger than 0.05, the PH assumption is established, and covariates are time-independent.

Since the PH and PR assumption results show all covariates (continues and category) for failure and repair data are time-independent, thus PHM and PRM are selected as a suitable model for analysis. For instance, the results of selecting effective covariates for PHM of DT.1 are shown in Table 5.

Table 4. The results of theoretical model for PH assumption for subsystem DT.1

Covariates	Rho	Chi2	Df	P-value
Temperature	−0.04	0.91	1	0.3395
Precipitation	0.008	0.07	1	0.7962
Shift	0.056	1.51	1	0.2196
Rock type	0.042	0.81	1	0.369
Number of service	0.063	1.27	1	0.259
Proportional loading	−0.05	1.26	1	0.262
Operation team	−0.009	0.04	1	0.838
Working place	0.038	0.75	1	0.385

Table 5. The results of PHM for hazard ratio (HR) and select effective covariates for DT.1

Covariates	HR	P-value	95% Conf. interval	
Temperature	1.005	0.426	0.993	1.017
Precipitation	0.891	0.038	0.800	0.994
Shift	0.929	0.216	0.827	1.043
Rock type	0.816	0.626	0.361	1.845
Number of service	0.056	0.000	0.047	0.075
Match with loader	1.212	0.049	1.000	1.469
Team	0.948	0.226	0.869	1.033
Working place	0.848	0.32	0.769	1.046

According to Table 5, the hazard ratio calculated for each covariate, the third column, shows the value of the z test statistic. The most important column is the last column, which determines the effective covariates. If the calculated p-value for the z-test is greater than 0.05, the null hypothesis will be accepted (covariate not affected); otherwise, the covariate is effective. According to Table 5, precipitation, the number of services, and match with loader are selected as effective covariates for DT.1. Finally, each component's reliability and maintainability functions are shown in Table 6 and Table 7, respectively. The MTBF and MTTR in the last columns were calculated according to Eq. (3) and (6) and using Wolfram Alpha software with mean values of the covariate.

3.2 Availability Importance Measure

After calculating RM characteristics and identification component interaction in a logical model, availability importance measure can be used to find the critical component of the system from an availability perspective. In this paper, we consider covariates that can

Table 6. Best-fit distribution for baseline, covariates, and MTTF calculation for failure data

Equipment	Baseline hazard rate (λ_{0s})		$\exp\left(\sum_{i=1}^{n} \alpha_i z_i\right)$	MTBF (Suitable)	MTBF (Unsuitable)
	Best fit	Parameter			
DT.1	Weibull	Shape = 2, Scale = 0.8	$Exp(-2.5Z_{NS} + 0.19Z_{ML} - 0.108Z_{P})$	2.2	26.41
DT.2	Weibull	Shape = 2.03, Scale = 0.57	$Exp(-2.6Z_{NS})$	1.82	23.55
DT.3	Weibull	Shape = 2.2, Scale = 1.7	$Exp(-2.6Z_{NS} - 0.2Z_{ML} + 0.108Z_{Sh})$	8.8	186.3
DT.4	Weibull	Shape = 2.1, Scale = 0.7	$Exp(-2.6Z_{NS} - 0.09Z_{Team} + 0.011Z_{T})$	2.2	25.7
DT.5	Weibull	Shape = 1.9, Scale = 2.2	$Exp(-2.6Z_{NS} + 0.14Z_{Team} + 0.02Z_{T})$	5.2	80.3
DT.6	Weibull	Shape = 1.9, Scale = 1.4	$Exp(-2.6Z_{NS})$	4.9	75.4
SH	Weibull	Shape = 1.9, Scale = 8.8	$Exp(-1.9Z_{NS} + 0.9Z_{MT} + 0.17Z_{T})$	2	15.12

Table 7. Best-fit distribution for baseline, covariates, and MTTR calculation for repair data

Equipment	Baseline repair rate (μ_{0s})		$\exp\left(\sum_{i=1}^{n} \alpha_i z_i\right)$	MTTR
	Best fit	Parameter		
DT.1	Weibull	Shape = 0.6, Scale = 10	–	15.05
DT.2	Weibull	Shape = 0.64, Scale = 1.7	$Exp(-0.28Z_{Sh})$	4.56
DT.3	Weibull	Shape = 0.67, Scale = 3.5	$Exp(-0.02Z_{T})$	8.22
DT.4	Weibull	Shape = 0.6, Scale = 4.8	–	7.22
DT.5	Weibull	Shape = 0.7, Scale = 1.7	$Exp(-0.023Z_{T})$	4.08
DT.6	Weibull	Shape = 0.7, Scale = 1.35	–	1.71
SH	Weibull	Shape = 0.6, Scale = 9.6	–	14.44

change the availability of the system in dynamic conditions. Therefore, critical components in different conditions should be identified. For this purpose, two scenarios for the number of times service (suitable ($Z_{NS} = 1$), unsuitable ($Z_{NS} = 3$)) is considered. For each scenario, availability importance measure (I_A), reliability-based availability importance measure ($I_{A,\,MTBF}$), and maintainability based availability importance measure ($I_{A,\,MTTR}$) are calculated using Eqs. (13), (15), and (16) then normalized by Eqs. (14),

(17) and (18) (NI_A- $NI_{A, MTBF}$- $NI_{A, MTTR}$), the calculation are represented in Table 8 and Table 9.

As shown in Tables 8 and 9, the list of components in decreasing order in the suitable condition is Sh., DT.6, DT.5, DT.3, DT.2, DT.4, and DT.1. The list of components in decreasing order in the unsuitable condition is Sh., DT.6, DT.3, DT.5, DT.2, DT.4, and DT.1. This indicates that various condition components may have different availability and correspondingly different importance from an availability point of view. From a reliability and maintainability perspective, each scenario has its own ordered list of components. From a reliability point of view in the suitable condition, the list of components in decreasing order is Sh., DT.2, DT.6, DT.5, DT.4, DT.3, and DT.1. However, in the unsuitable condition, again in descending order, the list of components is Sh., DT.2, DT.4, DT.1, DT.6, DT.5, and DT.3. Thus, various condition components have a different priority for resource allocation problems.

Table 8. Availability importance measure of mining fleet in the suitable condition

Equipment	Availability	I_A	NI_A	$I_{A, MTBF}$	$NI_{A, MTBF}$	$I_{A, MTTR}$	$NI_{A, MTTR}$
DT.1	0.128	0.0037	0.0036	0.000185	0.00342	0.000027	0.0028
DT.2	0.285	0.0045	0.0044	0.000501	0.00926	0.000200	0.0209
DT.3	0.517	0.0066	0.0065	0.000188	0.00347	0.000201	0.0210
DT.4	0.233	0.0042	0.0041	0.000339	0.00627	0.000103	0.0108
DT.5	0.560	0.0073	0.0072	0.000344	0.00636	0.000438	0.0458
DT.6	0.741	0.0123	0.0122	0.000483	0.00894	0.001385	0.1449
Sh	0.122	0.9738	0.9620	0.052014	0.96227	0.007202	0.7537

Table 9. Availability importance measure of mining fleet in the unsuitable condition

Equipment	Availability	I_A	NI_A	$I_{A, MTBF}$	$NI_{A, MTBF}$	$I_{A, MTTR}$	$NI_{A, MTTR}$
DT.1	0.637	0.0000008	0.0000008	7.22E−09	4.36654E−07	1.27E−08	7.3218E−07
DT.2	0.838	0.0000018	0.0000018	1.06E−08	6.44055E−07	5.50E−08	3.1802E−06
DT.3	0.958	0.0000071	0.0000071	1.54E−09	9.30580E−08	3.48E−08	2.0142E−06
DT.4	0.781	0.0000014	0.0000014	9.09E−09	5.49841E−07	3.23E−08	1.8691E−06
DT.5	0.952	0.0000062	0.0000062	3.55E−09	2.14521E−07	6.97E−08	4.0306E−06
DT.6	0.978	0.0000135	0.0000135	3.88E−09	2.34757E−07	1.71E−07	9.8948E−06
Sh	0.511	0.99999942	0.9999692	1.65E−02	9.99998E−01	1.73E−02	9.9998E−01

4 Conclusion

Availability improvement with minimum effort is an interesting issue. Dynamic operation environment is changes component technical characteristics and correspondingly component influence on system availability. Thus, to avoid wrong results, all influence factors on system characteristics must be identified. In this paper, the RAM characteristics of the system are analyzed using both time and operating condition variables (covariates). After that, using availability importance measure, component priority in various working conditions is identified. To illustrate the proposed methodology, a shovel-truck fleet that consists of one shovel and 6 dump trucks is selected. Availability importance measure in two different conditions (suitable and unsuitable) shows that DT.5 has the greatest effect in suitable conditions than DT.3. However, in an unsuitable condition, DT.3 has the greatest effect on system availability. Comparing reliability and maintainability-based availability importance measure for sh., DT.1, DT.2, and DT.4 show that, in the suitable condition, reliability improvement is the best strategy. However, in unsuitable conditions decreasing repair time is more appropriate.

References

1. Kumar, U.: Availability studies of load-haul-dump machines. In: Application of Computers and Operations Research in the Mineral Industry: 27/02/1989–02/03/1989. Society for Mining, Metalurgy and Exploration (1989)
2. Barabadi, A., Barabady, J., Markeset, T.: Maintainability analysis considering time-dependent and time-independent covariates. Reliab. Eng. Syst. Saf. **96**(1), 210–217 (2011)
3. Birnbaum, Z.W.: On the importance of different components in a multicomponent system. Washington Univ Seattle Lab of Statistical Research (1968)
4. Barabady, J., Kumar, U.: Availability allocation through importance measures. Int. J. Qual. Reliab. Manag. **24**(6), 643–657 (2007)
5. Gao, X., Markeset, T., Barabady, J.: Design and operational maintainability importance measures—a case study. Opsearch **45**(3), 189–208 (2008)
6. Gao, X., Barabady, J., Markeset, T.: Criticality analysis of a production facility using cost importance measures. Int. J. Syst. Assur. Eng. Manag. **1**(1), 17–23 (2010)
7. Wu, S., Coolen, F.P.: A cost-based importance measure for system components: an extension of the Birnbaum importance. Eur. J. Oper. Res. **225**(1), 189–195 (2013)
8. Nouri Qarahasanlou, A., Khalokakaie, R., Ataei, M., Ghodrati, B.: Operating environment-based availability importance measures for mining equipment (case study: sungun copper mine). J. Fail. Anal. Prev. **17**(1), 56–67 (2016). https://doi.org/10.1007/s11668-016-0205-z
9. Hoseinie, S.H., et al.: Reliability modeling of water system of longwall shearer machine. Arch. Min. Sci. **56**(2), 291–302 (2011)
10. Gorjian, N., et al.: The explicit hazard model-part 1: theoretical development. In: 2010 Prognostics and System Health Management Conference. IEEE (2010)
11. Ghodrati, B., Kumar, U.: Reliability and operating environment-based spare parts estimation approach: a case study in Kiruna Mine, Sweden. J. Qual. Maint. Eng. **11**(2), 169–184 (2005)
12. Kumar, D., Klefsjö, B.: Proportional hazards model: a review. Reliab. Eng. Syst. Saf. **44**(2), 177–188 (1994)
13. Gao, X., Markeset, T.: Design for production assurance considering influence factors. In: Proceedings of the European Safety and Reliability Conference (ESREL2007) (2007)
14. Kleinbaum, D.G.: Statistics in the Health Sciences: Survival Analysis. Springer, New York (2011)
15. Barabadi, A., Markeset, T.: Reliability and maintainability performance under Arctic conditions. Int. J. Syst. Assur. Eng. Manag. **2**(3), 205–217 (2011)

16. Tsarouhas, P.: Statistical techniques of Reliability, Availability, and Maintainability (RAM) analysis in industrial engineering. In: Ram, M., Davim, J.P. (eds.) Diagnostic Techniques in Industrial Engineering. MIE, pp. 207–231. Springer, Cham (2018). https://doi.org/10.1007/978-3-319-65497-3_8
17. Sherwin, D.J., Bossche, A.: The Reliability, Availability and Productiveness of Systems. Springer, Heidelberg (2012)
18. Nouri Gharahasanlou, A., et al.: Normalised availability importance measures for complex systems. Int. J. Min. Reclam. Environ. **32**(2), 109–122 (2018)

Industrial Equipment's Throughput Capacity Analysis

Ali Nouri Qarahasanlou[1], Ali Hazrati[2], Abbas Barabdi[3](✉), Aliasqar Khodayari[2], and Mehdi Mokhberdoran[4]

[1] Faculty of Technical and Engineering, Imam Khomeini International University, Qazvin, Iran
Ali_Nouri@eng.ikiu.ac.ir
[2] School of Mining, College of Engineering, University of Tehran, Tehran, Iran
{Hazrati.ali,Khodaiar}@ut.ac.ir
[3] Department of Engineering and Safety, UiT, The Arctic University of Norway, Tromsø, Norway
abbas.b.abadi@uit.no
[4] SGS, Tabriz, Iran
Mahdi.Mokhberdoran@sgs.com

Abstract. Throughput capacity (TC) is defined as the total amount of material processed or produced by the system in the given time. In practice, full capacity performance for industrial equipment is impossible because the failures are affected and cause a reduction. Therefore, failure interruptions, especially critical ones (bottlenecks), must be detected and considered in production management. From the point of production view, the bottleneck has the lowest production or performance. Most of the previous works used the availability and related importance measures as performance indicators and prioritization of subsystems. However, these measures cannot consider system production in their prioritization. This paper presents a bottleneck detection framework based on system performance and production capacity integration. The integrated approach is used to assess the loading and hauling subsystems of Golgohar Iron Mine, Iran. As a result of the analysis, the hauling subsystem identifies the system's bottleneck.

Keywords: Reliability · Maintainability · Throughput capacity · Mining Fleet

1 Introduction

Nowadays, the deliverability of systems is the main challenge of the competitive environment. Production performance analysis is a proposed tool for deliverability assessment in the production plants [1]. Capacity and availability are both important dimensions of the production performance of a production plant and mining equipment [2, 3]. In mining equipment, capacity is a function of equipment utilization and performance [3, 4]. Therefore, the availability and capacity analysis of mining equipment is the first step for assessing system throughput and detecting bottleneck(s) of the production process.

R. Karim et al. (Eds.): IAI 2021, LNME, pp. 99–111, 2022.
https://doi.org/10.1007/978-3-030-93639-6_9

Availability is the function of uptime and downtime, controlled by failures. The failure can be caused by production loss, company reputation, safety, environmental issues, maintenance costs, etc. [2]. In engineering systems, failures cannot be eliminated, but it is possible to mitigate the impact of failures with a better understanding of system behaviour and management.

Availability is defined as "the ability of an item to be in a state to perform a required function under given conditions at a given instant of time or over a given time interval, assuming that the required external resources are provided." Availability is a function of reliability (uptime), maintainability, and supportability (downtime) [2]. Therefore, calculating the system technical characteristics (RAMS) prerequisite for production performance analysis.

Quantitative Reliability, Maintainability, Availability, and Supportability (RAMS) analysis in mining equipment can be traced to the last 1980s. After that, various studies such as Loud-haul-dump machine in fleet and equipment level [5–7], drum shearer [8], powered supports [9], conveyor system [10], Crushing plant as a part of processing plant [11] had been carried out. Also, open-pit mine equipment's such as wagon drill [12], shovel [13], and truck [14, 15], have been analyzed in other studies using a statistical approach. Artificial intelligence techniques such as Genetic Algorithm (GA) [16, 17] and Machine Learning (ML) [18, 19] have been used for reliability and maintainability analysis and maintenance management of the mining equipment. In this paper, the statistical method is preferred due to the lack of data and user-friendly statistical procedures for managers and practitioners. In the mining operation, drilling, loading, and hauling equipment should interact with material production. Therefore, fleet or navigation level studies are needed to analyze the real situation. Recently, some studies have focused on RAM-based TC analysis and their equipment's or subsystems interactions in system networks [20, 21]. The studies at this level can reveal the production line bottleneck used for future planning and decision-making process on improving bottleneck and throughput capacity. The reviews present a narrow study carried out in the equipment level of the system in bottleneck recognition. Therefore, the present paper considers loading and hauling a fleet consisting of a shovel and trucks as a system in an open-pit mine. Since the primary purpose of the article is detecting bottleneck in system level, thus, performance characteristics of equipment (RAM) must be analyzed. After that, the TC of each subsystem based on availability, system configuration and capacity is predicted. Finally, subsystems prioritize and critical one is detected. The rest of the paper organized as follow:

Section 2 describes the proposed methodology, which is how to analyze the system's reliability, maintainability, and simulated TC. Section 3 analyzes a case study of Golgohar Iron Mine equipment by the proposed methodology. Finally, in Sect. 4, the conclusions of the paper are provided. This study assumed that:

- Each component, subsystem, and system has two states: working or failed.
- The effect of risk factors (covariates) such as environmental condition, operator skill, etc. not considered in the study.
- The equipment is repairable.
- Due to a lack of supportability data, the indicator is not considered in the study.

2 Methodology

Figure 1 illustrates the proposed approach flowchart. The methodology consists of four main steps that represent in detail as follows:

1. The identifying system, subsystem, and component boundaries and collecting required data in defined boundaries.
2. The identical and independent (iid) assumptions validation. Then select the best model for reliability and maintainability.
3. The estimation of reliability, maintainability, and availability characteristics.
4. Merging RAM characteristic and capacity at equipment level and simulating considering by system configuration, finally, predicting system and subsystem's TC.

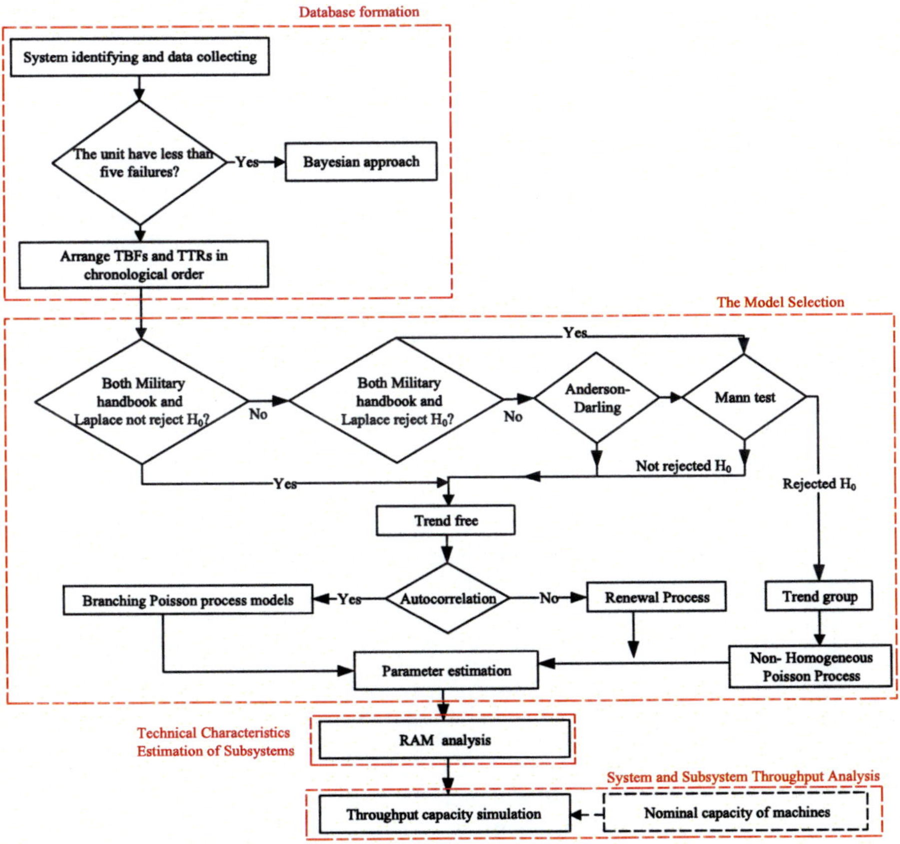

Fig. 1. TC analysis methodology

2.1 Boundary Identification and Data Collection

Preceding of data collection, system, subsystem, and component boundaries should be defined. These boundaries depend on the importance of the studied system, expert opinions, system configuration, management comments, data sources, and data collection process. Also, it must be preventing overlap in adjoining systems, subsystems, and components [20]. Sometimes, the study level (fleer or equipment) can be a useful factor in defining these boundaries [22].

When boundaries are defined, required data should be collected. Data can be gathered from many sources such as sensors, operator interfaces onboard equipment, historical operational and maintenance reports [23]. In addition, data collecting time intervals is a crucial issue. The data collecting time interval must be at least ten months [24]. The reliability and maintainability data is sorted as Time between Failures (TBFs) and Time to Repairs (TTRs) in chronological order. This data can be categorized into censored and complete groups. Failure occurring time is precisely not known in the censored data, but it is correctly known in the complete data [25].

2.2 Validation of Assumption of Identical and Independent Distribution (IID)

After the data collection, these data should be sorted into the required formatting, and the assumption of the Identical and Independent (iid) nature of data must be validated. Independency means that the data are free of trends and that each failure (repair) is independent of the preceding or succeeding failure (repair). Identically distributed data mean that all the data in the sample are obtained from the same probability distribution. In the non-repairable component, a failed component is replaced by the new one; therefore, the new component failure is independent of the previous component failure [26]. But in the repairable component, verification of the assumption that the data are iid should be done. Otherwise, completely wrong conclusions can be drawn [27].

Two common methods for validating the assumption of independent and identically distribution of sample are trend and autocorrelation test, respectively. If the assumption that the data are identical and independent is valid, the homogeneous Poisson Process (HPP) and Renewal Process (RP) must be fitted. Otherwise, the Non-Homogeneous Poisson Process (NHPP) is an appropriate model. For more information, see ref [25, 28, 29].

2.2.1 Trend Test

Trend tests can be categorized into two main groups; graphical and analytical. The graphical method is simple and does not require any calculations. The cumulative failures are plotted against the cumulative time in this kind of test. The straight line indicates trend free and convex or concave shows the trend in the data. The graphical method is strong when there are definite trends in the data. This solution may not be enough when slight trends are present and analytical tests should be performed [28].

In the analytical methods, the null hypothesis (H_0) is trend-free, and the alternative hypothesis (H_1) is a monotonic or non-monotonic (or both of them) trend. Four communal tests to trend analysis are the Laplace test, Military handbook test, Mann-Kendall test, and Anderson–Darling test. Depending on the nature of data, two or more tests must be performed. The null hypothesis in the military handbook and Laplace test is HPP. However, in the Mann-Kendall test is RP. Therefore, the Mann test must be performed after the rejection null hypothesis by both the Military handbook and Laplace test [25]. For more information about the tests as mentioned earlier, see [29]:

2.2.2 Autocorrelation

According to the proposed framework in Fig. 1, if there is no apparent trend, then the autocorrelation test needs to be carried out to check the independence between the TBF and TTR data. After sorting data in chronological order, ith failure against (i-1)'th failure is plotted. If all plot points generate a single cluster, then the data are independent, whereas the multi-clusters (two or more clusters) or a straight line lead to data dependency [28].

2.3 Reliability and Maintainability Characteristics Estimation

After selecting the best analysis model for the system's technical characteristics, parameter estimation and the goodness of fit test should be carried out. Weibull is the most popular distribution in the field of reliability engineering. Weibull distribution can model the early phase (decreasing failure rate), useful life (constant failure rate), and wear-out phase (increasing failure rate) of the system [7]. However, the lognormal distribution is suitable for mechanical component maintainability modeling [30]. The goodness of fit tests such as Anderson Darling (A-D), Kolmogorov-Smirnov (K-S), etc., can be used for best distribution selection through candidate distributions.

2.4 Throughput Capacity

TC is defined as; the amount of material that each system can process. When output at the system level is analyzed, the interaction between components must be considered [20]. For this purpose, the reliability block diagram (RBD) [26] and Fault tree (FT) [31] are two conventional methods. TC can be calculated using analytical and simulation techniques. The analytical approach is cheaper; on the other hand, the simulation method is more realistic [32].

To measure the product's availability, its component's availability must be calculated. However, to deal with this issue, the Norwegian oil and gas industry (Norwegian Technology Standards Institution, 1998) developed the Norsok Z-016 standard to assess the production availability of the system based on Fig. 2. This standard shows that system availability is the function of item availability and product availability is the function of system availability [1].

However, capacity performance must be considered an essential parameter to assess production performance. For instance, one has a production plant availability of 90 percent, the plant throughput capacity, and production rate may be less than desired due to the low capacity of items. Therefore, capacity performance at the item level must be considered. Figure 3 illustrates the production performance that consists of availability and capacity performance. It is noteworthy that throughput capacity can be described by production performance and production availability [2].

3 Case Study

Here we present a case study describing the proposed methodology. Golgohar complex is the first-largest iron mine in Iran. This complex is located in Sirjan, southeast of Iran. The complex consists of six mines, named by numbers, from one to six. Mine 1 is the largest and oldest mine in this complex and is operated by Arman Gohar Sirjan (AGS) Co. Golgohar mines are open-pit, so the shovel-truck fleet is used for raw material production.

This mine uses Liebherr shovels and three trucks: Terex, Caterpillar, and Komatsu. Dispatching priority in this mine has led to choosing Terex trucks for top levels and a combination of Komatsu and Caterpillar trucks for lower levels. The top-level (bench 10 to 12) using expert opinion and dispatching data shovel-truck fleet illustrated in Fig. 4 is selected as a case study.

This shovel-truck fleet is defined as a system. Each series and machine is defined as a subsystem and component. As previously noted, the illustrated system is working in the pre-stripping phase and on the top-level (bench 10 to 12) (Fig. 5). According to the mine production plan, this system will work about 500 h on the mentioned level. Therefore, analysis of this system during 500 h operation time is interesting. Table 1 depicts the production line corresponding codes as well as their functional capacity.

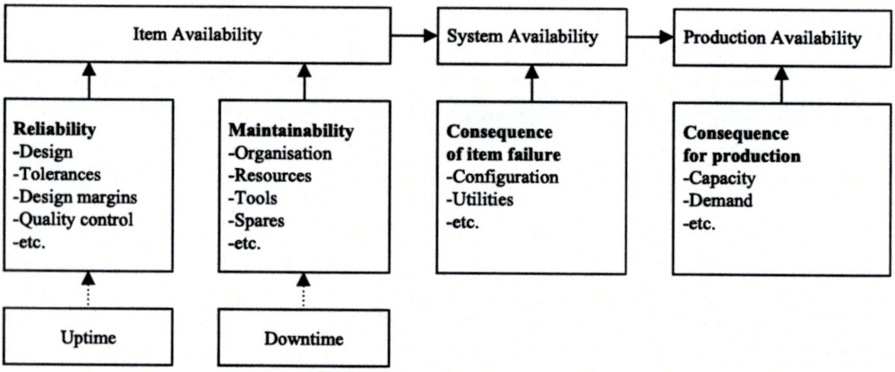

Fig. 2. Relationship between the availability of component and production availability [1]

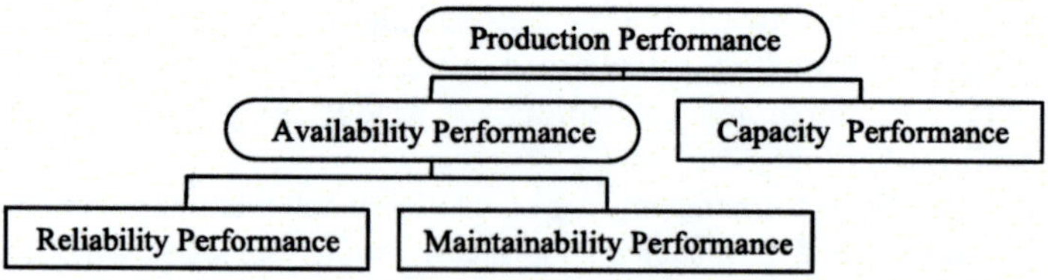

Fig. 3. Production performance concept, Adapted from [2]

After boundary identification, required data were collected, sorted, and classified in the form required for the analysis (i.e., TBF and TTR). This mine uses the SmartMine

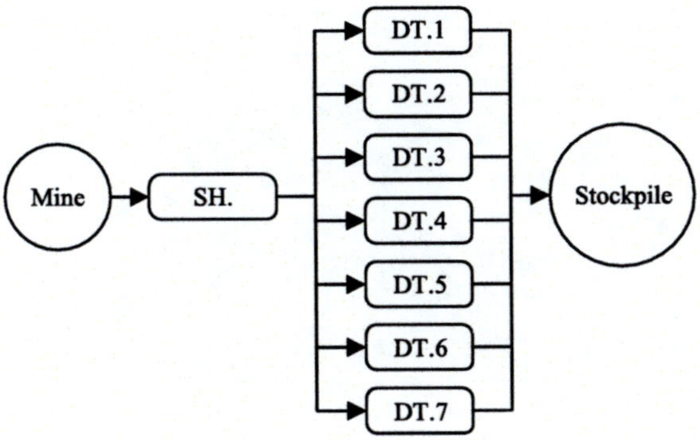

Fig. 4. Block diagram for shovel-truck fleet

system, so every moment equipment state (working or failed) automatically is recorded. Therefore, archived data in this data bank is a reliable data source. Data is used in this study were collected over 18 months.

The next step after collecting, sorting, and classifying the data is to validate the iid assumption of the nature of the data. As mentioned in Fig. 1, trend tests should be performed using two or more tests, according to trend behavior. Table 2 shows the p-value statistic of tests. The null hypothesis is considered in the confidence level of 95%. Thus, this assumption is rejected for less than 5% P-value.

Table 1. Mine equipment model and functional capacity

Row	Equipment	Model	Code	Mean capacity (m^3/h)
1	Shovel	Liebherr R9350	SH	420
2	Dump Truck	Terex-TR100	DT.1	53.8
3	Dump Truck	Terex-TR100	DT.2	53.3
4	Dump Truck	Terex-TR100	DT.3	54.5
5	Dump Truck	Terex-TR100	DT.4	53.4
6	Dump Truck	Terex-TR100	DT.5	53.7
7	Dump Truck	Caterpillar-777D	DT.6	57.7
8	Dump Truck	Caterpillar-777D	DT.7	60

NHPP (PLP) must be fitted for components that have a trend. Trend-free groups are subject to dependency analysis. Due to the scarcity of space, the DT.7 TTR autocorrelation test is illustrated in Fig. 6.

As can be seen, the first lag in the ACF graph is in the confidence level (95%), and the data has no evidence of any trend in the scatter plot. For all trend-free components, the dependency test shows the same result.

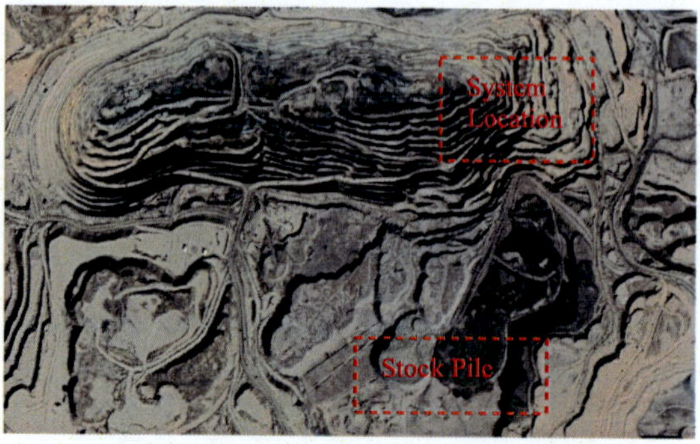

Fig. 5. Location of the selected system

Table 2. Computed p-value of trend tests

Eq.	Data	MIL.	Laplace	A-D	M-K	Result
SH	TBF	0.002	0	–	0.02	R*
	TTR	0.05	0.017	–	0.035	R
DT.1	TBF	0.2	0.042	0.029	0.078	NR**
	TTR	0.059	0.004	0.003	0.04	R
DT.2	TBF	0.076	0.38	–	–	NR
	TTR	0	0	–	0.007	R
DT.3	TBF	0.18	0.62	–	–	NR
	TTR	0	0	–	0.22	NR
DT.4	TBF	0.015	0.033	–	0.03	R
	TTR	0.67	0.74	–	–	NR
DT.5	TBF	0	0	–	0	R
	TTR	0	0.001	–	0.15	NR
DT.6	TBF	0.54	0.82	–	–	NR
	TTR	0.87	0.49	–	–	NR
DT.7	TBF	0.16	0.021	0.019	0.04	R
	TTR	0.22	0.42	–	–	NR

*R: Rejected
**NR: Not Rejected

The next step is selecting the best distribution through candidate distributions. For this purpose, the Anderson-Darling goodness of fit test is used. The distribution that has the lowest AD is chosen as the best distribution. The reliability and maintainability characteristics of the equipment have been calculated and presented in Table 3 and Fig. 7

using the software Minitab 18. As can be seen, Weibull and lognormal distributions are appropriate for reliability and maintainability modeling, respectively.

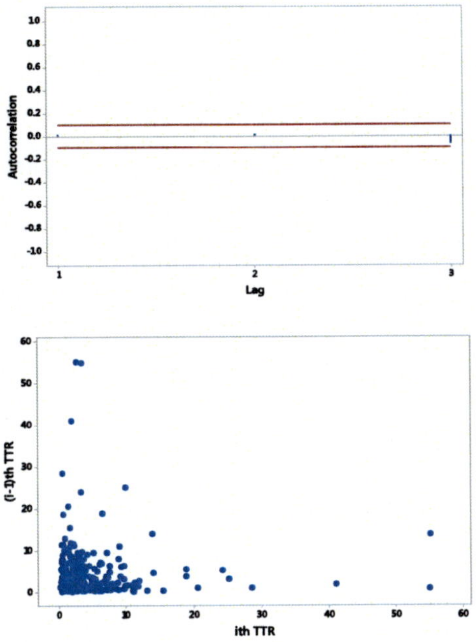

Fig. 6. Autocorrelation test for the DT.7 TTR

The logical relationship between components and components capacity is calculated for the TC analysis component technical characteristic. After that, the TC of the fleet is predicted using BlockSim 9 software for 500 h.

In the "Blocksim software" at the first step, simulate failure and repair characteristics of components, then incorporate the component's capacity into simulated times [33]. The failure and repair characteristics simulation technique works as the following; the first simulation yields first a random time to first failure, then a random time to first repair, then a random time to second failure, then a random time to second repair, and so on, until the chosen mission time ends. This sequence is repeated based on the number of simulations, yielding a different sequence. All the different sequences are stored each time. The number of simulations represents the number of different times to the first failure, the number of different times to first repair, and so on. The average of all times to the first failure is used as the time to the first failure. Similarly, the average of all times to first repair is used as the repair time for the first repair. The same process applies for the rest of the failures and repairs until the mission end time is reached [34].

The result of the simulation is shown in Table 4; the following are revealed:

- FCI is the contribution of equipment failure to system failure. Because of existing only one shovel in the system, shovel failure has the greatest influence on the system failure.

Table 3. Best-fit distribution for TBF and TTR data

Eq.	Reliability		Maintainability	
	Best-fit	Parameters	Best-fit	Parameters
Sh	PLP	Beta = 1.3; Eta = 140.4	PLP	Beta = 1.18; Eta = 18.36
DT.1	Weibull 3P.	Beta = 0.88; Eta = 12.08, Gamma = 0.44	PLP	Beta = 1.09; Eta = 10.31
DT.2	Weibull 3P.	Beta = 0.803; Eta = 9.8; Gamma = 0.437	PLP	Beta = 1.19; Eta = 16.93
DT.3	Weibull 2P.	Beta = 0.99; Eta = 19.9	Lognormal	LMean = 1.16; LStd = 1.31
DT.4	PLP	Beta = 0.89; Eta = 7.53	Lognormal 3P.	LMean = 1.08; LStd = 1.45; Location = 0.12
DT.5	PLP	Beta = 0.75; Eta = 2.27	Lognormal	LMean = 1.08; LStd = 1.48
DT.6	Weibull 2P.	Beta = 0.88; Eta = 27.94	Lognormal 3P.	LMean = 0.347; LStd = 1.25; Location = 0.028
DT.7	PLP	Beta = 1.08; Eta = 41.75	Lognormal	LMean = 0.68; LStd = 1.14

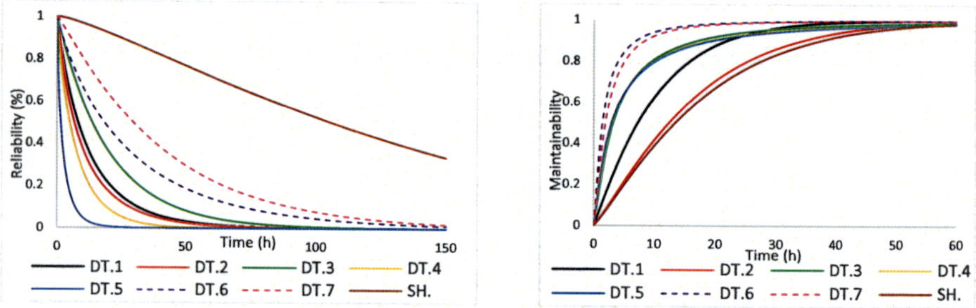

Fig. 7. Reliability and maintainability characteristics of the trucks and shovel.

- Excess Capacity is the additional amount a component could have produced while up and running.
- The utility is a proportion of time that available components can produce without blockage.

Because of low reliability and maintainability, Terex trucks have a weak performance (Table 4), while based on the result, Caterpillar trucks have higher strength and reliability. Therefore, a detailed study should be performed on Terex trucks at the equipment level. According to the simulation result, system throughput will be the sum of the truck's mean capacity (110039). Due to the high gap between loading and hauling subsystems, the hauling subsystem was identified as a bottleneck from the throughput capacity perspective. In other words, improving the truck's performance or increasing the number of trucks is needed.

Table 4. Throughput analysis result for individual equipment

Eq.	FCI (%)	MA* (%)	MC** (m3/h)	EC*** (m3)	Utility (%)
DT.1	0.21	69	15960	2617	85.92
DT.2	0.12	47	10399	2122	83.05
DT.3	0.5	76	18018	2700	86.97
DT.4	0.44	54.46	12279	2256	84.48
DT.5	0.35	30.2	6176	1920	76.28
DT.6	0.65	91.7	23294	3145	88.1
DT.7	0.36	90.32	23913	3182	88.25
SH	95.76	88.81	186457	49	99.97

MA*: Mean Availability, MC**: Mean Capacity, EC***: Exceed Capacity

4 Conclusion

The fleet level is more effective for the mining production system than the equipment level assessment for weakness point detection. Therefore, the right decision can be made for system improvement at the right time. In this paper, the reliability and maintainability characteristics of equipment were analyzed. After that, the system's throughput capacity using RM characteristics predicted the logical relationship between components and components capacity. The result shows that Terex trucks are tangibly weaker than Caterpillar trucks. Caterpillar truck's high performance is due to high reliability and maintainability (dashed lines in Fig. 7). The shovel is the newest equipment in the system, so; this equipment represents high reliability. In addition, poor maintainability of the shovel can result in in-site maintenance. Therefore, a detailed study should be performed on the feasibility of Terex trucks. From the bottleneck point of view, the high gap between loading and hauling subsystems (76418 m^3) demonstrated that improving trucks' performance or increasing trucks' number is necessary for throughput capacity improvement.

References

1. N Standard: Regularity management & reliability technology. Norwegian Technology Standards Institution, Oslo, Norway (1998)
2. Barabady, J., Markeset, T., Kumar, U.: Review and discussion of production assurance program. Int. J. Qual. Reliab. Manag. **27**(6), 702–720 (2010)
3. Moniri-Morad, A., Pourgol-Mohammad, M., Aghababaei, H., Sattarvand, J.: Capacity-based performance measurements for loading equipment in open pit mines. J. Central South Univ. **26**(6), 1672–1686 (2019). https://doi.org/10.1007/s11771-019-4124-5

4. Lanke, A.A., Hoseinie, S.H., Ghodrati, B.: Mine production index (MPI)-extension of OEE for bottleneck detection in mining. Int. J. Min. Sci. Technol. **26**(5), 753–760 (2016)
5. Kumar, U.: Availability studies of load-haul-dump machines. In: Application of Computers and Operations Research in the Mineral Industry: 27/02/1989–02/03/1989. Society for Mining, Metalurgy and Exploration (1989)
6. Kumar, U., Klefsjö, B., Granholm, S.: Reliability investigation for a fleet of load haul dump machines in a Swedish mine. Reliab. Eng. Syst. Saf. **26**(4), 341–361 (1989)
7. Balaraju, J., Govinda Raj, M., Murthy, C.: Estimation of reliability-based maintenance time intervals of Load-Haul-Dumper in an underground coal mine. J. Min. Environ. **9**(3), 761–770 (2018)
8. Hadi Hoseinie, S., Ataei, M., Khalokakaie, R., Ghodrati, B., Kumar, U.: Reliability analysis of drum shearer machine at mechanized longwall mines. J. Qual. Maint. Eng. **18**(1), 98–119 (2012)
9. Morshedlou, A., Dehghani, H., Hoseinie, S.H.: Reliability-based maintenance scheduling of powered supports in Tabas mechanized coal mine. J. Min. Environ. **5**(2), 113–120 (2014)
10. Simon, F., Javad, B., Abbas, B.: Availability analysis of the main conveyor in the Svea Coal Mine in Norway. Int. J. Min. Sci. Technol. **24**(5), 587–591 (2014)
11. Barabady, J., Kumar, U.: Reliability analysis of mining equipment: a case study of a crushing plant at Jajarm Bauxite Mine in Iran. Reliab. Eng. Syst. Saf. **93**(4), 647–653 (2008)
12. Rahimdel, M.J., Ataei, M., Khalokakaei, R., Hoseinie, S.H.: Maintenance plan for a fleet of rotary drill rigs/Harmonogram Utrzymania I Konserwacji Floty Obrotowych Urządzeń Wiertniczych. Arch. Min. Sci. **59**(2), 441–453 (2014)
13. Roy, S., Bhattacharyya, M., Naikan, V.: Maintainability and reliability analysis of a fleet of shovels. Min. Technol. **110**(3), 163–171 (2001)
14. Allahkarami, Z., Sayadi, A.R., Lanke, A.: Reliability analysis of motor system of dump truck for maintenance management. In: Kumar, U., Ahmadi, A., Verma, A.K., Varde, P. (eds.) Current Trends in Reliability, Availability, Maintainability and Safety. LNME, pp. 681–688. Springer, Cham (2016). https://doi.org/10.1007/978-3-319-23597-4_50
15. Morad, A.M., Pourgol-Mohammad, M., Sattarvand, J.: Reliability-centered maintenance for off-highway truck: case study of sungun copper mine operation equipment. In: Proceedings of the ASME International Mechanical Engineering Congress & Exposition (2013)
16. Vayenas, N., Peng, S.: Reliability analysis of underground mining equipment using genetic algorithms: a case study of two mine hoists. J. Qual. Maintenance Eng. **20**(1), 32–50 (2014). https://doi.org/10.1108/JQME-02-2013-0006
17. Peng, S., Vayenas, N.: Maintainability analysis of underground mining equipment using genetic algorithms: case studies with an LHD vehicle. J. Min. **2014**, 1–10 (2014). https://doi.org/10.1155/2014/528414
18. Taghizadeh Vahed, A., Ghodrati, B., Demirel, N., Hosseini Yazdi, M.: Predictive maintenance of mining machinery using machine learning approaches. In: Proceedings of the 29th European Safety and Reliability Conference, Hannover (2019)
19. Taghizadeh Vahed, A., Ghodrati, B., Hossienie, H.: Enhanced K-nearest neighbors method application in case of draglines reliability analysis. In: Widzyk-Capehart, E., Hekmat, A., Singhal, R. (eds.) Proceedings of the 27th International Symposium on Mine Planning and Equipment Selection—MPES 2018, pp. 481–488. Springer, Cham (2019). https://doi.org/10.1007/978-3-319-99220-4_40
20. Barabadi, A., Barabady, J., Markeset, T.: A methodology for throughput capacity analysis of a production facility considering environment condition. Reliab. Eng. Syst. Saf. **96**(12), 1637–1646 (2011)
21. Gharahasanlou, A., Ataei, M., Khalokakaie, R., Einian, V.: Throughput capacity analysis (case study: sungun copper mine). J. Fundam. Appl. Sci. **8**, 1531–1556 (2016)

22. Hoseinie, S., Al-Chalabi, H., Ghodrati, B.: Comparison between simulation and analytical methods in reliability data analysis: a case study on face drilling rigs. Data **3**(2), 12 (2018)
23. Barabady, J.: Reliability and maintainability analysis of crushing plants in Jajarm Bauxite Mine of Iran. In: Proceedings of the Annual Reliability and Maintainability Symposium. IEEE (2005)
24. Vagenas, N., Runciman, N., Clément, S.R.: A methodology for maintenance analysis of mining equipment. Int. J./Surface Min. Reclam. Environ. **11**(1), 33–40 (1997)
25. Garmabaki, A.H.S., Ahmadi, A., Mahmood, Y.A., Barabadi, A.: Reliability modelling of multiple repairable units. Qual. Reliab. Eng. Int. **32**(7), 2329–2343 (2016)
26. Rausand, M., Høyland, A.: System Reliability Theory: Models, Statistical Methods, and Applications, 396. Wiley, Hoboken (2003)
27. Kumar, U., Klefsjö, B.: Reliability analysis of hydraulic systems of LHD machines using the power law process model. Reliab. Eng. Syst. Saf. **35**(3), 217–224 (1992)
28. Louit, D.M., Pascual, R., Jardine, A.K.: A practical procedure for the selection of time-to-failure models based on the assessment of trends in maintenance data. Reliab. Eng. Syst. Saf. **94**(10), 1618–1628 (2009)
29. Barabadi, A., Garmabaki, A., Yuan, F., Lu, J.: Maintainability analysis of equipment using point process models. In: 2015 IEEE International Conference on Industrial Engineering and Engineering Management (IEEM). IEEE (2015)
30. Kline, M.: Suitability of the lognormal distribution for corrective maintenance repair times. Reliab. Eng. **9**(2), 65–80 (1984)
31. Verma, A.K., Ajit, S., Karanki, D.R.: Reliability and Safety Engineering, vol. 43. Springer, Heidelberg (2010)
32. Kawauchi, Y., Rausand, M.: A new approach to production regularity assessment in the oil and chemical industries. Reliab. Eng. Syst. Saf. **75**(3), 379–388 (2002)
33. ReliaSoft. Additional Analyses- ReliaWiki, ReliaSoft Publishing, USA (2019)
34. ReliaSoft. Repairable Systems Analysis Through Simulation-ReliaWiki, ReliaSoft Publishing, USA (2019)

The Effect of Risk Factors on the Resilience of Industrial Equipment

Abbas Barabadi[1] (✉), Ali Nouri Qarahasanlou[2], Adel Mottahedi[3], Ali Rahim Azar[4], and Ali Zamani[4]

[1] Department of Engineering and Safety, UiT, The Arctic University of Norway, Tromsø, Norway
abbas.b.abadi@uit.no
[2] Faculty of Technical and Engineering, Imam Khomeini International University, Qazvin, Iran
[3] Faculty of Mining, Petroleum and Geophysics Engineering, Shahrood University of Technology, Shahrud, Iran
[4] School of Mining Engineering, Tehran University, Tehran, Iran
ali.rahim@ut.ac.ir

Abstract. Recently, to evaluate the response of systems against disruptive events, the application of the resilience concept has been increased. Resilience depicts the system's ability to return to its normal operational status after the disruption. Various studies in the field of engineering and non-engineering systems have only considered systems' performance indicators to estimate resilience. Therefore, the impact of operating and environmental factors (risk factors) has been neglected. In this paper, the influence of the risk factors (rock type), as well as the system's performance indicators, are considered in the resilience estimation of the excavator system of Gol-E-Gohar Iron mine.

Keywords: Resilience · Reliability · Maintainability · Supportability

1 Introduction

Keeping the stable operation of systems is a challenge for engineering design. External or internal events like failures or disturbances affect the system operation, negatively. Systems based on their stability have a different reaction against failures. Some systems are vulnerable in case of disruption and lost their functionality. In the engineering domain, the ability of systems to withstand against the failure events and recover their performance in a suitable time is known as resilience [1]. At first, the concept of resilience was introduced by Holling in 1973 in the field of geological systems [2].

According to the necessary role of the resilience concept in the mitigation of the risk of disturbances, it has been used in various fields such as engineering [3, 4], social [5, 6], economic [7, 8], ecological [2, 9], socio-technical [10, 11], and socio-ecological [12, 13]. Newly, it has also been used to COVID-19 pandemic by many researchers [14–17]. For example, Barabadi et al. [14] evaluate the resilience of the health infrastructure

R. Karim et al. (Eds.): IAI 2021, LNME, pp. 112–127, 2022.
https://doi.org/10.1007/978-3-030-93639-6_10

systems (HIS) in Jajarm and Garme cities before and after the pandemic. Therefore, the resilience concept can be applied in each field.

There is a wide range of definitions for system resilience in the literature. Here are some of these definitions. Allenby and Fink [18] defined resilience as the system's ability to preserve its operations and structure in the presence of internal or external disturbances and to degrade gracefully when it must. Rose [19] defined resilience as the system's ability to maintain its functionally when a disruption occurs. Haimes [20] defined resilience as the system's ability to withstand a major disturbance within acceptable degradation parameters and to recover with a suitable time and reasonable costs and risks. Moreover, the number of resilience definitions have been presented for more specific domains such as engineering, economy, social, ecological, etc. In the engineering domain, resilience is defined as the system's ability to predict, absorb, adapt, and/or quickly recover from a disruptive event [1]. The given definitions highlight the ability of the system to resist against the failures and absorb failures adverse consequences. It should be noted, the pre-failure (preparedness) and post-failure (recovery) activities are both essential in the resilience concept [3]. These activities make the system reliable, supportable, flexible, adaptable, and maintainable.

In Fig. 1, the schematic view of the system resilience is shown. As can be seen, the performance level of the system degraded after the disruptive event at t_e. Then, the system performance level reached to lowest value at t_d, and remained at this level until t_s. After commence of the recovery activities, this level increased and reached to the new steady state. It can be either close to or higher than the initial state of the system [21]. It depends on the quality and quantity of the pre- and post-failure activities.

There are many approaches to resilience analysis. Hosseini et al. [1] classified the resilience analysis approach into the qualitative and quantitative groups (see Fig. 2). Qualitative approaches, which are divided into the conceptual frameworks and semi-quantitative indices, analysed the resilience of the system without any quantifications. Qualitative approaches are often used in the field of non-engineering systems. Quantification approaches, which are classified into general measures and structural based models, are suitable to apply in the field of engineering systems.

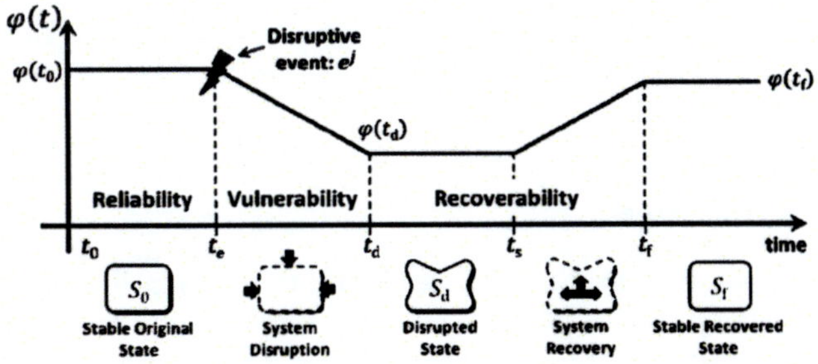

Fig. 1. System performance and state transition to describe resilience [1].

The structural based models are divided into optimization, simulation, and fuzzy models. These models examine how the structure of a system affects its resilience. In

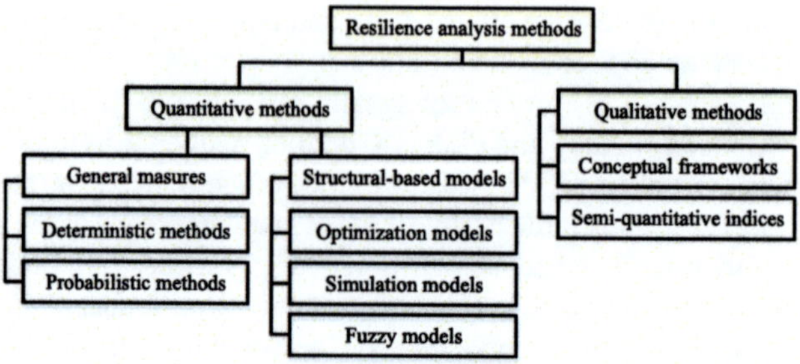

Fig. 2. Resilience analysis approaches [1].

these models, system behaviour must be observed and the characteristics of the system must be modelled or simulated. General measures are deterministic and probabilistic metrics. In these approaches, the system performances are compared before and after the failure event [1]. Deterministic measures never used uncertainties (like the probability of system repair) in order to analyse the system resilience. While probabilistic measures consider the uncertainties. Here, some of the presented resilience analysis models in the field of engineering are described in Table 1.

Among these models, Rød et al. model has more capability to use. It is probabilistic and time-dependent. It considers the resilience of the owner organization of the system as well as system performance indicators. Therefore, in this paper, this model is used for resilience analysis. However, none of these models considers the impact of operational factors on system resilience. These factors, which have different origins, impact the system reliability, maintainability, and supportability [22]. Therefore, in this study, the impact of these influence factors is considered in the resilience analysis process.

2 Resilience Analysis Methodology

In the present paper, the resilience analysis is based on the Rød et al. model. Hence, system reliability, maintainability, and supportability (RMS) analysis should be carried out. It must be mentioned these analyses can be conducted considering the effect of risk factors. For this aim, five steps are considered, which are described in the following (see Fig. 3).

2.1 Database Establishment

A database, including time between failure (TBF), time to repair (TTR), and time to delivery (TTD) data should be collected from the accessible sources. Simultaneously, the most critical operational factor should be identified. Afterward, the collected database must be segmentation based on the identified risk factor.

Table 1. Resilience analysis models.

Author(s)	Model	Model description	
Bruneau et al. [23]	$R=\int_{t_0}^{t_1}[100-Q(t)]dt$	R: Resilience reduction $Q(t)$: System performance function t_0: Disruption initiation time t_1: Recovery actions stoppage time	- It is deterministic - The initial system performance level is considered as 100%
Rose [19]	$\psi=\frac{\%\Delta DY^{max}-\%\Delta DY}{\%\Delta DY^{max}}$	ψ: System resilience $\%\Delta DY$: Difference in non-disrupted and expected disrupted system performance $\%\Delta DY^{max}$: Difference in non-disrupted and worst case disrupted system performance	- It is deterministic - It measures the ratio of the avoided drop in system output and the maximum possible drop in system output
Orwin and Wardle [24]	$\psi=\left[\frac{2\lvert D_0\rvert}{\lvert D_0\rvert+\lvert D_x\rvert}\right]-1$	ψ: System resilience D_0: Refers the maximum intensity of absorbable force without perturbing the system function D_x: Refers to the magnitude of the disturbance effect on safety at time (t_x)	- It is deterministic - Resilience can take the value between 0 and 1 - When the magnitude of the disturbance's effect is equal to zero ($D_x = 0$), the maximum resilience can be obtained
Tierney and Bruneau [25]	$R=\frac{\int_{t_0}^{t_1}[Q(t)]dt}{100(t_1-t_0)}$	R: Resilience reduction $Q(t)$: System performance function t_0: Disruption initiation time t_1: Recovery actions stoppage time t_1-t_0: System recovery duration	- It is deterministic - It is time-dependent - The initial system performance level is not considered as 100%
Cimellaro et al. [26]	$\psi=\alpha\int_{T_{LC}}\frac{Q_1(t)}{T_{LC}}dt+(1-\alpha)\int_{T_{LC}}\frac{Q_2(t)}{T_{LC}}dt$	ψ: System resilience $Q_1(t)$: System services quality before the disruption, $Q_2(t)$: System services quality after the disruption T_{LC} : Control time of the system α : Weighting factor representing the importance of pre- and post-failure activities qualities	- It is deterministic - It considers both pre- and post-failure activities - It considers the importance levels (weights) of the pre- and post-failure activities
Ayyub [27]	$\psi=\frac{T_i+F\Delta T_f+R\Delta T_r}{T_i+\Delta T_f+\Delta T_r}$	ψ: System resilience T_i: Time to incident, T_f: Time to failure T_r: Time to recovery $\Delta T_f = T_f - T_i$: Duration of failure $\Delta T_r = T_f - T_f$: Duration of recovery F: Failure profile R: Recovery profile	- It is probabilistic - It is time-dependent - It considers both pre- and post-failure activities - The failure profile is a measure of robustness and redundancy - the recovery profile measures recoverability
Youn et al. [28]	$\psi=R+\rho=R+\kappa.\Lambda_P.\Lambda_D.(1-R)$	ψ: System resilience R: System reliability ρ: System restoration κ: Probability of successful recovery event Λ_P: Probability of correct Prognosis event Λ_D: Probability of correct diagnosis event	- It is probabilistic - It is time-independent - It considers both pre- and post-failure activities - Resilience can take the value between 0 and 1
Rød et al. [6]	$\psi=R+\Lambda(1-R)$ $=R+\left[\prod_{i=1}^{4}\beta_i\right](1-R)$	ψ: System resilience R: System reliability β_1: System maintainability β_2: Owner organization resilience β_3: Prognostic and health management (PHM) efficiency β_4: System supportability	- It is probabilistic - It is time-dependent - It considers both pre- and post-failure activities - Resilience can take the value between 0 and 1 - It considers the system resilience as a function of system reliability and recoverability
Sarwar et al. [29]	$\psi=R+\mu[V,M]$	ψ: System resilience R: System reliability μ: System recoverability function V: System vulnerability M: System maintainability	- It is probabilistic - It is time-dependent - It considers both pre- and post-failure activities - Resilience can take the value between 0 and 1
Najarian and Lim [30]	$r=r_1\lambda_1+r_2\lambda_2+r_3\lambda_3$, $\sum_{i=1}^{3}\lambda_i=1 \text{ and } \lambda_i\geq 0 \text{ for } i=1,2,3$	r: System resilience r_1: System absorptive component r_2: System adaptability component r_3: System recovery component λ_i: Weight of the ith component	- It is time-dependent - It considers both pre- and post-failure activities - Resilience can take the value between 0 and 1

2.2 Selection of the Best Fit Statistical Model

After data collection, to pick the best fit model, the assumption of the independent and identically distributed nature (iid) of data should be judged. For this aim, trend tests include Military handbook (MIL), Laplace, Anderson-Darling (A-D), and Mann-Kendall (M-K)) tests should be adopted. Moreover, autocorrelation tests include Graphical method and autocorrelation function (AFC) should be performed. Here, to conduct trend and serial correlation tests, the represented algorithm in Fig. 3 is suggested. For more information about trend and autocorrelation tests refer to [31, 32].

2.3 RMS Analysis

Here, based on the results of the previous step, if there is any sign of the presence of the trend among the data, then the nonhomogeneous models like the Power Law Process (PLP) should be applied. If there is autocorrelation in data and trend test results do not confirm the potential of the presence of the trend among the data, then the Branching Poisson Process (BPP) models can be utilized. Furthermore, if there is no indication of the existence of trend and autocorrelation among the data, the classical distribution models such as normal or lognormal models can be used (see Fig. 3). For more details refer to [8–10].

2.4 Estimation of the Management Indicators

In this step, the PHM efficiency of the system and the organization's resilience should be estimated.

2.5 Resilience Analysis of the System and Subsystems

Finally, using the obtained results from the previous steps, the resilience of a subsystem of the series system with n subsystems can be analysed as follow [3]:

$$\psi_i(t) = R_i(t) + \Lambda_i(t)(1 - R_i(t)) \tag{1}$$

The resilience of the series system can be analysed as follow:

$$\psi(t) = \prod_{i=1}^{n}[R_i(t) + \Lambda_i(t)(1 - R_i(t))] \tag{2}$$

3 Case Study

The mining industry is one of the most important sectors among the industries. It is consisted of many complex processes like ore mining, ore processing, and so on, to supply the raw material of other industrial sectors. This field consists of many critical systems like the excavator system. Out of schedule stoppage of this system will lead to mine production reduction and create several problems for the mine management. The application of the resilience concept seems necessary for this sector.

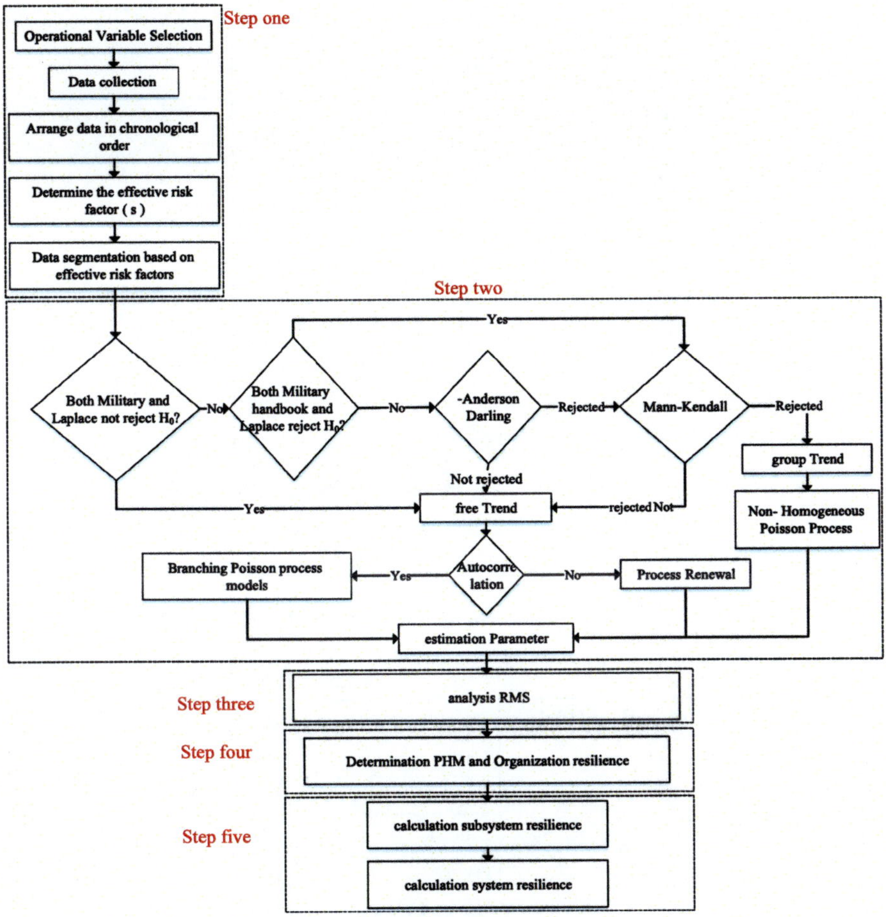

Fig. 3. Resilience analysis algorithm.

In this work, Gol-E-Gohar Iron mine is selected as a case study. This mine, with six mining sites, is one of the largest producers of Iron with and supplies the raw material of some industries like automobile manufacturing. The estimated deposit of this mine is about 1135 million tons. It is situated 55 km southwest of Sirjan between 551150E and 551240E longitudes and 29,130N and 29,170N latitudes at an altitude of 1750 m above sea level (see Fig. 4), and surrounded by 2500 m height Mountains [34]. The excavator system of the mining site No.1 is selected for resilience analysis. The excavator system consists of six series subsystems. The characteristics of this system and its flowchart are presented in Table 2 and Fig. 5, respectively.

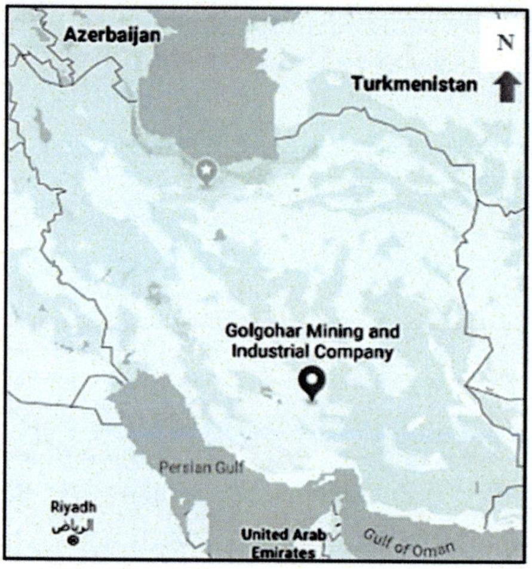

Fig. 4. Gol-E-Gohar Iron mine location.

Table 2. Excavator system characteristics.

System model	Series subsystems	Codes
Caterpillar 390DL	Boom	Bo
	Cabin	Ca
	Engine	En
	Electric	El
	Hydraulic	Hy
	Undercarriage	Un

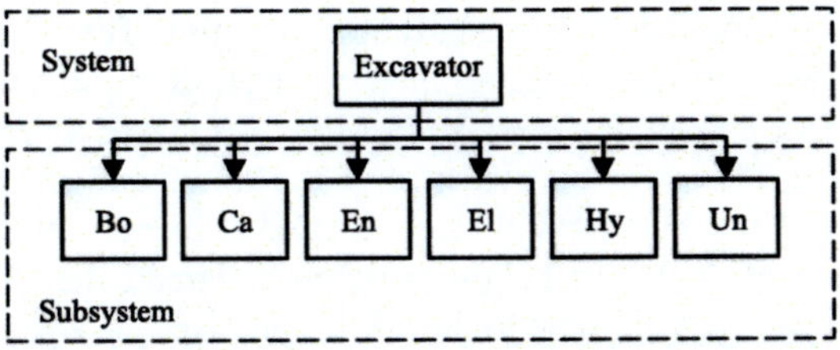

Fig. 5. Flowchart of the excavator system.

4 Results and Discussion

According to Sect. 2, a database of TBFs, TTRs, and TTDs of the excavator system and its subsystems were collected. These data belonged to a period of 24 months from January 2016 to December 2018. Moreover, rock type was recognized as the most important factor that impacts the failure of the excavator system. In the field of mining engineering, rock type is such operational factors. This factor can be divided into ore and waste rocks, based on the hardness and specific gravity variations. For example, high hardness of rock can damage the teeth of the bucket, and the high specific gravity of rock can put excessive pressure on the engine. The collected database (only TBFs) was segmented based on the rock type. Some parts of the collected data are displayed in Table 3. Afterward, the iid assumption was evaluated for TBFs, TTRs, and TTDs. For checking the data autocorrelation, the Graphical method is used. In the graphical method, the autocorrelation test was carried out graphically by plotting the ith TBF (TTR) against (i-1)th TBF (TTR). If the plotted points do not follow a special trend and randomly scattered without any clear pattern, it can be inferred that the data are free from serial correlation or independent [35]. For example, the results of the autocorrelation test for the TBFs data of the Bo subsystem is shown in Fig. 6. As can be seen, the TBFs data of the Bo subsystem has no autocorrelation.

The results of the trend tests for failure, repair, and delivery data are illustrated in Tables 4 and 5, respectively. Based on the results, in the ore segment, TBFs of El, En,

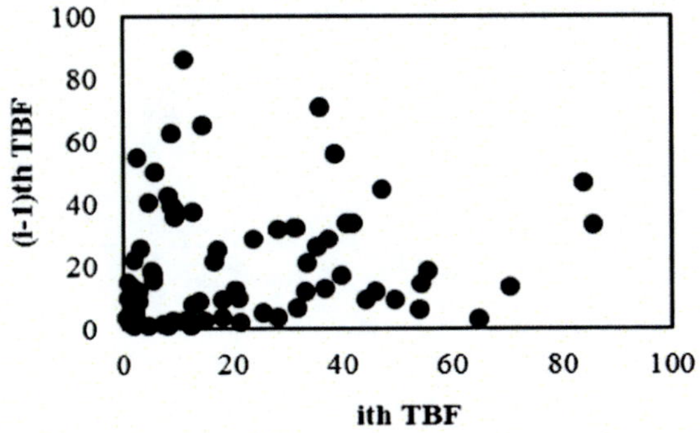

Fig. 6. Autocorrelation test for the TBFs data of the Bo subsystem.

Hy, Ca, and Bo subsystems had not trend and autocorrelation. Then classical distribution models were used for their modelling. As there were signs of the trend in TBFs of the Un subsystem, the PLP model was used for it. Moreover, based on the obtained results in the waste segment, there was no evidence of the trend, and autocorrelation in the TBFs of El, En, and Bo subsystems, thus classical distribution models were used for them. While the TBFs of Ca, Hy, and Un subsystems had trend, then the PLP model was used for these subsystems. According to the TTRs data, Bo, Ca, Hy, and El subsystems had no trend and autocorrelation. Then classical distribution models were used for their modelling. While the TTRs of En and Un subsystems had trend, then the PLP model was used for these subsystems. It must be mentioned, because of the same model, repair and, spare parts for the excavator subsystems, the supportability analysis was performed only for the entire excavator system. In Figs. 7, 8 and 9 and Table 6, the best-selected models for the subsystems are presented (Fig. 10).

Table 3. Samples of the collected database.

Reliability database				Maintainability database		Supportability database	
Ore section		West section					
TBF (Hr)	System	TBF (Hr)	System	TTR (Hr)	System	TTD (Hr)	System
160.45	1	6.56	1	0.33	1	0.80	1
54.17	1	40.85	1	1.20	1	31.11	1
5.86	1	203.00	1	0.61	1	0.33	1
143.31	1	29.78	1	0.22	1	1.20	1

Based on Fig. 3, After RMS analysis, management indicators should be determined. According to Rød et al. [3], the values of PHM efficiency and organization resilience can be considered as the constant values. In this study, the adopted values by Rød et al. were considered for PHM efficiency and organization resilience (see Table 7).

Table 4. The result of the trend test for TBFs data.

Index	Subsystem	Data		MIL	Laplace	A-N	M-K	Result
Reliability	Bo	TBF of ore	Test statistic	920.53	0	6.13	–	No Trend
			P-value	0.409	0.709	0.001	–	
		TBF of west	Test statistic	418.24	0.99	–	–	No trend
			P-value	0.97	0.321	–	–	
	Ca	TBF of ore	Test statistic	47.98	1.02	–	–	No trend
			P-value	0.354	0.307	–	–	
		TBF of west	Test statistic	31.86	1.99	–	−2.084	Trend
			P-value	0.025	0.046	–	0.018	
	En	TBF of ore	Test statistic	126.86	3.59	–	−4.238	No trend
			P-value	0.001	0	–	0.13	
		TBF of west	Test statistic	208.19	0.07	–	–	No trend
			P-value	0.811	0.946	–	–	
	Hy	TBF of ore	Test statistic	179.18	−0.34	–	–	No trend
			P-value	0.839	0.733	–	–	
		TBF of west	Test statistic	242.48	−3.03	–	3.55	Trend
			P-value	0	0.002	–	0.019	
	El	TBF of ore	Test statistic	86.41	1.01	–	–	No trend
			P-value	0.504	0.314	–	–	
		TBF of west	Test statistic	71.99	1.18	–	–	No trend
			P-value	0.66	0.24	–	–	
	Un	TBF of ore	Test statistic	135.94	4.33	–	−4.48	Trend
			P-value	0	0	–	0.036	
		TBF of west	Test statistic	158.58	5.71	2.68	−4.162	Trend
			P-value	0	0	0.04	0.0157	

Table 5. The result of trend test for the TTRs and TTDs data.

Index	System or Subsystems		MIL	Laplace	A-N	M-K	Result
Maintainability	Bo	Test statistic	1295.05	−0.64	–	–	No Trend
		P-value	0.19	0.523	–	–	
	Ca	Test statistic	119.06	0.64	–	–	No trend
		P-value	0.986	0.52	–	–	
	En	Test statistic	238.65	3.12	–	−3.73	Trend
		P-value	0.07	0.002	–	0.0092	
	Hy	Test statistic	343.43	−1.29	–	–	No trend
		P-value	0.161	0.199	–	–	
	El	Test statistic	186.16	1.34	6.77	–	No trend
		P-value	0.566	0.179	–	–	
	Un	Test statistic	455.59	2.8	–	−3.129	Trend
		P-value	0.024	0.005	–	0.009	
Supportability	Excavator	Test statistic	2680.97	−0.65	–	–	No trend
		P-value	0.446	0.516	–	–	

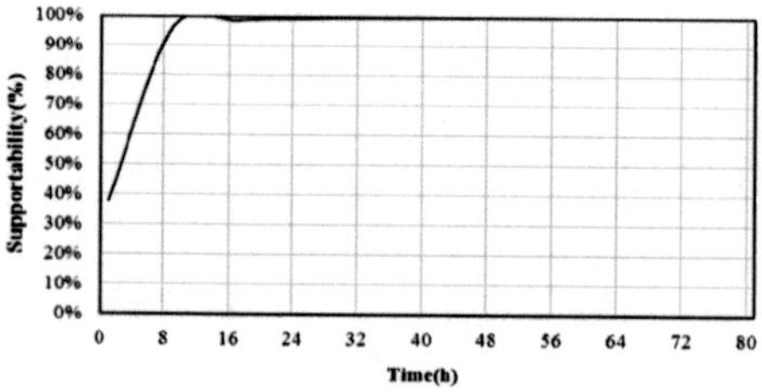

Fig. 7. Excavator system supportability analysis results for 81 h of activities.

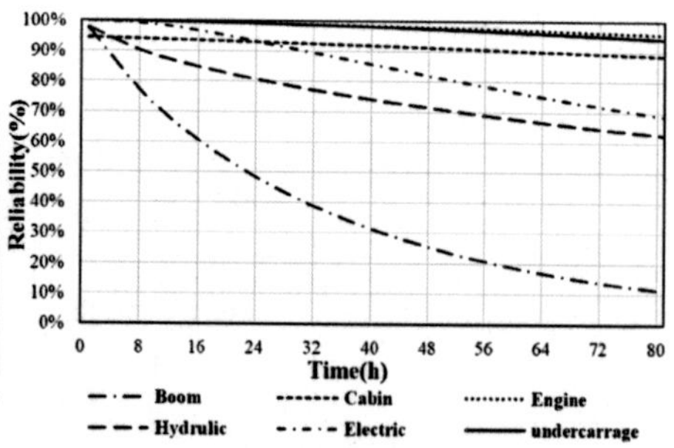

Fig. 8. Excavator subsystem reliability analysis results for Ore section for 81 h of activities.

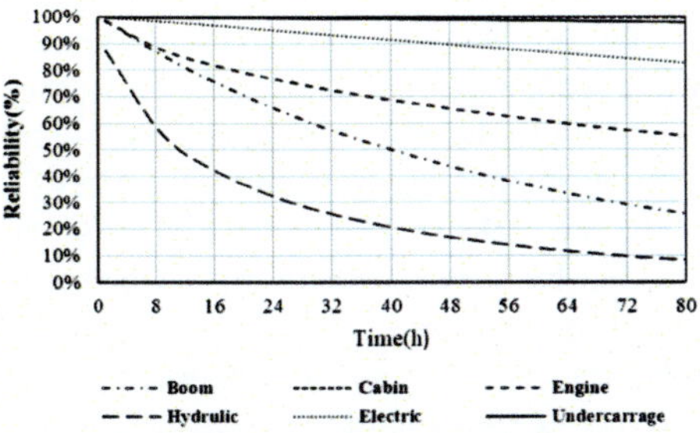

Fig. 9. Excavator subsystem reliability analysis results for waste section for 81 h of activities.

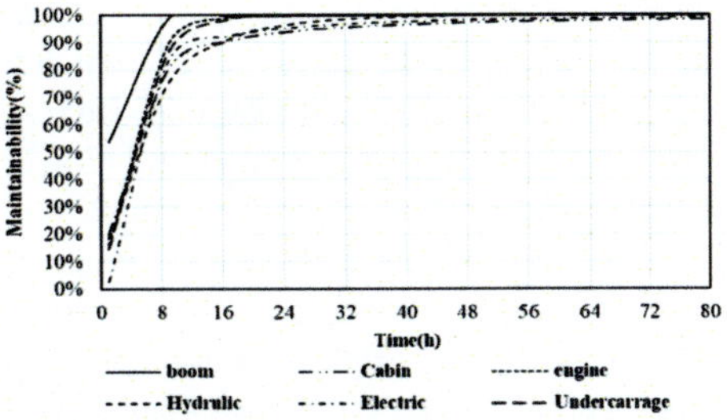

Fig. 10. Excavator subsystem maintainability analysis results for 81 h of activities.

Table 6. Best-fitted models for the subsystems reliability, maintainability, and supportability.

Subsystems		Reliability				Maintainability	
		ore		West			
Bo	Model	weibull-3P		weibull-3P		loglogistic-3P	
	parameter	β	0.905	β	0.975	β	- 0.20717
		η(hr)	33.69	η(hr)	57.77	η(hr)	0.51
		γ(hr)	0.3706	γ(hr)	0.315	γ(hr)	0.121
Ca	Model	Normal		PLP		Log logistic	
	parameter	mean	333.72	β	1.633	β	1.069
		Std	209.66	η(hr)	3170	η(hr)	0.778
		-	-	γ(hr)	-	γ(hr)	-
En	Model	PLP		weibull-3P		PLP	
	parameter	β	1.448	β	0.664	β	1.173
		η(hr)	692.35	η(hr)	172.2	η(hr)	4.4146
		γ(hr)	-	γ(hr)	0.633	γ(hr)	-
Hy	Model	wiebull-3P		PLP		weibull-3P	
	parameter	β	0.6437	β	0.66	β	0.8423
		η(hr)	260.61	η(hr)	19.93	η(hr)	5.887
		γ(hr)	0.154	γ(hr)	-	γ(hr)	0.10132
El	Model	lognormal		weibull-2P		Lognormal-3P	
	parameter	LMe	4.977	β	1.078	β	0.555
		LSt	1.207	η(hr)	371.5	η(hr)	1.542
		-	-	γ(hr)	-	γ(hr)	0.912
Un	Model	PLP		PLP		PLP	
	parameter	β	1.588	β	1.778	β	1.154
		η(hr)	469.69	η(hr)	652.4	η(hr)	4.866
		γ(hr)	-	γ(hr)	-	γ(hr)	-
System	Supportability						
Excavator	Model	Loglogistic-3P					
	parameter	β			0.275		
		η(hr)			0.603		
		γ(hr)			0.018		

Table 7. The considered values for PHM efficiency and organization resilience [3].

Parameters	Values
Organization resilience	0.85
PHM efficiency	0.75

Finally, using Eqs. 1 and 2, the resilience of the excavator system and its subsystems (in the ore and waste rocks) for 81 h of activities were analysed. The results are shown in Figs. 11, 12 and 13. For example, the resilience of the Hy subsystem (see Figs. 11 and 12) in the waste rocks will be notably lower than its resilience in the ore rocks. After 81 h of activities, the Hy subsystem resilience in the ore rocks will be reached to 86%, but it will be 66% in the waste rocks. Because the hardness and specific gravity of waste rocks are more than ore rocks, which can damage the excavator subsystems like the Bo subsystem. These differences indicate the influence of rock type on system resilience. Moreover, as can be seen, the impact degree of this factor depends on its direct contact with the subsystem. For example, the impact of rock type on the Ca subsystem is lower than the impact on the Hy subsystem.

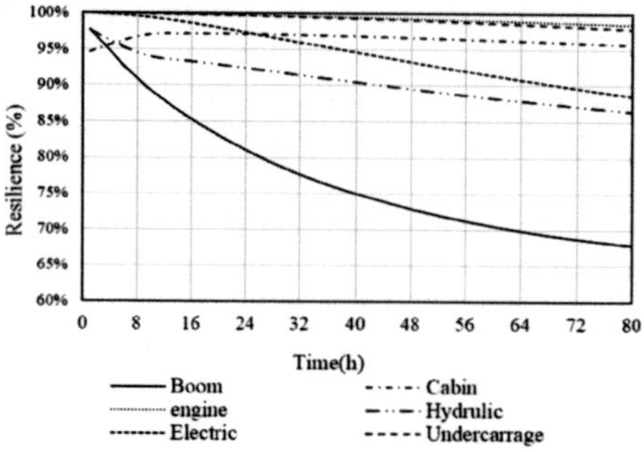

Fig. 11. Excavator subsystem resilience analysis results for ore section for 81 h of activities.

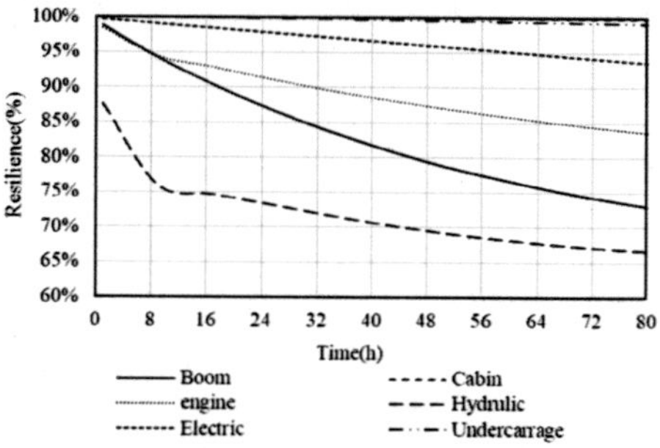

Fig. 12. Excavator subsystem resilience analysis results for waste section for 81 h of activities.

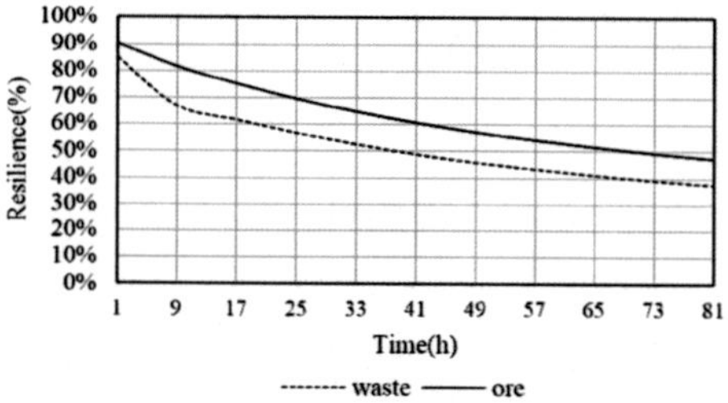

Fig. 13. Excavator system resilience analysis results for ore and waste sections for 81 h of activities.

5 Conclusion

Systems based on their stability have a different reaction against failures. Some systems are vulnerable in case of disruption and lost their functionality. In the engineering domain, the ability of systems to withstand against the failure events and recover their performance in a suitable time is known as resilience. In this work, the resilience concept was used in the field of mining industry, and the excavator system of Gol-E-Gohar Iron mine was considered as a specific case study. In this paper, the influence of the risk factors on the system resilience was considered. The rock type was identified as the risk factor that affects the resilience of the excavator. Therefore, by segmentation of the collected database into the ore and waste rocks groups, the impact of rock type on the excavator resilience were evaluated. This factor affects the reliability indicator. The results emerge the importance of consideration of the risk factors in the resilience estimation process. As shown, the resilience of the excavator system can be varied based on the rock type. It can be due to the high hardness and density of the waste rock compared to the ore rock. It must be mentioned that the mining industry is an essential industry that supplies the raw material of other sectors. Thus, the application of the resilience concept with consideration of risk factors will improve its overall functionality.

References

1. Hosseini, S.M., Barker, K., Ramirez-Marquez, J.E.: A review of definitions and measures of system resilience. Reliab. Eng. Syst. Saf. **145**, 47–61 (2016)
2. Holling, C.S.: Resilience and stability of ecological systems. Annu. Rev. Ecol. Syst. **4**, 1–23 (1973)
3. Rød, B., Barabadi, A., Gudmestad, O.: Characteristics of arctic infrastructure resilience: application of expert judgement. Presented at the 26th International Ocean and Polar Engineering Conference, Rhodes, Greece, 26 June–2 July (2016)
4. Najarian, M., Lim, G.J.: Design and assessment methodology for system resilience metrics. Risk Anal. 1–14 (2019). https://doi.org/10.1111/risa.13274
5. Amico, A.D., Currà, E.: The role of urban built heritage in qualify and quantify resilience. Specific issues in Mediterranean City. Procedia Econ. Finan. **18**, 181–189 (2014). https://doi.org/10.1016/S2212-5671(14)00929-0
6. Zhang, N., Huang, H.: Resilience analysis of countries under disasters based on multisource data: resilience analysis of countries under disasters. Risk Anal. **38**(1), 31–42 (2018). https://doi.org/10.1111/risa.12807
7. Vugrin, E.D., Warren, D.E., Ehlen, M.A.: A resilience assessment framework for infrastructure and economic systems: quantitative and qualitative resilience analysis of petrochemical supply chains to a hurricane. Process Saf. Prog. **30**(3), 280–290 (2011). https://doi.org/10.1002/prs.10437
8. Pant, R., Barker, K., Zobel, C.W.: Static and dynamic metrics of economic resilience for interdependent infrastructure and industry sectors. Reliab. Eng. Syst. Saf. **125**, 92–102 (2014). https://doi.org/10.1016/j.ress.2013.09.007
9. Müller, F., et al.: Assessing resilience in long-term ecological data sets. Ecol. Ind. **65**, 10–43 (2016). https://doi.org/10.1016/j.ecolind.2015.10.066
10. Omer, M., Mostashari, A., Lindemann, U.: Resilience analysis of soft infrastructure systems. Procedia Comput. Sci. **28**, 873–882 (2014). https://doi.org/10.1016/j.procs.2014.03.104

11. Cook, A., Delgado, L., Tanner, G., Cristóbal, S.: Measuring the cost of resilience. J. Air Transp. Manag. **56**, 38–47 (2016). https://doi.org/10.1016/j.jairtraman.2016.02.007
12. Brown, E.D., Williams, B.K.: Resilience and resource management. Environ. Manag. **56**(6), 1416–1427 (2015). https://doi.org/10.1007/s00267-015-0582-1
13. Meng, F., Fu, G., Farmani, R., Sweetapple, C., Butler, D.: Topological attributes of network resilience: a study in water distribution systems. Water Res. **143**, 376–386 (2018). https://doi.org/10.1016/j.watres.2018.06.048
14. Barabadi, A., Ghiasi, M.H., Nouri Qarahasanlou, A., Mottahedi, A.: A holistic view of health infrastructure resilience before and after COVID-19. ABJS, vol. 8, no. Covid-19 Special Issue (2020). https://doi.org/10.22038/abjs.2020.47817.2360
15. Blanton, R.E., et al.: African resources and the promise of resilience against COVID-19. Am. J. Trop. Med. Hyg. **103**, 539 (2020). https://doi.org/10.4269/ajtmh.20-0470
16. Barton, M.A., Christianson, M., Myers, C.G., Sutcliffe, K.: Resilience in action: leading for resilience in response to COVID-19. BMJ Leader, p. leader-2020-000260, May 2020. https://doi.org/10.1136/leader-2020-000260
17. Trump, B.D., Linkov, I.: Risk and resilience in the time of the COVID-19 crisis. Environ. Syst. Decis. **40**(2), 171–173 (2020). https://doi.org/10.1007/s10669-020-09781-0
18. Allenby, B., Fink, J.: Toward inherently secure and resilient societies. Science **390**, 1034–1036 (2005)
19. Rose, A.: Economic resilience to natural and man-made disasters: Multidisciplinary origins and contextual dimensions. Environ. Hazards **7**(4), 383–398 (2007). https://doi.org/10.1016/j.envhaz.2007.10.001
20. Haimes, Y.Y.: On the definition of resilience in systems. Risk Anal. **29**(4), 498–501 (2009)
21. Henry, D., Emmanuel Ramirez-Marquez, J.: Generic metrics and quantitative approaches for system resilience as a function of time. Reliab. Eng. Syst. Saf. **99**, 114–122 (2012). https://doi.org/10.1016/j.ress.2011.09.002
22. Barabadi, A., Barabady, J., Markeset, T.: A methodology for throughput capacity analysis of a production facility considering environment condition. Reliab. Eng. Syst. Saf. **96**, 1637–1646 (2011)
23. Bruneau, M., et al.: A framework to uantitatively assess and enhance the science the seismic resilience of communities. Earthq. Spectra **19**(4), 733–752 (2003)
24. Orwin, K.H., Wardle, D.A.: New indices for quantifying the resistance and resilience of soil biota to exogenous disturbances. Soil Biol. Biochem. **36**, 1907–1912 (2004)
25. Tierney, K., Bruneau, M.: Conceptualized and measuring resilience. TR News **250**, 14–17 (2007)
26. Cimellaro, G.P., Reinhorn, A., Bruneau, M.: Seismic resilience of a hospital system. Struct. Infrastruct. Eng. **6**, 127–177 (2010)
27. Ayyub, B.M.: Systems resilience for multihazard environments: definition, metrics, and valuation for decision making. Risk Anal. **34**(2), 340–355 (2014)
28. Youn, B.D., Hu, C., Wang, P.F.: Resilience-driven system design of complex engineered systems. J. Mech. Des. **133**, 1–15 (2011)
29. Sarwar, A., Khan, F., Abimbola, M., James, L.: Resilience analysis of a remote offshore oil and gas facility for a potential hydrocarbon release: resilience analysis of a remote offshore operation. Risk Anal. **38**(8), 1601–1617 (2018). https://doi.org/10.1111/risa.12974
30. Najarian, M., Lim, G.J.: Design and assessment methodology for system resilience metrics. Risk Anal. **39**(9), 1885–1898 (2019). https://doi.org/10.1111/risa.13274
31. Garmabaki, A.H.S., Ahmadi, A., Block, J., Pham, H., Kumar, U.: A reliability decision framework for multiple repairable units. Reliab. Eng. Syst. Saf. **150**, 78–88 (2016)
32. Garmabaki, A.H.S., Ahmadi, A., Mahmood, Y.A., Barabadi, A.: Reliability modelling of multiple repairable units: reliability modelling, multiple repairable systems. Qual. Reliab. Engng. Int. **32**(7), 2329–2343 (2016). https://doi.org/10.1002/qre.1938

33. Hoseinie, S.H., Ataei, M., Khalokakaie, R., Ghodrati, B., Kumar, U.: Reliability analysis of drum shearer machine at mechanized longwall mines. J. Qual. Maint. Eng. **18**(1), 98–119 (2012)
34. Rezaei, M., Monjezi, M., Yazdian Varjani, A.: Development of a fuzzy model to predict flyrock in surface mining. Saf. Sci. **49**(2), 298–305 (2011). https://doi.org/10.1016/j.ssci.2010.09.004
35. Basiri, M., Sharifi, M., Ostadi, B.: Reliability and risk assessment of electric cable shovel at Chadormalu iron ore mine in Iran. Int. J. Eng. **33**(1), 170–177 (2020). https://doi.org/10.5829/ije.2020.33.01a.20

Analysis of Systematic Influences on the Insulation Resistance of Electronic Railway Interlocking Systems

Judith Heusel(✉) and Jörn C. Groos

German Aerospace Center (DLR), Lilienthalplatz 7, 38108 Braunschweig, Germany
{judith.heusel,joern.groos}@dlr.de

Abstract. Railway interlocking systems take the role of managing train traffic by locking routes and are responsible for light signalling and switch positioning. Their correct operation is highly relevant for safety as they prevent trains from taking conflicting paths which could lead to severe accidents. In addition, as interlocking systems organise traffic at infrastructure nodes, their failure can have considerable negative impact on train delays and profitability.

Electronic interlocking systems in Germany are isolated electrical systems. Insulation faults can cause safety-critical unintended interlocking behaviour and are responsible for a significant amount of interlocking-related delay minutes. The goal of the present research is to prevent system failures by early detecting abnormal behaviour of the insulation resistance and by localising elements related to incipient insulation faults.

In Germany specific measurement devices continuously observe the insulation resistance of the entire electrical interlocking system. If the observed insulation resistance drops below a defined threshold an alarm is raised. As part of a research project and proof-of-concept, the monitoring device of the electronic interlocking at Plattling, Bavaria, has been equipped with an additional data logger to record the observed insulation resistance for a period of 18 months. Furthermore, status messages of the interlocking control system are available for a period of 20 days. Observed systematic influences on the insulation resistance such as weather conditions and interlocking operations are presented. Time-series modelling of insulation resistance as well as classification with regard to normal/abnormal behaviour based on weather condition and railway control notifications, respectively, give promising results.

Keywords: Railway infrastructure · Electronic interlocking · Condition monitoring · Time series modelling · Insulation faults

1 Introduction

In Germany, the entire electrical system of an interlocking is isolated against earth.

Insulation faults that appear in one point of the isolated system remain uncritical but have to be eliminated timely since a second insulation fault becomes safety-critical:

R. Karim et al. (Eds.): IAI 2021, LNME, pp. 128–140, 2022.
https://doi.org/10.1007/978-3-030-93639-6_11

Unintended switch repositioning can be a consequence, for instance. A crucial part of the isolated system is the cable installation of the interlocking. Insulation faults at these cables are known to cause an essential part of interlocking-related disturbances of the railway system in Germany [1]. In case of an emerging insulation fault the localisation of one or multiple failure-causing elements would be of great benefit because a reduction of the consumed time until the error is found and repaired reduces delay minutes as well as monetary and reputational loss.

Until now the insulation resistance of electronic interlocking systems is monitored not to fall below a threshold by raising alarms in case the threshold is not reached. With a test-setting established in cooperation with Bender and DB Netz AG at the electronic interlocking in Plattling, the potential of a condition monitoring approach for the electrical installation using the complete course of the insulation resistance is evaluated. To this end, systematic influences on the insulation resistance course are examined. For a detailed description of the test setting see [1]. To our knowledge, this use case is novel to the railway industry and there is no precedent related published work and research.

This paper is organised as follows: Sect. 2 gives an overview of the available data used for the analysis, its origin and its characteristics. In Sect. 3 results of a first explorative data analysis are presented which are refined and used to model the insulation resistance behaviour of the interlocking in dependence of the different examined parameters in Sect. 4. Results are discussed in Sect. 5.

2 Data Description

The available data for this analysis comprises insulation resistance measurements, notifications from the operator desk of the interlocking and weather data. The data from the different sources covers different time intervals and each source provides a particular time resolution. The data is described in more detail in the following subsections.

2.1 Insulation Resistance Data

To ensure a minimum insulation resistance, each electronic interlocking in Germany is equipped with a device that measures the insulation resistance constantly (typically at an effective rate of 1–2 Hz) and raises an alarm if it falls below the minimum threshold of 30 kOhm. Besides raising alarms, this data has neither been stored nor analysed systematically in the past. For a research project a data logger has been installed in the electronic interlocking of Plattling and insulation resistance data has been collected between 28 January 2015 and 30 August 2016. For the available data set, the insulation resistance typically takes values between 500 and 2000 kOhm; the median value amounts to 890 kOhm. The values show a large variation over time and several time-series characteristics can be observed (see Fig. 1).

2.2 Data of Interlocking Operation

This data set covers all activities executed by the interlocking within a time interval of 20 days (20 April 2015 to 4 May 2015). It contains almost 1.5 million notifications about activities. This corresponds to an average of 70 notifications per minute, whereas multiple notifications per second appear regularly. The interlocking manages 422 elements of which 210 are located at the station in Plattling. The element types are for example switches, signals, notifications of routes and track sections.

2.3 Weather Data

In addition to the available data related to the electronic interlocking, weather data is used for data analysis. This data is provided by the German Weather Services (Deutscher Wetterdienst, DWD) and freely available content. The utilised data has been measured at the weather station nearest to the interlocking which is located in Metten, about 8 km from Plattling. Meteorological parameters are available in a 10-min-resolution and cover the temperature in degree Celsius at 2 m and 5 cm height above ground, the relative air humidity at 2 m height, the precipitation in mm and the sunshine duration in hours. The weather data is available for the complete time interval covered by the insulation resistance data collection.

3 Explorative Data Analysis

The insulation resistance values show a large variation over time. A first analysis suggests different influences on its development, containing both meteorological parameters and interlocking operations. Figure 1 depicts a typical course of the insulation resistance over several days.

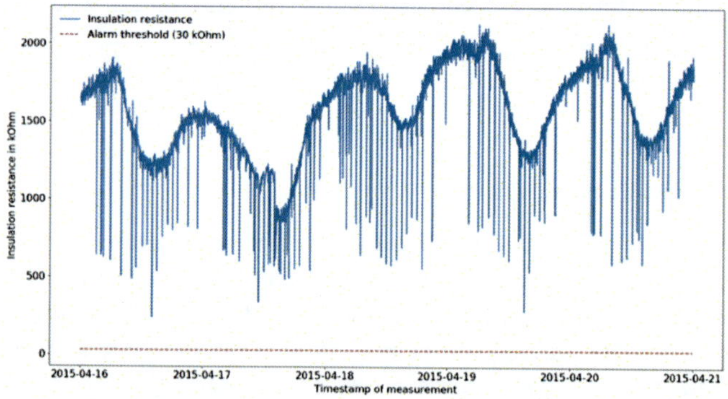

Fig. 1. Typical course of the insulation resistance.

There are two main observations that can be made immediately. First, the values show a long-term variation over the course of a day and trends concerning the development over multiple days. Roughly speaking, the insulation resistance time series takes a day course which is increasing in the early morning hours, decreasing to the evening and then increasing again. Second, the insulation resistance takes short-term, sudden variations: It drops repeatedly over the day, sometimes by large percentages of its initial values, and recovers after a short time period (several minutes).

Concerning the long-term variations, the first striking observation comes to eye by comparing temperature to insulation resistance behaviour. Figure 2 illustrates the day courses of temperature and insulation resistance. The comparison of the two curves shows timewise negatively correlated courses and timewise a time-lagged version of this behaviour. More precisely, at turning points of the temperature, e.g., at a change from decreasing to increasing temperature (for example on 19 April at around 2 p.m.) the insulation resistance keeps decreasing and starts increasing with a time shift (here not before 4 p.m.). This is interpreted as a cooling or warming period, respectively.

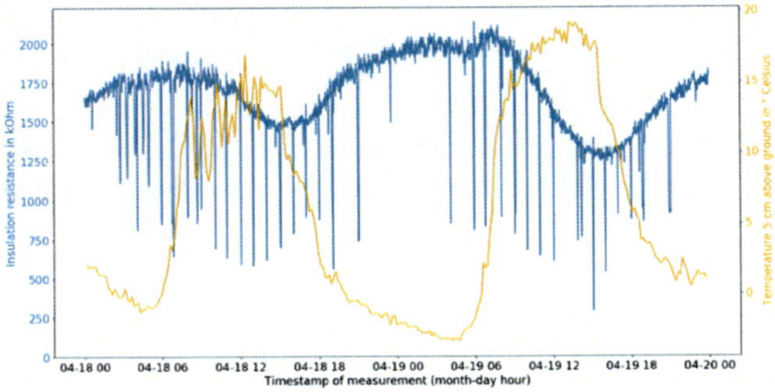

Fig. 2. Anti-cyclical behaviour of temperature and insulation resistance.

Many insulation materials are known to change their insulation behaviour as temperature changes and temperature corrections are commonly used [2]. The temperature dependence of insulation resistance is therefore physically reasonable and analysed in more detail in Sect. 4.

The negative correlations with temperature are superimposed by a decrease in values during time periods with precipitation. As the precipitation ends, the values start increasing at long-term range. This behaviour can be observed in Fig. 3 and is interpreted as a decline in insulation property by penetrating humidity, possibly caused by defective seals or cables, which diminishes as the soil dries. A time lag between the occurrence of precipitation and a decrease of insulation resistance may be explained by deviations in precipitation between weather station and asset or by the time taken by soil moistening.

Underlying to the day-course variation there is a long-term development of the mean value per day around which this oscillation takes place. The long-term development shows a high autocorrelation for a time lag of 1 day, particularly at dry conditions, a negative correlation to the temperature difference between the means of the actual to the previous day, and a negative correlation with the precipitation amount of the day. This is further discussed in Sect. 4.1.

Returning to the second phenomenon, the suddenly appearing drops are assumed to originate from interlocking operation: It is observed that they arise at similar timestamps of each day. This behaviour is illustrated in Fig. 4. These timestamps correlate with departure times of mostly long-distance trains (inter-city and inter-city express in

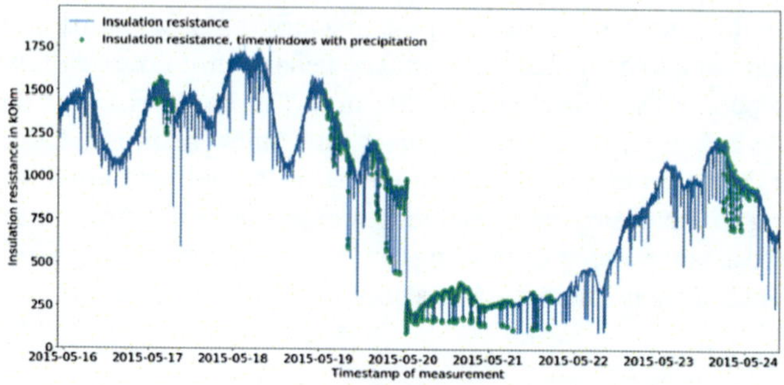

Fig. 3. Negative correlation between precipitation and insulation resistance.

Germany), whereas at departure times of regional trains typically no drops occur. Additionally, the timestamps differ e.g. from weekdays to weekend according to timetable changes.

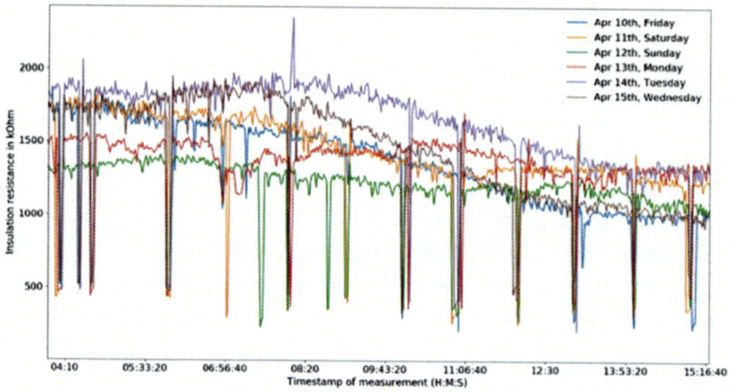

Fig. 4. Time-of-day correlation of drops for several subsequent days.

In the next chapter the hypotheses obtained in the explorative data analysis are investigated in more detail.

4 Time Series Modelling

With regard to detecting abnormal behaviour of the insulation resistance caused by emerging interlocking failures, e.g. by incipient insulation faults, one goal is to identify other causes for variations in the insulation resistance values and to understand the systematic influences on its development.

4.1 Modelling of Weather Influences

Before analysing influences of meteorological parameters, the insulation resistance data is pre-processed in order to reach a consistent time resolution and to eliminate influences coming from other sources. Since the sudden drops of the insulation resistance are supposed not to depend on weather conditions but on interlocking operation (e.g. due to high time of day correlation between different days, time correlations to train timetables and weekday to weekend differences), these outliers are removed by an algorithm before averaging over 10-min-intervals that correspond to the time resolution of the weather data.

First of all, in Sect. 3 it has been observed that the insulation resistance and the temperature curves show a roughly anti-cyclical behaviour over the day. Switching to a larger perspective by considering all measurements of the campaign, the empirical distributions of the insulation resistance in dependence on the temperature show a systematic behaviour: The empirical distributions belonging to higher temperatures tend to dominate the distributions belonging to lower temperatures (a real-valued probability distribution P_1 stochastically dominates the distribution P_2 if the corresponding cumulative distribution functions F_1 and F_2 fulfill $F_1(x) \geq F_2(x)$ for all $x \in \mathbb{R}$, such that there is at least one $x \in \mathbb{R}$ for which the inequality is strict [3]).

Figure 5 depicts the empirical distribution of the insulation resistance (values of 10-min means without sudden drops) in dependence on the measured temperature at 5 cm above ground, for bins of 4 °C. Although the domination does not hold for every pair of distribution functions, a clear systematic influence becomes visible. This behaviour is both observed at dry and at wet conditions.

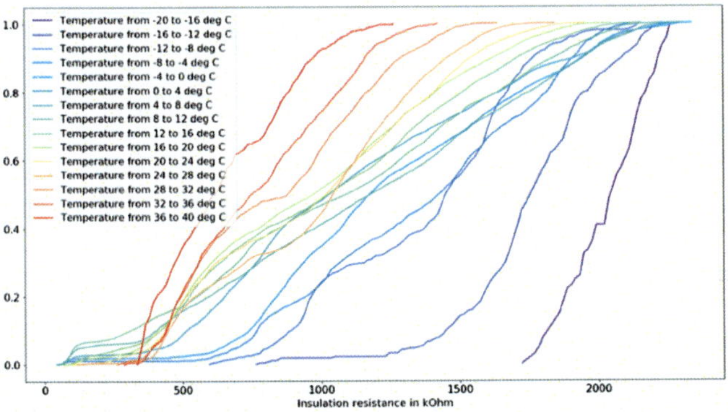

Fig. 5. Empirical distribution function of the insulation resistance in dependence on the temperature at 5 cm above ground, per bin of 4 °C.

The negative correlation with temperature is once more illustrated in Fig. 6 for a finer resolution of temperature bins: The figure shows the mean of the pre-processed 10-min-resolved insulation resistance values for temperature bins of 1 °C (blue solid line) and the intervals *[mean - standard deviation, mean + standard deviation]* per temperature bin in vertical orange lines. The input data for this figure covers the whole period of data collection (including measurements taken at presence of precipitation).

However, Fig. 6 also shows that the empirical standard deviation of the measured values in a temperature bin is quite large, especially in common temperature ranges, and comparable to the span of the insulation resistance of one whole day. This can even be

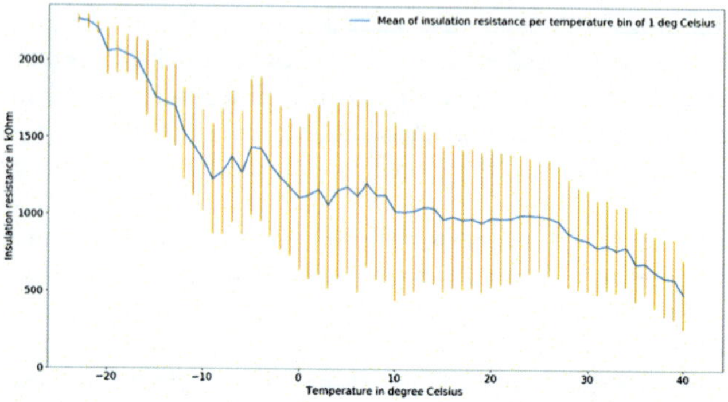

Fig. 6. Empirical mean and standard deviation of the insulation resistance per temperature bin of 1 °C

found if the temperature bins are chosen to be of a length of 0.5 °C. A higher resolution is not reasonable because the temperature measured at a weather station about 8 km away from the asset does not reflect the asset temperature with such a high accuracy. Even at long dry periods the temperature value is therefore not appropriate as single variable to model the insulation resistance development and moreover does not cover all systematic behaviour; this results from the analysis of the time series of normalised values per temperature bin. Note that at rarely occurring temperatures (partly appearing just within a narrow time span of the observed period) the standard deviation is smaller: Multiple measurements in small time distance to each other are not independent when the time series is modelled to be autoregressive (see next paragraph). As already observed in Sect. 3, the daily variation of the insulation resistance shows in large parts a clear opposite development to the temperature but in order to determine the mean value around which the oscillation takes place the history of its development has to be taken into account. As will be described in the subsequent paragraph, an autoregressive model is therefore required to come up to model the insulation resistance course adequately.

From a larger time perspective, considering the long-term development of the insulation resistance represented by its mean per day, a clear autocorrelation respective to a time lag of 1 (day) is found. The development of the mean, especially at days without precipitation, is mostly explained by the previous day's insulation resistance mean and the temperature difference to the previous day. As an example, the variation of the mean of the last 24 h (evaluated just once per 24 h) where the last precipitation dates back at least 48 h is explained by the variations of the mean of the previous 24 to 48 h and the temperature difference by an R^2 value of 0.91. Particularly for dry conditions the average around which the insulation resistance is oscillating is mainly driven by its own historical values and not by a change in temperature values (see Fig. 7) – removing the mean of the previous day as predictor variable in the model and replacing the temperature difference by the mean temperature per day yields poor results. Similar to the mean, other parameters like maximum and minimum value of the insulation resistance per day can be modelled by linear regression in dependence of the corresponding values of the previous day and temperature/precipitation values with a good performance.

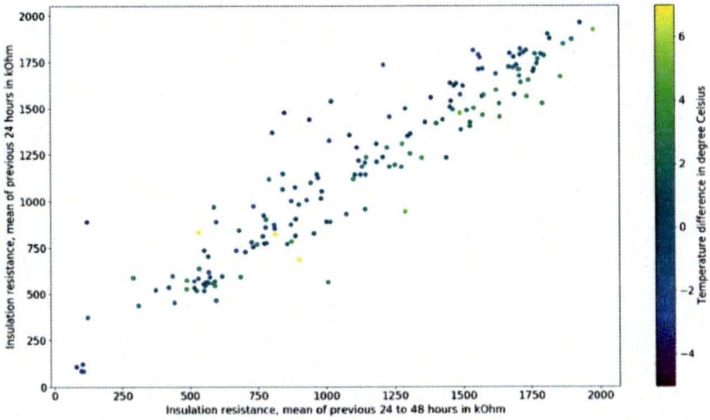

Fig. 7. Day mean of insulation resistance in dependence of the mean of the previous day and the temperature difference (day mean) to the previous day

Returning to the daily variation of the insulation resistance, an autoregressive time series model is required, involving additionally exogenous variables like temperature or precipitation values. A widely-used model is the autoregressive (AR) model, which assumes a linear relationship of the actual value of a time series R_t and its p predecessors, disturbed by a residual random variable ε_t:

$$R_t = \sum_{i=1}^{p} a_i R_{t-i} + \varepsilon_t$$

The random errors $(\varepsilon_t)_{t \geq 1}$ are centred and independent or at least uncorrelated random variables (see [4]), sometimes assumed to be identically standard normal distributed. Since for the weather data and the pre-processed insulation resistance data there is one measurement per 10 min available, t corresponds to an integer denoting the t-th 10-min time window.

The AR model is further developed and adapted to take into account several information gains from the explorative analysis. First, it is found that the insulation resistance's deviation from its mean (previous 24 h) can be adequately described by the corresponding mean deviation of the temperature multiplied by the fraction of the insulation resistance to temperature span (difference between maximum and minimum of the previous 24 h). Second, the time lags at different times of the day are included in the model by accounting for the time of day: for a time resolution of 10 min there are 144 time windows per day. More precisely, the insulation resistance is modelled by

$$R_t = a_{t \bmod 144} \cdot X_t + b_{t \bmod 144} \cdot \frac{1}{144} \cdot \sum_{i=1}^{144} R_{t-i} + \varepsilon_t.$$

As described above, the variable X_t is defined as

$$X_t = \left(T_t - \frac{1}{144} \cdot \sum_{i=1}^{144} T_{t-i} \right) \cdot \frac{\max\limits_{1 \leq i \leq 144} R_{t-i} - \min\limits_{1 \leq i \leq 144} R_{t-i}}{\max\limits_{1 \leq i \leq 144} T_{t-i} - \min\limits_{1 \leq i \leq 144} T_{t-i}}$$

The parameters of the model that have to be determined are given by a_j, b_j for $1 \leq j \leq 144$. Optionally an intercept c_j can be included, as well as a term accounting for the amount of precipitation. The model also yields good results if the coefficients b_j are dropped. The problem is reduced to fitting a linear regression for each j and is performed after having randomly applied a train-test split with approximately a third of the measurements (in packages of consecutive measurements lasting one week) belonging to the training and two thirds belonging to the test set. Only measurements taken at moments where the last precipitation dates back at least 24 h are considered for both training and test set. Figure 8 shows the predicted course of the insulation resistance for several days in May 2016.

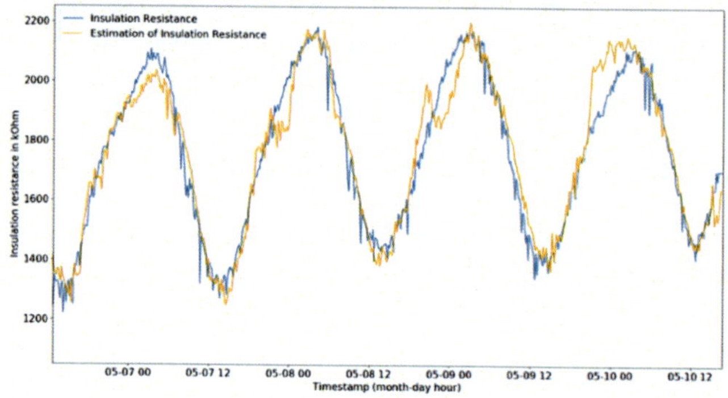

Fig. 8. Measured (blue) and predicted (orange) values of the insulation resistance, for time windows of 10 min.

The coefficient of determination equals 0.97 on the test set and 0.96 on the training set. Systematic behaviour is captured and systematic errors provided by other models have been eliminated. The above model can be used to predict the insulation resistance if the insulation resistance history as well as the temperature are known; in addition, assuming that the residuals ε_t are independent and identically normal distributed, performing a linear regression allows for determining both confidence and prediction intervals for new measurements, given the values of the independent variables. In fact, the above result shows that the variation of the insulation resistance over the day at dry conditions at a 10-min time resolution can be explained to a very high degree by the variation of temperature and historical values of mean, minimum and maximum of temperature and insulation resistance, respectively.

Capturing and quantifying the effect of precipitation in moistening or drying periods is content of current research. First results have been obtained but are beyond the scope of this paper.

4.2 Interlocking Operation

To analyse interlocking operation-related behaviour of the insulation resistance which is presumably represented by the sudden drops observed in the data as illustrated in Fig. 1

and Fig. 4, some data preparation is necessary. In the raw data the activity notifications are encoded as 32-bit vectors with groups of them representing a certain condition aspect of the element. Thus, one condition of a field element itself consists of multiple components. As data preparation step, these 32-bit conditions are recoded to only retain relevant information: groups only taking one value and indices having no meaning are dropped; those taking multiple values are reduced to one entry. Taking all the up-to-date conditions of all elements from the complete list of element status changes, a condition vector of the whole system is created for each timestamp and after each change of the system. This vector reflects the condition of the whole system and contains all elementary conditions of all elements. It has overall 1440 entries for the 422 elements, i.e., in average 3 to 4 components per element, each taking individual co-domains.

Since there is a large gap between the time resolutions of insulation resistance measurements and timestamps of interlocking operation activities, the analysis is challenging. Moreover, the problem is high-dimensional and with more than 10^{400} possible vector states. The preparation of the insulation resistance data mainly concerns the classification of the measurements. An algorithm identifies begin and end of the drops in terms of first abnormal and first normal measurements, respectively, as well as abnormal measurements which are preceded by abnormal ones and normal measurements preceded by normal ones. Here different approaches can make use of different information coming from the data. To illustrate this, consider the first timestamp at which a drop appears in the insulation resistance data. The elements changing their status between the last normal and the first low measurement are suspicious to take part in causing the drop. But the vectors appearing between these two measurements cannot be uniquely assigned to a normal or abnormal behaviour of the insulation resistance. In contrast, vectors describing the system status between two low measurements can indeed clearly be assigned to an abnormal measurement. But elements changing their status within this time interval are unlikely to cause the drop since their changes do not affect the low value of the measurement (where it is assumed that the measurement remains low between two low measurements). Between the last low and the first recovered measurement, the situation is similar to the situation between the last normal and the first low measurement.

Concerning the information contained in the changes between normal and abnormal measurements, examining the data for one-to-one correlations between drops and changes of one specific element yields no positive result. In fact, the intersection of the set which contains elements that change their status between the last normal and first abnormal measurement and the set containing those changing between the last abnormal and first recovered measurement, respectively, is frequently empty. Thus, for accessing time correlations between a certain state of the system and appearing drops, it is necessary to involve multiple elements where the sizes of the subsets of involved elements is unknown. Examining frequent combinations of two elements, one of each set, in general shows that the state combination observed in abnormal measurements likewise often appears under normal conditions. The problem becomes quickly large because there appear up to 150 changes between two insulation measurements. Further analysis requires machine learning methods due to the high complexity and number of possible subsets that have to be examined.

Since the predictors in form of the elementary conditions of the status vectors are categorical and one goal of the model is to attain information about involved elements and conditions causing sudden drops, the model used to predict abnormal behaviour is a decision tree. Decision trees can be used for classification as well as for regression purposes. A decision tree sequentially splits the domain of the independent variables, in each step dividing it into two subsets. For obtaining the class or regression value of one sample a tree is followed by deciding at each split to which subset it belongs until a leaf (end point) is reached. If the domain of an independent variable is numerical, the split is axis-parallel. In this way the domain D is split into pairwise disjoint subsets which correspond to the leaf nodes: $D = \bigcup_{i=1}^{m} D_i$. If the tree is grown for regression, the regression value assigned to a new sample $x \in D_i$ is the average of the dependent variables' values taken by the training data points in the subset D_i. If the objective is classification and the classes are given by $C_j, j \leq n, x \in D_i$ is assigned to the class j for which the number of training samples that belong to $D_i \bigcap C_j$ is maximal. The splits are performed to minimise a pre-defined target value (cost function), e.g. mean squared error for regression; misclassification rate, entropy or Gini index for classification purposes [5]. Overall, decision trees are a very illustrative tool that allows for traceability of decision making.

Two different approaches are presented to predict the sudden drops by the knowledge of interlocking status messages. The first is a classification approach using a decision tree. For training, only the status vectors that are definitely assigned to a normal/abnormal measurement, respectively, are used (i.e., they are both followed and preceded by a normal respective an abnormal one). The normal class is assigned to the value 0 and the abnormal class to the value 1. From the whole set of these vectors, only 40,000 corresponding to a time-interval of about 12 h are used for training.

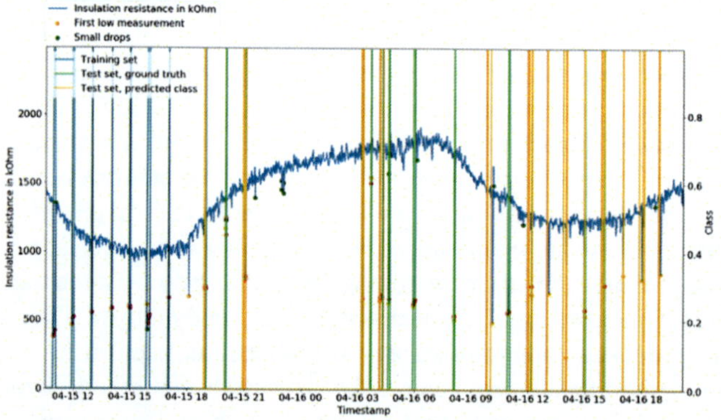

Fig. 9. Classification of condition vectors. Red dots symbolise more than one subsequent abnormal measurement.

As Fig. 9 shows, the decision tree classifier succeeds in classifying vectors appearing in temporal proximity to drops that consist just of one insulation resistance measurement (and therefore were not part of the set defining the ground truth) to be abnormal (see e.g. on 16 April at 10 a.m. as well as at 1 p.m. and 2 p.m.). Anyway, this approach misses a lot of information by not taking into account drops consisting only of one low measurement which can be a reason for the false negatives e.g. on 16 April at 5 and 8 p.m.

The second approach includes all vectors but passes to a fuzzy classification by assigning values from [0, 1] to the vectors. The vectors that are clearly identifiable as normal or abnormal (those used in the classification described in the previous paragraph) are again attributed by the values 0 and 1. Those which are not uniquely identifiable cases (e.g. between two measurements of which one is normal and one abnormal or before a measurement only showing a small difference to the preceding cleaned 10-min mean) are assigned to a heuristically determined probability of belonging to an abnormal measurement. For the former case, this is done by weighting with the fraction of time-difference to the normal measurement divided by the time-difference between the precedent and next measurement, for the latter by including the relative difference of the drop to the normal measurement. Then a decision tree for regression is trained with only 20,000 vectors at the beginning of the period. The begin is clipped because it takes some time until the majority of the elements has changed their status for the first time and a meaningful vector is created. The training values are depicted in Fig. 10 in teal dashed lines and comprise only 6 drops.

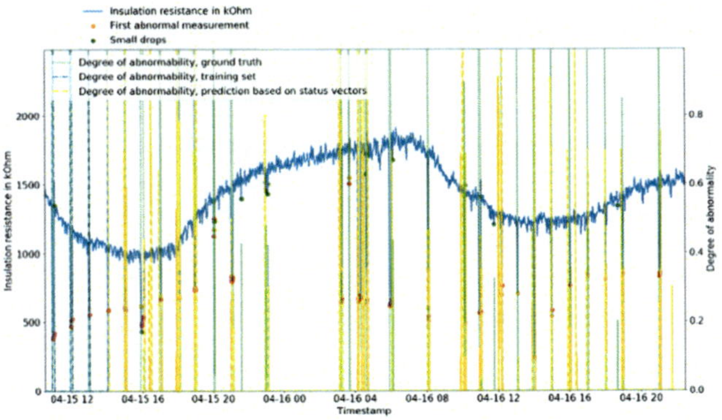

Fig. 10. Prediction of fuzzy class affiliation of the condition vectors (four drops out of six of the training set are visible).

The algorithm can nevertheless detect drops of the insulation resistance on basis of the condition vectors representing the interlocking element status. Depending on a threshold dividing the fuzzy classes in normal/abnormal, a true positive rate (sensitivity) of 33% is reached on the test set, with a specificity of 93%. It has to be taken into account that false positive predictions often occur in very small time distance to a drop as well as false negative predictions often belong to a drop where a part of the vectors has been correctly classified as abnormal. Other approaches and improvements of the performance are content of ongoing research but the first results already give a strong hint on a dependence of the sudden drops on interlocking activities. Examining the tree splits can be exploited to obtain suspicious elements for causing the strong drops of the insulation resistance.

5 Discussion

The analysis of the insulation resistance values of the test set in Plattling presented in this paper includes correlations of changes of the insulation resistance both with meteorological parameters and with interlocking operation. A first investigation already provides interesting and promising results. The approach shows potential to monitor the condition of the electrical installations of electronic interlocking systems and to detect abnormal behaviour as well as elements that are suspicious to be at the responsible for insulation faults.

The insulation resistance seems to be affected by temperature and precipitation which both are observed to negatively correlate with the insulation resistance magnitude. For the temperature-driven behaviour, a good performing model is found; first results on quantifying the precipitation-related behaviour are obtained but have to be improved. In particular, since the weather station providing the meteorological data is about 8 km away from the interlocking in this study, there can be deviations between the local weather conditions that hinder the analysis. Future research would benefit from local weather sensors at the asset.

In addition to that, using data acquired at the interlocking operator desk suggests a relationship between sudden drops of the insulation resistance and certain states of the interlocking elements. This allows drawing conclusions on involved elements, which has to be further investigated in future.

Until now there has been only one measurement campaign with data coming from one specific interlocking. Future research should comprise data from other interlocking systems as well. The next generation of measurement devices including a data logger is currently under field testing and will allow gathering data from further electronic interlockings in the future.

Acknowledgments. We want to thank Bender GmbH&Co.KG and DB Netz AG for offering us the opportunity to install a test setting, collect data and enable us to do this research.

References

1. Groos, J.C., Zhang, X., Linder, C.: The relevant influences on the insulation resistance of electronic interlocking cable systems. In: SIGNAL + DRAHT, vol. 110, no. 5, pp. 17–24 (2018)
2. Zurek, S., El-Rasheed, A., Ohlen, M.: Individual temperature correction (ITC) for insulation resistance measurements. In: INSUCON - 13th International Electrical Insulation Conference (INSUCON), pp. 1–5 (2017)
3. Heathcote, A., Brown, S., Wagenmakers, E.J., Eidels, A.: Distribution-free tests of stochastic dominance for small samples. J. Math. Psychol. **54**(5), 454–463 (2010)
4. Akaike, H.: Fitting autoregressive models for prediction. Ann. Inst. Stat. Math. **21**(1), 243–247 (1969). https://doi.org/10.1007/BF02532251
5. Murphy, K.P.: Machine Learning. A Probabilistic Perspective. MIT Press, Cambridge (2013)

Study on the Condition Monitoring Technology of Electric Valve Based on Principal Component Analysis

Renyi Xu, Minjun Peng, and Hang Wang(✉)

Harbin Engineering University, Harbin, Heilongjiang, China
heupmj@hrbeu.edu.cn

Abstract. Valves are the most diverse and widely used general purpose mechanical equipment in nuclear power plants. During its use, due to the prolonged exposure to vapors, oils and radioactive liquids, the valve inevitably ages and fails. Furthermore, the safety and reliable operation of nuclear power plants are adversely affected. Therefore, it is of great significance to monitor the running state of valves in real time and find the hidden danger in time to ensure the safe operation of nuclear power plants. Based on this, this study built an experimental bench that can simulate the valve leakage fault, and used acoustic emission sensor to measure the operation data of the valve under normal and leakage conditions. And the intelligent condition monitoring technology of the valve is preliminarily discussed with artificial intelligence algorithm PCA as the monitoring means. So as to provide the technical basis for the follow-up valve maintenance and fault diagnosis. It is found that the PCA algorithm can be used to monitor the running state of the valve, and the leakage state of the valve can be found in time. Therefore, it has certain reference significance for the follow-up maintenance treatment.

Keywords: Electric valve · Acoustic emission sensor · PCA

1 Introduction

With the continuous development of artificial intelligence technology and big data theory, the digitization and informatization degree of each system and equipment of nuclear power plants (NPPs) is constantly improving, and the trend of intelligent operation is gradually developing. However, due to the complexity and potential danger of nuclear power plants, the security of intelligent operation has always been highly concerned by the public. In the nuclear power plants, from the simplest chopping device to the complex automatic control system, the valve types and specifications are numerous, the corresponding valve failure is also the most common, among which the valve leakage and jam problems are the most common. Based on this consideration, this paper takes the valve leakage fault as the research object to deeply analyze the monitoring mechanism.And the intelligent condition monitoring technology of the valve is preliminarily discussed with artificial intelligence algorithm PCA as the monitoring means. So as to provide the

R. Karim et al. (Eds.): IAI 2021, LNME, pp. 141–151, 2022.
https://doi.org/10.1007/978-3-030-93639-6_12

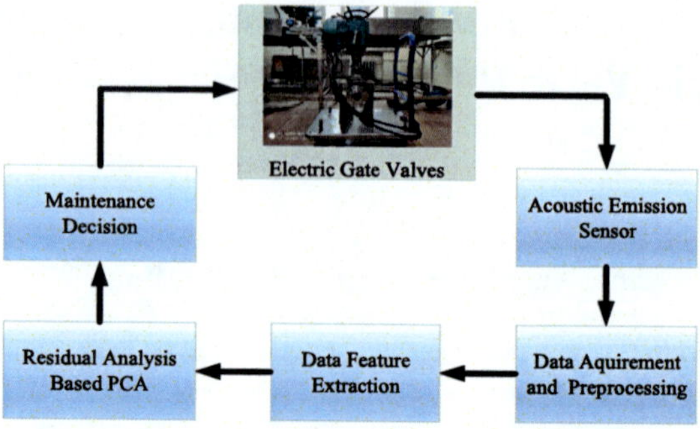

Fig. 1. Research route of electric valve condition monitoring technology.

technical basis for the follow-up valve maintenance and fault diagnosis. Figure 1 shows the technical route studied in this paper.

As shown in Fig. 1, this study takes the LFZ10-18W electric gate valve as the research object, and the valve leakage fault as the monitoring object. Acoustic emission sensor is used to collect the operation data under the condition of valve leakage to perceive the operation state of the valve. By analyzing the response of different characteristic parameters of acoustic emission sensor to leakage fault, the characteristic extraction of valve leakage signal is carried out. Finally, the residual analysis is carried out by PCA algorithm to realize the on-line monitoring of valve leakage fault.

2 Experiment

2.1 Electric Valve Test Bench

In this study, PCA algorithm was used to monitor the operation state of the electric valve. In the online monitoring process of the electric valve, the most critical step was to obtain the operation data of the valve under normal state and fault state. Therefore, a set of test bench that can simulate the leakage fault of the electric gate valve is built. In conjunction with the specially designed external sensing system and relevant test schemes, process parameters such as relevant pressure, differential pressure and flow are measured to collect information and data that can reflect the status of the electric gate valve, and serve as the research basis for the online monitoring of the electric valve.

The structure of the test bench designed in this study is shown in Fig. 2. The type of electric gate valve for analysis is LFZ10-18W straight screw gate valve, driven by squirrel-cage coil motor. Gate valves have a diameter of 50 mm and are truncated by rigid single gate and connected by flanged pipe with nominal pressure of 2.5 MPa.

As shown in Fig. 2, both the front and back ends of the electric gate valve are equipped with a front ball valve, and a static pressure gauge with a measuring range of 0–2.5 MPa is set before and after it. A differential gauge with a range of 0–500 kPa is also provided on the electric gate valve. The valve is connected with the system loop by a hose with a diameter of 50 mm and connected with the valve by a flange. Figure 3 below shows the

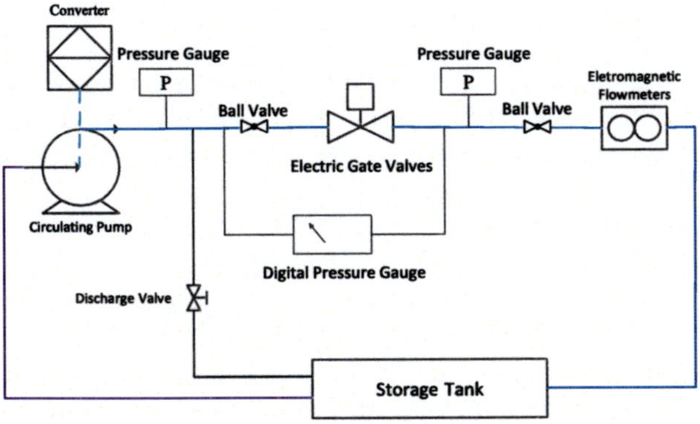

Fig. 2. Schematic diagram of on-line monitoring test bench of electric valve

fault gate valve used in the experiment. Flow measurement valves and downstream ball valves are installed in the main channel. Finally, an electromagnetic flowmeter with a measuring range of 0.2 m^3/h– 4 m^3/h is set in the downstream. The flow measuring valve can be used to measure the flow of the pipeline less than the range of the electromagnetic flowmeter. In order to prevent the pump from being damaged when the main pipeline is completely cut off, a front discharge valve is installed near the pump outlet to direct the working medium back to the storage tank. The diameter of the main pipe is 50 mm. It relies on the vertical circulation pump to provide the driving pressure. After pumping water from the storage tank, it is pressurized by the pump to provide the pipe circulation. The water tank is an open container with a maximum storage capacity of about 3 m^3.

Fig. 3. Field diagram of faulty electric valves

2.2 Setting of Electric Valve Test Parts

Electric valve is in direct contact with the working medium in the process of use, and the control function of the fluid pipeline is realized by cutting or connecting the pipeline. Valves are prone to aging and failure when operating conditions change or improper use.

Since valve leakage fault accounts for a large proportion of valve faults, this study selected valve leakage as a typical fault mode. The appearance of valve leakage fault is generally due to the valve crack and with the extension of time, the crack gradually expanded. The occurrence of cracks mainly has the following reasons. First, the valve internal lattice is uneven, and there are material defects. Second, uneven impact of fluid or installation defects lead to uneven valve force, and then mechanical cracks. The third is the corrosion leakage caused by fluid corrosion and radioactive material irradiation. Therefore, in order to simulate the leakage fault of the electric valve, the test piece of the electric valve as shown in Fig. 4 is designed in this study.

Fig. 4. Simulated fault parts of electric valves

As shown in Fig. 4, three holes M3, M5 and M10 are opened on the valve body in front of the valve plate to simulate the leakage failure of the valve. During the experiment, the degree of leakage outside the valve was controlled by the number of turns of the rotating screw to simulate the operation state of the electric valve in different leakage states.

2.3 Sensing and Measurement of Electric Valve Test Bench

In the process of on-line monitoring of electric valve, it is very important to select and collect parameters that can represent the running state of electric valve. As shown in Fig. 2, static pressure of the front and rear pipelines of the valve can be collected and measured through the static pressure gauge on the test bench, the pressure difference before and after the valve can be measured by a pressure gauge, The leakage of the valve can be calculated by flow measurement valve and electromagnetic flowmeter. However, it is not enough to completely represent the running state of the electric valve only by the signals such as flow rate and pressure, so it is necessary to use other detection methods to measure the characteristic parameters under the leakage state. Since the acoustic emission method is to measure the transient stress wave on the surface of the valve body when leakage occurs without any external drive, it is a good non-destructive testing method for valve leakage. Therefore, this study uses acoustic emission technology to conduct intelligent nondestructive testing on the valve. As shown in Fig. 5, the layout of the acoustic emission sensor of the electric valve is shown. It can be seen from the diagram that the acoustic emission sensor is positioned in the center of the valve body. This is mainly due to the fact that when the electric gate valve fails to leak, the acoustic emission signal is triggered by the turbulence field generated near the leakage hole. Therefore, acoustic emission sensors need to be installed as close as possible to the sound source.

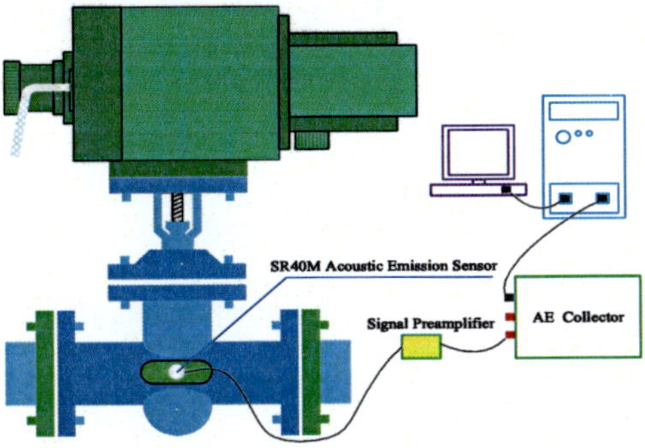

Fig. 5. Arrangement of acoustic emission sensors on electric gate valves

The time characteristic parameters of acoustic emission signal refer to a series of statistics in the time dimension of acoustic emission waveform that can represent the signal properties, including ringing counts, amplitude, duration, energy, and root mean square (RMS), average signal level (ASL). The on-line monitoring research of electric valve is based on the analysis of these characteristic parameters.

3 Principal Component Analysis

Principal component analysis (PCA) is a multivariate statistical method. It uses the historical data obtained from the normal operation of the system to establish the principal component model, compress the high-dimensional information and project it into the low-dimensional feature subspace, so as to describe the characteristic information of the original data space with fewer principal component variables in the low-dimensional space. It can synthetically consider the correlation among residuals, so it can effectively avoid the false alarm problem which is easy to occur in the residuals threshold method.

Assuming that the number of effective measurement points in a subsystem is n, n groups of residual variables can be generated when running synchronously with the simulation model. Take m values of this group of residual variables as the residual sample data needed for the principal component model modeling, and get a m × n matrix X^l. The steps for establishing the principal component model are as follows:

(1) Standardized treatment
Because the dimensions of the original data are different, in order to eliminate the influence of dimensional factors, the original data is first standardized. Let X be the processed matrix, then:

$$x_{ij} = \frac{x_{i,j}{}^{\iota} - \overrightarrow{x}_j{}^{\iota}}{\sigma_j}(\mathrm{i} = 1, 2, \ldots, \mathrm{m}; \mathrm{j} = 1, 2, \ldots, \mathrm{n}), \tag{1}$$

Where, the mean is $\overrightarrow{x}_j{}^{\iota} = \frac{1}{n}\sum_{i=1}^{n} x_{i,j}{}^{\iota}$, the standard deviation is

$$\sigma_j = \sqrt{\frac{1}{n-1}\sum\nolimits_{i=1}^{n} (x_{i,j}{}^{\iota} - \overrightarrow{x}^{\iota})^2} \tag{2}$$

(2) For standardized X, find its covariance matrix Cov(X)

$$\mathrm{Cov(X)} = \frac{X^T \mathrm{X}}{n-1} \tag{3}$$

(3) Find the eigenvalues of Cov(X) and the corresponding eigenvectors

$$\mathrm{Cov(X)} p_i = \lambda_i p_i \tag{4}$$

Sort the obtained eigenvalues from large to small: $\lambda_1, \lambda_2, \ldots, \lambda_n$.The corresponding eigenvectors are respectively: $p_1, p_2, \ldots, p_n$.

(4) Determine the number of pivot entries A
Determine the required number of pivot elements according to the cumulative contribution rate Q_m of pivot elements,$\mathrm{Q_m}$ is calculated by the following formula:

$$Q_m = \frac{\sum_{i=1}^{A} \lambda_i}{\sum_{i=1}^{n} \lambda_i} \tag{5}$$

(5) Get the calculation model
$p_1, p_2, \ldots, p_n$ is called the load vector, $p_{n\times A} = [p_1, p_2, \ldots, p_n]$. is called the load matrix. Span{P} represents the principal element subspace of the principal element model.

$$T_{m\times A} = [t_1, t_2, \ldots, t_A] = XP_{n\times A} \tag{6}$$

(6) Calculate control limits for statistics
PCA model is used for real-time analysis of data. Generally, statistics T^2 is used for projection on principal component space and statistics SPE is used for hypothesis testing for projection on residual subspace.

4 Results and Discussion

This experiment needs to measure the signals and parameters of the electric gate valve under normal operation and leakage failure. By setting the leakage phenomenon of electric valve into the environment, the relationship between acoustic emission parameters and leakage quantity of electric valve is studied when the flow is stable. In the experiment, at a certain circulating pump frequency and valve opening, the acoustic emission measurement system was started when the water completely filled the pipeline and the flow rate was stable. Then slowly turn the screw to increase the leakage gradually. Different pump frequencies and valve openings represent different operating conditions of electric valves. In each experiment, the characteristic parameters obtained included the frequency of the circulating pump, the actual opening of the electric valve, the pressure difference between the front and rear ends of the valve, fluid flow, and five acoustic emission signals, namely amplitude, ringing count, energy, RMS, and ASL. Figure 6 and Fig. 7 below show the variation trend of amplitude, ringing count, RMS and ASL in a certain aging process when the circulating pump frequency is 30 Hz and the valve opening degree is 35 s.

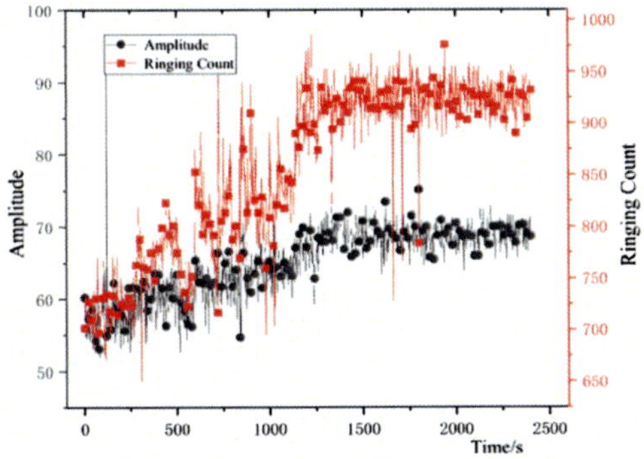

Fig. 6. The variation trend of amplitude and ringing count at frequency 30 Hz and valve opening degree 35 s

Through comprehensive analysis of the characteristic parameters obtained in the leakage process, it can be known that, At a certain pump frequency and valve opening, as the leakage volume increases from small to large, the amplitude of acoustic emission parameters, ringing count, RMS and ASL all show approximately the same trend: When the leakage quantity is less than a certain value, the size of relevant characteristic parameters is approximately unchanged. When the leakage quantity exceeds this value, the parameters show an obvious change trend. After the leakage volume increases further to a critical value, the parameter remains approximately unchanged again. This is because when the leakage amount is very small, the leakage has little effect on the flow of working medium in the pipeline. When the leakage volume is very large, the pipeline flow is no longer the turbulent state in the pipe, but is completely opened. However, when the leakage amount is between these two critical values, the turbulence effect in the

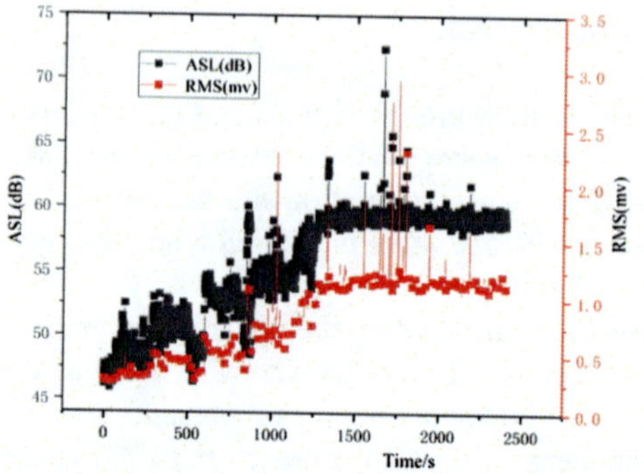

Fig. 7. The variation trend of RMS and ASL at frequency 30 Hz and valve opening degree 35 s

tube is obvious, and also the acoustic emission measurement parameters have a positive correlation with the size of the leakage amount.

At the same time, principal component analysis is used to monitor the residual error of the characteristic parameters in real time. The monitoring results of PCA are shown in Fig. 8.

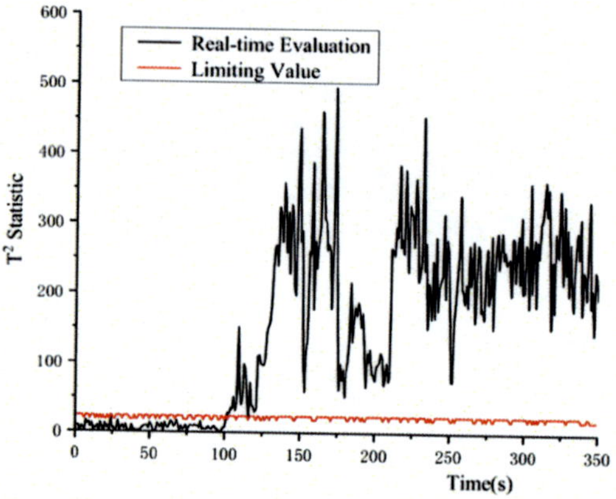

Fig. 8. Monitoring results of PCA at frequency 30 Hz and valve opening degree 35 s

As can be seen intuitively from Fig. 8, the calculated value of the T^2 statistic exceeds the limit value at 103 s. When at 60 s, the valve fails by turning the screw, valve leaks can be quickly identified.

Figure 9 and Fig. 10 below show the variation trend of amplitude, ringing count, RMS and ASL in a certain aging process when the circulating pump frequency is 35 Hz and the valve is opened for 30 s. As shown in the figure, the same conclusion can be reached.

Also Fig. 11 show the monitoring results of PCA at frequency 35 Hz and valve opening degree 30 s. As can be seen from Fig. 11, valve leakage can also be detected quickly through PCA monitoring.

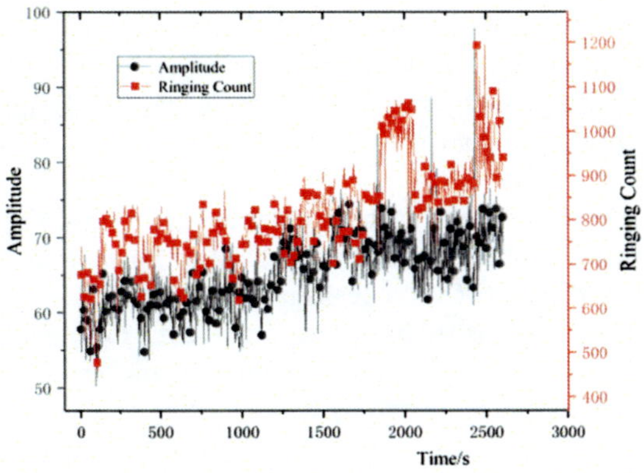

Fig. 9. The variation trend of amplitude and ringing count at frequency 35 Hz and valve opening degree 30 s

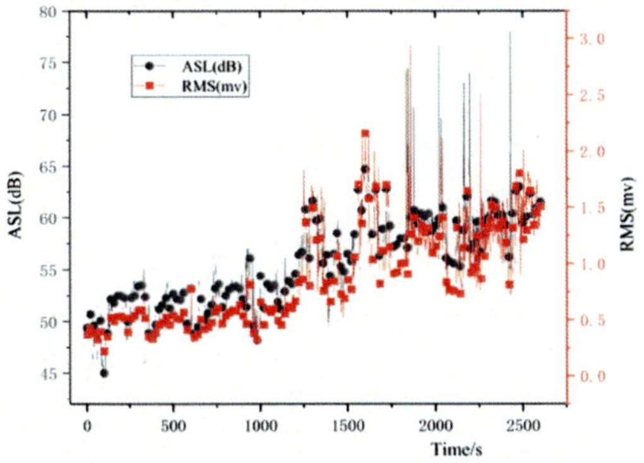

Fig. 10. The variation trend of RMS and ASL at frequency 35 Hz and valve opening degree 30 s

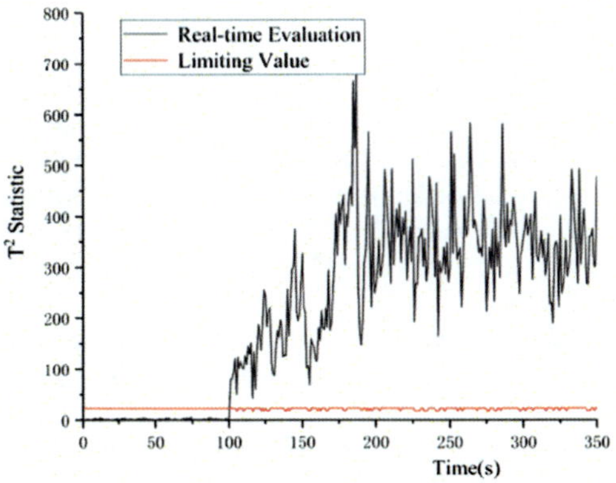

Fig. 11. Monitoring results of PCA at frequency 35 Hz and valve opening degree 30 s

In conclusion, the PCA model established in this paper can monitor the running state of the valve and find the leakage state of the valve in time. The effectiveness of the proposed method is proved by the actual valve leakage experiment. Therefore, the following conclusions can be drawn:

(1) The characteristic parameters such as amplitude, ring count, RMS and ASL in are signal can better reflect the operating state of valve, and can be used as the characteristic information of valve leakage detection.
(2) After PCA model is established based on normal data, the model can realize timely detection of valve leakage fault based on square predict error (SPE) and Hotelling (T^2) statistics, which is also proved by experiments.

Acknowledgments. Our thanks to sponsors of IAI2020 Conference for their intellectual and financial support. The authors greatly appreciate the support from Young Talent Program of China National Nuclear Corporation: Research on multi-strategy intelligent fault diagnosis technology for important nuclear power equipment. Project Number: KY90200210007.

The authors greatly appreciate the support from Fundamental Science on Nuclear Safety and Simulation Technology Laboratory, Harbin Engineering University, China.

References

1. Lee, S.G., Park, J.H., Yoo, K.B., et al.: Evaluation of internal leak in valve using acoustic emission method. Key Eng. Mater. **48**(1), 326–328, 661–664 (2006)
2. Wold, S., Geladi, E.K.: Principal component analysis. Chemometrics Intell. Labor. Syst. **13**(2), 37–52 (1987)
3. Bakshi, B.R.: Multiscale PCA with application to multivariate statistical process monitoring. AIChE J. **44**(7), 1596–1610 (1998)
4. van de Loo, P.J., Ronemeijer, D.K.: Acoustic emission tank testing: how to discriminate between the onset of corrosion and further stages of degradation. In: European Conference on Acoustic Emission Testing, TUA Austria, Wien, 6–8 May 1998
5. Mallat, S.: Zero-crossings of a wavelet transform. IEEE Trans. Inf. Theory **37**(4), 1019–1033 (1991)
6. Lee, J.H., Lee, M.R., Kim, J.T., et al.: Analysis of acoustic emission signals for condition monitoring of check valve at nuclear power plants. Key Eng. Mater. 144–149 (2004)
7. Shukri, I.N.B.M., Mun, G.Y., Ibrahim, R.B.: A study on control valve fault incipient detection monitoring system using Acoustic Emission technique. In: 2011 3rd International Conference on Computer Research and Development (ICCRD), vol. 4 (2011)
8. Silva, L.R., Deschamps, C.J.: Modeling of gas leakage through compressor valves. Int. J. Refrig. **53**, 195–205 (2015)
9. Lee, M.R., Lee, J.H., Kim, J.T., et al.: Condition monitoring of a nuclear power plant check valve based on acoustic emission and a neural network. J. Pressure Vessel Technol. **127**(3), 230–236 (2005)
10. Kaewwaewnoi, W., Prateepasen, A., Kaewtrakulpong, P.: Investigation of the relationship between internal fluid leakage through a valve and the acoustic emission generated from the leakage. Measurement **43**(2), 274–282 (2010)

11. Wise, B.M., Ricker, N.L., Veltkamp, D.F., et al.: A theoretical basis for the use of principal component models for monitoring multivariate processes. Process Control Qual. **1**(1), 41–51 (1990)
12. Tian, Q., Xiang, Z.: The applicability of hawkins statistic for PCA modelling. ICICIP **31**(2), 436–441 (2011)

Multivariate Alarm Threshold Design Based on PCA

Yue Yu, Minjun Peng, Hang Wang(✉), and Zhanguo Ma

Fundamental Science on Nuclear Safety and Simulation Technology Laboratory, Harbin Engineering University, Harbin, Heilongjiang, China
{yueuy,heupmj,heuwanghang,mazhanguo}@hrbeu.edu.cn

Abstract. Now, the alarm system is the first line of defence for nuclear power plants. however, there is a serious problem called alarm overloading. Under the steady state of the nuclear power plant operation, there are too many nuisance alarms due to noise and fluctuations. In order to reduce the number of the nuisance alarms and improve the alarm accuracy in fault conditions, a multivariate alarm threshold design method based on principal component analysis (PCA) is proposed. In this method, a second threshold is added to the alarm threshold which can eliminate the number of false alarms effectively. The PCA model is established using actual operating data from nuclear power plants. Then, according to the actual data sample, the multivariate threshold is set to eliminate the number of the nuisance alarm, and the simulated accident data is used to analyse the alarm capability of the threshold in accident mode. The simulation results show that the alarm is triggered immediately under the accident and the nuisance alarm in the steady state can be effectively suppressed.

Keywords: Alarm system · Multivariate threshold · PCA · Nuisance alarm · Alarm overloading

1 Introduction

The alarm system is the first protective layer of the nuclear power plant [1]. Its operational performance will affect the safety and economy of the operation of the nuclear power plant to a certain extent. With the development and application of digital I&C system (digital instrument and control system), the number of sensors increases and it helps to obtain more operating status data of the system equipment. But more alarm detection variables are introduced, and increase the operating pressure of the operator.

Every year, the world will lose millions or even billions of dollars due to unexpected failure, operational errors and other reasons. In the actual industrial process, the system and the sensor will be affected by the complex environment, and itself, which often produces excessive nuisance alarms. A nuisance alarm is defined as one that does not require a specific action or response from operators [2]. Obviously, it can be seen from Table 1 that there is a certain difference between the actual industrial statistical alarm results and the EEMUA standard [3].

R. Karim et al. (Eds.): IAI 2021, LNME, pp. 152–162, 2022.
https://doi.org/10.1007/978-3-030-93639-6_13

At present, the single-variable threshold is widely used in nuclear power plant alarm. The designer usually determines the alarm threshold through the empirical judgment. Many scholars want to design advanced single-variable threshold to optimize alarm system. Zhu adopted a dynamic threshold management method based on Bayesian-estimation to ensure the accuracy of alarm signals [4]. Mezache applied fuzzy neural network and genetic algorithm to the training of threshold estimation [5].

The above researchers used methods such as adaptive threshold, filter and dynamic threshold and so on to design the alarm threshold. However, they focus on the single-variable threshold design, which is usually independent of other related process variables, without considering the correlation between the variables in the industrial process. Those design approaches may bring negative effects.

Therefore, scholars began to pay attention to the multivariate alarm threshold. Bristol proposes an adaptive alarm threshold designed for different working conditions, which increases the range of the model [6]. Zang found the missed alarm rate and false alarm rate by calculating the probability density function of multi-dimensional space [7]. But the calculation of multivariable joint probability density function is very complex.

PCA method has been most applied in the fault detection and identification of sensors in various industrial processes in the past few years. Li applied PCA to the sensor monitor to ensure the accuracy of monitoring information [8]. A comprehensive PCA method is proposed in her research. However, PCA is rarely used in alarm threshold designs. At the same time, the second threshold is added to ensure the accuracy of the alarm under accident modes [9]. Different from the traditional sensor fault diagnosis, this paper mainly uses the characteristics of PCA to design multivariate alarm threshold to reduce the steady-state nuisance alarm.

Table 1. The comparison between actual industry and the EEMUA standard

Type	EEMUA	Oil-Gas	Petro-Chem	Power
Average alarms/day	144	1200	1500	2000
Peak alarms/10 min	10	220	180	350
Average alarms/10 min	1	6	9	8

This method is more suitable for alarm system than traditional and single-variable threshold. Firstly, the multivariate alarm threshold design mainly considers the correlation between the variable and system designs of the multivariate alarm threshold. It can effectively suppress the false alarm caused by fluctuation compared with traditional threshold. Furthermore, PCA method can achieve "de-noising" and "de-redundancy" of original data, and extend the scope of application. Moreover, a second threshold is added relative to the traditional PCA to eliminate the number of false alarms and reduce false alarms. In a word, this method can obtain a better solution at a less cost.

According to the industry standard ANSI/ISA-18.2, “an alarm system is the collection of hardware and software that detects an alarm state, communicates the indication of that state to operators, and records changes in the alarm state [10].” The general idea and method of an alarm system is to compare the real-time measurements of analogy variables with the high or low alarm threshold. Alarm threshold is mainly designed for the pair of indicators of false alarm and missed alarm [11]. As a result, false alarms (missed alarms) are possibly present, showed as the star (circle) points in Fig. 1.

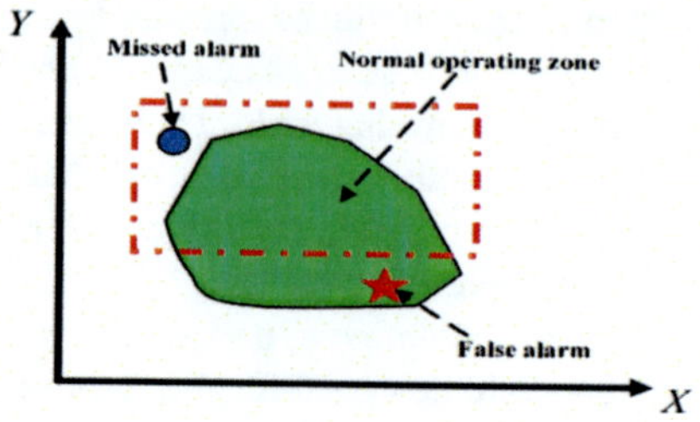

Fig. 1. Schematic diagram of normal operating zone with isolated alarm threshold.

However, false alarm rate and missed alarm rate are a pair of contradictory reference indicators, which greatly increase the complexity of alarm threshold design. Meanwhile, these types of nuisance alarms are typically generated due to random noise and/or disturbances on the process variables configured with alarms. Sequentially, it also brings another troublesome to the design of the alarm threshold.

At present, the single-variable threshold is widely used but not suitable for modern complex industrial conditions. Therefore, this paper proposes the multivariate threshold. Meanwhile, the second threshold is added to improve the PCA model.

2 Multivariate Threshold and Second Threshold Design

In this section, a PCA-based multivariate threshold and second threshold operator are proposed. In this method, PCA modelling procedures and multivariate threshold in the nuclear power plant are exhibited respectively in detail. Meanwhile, a false alarm reducing methodology is also presented for the alarm system using second control limit. The performance of alarm model is greatly improved by the introduction of false alarms reducing method.

2.1 PCA Theoretical Basis

PCA is a method that transforms a set of correlated variables into a small set of new uncorrelated variables and retains most of the information of the original data at the same time. The variables of modern engineering systems are usually multi-dimensional and have a certain correlation, so the high-energy dimensions can be selected from the new coordinate space to replace the original data with high redundant information. The process of data analysis and alarm system using PCA is as follows (Fig. 2).

The sample matrix $X^0_{n \times m}$ should be normalized first to eliminate the influence caused by diverse magnitudes of variables in $X^0_{n \times m}$. The covariance matrix of the normalized data matrix $X_{n \times m}$ is $C_{m \times m}$. The eigenvalue and eigenvector can be solved as:

$$U = [P_1, P_2, \ldots, P_m]$$

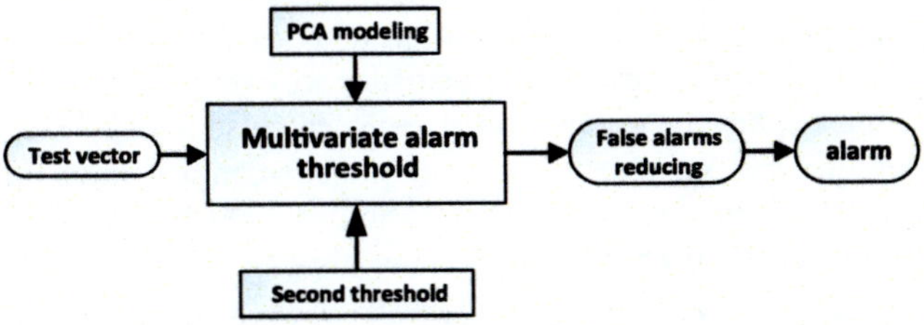

Fig. 2. The integrate multivariate alarm threshold framework in this paper

$$\lambda = [\lambda_1, \lambda_2, \ldots, \lambda_m] \tag{1}$$

There are various criteria to determine the number of PCs in a PCA model. Then we adopt CPV percentage as the selection criterion to determine the number of PC. It is defined as:

$$CPV = \frac{\sum_{i=1}^{i=k}\lambda_i}{\sum_{i=1}^{i=m}\lambda_i} \times 100\% \tag{2}$$

The eigenvalues corresponding to the eigenvectors describe that how much information each PC contains. The cumulative percent variance (CPV) percentage represents that the variability of selected PCs accounts for all the variability in the original data matrix.

As explained above, we can note that PCA divides the original data matrix X into two parts: one is the model estimation matrix P, which contains information of system variation; the other is residual matrix V, which contains noise or model error information.

2.2 Multivariate Threshold Design

After PC is determined by the foregoing steps, the following step for PCA technique is multivariate alarm threshold design. The Q statistic and T^2 statistic.

Q statistic can be defined as:

$$Q = x(I - P_k {P_k}^T)x^T \leq Q_\alpha \tag{3}$$

T^2 statistic can be defined as:

$$T^2 = t_i \Lambda^{-1} t_i^T = xP_k \Lambda^{-1} {P_k}^T x^T \leq T_\alpha^2 \tag{4}$$

Λ in the Eq. (4) is the eigenvalue vector λ_k. Q_α and T_α^2 in Eq. (3) and (4) are the corresponding confidence limits for Q and T^2 statistics respectively.

It can be seen from the above calculation formula that the T^2 statistics are based on matrix P and eigenvalues λ_k. Therefore, T^2 statistics are overly sensitive to the existence of small eigenvalues in the eigenvalues, because the calculation of T^2 statistics uses the reciprocal of the first k eigenvalues. But the Q statistic can just solve this problem. It can be seen from the formula that the Q statistic describes the main variation of the test vector in the residual space. It quantifies the distance of the vector falls from the PC model. Since the PC is not used in the Q statistic calculation, the Q statistic is not sensitive to the noise, the selected principal element eigenvalues of which are too small. The Q statistic analyses the variation of some smaller eigenvalues in the residual space.

Supposed that a test vector x is expressed as $x = [x_1, x_2 \ldots x_m]$, m is the number of variables in$x$. And m is also the corresponding number of parameters in the PCA model. The contribution of variable x_i to the total variation in the residual space is defined as:

$$Q_{x_i} = \frac{\|x_i(I - PP^T)\|}{\|x(I - PP^T)\|} = \frac{e_i^2}{e^2} \times 100\% = \frac{e_i^2}{e_1^2 + e_2^2 + \ldots + e_m^2} \times 100 \quad (5)$$

The sum of contributions of all m variables in the residual space is equal to the Q statistic of x. In accordance with the contribution of each variable, alarm can be located, since a large contribution on x_i usually means a faulty state on variable i.

The contribution of variable x_i to the total variation in PC space can be calculated as the following steps.

(1) Calculate the contribution of x_i to score vector t_j:

$$CR_{j,xi} = \frac{t_j p_{j,i}}{\lambda} xi \ (i = 1, 2, \ldots, m) \quad (6)$$

where $p_{j,i}$ is the *ith* element of eigenvector p_j.

(2) Calculate the contribution of x_i to T^2 statistic:

$$T_{x_i}^2 = \sum\nolimits_{j=1}^{k} CR_{j,x_i} = \sum\nolimits_{j=1}^{k} \left(\frac{t_j p_{j,i}}{\lambda_j} x_i \right) (i = 1, 2, \ldots, m) \quad (7)$$

2.3 Second Threshold Design

Supposed that false alarm probability for T^2 and Q statistics under steady state is α.In accordance with the statistical experience in process industries, the commonly used value for α is between 0 and 0.05.

We further consider a basic observation unit with n sample points. If T^2 and Q statistics of the n sample points are independent with each other, then T^2 and Q statistics approximately obey Bernoulli distribution [12]. If so, the probability distribution of T^2 and Q statistics in each basic unit can be approximately expressed as:

$$b(s; n, \alpha) = C(n|s)\alpha^s(1 - \alpha)^{n-s} \quad (8)$$

Where, s represents the number of false alarms in an observation unit. According to Eq. (8), we can get the probability distribution that the number of false alarms of the test sample statistics in the window does not exceed s as:

$$F(s; n, \alpha) = \sum\nolimits_{i=0}^{s} b(s; n, \alpha) = \sum\nolimits_{i=0}^{s} C_n^i \alpha^i (1 - \alpha)^{n-i} \quad (9)$$

Another probability value β is defined to make Eq. (9) satisfy the following inequality:

$$F(s; n, \alpha) = \sum_{i=0}^{s} C_n^i \alpha^i (1-\alpha)^{n-i} \le \beta \quad (10)$$

If the foregoing confidence limits (Q_α and T_α^2) are called the first multivariate threshold, this new confidence limit s can be called the second threshold for T^2 and Q statistics in this paper.

The specific meaning for the second threshold s is explained as follows. If T_α^2 or Q_α statistic is beyond the first confidence limit (T_α^2 or Q_α) at current testing time j. then the previous n samples will be further analysed (including testing sample x_j). We can see that it is exactly a basic observation unit. During the n testing samples, if the number of false alarms for T^2 or Q statistics is more than the second threshold s, then the current testing sample x_j is regarded as a true faulty state. Otherwise it will be treated as a false alarm and will be ignored.

Usually β is set between 0.95 and 0.99 according to the statistical experience in process industries. The above is the theoretical analysis process of false alarm elimination based on statistical analysis.

2.4 Alarm Design Framework

Sample data of the system are collected for PCA model training. According to the specific process of PCA modelling, the control limits of T^2 and Q statistics were obtained after the sensor sample data which were trained. According to the projection matrix and residual space projection matrix of PCA model, the projection analysis was performed on the test vector, and the corresponding T^2 and Q statistics of the test vector were obtained.

In this paper, both T^2 and Q statistics are used to monitor the status of the sensor, and the changes in the principal component space and residual space are monitored simultaneously. When statistics exceed the limit, the system enters the false alarm elimination stage. In the phase of false alarm elimination, this paper adopts the second threshold based on statistics to eliminate the number of false alarms and the flowchart of method is shown in Fig. 3.

3 Simulation Verification of the Method

This chapter is divided into two parts. Section 3.1 tests the effect of the nuisance alarm suppression, using actual data and simulation data respectively. Section 3.2 uses simulation data to simulate the accident and tests the accuracy of the model alarm.

3.1 Nuisance Alarm Suppression Test

3.1.1 Actual Data Test

The actual data comes from the normalized operational data of the nuclear power plant. The results of test are shown in Fig. 4 and Fig. 5 below. As can be seen from Fig. 4, there is nearly no false alarm in the T^2 statistics. Due to the noise, The Q statistic fluctuates, but it is in a normal state after being controlled by the second threshold. Test results show that the multivariate threshold and the second threshold can effectively reduce the nuisance alarm caused by noise and disturbance, and the number of nuisance alarms is decreased from 52 to 0. From the perspective of the CPV of T^2 and Q, it can be seen that the CPV of the various variables in Fig. 5 is basically the same. This can also indicate that the test data is stable without exceptions.

3.1.2 Simulation Data Test

The was performed using the simulation data, and the results are shown in Figs. 6 and 7. As can be seen from Fig. 6, the nuisance alarm is well suppressed under steady state. There is basically no nuisance alarm throughout the process, probably because the simulation data is smoother and less noisy than the actual data.

It can be more clearly seen by comparing Fig. 5 with Fig. 7 that the contribution rates of different variables are inconsistent. Because the simulation data is relatively smooth, resulting in poor anti-interference ability after training, the contribution rate of the test results is evenly not distributed in Fig. 7. The contribution rate represents the degree of interference of external influences on each variable during the test. The greater the contribution rate are, the greater the degree of interference with the corresponding variables become.

It can be seen from the experimental results that the multivariate threshold based on PCA has an excellent performance. Firstly, the multivariate threshold is designed relative to the single variable threshold, and the multivariate is used as a whole for threshold judgment. Secondly, compared with the traditional single-variable threshold design, it can effectively reduce the number of the nuisance alarm caused by environmental factors such as noise, which greatly reduces the operator's operating pressure. Finally, it is possible to judge variables which are weak to interference by the contribution rate of each variable, and help the operator to analyse the operating state of the nuclear power plant more reasonably.

3.2 Alarm Capability Test for Simulated Fault

This section uses simulation data to test the accuracy and timeliness of multivariable alarm thresholds. Through the characteristics of "denoising", "de-redundancy" and "dimension reduction" of the PCA method, the number of false alarms can be reduced to some extent. However, this method does not fundamentally solve the alarm overload problem in the event of an accident. This test only considers the accuracy of alarm and timeliness of multivariate alarm thresholds in the accident mode.

In order to verify the model's fault detection and alarm capabilities, a LOCA accident was added at 80s. As can be seen from Fig. 8, both T^2 statistics and Q statistics can detect accident immediately after the accident occurs. The reason why Q statistic triggers an alarm faster than T^2 statistics in Fig. 9 is that the Q statistic is more sensitive than the T^2 statistic. When T^2 statistic or Q statistic exceeds its control limit, the second threshold is triggered to determine whether to alert.

The contribution rate of variables can be used to determine which variables dominate the fault. It can be seen from Fig. 9 that the 5th and 55th (steam generator water level and inside the containment of radioactive) variables have the highest contribution rate. The reason for this phenomenon is that the contribution rate is determined by the degree of change of the variable. And these two variables are basically 0 under normal conditions, so their changes are more drastic, so their contribution rate is also higher than other variables.

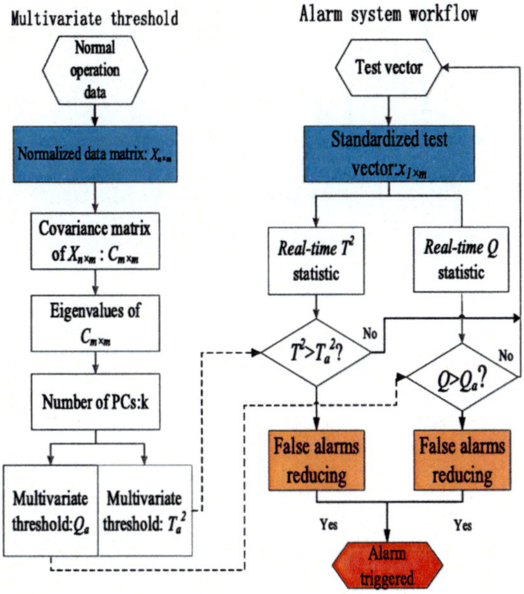

Fig. 3. The threshold flow chart of alarm system with PCA

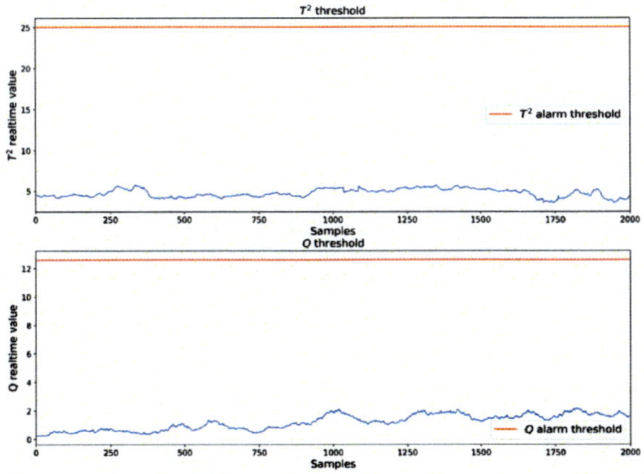

Fig. 4. The real-time data of T^2 and Q with actual data

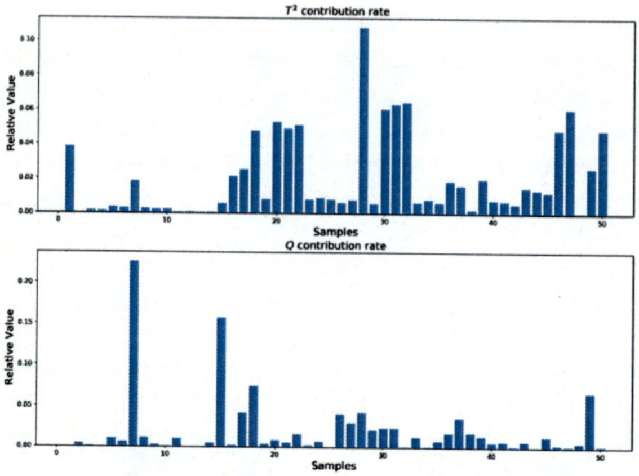

Fig. 5. The contribution rate of T^2 and Q with actual data

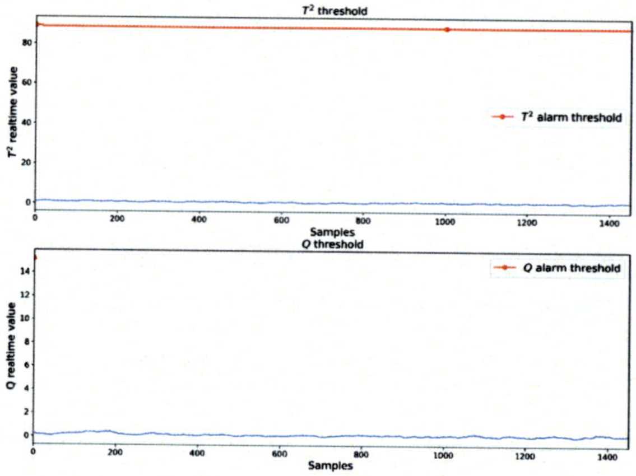

Fig. 6. The real-time data of T^2 and Q with simulation data

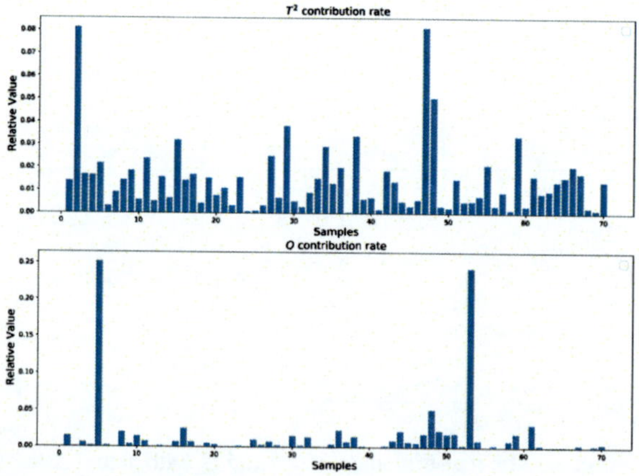

Fig. 7. The contribution rate of T^2 and Q with simulation data

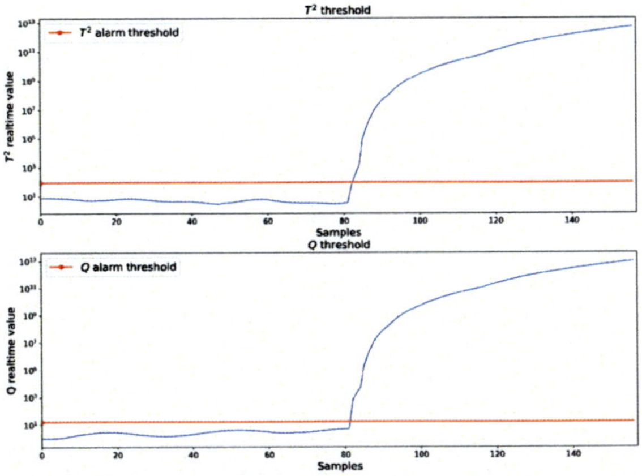

Fig. 8. The real-time data of T^2 and Q with fault data (logarithmic coordinates)

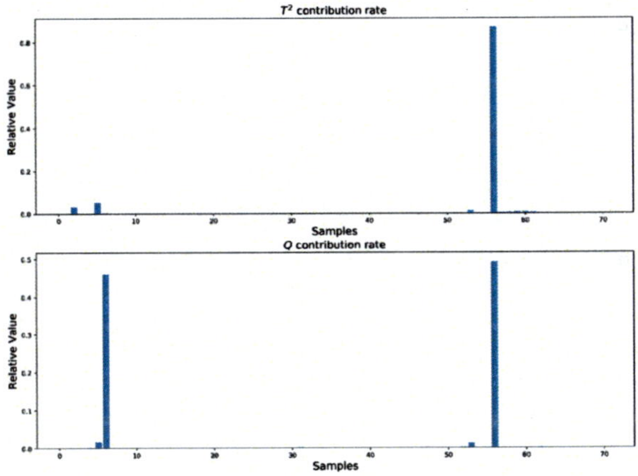

Fig. 9. The contribution rate of T^2 and Q in 85 s

4 Conclusion

This paper designs a multivariate threshold for PCA and adds a second threshold. The model was tested using actual data and simulation data. The test includes nuisance alarm suppression and alarm performance analysis. From the test results, compared with the traditional threshold, it can be seen that the threshold has a good effect on the nuisance alarm suppression under steady state, and it can also perform an accurate and effective alarm in the event of an accident.

However, due to the shortcomings of the PCA model, the model still needs to be improved and optimized. For example, the alarm overload problem in the event of an accident cannot be well solved. At the same time, there still have been a small number of nuisance alarms existing in the test results. Therefore, in the subsequent work, it is necessary to analyse and classify the alarm information under the accident to reach the anticipated result of suppressing the alarm overload. The alarm data is classified to optimize the performance in the accident mode. On this topic, further research is needed.

References

1. Stauffer, T., Clarke, P.: Using alarms as a layer of protection. Process Saf. Prog. **35**(1), 76–83 (2016)
2. Wang, J., Yang, F., Chen, T., Shah, S.L.: An overview of industrial alarm systems: main causes for alarm overloading, research status, and open problems. IEEE Trans. Autom. Sci. Eng. **13**(2), 1–17 (2015)
3. Noyes, J.: Alarm systems: a guide to design, management and procurement. Eng. Manag. **9**(5), 226 (1999)
4. Zhu, J., Shu, Y., Zhao, J., Yang, F.: A dynamic alarm management strategy for chemical process transitions. J. Loss Prev. Process Ind. **30**, 207–218 (2014)
5. Mezache, A., Soltani, F.: A novel threshold optimization of ML-CFAR detector in weibull clutter using fuzzy-neural networks. Signal Process. **87**(9), 2100–2110 (2007)
6. Bristol, E.H.: Improved process control alarm operation. ISA Trans. **40**(2), 191–205 (2001)
7. Hao, Z., Hongguang, L.: Optimization of process alarm thresholds: a multidimensional kernel density estimation approach. Process Saf. Prog. **33**(3), 292–298 (2014)
8. Wei, L., Minjun, P., Yongkuo, L., Shouyu, C., Nan, J., Hang, W.: Condition monitoring of sensors in a NPP using optimized PCA. Science and Technology of Nuclear Installations, 2018(PT.1), pp. 1–16 (2018)
9. Wei, L., Minjun, P., Qingzhong, W.: False alarm reducing in PCA method for sensor fault detection in a nuclear power plant. Ann. Nucl. Energy **118**, 131–139 (2018)
10. Takai, T., Kutsuma, Y., Ishihara, H.: Management of alarm system for process industries. In: 2012 Proceedings of SICE Annual Conference (SICE), pp. 688–692 (2012)
11. Zhu, Q.X., Gao, H.H., Xu, Y.: A survey on alarm management for industrial processes. Zidonghua Xuebao/Acta Automatica Sinica **43**(6), 955–968 (2017)
12. Chen, T., Martin, E., Montague, G.: Robust probabilistic PCA with missing data and contribution analysis for outlier detection. Comput. Stat. Data Anal. **53**(10), 3706–3716 (2009)

Evaluation of Contact-Type Failure Using Frequency Fluctuation Caused by Nonlinear Wave Modulation Utilizing Self-excited Ultrasonic Vibration

Takashi Tanaka(✉), Yasunori Oura, Syuya Maeda, and Zhiqiang Wu

University of Shiga Prefecture, 2500 Hassaka, Hikone 522-8533, Shiga, Japan
{tanaka.ta,oura,wu.z}@mech.usp.ac.jp, of22smaeda@ec.usp.ac.jp

Abstract. This study concerns the evaluation method of contact-type failure, which is difficult to detect by linear ultrasonic because of transmission of ultrasonic at the contact surfaces utilizing frequency fluctuation caused by nonlinear wave modulation. When the contact-failure is vibrated by low-frequency vibration, the amplitude and phase of ultrasonic in the vicinity of the failure area is fluctuated in synchronization with low-frequency vibration (nonlinear wave modulation). This amplitude of amplitude fluctuation and phase fluctuation is the evaluation index of contact-type failure. However, the change in this index depends on the viscous damping of the structure. In the nonlinear wave modulation utilizing self-excited ultrasonic vibration, the frequency of ultrasonic vibration is fluctuated by low-frequency vibration. The frequency fluctuation may be the evaluation index of contact-type failure independent of the viscous damping of the structure. In this paper, the experimental verification using beam structure is done. Firstly, the self-excitation technique of ultrasonic natural vibration using feedback control is introduced. Secondarily, it is shown that the amplitude of frequency fluctuation is the evaluation index of contact-type failure by experiment. The uniform beam structure with simulated failure, which is simulated the contact-type failure is used. The vibration characteristics of this structure and the self-excited ultrasonic vibration excited by feedback control can measure easily. Finally, the viscous damping dependency of the frequency fluctuation is investigated. As the result, it is shown that the amplitude of frequency fluctuation is the evaluation index independent of viscous damping.

Keywords: Detection · Contact-type failure · Ultrasonic · Self-excitation · Nonlinear wave modulation

1 Introduction

The deterioration of infrastructure is a serious problem in many countries. To keep the performance of infrastructure and guarantee the user's safety, the maintenance is very important. The sensing techniques of physical quantity and the detection and evaluation

R. Karim et al. (Eds.): IAI 2021, LNME, pp. 163–175, 2022.
https://doi.org/10.1007/978-3-030-93639-6_14

technique of failure condition based on physical quantity are necessary to schedule maintenance and keep the health of infrastructure [1, 2].

The monitoring systems using piezo-electric material are developed by many researchers [3, 4]. Piezo-electric materials can use as sensor and actuator. Therefore, the smart structure using piezo-electric material can realize vibration control of structure. The authors have developed the active monitoring system using vibration control of piezo-electric material.

It is difficult to detect the contact-type failure e.g. fatigue crack, delamination of composite material and debonding. To detect contact-type failure, the detection method based on nonlinear wave modulation was proposed [5]. Nonlinear wave modulation is the phenomenon caused by fluctuation of the contact-condition of failure. The authors consider that the essence of nonlinear wave modulation is the fluctuation of natural frequency caused by fluctuation of contact stiffness by low-frequency vibration. The natural frequency is independent parameter of viscous damping. Thus, the fluctuation of natural frequency caused by nonlinear wave modulation may be independent of viscous damping.

Thus, the authors study the detection method of fluctuation of natural frequency by ultrasonic vibration generated by the local feedback control (self-excited ultrasonic vibration) by time response analysis using SDOF model nonlinear wave modulation [6]. Firstly, the SDOF model of nonlinear wave modulation is introduced. From the analysis result, the problem that the conventional failure index using the amplitude of amplitude fluctuation and phase fluctuation depends on viscous damping. Secondarily, the concept of detection of contact-type failure based on nonlinear wave modulation utilizing self-excited ultrasonic vibration was proposed. The local feedback control is the excitation control using sensor and actuator which locations are the same. Self-excited vibration at the natural frequency of the structure is generated automatically by the local feedback control. Therefore, the excitation frequency changes, chasing with the change of natural frequencies. As the result, the frequency modulation of excitation signal in synchronization with low-frequency vibration caused by nonlinear wave modulation occurs. Therefore, it is confirmed that the amplitude of frequency demodulation is the novel evaluation index of the level of contact-type failure.

In this paper, the independence of viscous damping is clarified. Firstly, the analog circuit of local feedback control is introduced. The self-excited ultrasonic vibration using introduced circuit is generated at the natural frequency. Secondarily, the new failure index using frequency fluctuation is proposed for the consideration about the relationship between fluctuations of the contact condition of failure and frequency fluctuation. Finally, the damping independence of evaluation index is investigated. It is clarified that the evaluation index using frequency demodulation is independent of viscous damping.

2 Nonlinear Wave Modulation Utilizing Self-excited Ultrasonic Vibration

In this section, the self-excited ultrasonic vibration and overview of nonlinear wave modulation are introduced. Firstly, the exciting technique of ultrasonic natural vibration by local feedback control is explained. Further, the analog circuit for self-excitation is shown. Next, the nonlinear wave modulation is explained by means of physical model. It is shown that the essence of nonlinear wave modulation is fluctuation of natural frequencies. Finally, it is shown that the frequency fluctuation of excitation signal is occurring by the nonlinear wave modulation utilizing self-excited ultrasonic vibration.

2.1 Self-excited Ultrasonic Vibration Excited by Local Feedback Control

In this subsection, the generation mechanism of the self-excited ultrasonic vibration and the analog circuit to realize are introduced. The authors proposed the exciting technique of natural vibration by local feedback control [7].

Self-excited vibration is the vibration generated by non-oscillatory excitation. The generation condition of self-excited vibration is that the gain value is higher than 0dB at the frequency which phase is $-180°$. The self-excitation at natural frequency can realize to make this condition at natural frequencies of feedback control.

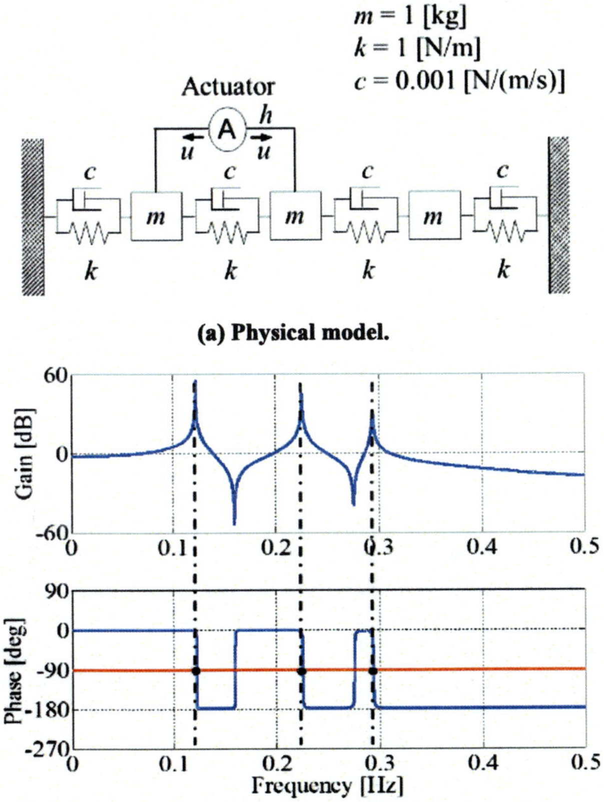

(b) Transfer function from force *u* to relative displacement *h*.

Fig. 1. Linear and 3 degree of freedom system.

Firstly, the vibration characteristics of common structure shown in Fig. 1 is expressed. Figure 1(a) is the physical model of linear 3 degree of freedom model system which boundary condition of both edges is fixed condition. In this model, the vibration characteristics measuring by the sensor and the actuator in the same arrangement are shown in Fig. 1(b). The phase is −90° at all natural frequencies. Therefore, the self-excited vibration occurs at natural frequencies by the feedback control using the sensor and the actuator in the same arrangement (local feedback control) which is designed that the phase of the controller is −90° at all frequencies. When the natural frequency varies, the frequency of the self-excited vibration occurring by local feedback control changes following variation of natural frequency automatically (automatic following of natural frequency).

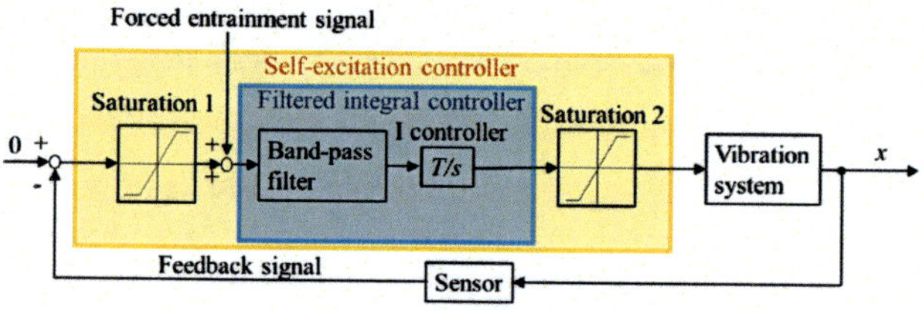

Fig. 2. Block diagram local feedback control using integral negative feedback control.

Figure 2 shows the block diagram of local feedback control in this paper. The integral negative feedback control with band-pass filter is used as the control which phase is −90° in wideband frequency range. The gain of integral negative feedback control is a high gain at low frequency range. Therefore, the 2-order Butterworth high-pass filter is used to suppress high gain in the low frequency range. Lastly, the 1-order Butterworth low-pass filter is used to be reduced noise at high frequency.

2.2 Nonlinear Wave Modulation Utilizing Self-excited Ultrasonic Vibration

In this subsection, the overview of nonlinear wave modulation utilizing self-excited ultrasonic vibration is introduced. Firstly, the mechanism of nonlinear wave modulation is explained. Next, the frequency modulation, which occurred when the self-excitation is used for ultrasonic excitation, is explained.

The nonlinear wave modulation [5, 6] is the phenomenon caused by contact acoustic nonlinearity. Figure 3 shows the conceptual illustration of nonlinear wave modulation. When the structure with contact-type failure is vibrating at low-frequency, the contact interface of failure is tapping and clapping. When the ultrasonic vibration of structure excites, the amplitude and phase of ultrasonic vibration modulates in synchronization with low-frequency vibration. When the natural frequency of ultrasonic vibration is enough larger frequency of stiffness fluctuation, expressed by the linear and time-varying transfer function model shown in Fig. 4(a). The blue line is the transfer function in the condition of non-fluctuation of local stiffness. The green line is the transfer function when the load force applied. The red line is the transfer function when the compressing force applied.

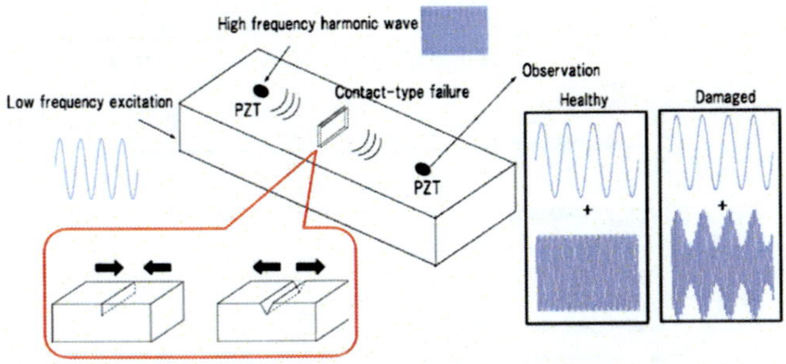

Fig. 3. Conceptual illustration of nonlinear wave modulation [6].

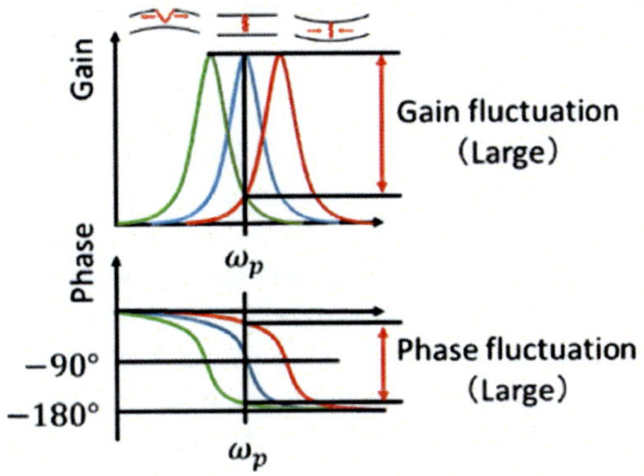

(a) In the case of small damping.

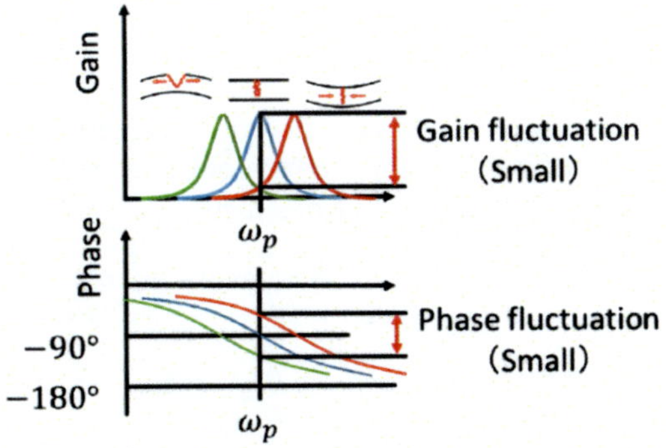

(b) In the case of large damping.

Fig. 4. Time-varying model of nonlinear wave modulation [6].

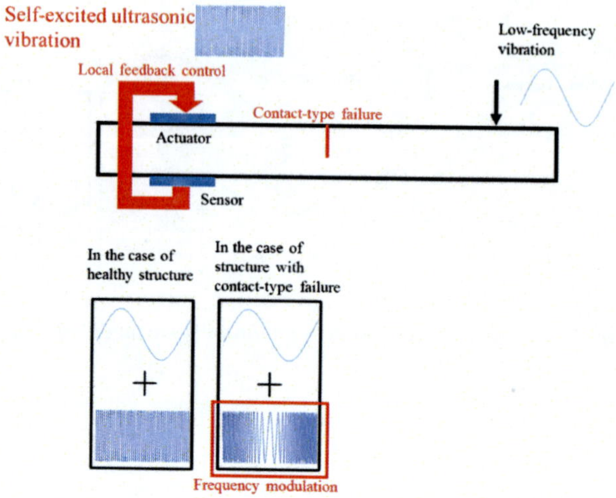

Fig. 5. Illustration of nonlinear wave modulation utilizing self-excited ultrasonic vibration.

The vibration characteristic in the case of large viscous damping is shown in Fig. 4(b). The peak of gain decreases and the inclining of phase slope is gentle. As the result, the fluctuation of gain and phase depends on viscous damping.

The authors focused that the nature of the nonlinear wave modulation is fluctuation of natural frequency [6] shown in Fig. 5. In the case of nonlinear wave modulation utilizing ultrasonic vibration driven by self-excitation (self-excited ultrasonic vibration), the frequency modulation is occurred because of automatic following of natural frequencies. Natural frequency is only determined by mass and stiffness. Thus, the failure development index calculated from frequency modulation is independent of the viscous damping.

3 Experimental Setup

In this section, the experimental setup and vibration characteristic of the structure is introduced. Firstly, the experimental device and analog circuit of local feedback control is introduced. Next, the vibration characteristic of structure and open loop transfer function is shown.

3.1 Experimental Device

Firstly, the experimental setup is shown in Fig. 6. The uniform beam fixed at both ends is used. The material of the beam is hot rolled general structural steel (SS400 in Japanese Industrial Standard (JIS)). The length, width and depth of beam is 3000 mm, 50 mm and 5 mm. The piezo-electric patches (Fuji Ceramics Corp., C-6, 30 mm × 20 mm) for control are provided at 600 mm. Exciter (Akashi Corp., 840-342) for excitation at low-frequency vibration is provided at 150 mm. The bipolar power amplifier (MESS-TEK Corp., M-2682) is used to supply the control signal to the piezo-electric patch. The control signal and open circuit voltage of piezo-electric for measurement are recorded by oscilloscope. The sampling frequency and measurement time are 1 MHz and 10 s. Thus, the polyurethane rubber sheet, of which length, width and depth are 100 mm × 50 mm × 5 mm, as damping material is adhered to the excitation point of both faces of the beam.

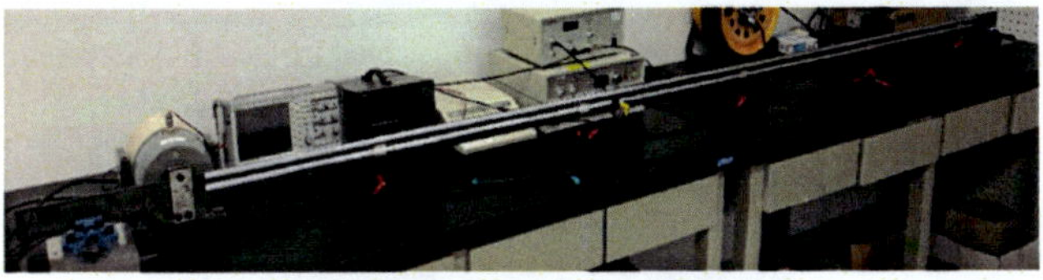

(a) Experimental apparatus.

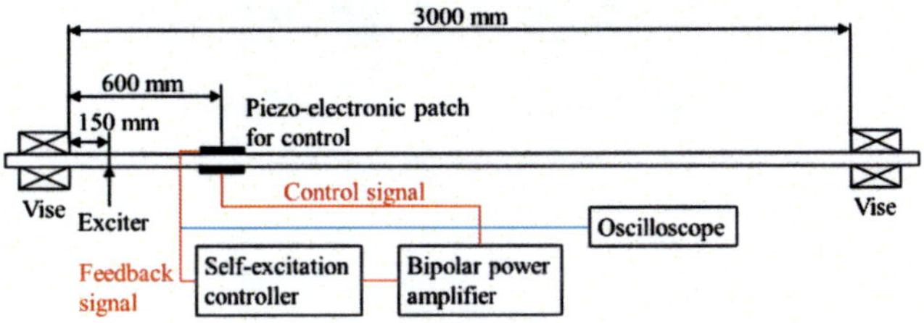

(b) Experimental device.

Fig. 6. Experimental setup.

Figure 7 shows the vibration characteristic of the beam measured by piezo-electronic patches for control. The blue line is the vibration characteristic of the beam without rubber sheet. Red line is the vibration characteristic of beam with rubber sheets. The natural frequencies are changed caused by addition of the damping material. The peaks of gain decrease and the reclining of phase curve is gentle by damping effect.

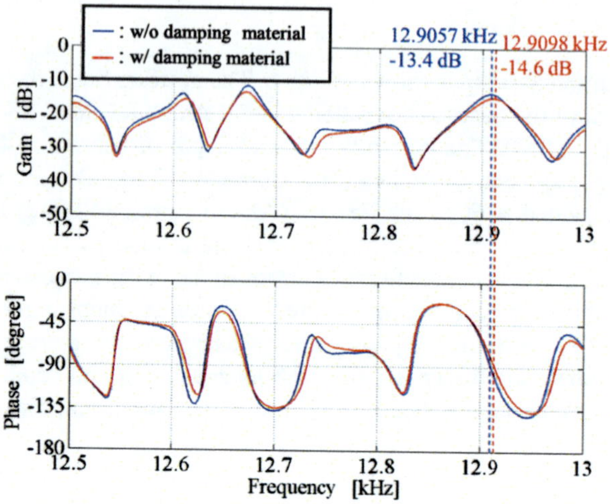

Fig. 7. Vibration characteristic of the beam measured by piezo-electronic patches.

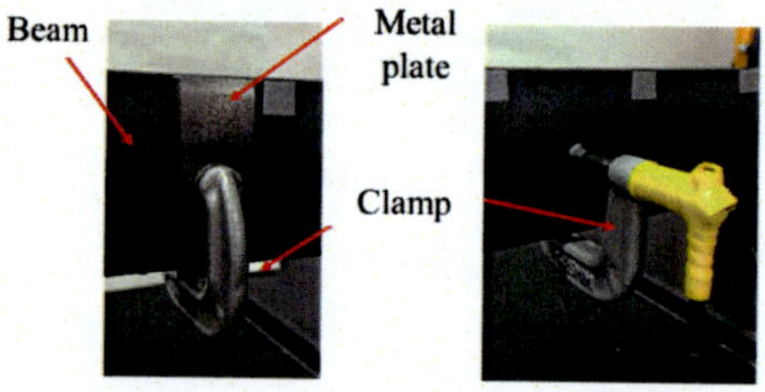

Fig. 8. Apparatus of simulated failure.

The real contact-type failure is difficult to control the failure level. In this experiment, the simulated failure of contact-type failure shown in Fig. 8 is used. This simulated failure consists of a metal plate and clamp shown in Fig. 8. The contact area between the metal plate and beam is fluctuated in synchronization with the amplitude of low-frequency vibration. This simulated failure is set at 2200 mm. When the voltage of the exciter is increase, the fluctuation range of the contact area of the simulated failure increase. In this paper, the voltage change of the exciter is regarded as the development of contact-type failure.

3.2 Analog Circuit of Local Feedback Control

In this subsection, the design of analog circuit is expressed. The circuits to realize local feedback control is shown in Fig. 9. Figure 9(a) is saturation circuit which is saturation 1 of Fig. 2. The saturation voltage is ±3V. Adder circuit is used to add the entrainment signal. Passive high-pass filter and high-pass filter is used as band-pass filter. Cut off frequency of high-pass filter and low- pass filter is 106 Hz and 1 MHz. The amplifier and saturation circuits are used as saturation 2 of Fig. 2. The amplifier element is necessary to coordinate the gain of open loop transfer function. The voltage followers are used connection between each circuits.

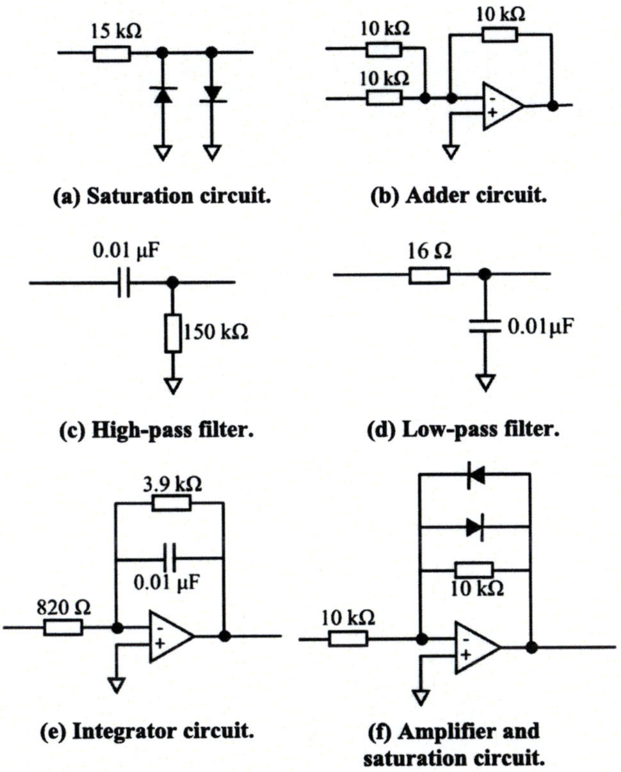

Fig. 9. Analog circuit of local feedback control.

4 Experimental Results

4.1 Experiment of Self-excitation and Detection of Contact-Type Failure

Firstly, the self-excited vibration generated by the local feedback control is expressed. Figure 10 shows the feedback signal excited by local feedback control. Figure 10(a) shows the result of the experiment using the beam without damping material. The period of signal is 7.7479 ms. This frequency is 12.9057 kHz, which means the one of peak frequencies of blue line of Fig. 7. Figure 10(b) shows the result of the experiment with damping material. The period of signal is 7.7460 ms. This frequency is 12.9098 kHz, which means the one of peak frequencies of red line of Fig. 7. From these results, the self-excited ultrasonic vibration is generated at the natural frequency of the beam automatically.

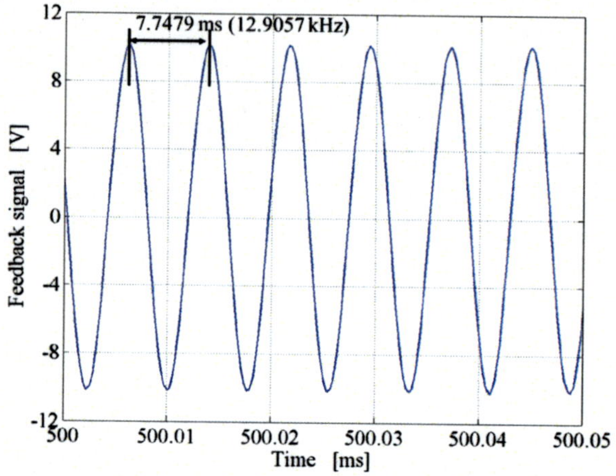

(a) In the case of the beam without the damping material.

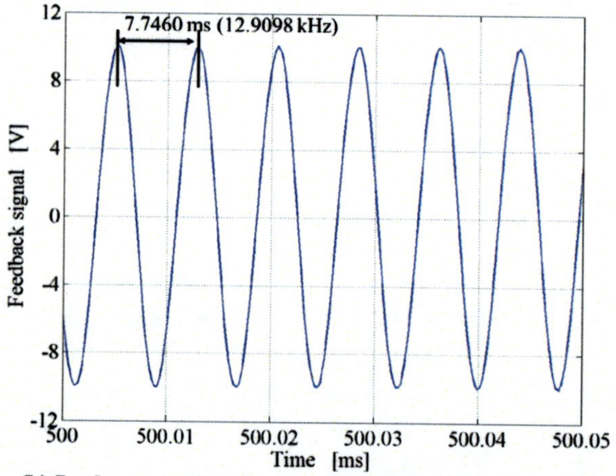

(b) In the case of the beam with the damping material.

Fig. 10. Feedback signal when the beam is excited by local feedback control.

Secondly, it is shown that the frequency modulation caused by contact-type failure is generated. In this experiment, the damping material is not used. The excitation frequency of the exciter is 8.6 Hz, which is the 2nd natural frequency of the beam. The short time

Fourier transformation (STFT) is used to demodulate frequency modulation. The time resolution is 0.002 s. The width of the window is 0.01 s. Thus, the zero-padding is used to improve apparent frequency resolution. The data length is extended to 200 s by zero-padding. As the result of zero padding, the apparent frequency resolution is 0.005 Hz.

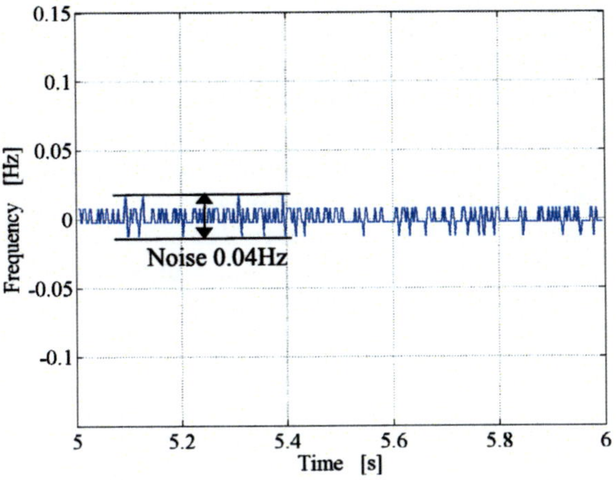

(a) In the case of the input voltage of the exciter is 0 $V_{p\text{-}p}$.

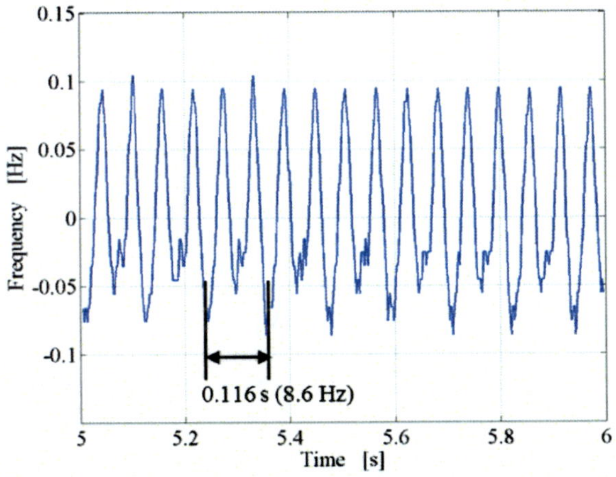

(b) In the case of the input voltage of the exciter is 1.2 $V_{p\text{-}p}$.

Fig. 11. The frequency fluctuation generated to the feedback control signal.

Figure 11 shows the frequency demodulation signal when the input voltage of the exciter is 0 $V_{p\text{-}p}$ and 1.2V $_{p\text{-}p}$. This figure is the result of experiment using the beam without damping material. The frequency modulation is not generated when the voltage of the exciter is 0 $V_{p\text{-}p}$ (Fig. 11(a)). The noise caused by the measurement noise is

0.04 Hz. On the other hand, the frequency fluctuation is generated when the voltage of the exciter is 1.2 $V_{p\text{-}p}$ (Fig. 11(b)). The fluctuation frequencies are 8.6 Hz and 17.2 Hz which mean the same and two times the low-frequency. From these results, the frequency modulation caused by contact-type failure is generated.

4.2 Independence of Viscous Damping

In this subsection, the independence of viscous damping is investigated. Figure 12 shows the relationship between amplitude of frequency fluctuation and the input voltage of the exciter. The blue points are the result in the case of beam without damping material. The red points are the result in the case of beam with damping material. From this result, the amplitude of frequency fluctuation increases in synchronization with the development of contact-type failure. Thus, the increase lines may be much the same in the regard to the volume of noise of frequency demodulation. The differences are in the range of noise, which measured from the result when the voltage of exciter is 0 $V_{p\text{-}p}$. The value differences between the frequency fluctuation in the case of the beam without and with damping material are within this error range. As this result, the frequency fluctuation is approximately independent of viscous damping.

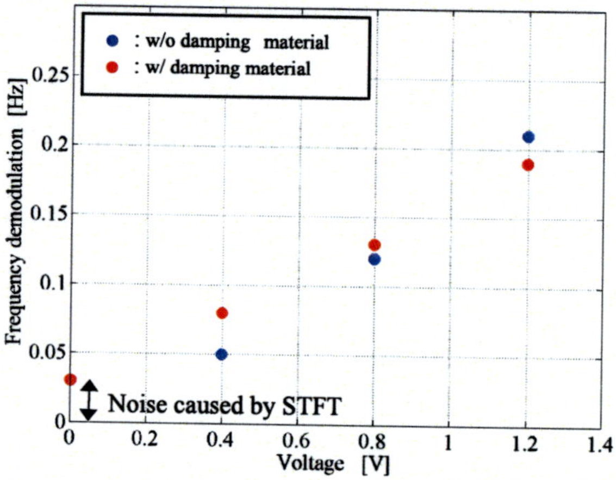

Fig. 12. Independency of viscous damping

5 Conclusions

In this paper, the detection method of contact-type failure based on nonlinear wave modulation utilizing self-excited ultrasonic vibration is investigated by the experiment. The summary of results is as follows.

(1). The analog circuit of local feedback control is made. The excitation experiment using procedure circuit is done. As the result, the self-excited ultrasonic vibration is generated at the natural frequency of the beam automatically.

(2). It is confirmed that the detection of contact-type failure based on nonlinear wave modulation utilizing self-excited vibration is possible to evaluate the development of failure. From the experimental result, the frequency modulation of the feedback signal caused by nonlinear wave modulation is observed. Therefore, the amplitude of the frequency fluctuation increases in the synchronization with the input voltage of the exciter, which simulates the development of contact-type failure.

(3). The independence of viscous damping of amplitude of frequency fluctuation is investigated. From the experiment result using the beam without and with damping material, the amplitude of frequency fluctuation is approximately independent of viscous damping.

These results suggested that the amplitude of frequency fluctuation caused by nonlinear wave modulation utilizing self-excited ultrasonic vibration is the novel index to evaluate the development of contact-type failure that is not affected by the change of the viscous damping.

Further experiments in the case of high amplitude of low-frequency vibration are needed in order to clarify the independence of viscous damping.

Acknowledgments. This research was supported by JSPS KAKENHI Grant Number 18K13716.

References

1. Maaskant, R., Alavie, T., Measures, R.M., Tadros, G., Rizkalla, S.H., Guha-Thakurta, A.: Fiber-optic bragg grating sensors for bridge monitoring. J. Qual. Maintenance Eng. **15**(2), 127–150 (1997)
2. Iba, D., et al.: Printed gear sensor for health monitoring (Development of three-axis laser printer for conductive ink and evaluation of laser-sintered electric circuits). J. Adv. Mech. Design Syst. Manuf. **11**(6), JAMDSM0090, pp. 1–10 (2017)
3. Mei, H., Haider, M.F., Joseph, R., Migot, A., Giurgiutiu, V.: Recent advances in piezoelectric wafer active sensors for structural health monitoring applications. Sensors **19**(2), 383 (2019)
4. Huan, Q., Chen, M., Su, Z., Li, F.: A high-resolution structural health monitoring system based on SH wave piezoelectric transducers phased array. Ultrasonics **97**, 29–37 (2019)
5. Masuda, A., Aoki, J., Shinagawa, T., Iba, D., Sone, A.: Nonlinear piezoelectric impedance modulation induced by a contact-type failure and its application in crack monitoring. Smart Mater. Struct. **20**(2), 1–11 (2011)
6. Tanaka, T., Oura, Y., Maeda, S.: Detection method of contact-type failure based on nonlinear wave modulation utilizing ultrasonic vibration driven by self-excitation. In: Ball, A., Gelman, L., Rao, B.K.N. (eds.) Advances in Asset Management and Condition Monitoring. SIST, vol. 166, pp. 79–89. Springer, Cham (2020). https://doi.org/10.1007/978-3-030-57745-2_8
7. Tanaka, T., Nakamura, N., Oura, Y., Kurita, Y.: Measurement of natural vibration of acoustic space by umdamped vibration using decentraried control. In: Proceedings of ICSV 2018 (2018)

Check for updates

Bearing Lubricant Corrosion Identification Through Transfer Learning

Richard Bellizzi[1(✉)], Jason Galary[1], and Alfa Heryudono[2]

[1] Nye Lubricants, Inc., 12 Howland Road, Fairhaven, MA 02719, USA
{richard.bellizzi,jason.galary}@fuchs.com
[2] Department of Mathematics, UMass Dartmouth, Dartmouth, MA 02747, USA
aheryudono@umassd.edu

Abstract. Finding a fast, automated, and more accurate method of inspection and qualification of corrosion in bearing races is one challenge in the lubricant industry. The goal is to inspect a section or multiple sections of a bearing and determine whether corrosion is present or not. The lubricant industry uses the EMCOR method (ASTM D-6138) and the Standard Corrosion methods (ASTM D-1743 and D-5969) to test how lubricants interact within bearings under corrosive conditions. Several factors that create difficulties when integrating such methods include limited available data sets, costly sample generation, uncertainties in visual observations, and bias in finite ratings. Moreover, for smaller companies, limited resources are considered additional development constraints.

Recent advancements in machine learning algorithms allow smaller companies to incorporate computational methodologies as supporting roles, such as developing solutions for inspection and quantifying corrosion on bearings. This paper designs a solution model using the limited amount of 'real' data from in-house testing coupled with Convolutional Neural Networks (CNN) and Transfer Learning (TL). We show that conflicting results caused by human error factors become mitigated through a more reliable and computational-based method. We also demonstrate a repeatable and accurate method for visualizing and classifying corrosion on bearings utilizing CNN and TL to bridge the gap in technology in the lubricant industry. The method can aid smaller companies in a digital transformation transition and provide more insight into products and improve development capabilities.

Keywords: Convolutional Neural Networks (CNN) · Transfer Learning (TL) · Feature extraction · Lubricants · Bearing corrosion · Machine vision

1 Introduction

Product testing is a common occurrence in most industries. The lubricant industry is no different though it may be lacking some analysis methods that are currently available. Historically, various corrosion tests generate bearing samples, and a technician inspects the resulting samples for corrosion. This paper shows that a simple classifier system recognizes and categorizes corrosion accurately and consistently on bearings. With a

R. Karim et al. (Eds.): IAI 2021, LNME, pp. 176–188, 2022.
https://doi.org/10.1007/978-3-030-93639-6_15

system like this in place, human error becomes reduced, allowing for reliable interpretation of test results. Convolutional Neural Networks (CNN), among other machine learning methods, require a substantial amount of resources to generalize well. Transfer learning bridges the gap between the resources generally required versus those typically available. Today, public data sets are becoming more common. Using these data sets as a resource for transfer learning is feasible if a developer correlates a data set to their potential test outputs. Leveraging the historical data saved for future analysis provides the developer with the required secondary set. Internal data, combined with a public data set consisting of similar features, set the base architecture for a functioning classifier.

Combining these known methods and applying them to industry-specific knowledge such as testing like this creates added value for these test outputs. The outcome is a working model for corrosion identification and a layout for other researchers to develop bearing corrosion models for similar applications. This example presents that by collaborating with public data sets and a component of internal research, laboratories should start to adopt these integrated methods to advance research, analysis, and development. Data pervades every industry, and companies can leverage this by applying domain-specific knowledge to find a data set with shared characteristics and then implement the appropriate methods. The applied methodologies can vary depending on the data, but the CNN and TL methodology was appropriate for this work. Adopting external data for internal uses allows for existing platforms to grow and improve research in various industries. These methods laid out and applied to the lubricant industry act as examples of situations where researchers can benefit from similar implementations.

1.1 Bearing Corrosion

Corrosion is a significant consideration in lubricant development for bearings, and there are various testing methodologies to simulate corrosion within the industry. Various chemicals, salts, or other environmental interactions, cause the inner surface of the bearing to degrade and weaken over time as exposure to these conditions persists. Lubricants help prevent or slow down this process by protecting the surface of the bearing while also reducing friction and wear. With bearings used in many applications, an improved method for inspecting the lubricants' capability to protect against corrosion helps advance the development process, further improving products. A deeper understanding of this interaction provides research scientists with an accurate estimate of the chemical reactions occurring in these harsher environments. In turn, this type of development leads to higher performing products that customers can then leverage within new product advancements.

Understanding corrosion is the end goal for the model, and to accomplish that, it needs to be able to recognize the various visual conditions that corrosion produces. One example of visible wear is the oxidation layer that forms when the exposed metal encounters the environment. Oxidation is a process where the metal surface becomes damaged to the point that the environment starts to react with any exposed metal. The reaction produces a fracturing on the top layer of metal, where visibility depends on the severity of the wear. Since bearings help to transfer motion, there is constant exposure to various loading conditions. This loading accelerates the corrosion process. Stainless steel is a durable material that is often the standard in bearing manufacturing. However, stainless steel is not used just in bearing manufacturing, as there are various processes regarding stainless-steel development.

Several methods exist that are the standard for determining the corrosion prevention characteristics of lubricants within the industry. A static corrosion test and a dynamic EMCOR corrosion test are the two core methodologies that Nye Lubricants Inc. uses [1, 2]. These methods are ASTM standards with ASTM D-1743 [2] as a static test and ASTM D-6138 [1] as a dynamic test. The methods listed are accessible through the ASTM International website. These tests both induce conditions that expose bearings to a corrosive environment. As each name suggests, one test has the bearing in a static environment resting in the corrosive solution, and the second test has a bearing in a dynamic environment with rotation occurring in timed intervals. After completion of these tests, disassembled bearing samples allow for further evaluation. These are the samples that are generated internally for various products. These samples are currently manually inspected according to both ASTM methods. The condition of the bearings post-test, inspected by human technicians, shows visible defects typically, which means the analysis does benefit from the methods listed in this paper due to the similar capabilities of Computer Vision Systems.

2 Background

2.1 Machine Vision

Industries have been using Machine Vision for quality inspection everywhere. Whether it is manufacturing facilities where machine vision provides insurance for repeatability or its medical facilities where precision is required, machine vision provides companies with automated tasks and improved functionality. Nye Lubricants (Nye) has incorporated a similar methodology by implementing a camera setup that can capture photos of the bearings after testing finishes. Nye calls this system the Bearing Corrosion Analyzer (BCA). This system allows for the bearing surface to be digitized and further analyzed. This feature of the technology is significant as it continuously provides new data, further improving the model. The continuous improvement process that this type of implementation provides allows companies to gain a competitive edge. The task becomes automated through the camera interface and the software system behind it. As proven by many in other industries, the methods are capable, but it requires the proper implementation to function correctly.

One crucial change that machine vision provides is an improved visual acuity over that of a human. The camera in question is a Basler monochrome product, and the resolution capabilities are an improvement over human inspection methods. This improvement makes the inspection more critical as it can recognize minor details more repeatably. The enhanced detail pushes for further improvements in lubricant formulation and raises the bar for analysis capabilities. The increased scrutiny then provides insight for researchers to meet these new demands introduced by integrating machine vision into the lubricant industry standards. Customers typically appreciate the improved inspection methodology since it further quantifies which product to select when comparing similar possibilities. Adding images as an additional output from some of these tests provides new ways to perform post-test analyses and provide deliverables to customers. Overall, the addition of a machine vision system raises the expectations and the capabilities for most processes.

2.2 Deep Learning

Image classification and recognition is not a new topic. Various industries have been implementing Deep Learning in numerous applications to make predictions, classify images, or automate workflows. A Convolutional Neural Network (CNN) is a well-known Deep Learning Method, and it has been applied and integrated into various software packages. Examples include the image classifier system, ImageNet [4], which is integrated into various languages already as it was influential in the rise of AI. These models represent methods and approaches that have proven efficient in numerous use cases. The lubricant industry still has methodologies that rely on human vision inspection techniques, but Nye developed its system to capture photos of bearings after testing in harsh environments. These images provide an essential tool when considering the use of CNN architectures. These architectures are robust at image classification, as has been proven with some of the examples mentioned. The drawback is the amount of data that is required to train a system effectively. Not many smaller companies have the optimal amount of domain data nor the resources at their disposal to obtain it. Big data is the fuel for AI, and obtaining data sources that fit this criterion is a challenge depending on the company's technological developments [6, 7].

Data generated for this work used the machine vision system developed by Nye. The tested bearings underwent scanning in the BCA so that the digital images could be cropped, modified, and manipulated in other ways to work with a CNN. However, as big data is defined, the amount of data generated from this system does not meet those requirements. The number of samples needed to create such a data set is too costly for most smaller companies, and this is the significant restriction holding these companies back from utilizing AI and other machine learning applications to their fullest. Any delays in fully leveraging internal experimental data provide more time for competitors to push further ahead. Given these motivations, the machine learning industry has provided a methodology to help bridge this gap using newer techniques like Transfer Learning [6, 7].

2.3 Transfer Learning

Transfer Learning (TL) is a methodology within Deep Learning that integrates domain data sets with networks already trained on big data. This combination generates a more accurate model than what a smaller data set provides on its own. Using new data sets provides flexibility where the initial data set provides the more extensive set for training, and the other is a smaller, more topic-specific data set for transfer onto the topic at hand. Methods like these help adapt image classifications to new fields of data. For example, modifications to the ImageNet [4] classifier adjust the system to a more specialized classification task, say to go from the thousands of classes in the original data set to a specialized set for classifying cats and dogs. This example, often used to introduce TL to new users, provides an excellent template for applying TL to new data. The transfer allows the CNN to extract features from the original set that the new data also contains so that the model does not have to learn features from scratch on the small data set. This extraction accelerates the process of training new data and provides a platform for smaller companies to utilize their data [6, 7].

The data generated from the Bearing Corrosion Analyzer (BCA) provided a data set for transference. With TL as one of the core methods used, a drastic reduction in the number of images required makes the sample generation a realistic goal for Nye. The data from this system is output as an image that is then fed into a pre-trained CNN for retraining on a new classification task. The lubricant industry has various methods for generating the corrosion samples needed for this work. The addition of these test methods into this new data set allows the model to have more robustness to different types of corrosion testing. The created TL data set combines with the external data set for use in the final model selection with these characteristics.

3 Data and Methodology

3.1 Experimental Goals

Stainless steel is one of the more common materials used in bearing manufacturing. While systems exist that inspect bearings for manufacturing flaws, they have not yet adapted to inspect the lubricants' preventative capabilities on stainless-steel surfaces. Observations in numerous manufacturing environments show some of the characteristics of stainless steel, and it is one of these scenarios that allow for a transfer from one stainless steel data set to another. Cold-rolled stainless steel manufacturing contains various defects that classification systems can recognize and categorize. The data set used as the core training set exists due to work in this cold-rolled manufacturing environment. Northeastern University developed a labeled data set for six different defects related to cold-rolled stainless steel manufacturing [5, 8].

Since bearing composition contains similar materials and after visual inspection of the data set, the similarities between bearing corrosion and stainless-steel manufacturing defects are enough to consider transfer learning. Combining data sets allows for the initially trained network to transfer features over to the bearing corrosion data. Overall, using these methods, a system is implemented with a reduced cost and time factor than considering the same methods without using intermediate data sets. As safety and other conditional requirements increase in severity, an improved method for analyzing products becomes necessary rather than a luxury. Implementing a system like this provides the foundation to build a classifier that can meet these requirements sooner than expected.

3.2 Data Generation

The subject expertise internally at Nye helped inspect various public data sets, and the final decision for a set yielded the Northeastern University (NEU) stainless steel set as the optimal choice. Each image is a greyscale 200 $\times$ 200 pixel picture of a defect example. Upon visual inspection, the sets shared similarities in the greyscale nature of the images. The NEU set contained six different classes of defects from the stainless-steel manufacturing processes. Of these six, several defects visually share some characteristics with corrosion defects. Some comparison images are presented below in Fig. 1. The patches defect consists of blotching like the corrosion seen in some of the more severe cases, while the scratches defect mimic some of the ball track types of corrosion sometimes seen in bearings. These similarities, paired with the image contrast nature of the images, make the NEU data set an excellent match for transferring weights from one network to another. The set consisted of 1800 total images with 300 images from each of the six defect classes. Applying various augmentations to supplement the original photos helps improve the data set by increasing the total data set size [5].

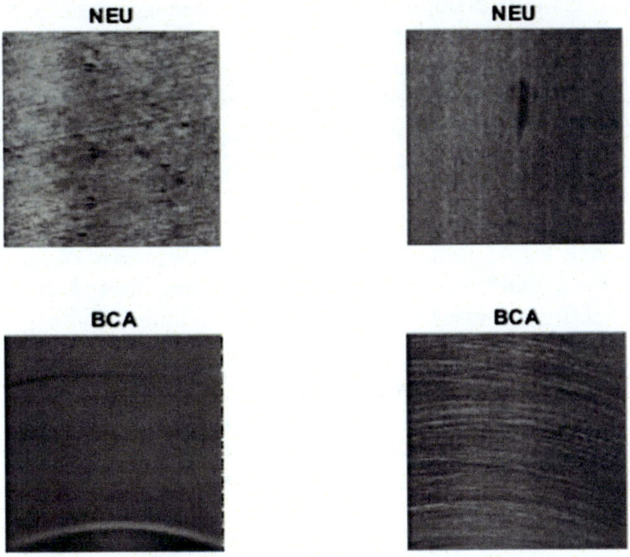

Fig. 1. Comparative sample images

Image augmentation methods are standard practice when adding robustness into a Deep Learning model. Techniques such as rotating the image, flipping the image, introducing noise, or some smudging methods all help provide the model with a variety of 'noisy' photos. These methods allow the system to be robust enough to handle new images that might not be the same as some training data. Introducing new images is a common issue with supervised learning, where the robustness of the model determines how well it can adapt to outlier examples in the data. Also, this work uses these methods to increase the data set size so that not only will the final model have more robustness to noise, but it also has more data for learning. The final model would never see any of these augmented photos in an application. However, the feature extractions that occur through some of those noisier images add to the capability for finding new areas of classification [6, 7].

The various augmentations implemented can vary depending on the types of images used. Since these images are square and greyscale, the system used four augmentations to increase the overall size of the data. Augmentations apply changes to the images, including flipping around the x-axis and y-axis and introducing some Gaussian Noise to the images (seen as white spots). The final transformation consisted of inverting the contrast of the image to minimize the effect of lighting fluctuations. With these four augmentations, the final NEU data set consisted of 9,000 images with about 1,500 images accessible for each defect [6].

The BCA images were captured on the machine vision system and then assembled into two different classes. Cropping the entire image of the bearing surface into several windows provided examples of isolated corrosion samples. It took 103 bearings to generate the final 721 sample images. These images were also augmented with the same methods to bring the total number of samples to 3605. The two classes include a 'clean' scenario with around 1175 examples and a 'corroded' scenario with 2430 samples. There are some images with minimal spotting occurrences that are considered corrosion, and then there are cases where the corrosion takes up most of the image. Some images also contain a wear track that is only sometimes factored in as corrosion. Analyzing these examples of corrosion required some experience in judging corrosion, which the technicians provided. From each cropped image, examples of corrosion were identified and labeled by technicians. It is beyond the scope of this initial work, but adding additional classes to the set might help introduce more variety into the analysis. However, since the ASTM method only calls for corrosion, the application remained consistent with these guidelines. It may be the basis for future work to expand the classes beyond the two simple cases of corrosion versus no corrosion.

3.3 Development Methods and Hardware

Acquiring these data sets provided the fuel Nye needed to implement some of these methods. Setting up a functioning CNN architecture for TL requires the appropriate software tools. There are numerous options, but for this work, MATLAB by MathWorks was the selection due to compatibility with the company's existing software setup as well as the teachings of the University of Massachusetts Dartmouth, who partnered on this work. Within this platform, a Deep Learning toolbox provides core frameworks for establishing methods that require CNN architectures. More specifically, the trainNetwork command is a function that allows the user to import various data types and set individual options to perform training on whatever supplied data is available [6].

One of the more well-known optimizer methods, the Stochastic Gradient Descent Method (SGDM), provided the optimizer for training the NEU system [3]. A typical Gradient Descent Method (GDM) utilizes the gradient calculation to determine the general direction for stepping towards the minimum of the loss function. However, gradient calculation using all data simultaneously in each iteration can be costly, particularly for problems involving large data sets. The SGDM and its variants, which can be traced back to Robbins and Monro's work in the 50s, use an unbiased estimate of the gradient [3]. It has a very low iteration per cost since only one sample is chosen randomly from the data set. Although the convergence is not as direct as typical GDM, the low computational cost makes it attractive. The initial training used only the bare minimum in options, and the model was not performing at a high enough accuracy, which prompted using some additional options [6].

Something that is always a concern when dealing with machine learning methods is the issue of over-fitting. During the early stages of development, the model appeared to be doing just that, which prompted the addition of L2 regularization [6]. This option allows the model to adapt and adjust its weights during training using a regularization term to help manage over-fitting issues. After the inclusion of these options, the model performance became more stable and consistent. Experimentation with these options was necessary to help adapt to the industry-specific data like the images generated from the BCA system. A Dell Precision tower system with a 10-core Intel Xeon 2.2 GHz processor and an Nvidia GeForce GTX 1660 Ti GPU made up the hardware for training the model. Once a final network architecture proved its capability, the model underwent training for thousands of iterations. Overall training took about eight hours to run for the NEU training and then an additional three and a half hours for transferring to the BCA data.

Table 1. Model layers

1	Image Input	200 × 200 × 1 images
2	Convolution	5 4 × 4 × 1 convolutions
3	Batch Normalization	5 channels
4	ReLU	ReLU
5	Max Pooling	4 × 4 max pooling
6	Convolution	20 4 × 4 × 5 convolutions
7	Batch Normalization	20 channels
8	ReLU	ReLU
9	Max Pooling	4 × 4 max pooling
10	Convolution	40 4 × 4 × 20 convolutions
11	Batch Normalization	40 channels
12	ReLU	ReLU
13	Max Pooling	4 × 4 max pooling
14	Convolution	80 4 × 4 × 40 convolutions
15	Batch Normalization	80 channels
16	ReLU	ReLU
17	Max Pooling	4 × 4 max pooling
18	Convolution	40 4 × 4 × 80 convolutions
19	Batch Normalization	40 channels
20	ReLU	ReLU
21	Fully Connected	6 fully connected layer
22	Softmax	softmax
23	Classification Output	'Cr' and 5 other classes

The network trained on the NEU data set is named the StainlessNet, with convergence shown in Fig. 2. The 23 layers, shown in Table 1, provide the trained network to transfer and fit the new training data from the BCA system. The fully connected layer is converted from an output of six to two so that the classification output can now predict the two new classes, clean or corroded. Techniques like this are typical when working with transfer learning. The concept of taking the weights from the initial training and using them as a starting point for new input data is how TL functions, and that is what allows smaller sets of data to be valuable. The class changes also allow the layers to adapt their prediction outcomes to the new data [6, 7].

4 Results

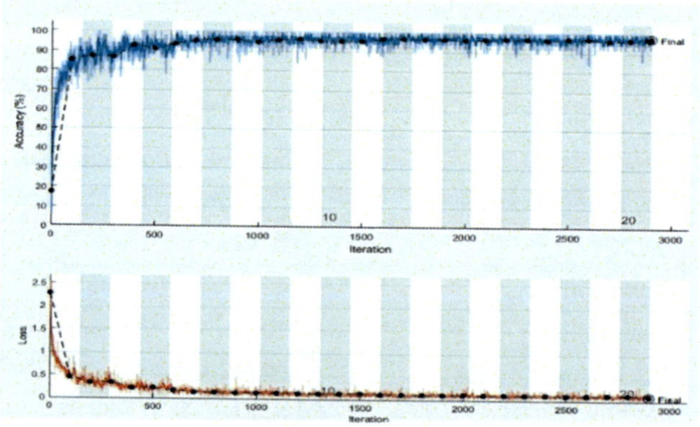

Fig. 2. NEU dataset convergence

After all the initial work to determine the optimal parameters, training for the final model used a 90/10 split from the complete data set. Another 10% of the data provided the validation set for training. Overall the data split is as follows: 80% used for training the model, 10% used for validation during training, and 10% for testing the trained model. The training accuracy for the NEU data set averages around 97% accuracy. The loss of the model after training was around 0.0821. This level of error is acceptable for Nye, given the source of the data and the size. Model testing used the test data withheld from training to confirm that it has not over-fit and can still maintain its accuracy. The results of that test, in Table 2, show the confusion matrix with the overall accuracy averaging around 96%. It is challenging to introduce entirely new data to the NEU-trained model since most of the images obtained were from manufacturing environments, and there was no way internally to generate utterly new data for the model. Due to this, the resulting model was enough to pursue the transfer learning portion.

Applying transfer learning from the NEU data to the BCA data required several adjustments during development. Since the BCA test is looking for corroded samples, the number of clean samples provided was insufficient to avoid a data imbalance. This data imbalance resulted in some flat-lining during some of the initial transfer training. The

Table 2. Confusion matrix for NEU test data

Class	Predicted classes					
	1	2	3	4	5	6
1	154	0	0	0	0	0
2	0	145	1	7	0	14
3	0	0	142	0	0	0
4	1	4	0	154	0	1
5	0	0	0	0	155	1
6	0	3	0	0	0	118

training accuracy would start around 70% and only improve to 82% accuracy. While this is not terrible given the limited amount of data, new clean images were extrapolated from the existing set to build more balance into the data set. Once retrained with the updated data set, the model converted to a higher level of accuracy. The BCA data set followed a similar training-validation-testing split as the NEU set, ensuring that the maximum allowable data is used during training while still maintaining testing capabilities [6, 7].

Future work seeks to overcome this convergence plateau that is still present by continuously improving the data set with newly generated images. With the adjustments finalized, the final transfer model achieved a training accuracy of around 93%, with a loss of around 0.164. Table 3 shows the test data results as a confusion matrix for the corroded and clean test examples. It takes a few epochs for the model to converge to just under this level, and then it oscillates with slight improvements to the accuracy until training completes. Even though it converges, the model did gain some benefit from incorporating the learning rate parameter. Adding these options allowed the model to improve in accuracy, albeit only several points higher. As this is the first model that Nye has implemented, its accuracy is acceptable, given that continuous improvements through development help adapt further.

Table 3. Confusion matrix for BCA test data

Class	Predicted clean	Predicted corroded
Clean	112	25
Corroded	8	215

5 Conclusions

Given that CNN architectures are well known to work on image data, it is no surprise that this model can classify the defects from the Northeastern University data set.

Transferring those features from manufactured stainless steel to curved stainless steel bearing surfaces can improve the efficacy of analyzing bearing corrosion in the lubricant industry. The first phase of this work was to confirm and implement a method to produce accurate classifications for corrosion on bearings. With a test accuracy of around 91–92%, the model can provide these insights. It is helpful to still grade the bearings with the technician's manual method as the accuracy continues to improve, but using this model to supplement the analysis helps introduce customers and the industry to this updated output. Overall it is confirmed that CNN architectures paired with transfer learning can function even in smaller companies given carefully matched data sets and proper implementations.

The current model can classify corrosion on bearings with about 91% accuracy. The confusion matrix for the transfer model in Table 3 seems to misclassify some samples as containing corrosion while the technician classified them as clean. After investigating it further, the findings showed that this is due to some discrepancies in how the technicians classify corrosion samples. Previously mentioned, the ball wear-track corrosion effect shown in Fig. 3 shows how the wear follows a line path around the circumference of the bearing. Depending on the intensity of this wear track, technicians might classify it as corrosion, or they might not. This specific classification is an outlier amongst the data, and introducing more samples may help improve the misclassification of these in the future. Another possibility of overcoming these errors would be creating an additional class to separate these outliers into a separate class for further analysis.

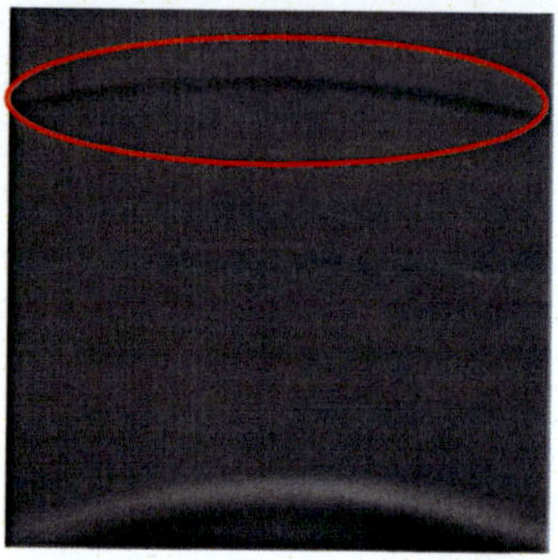

Fig. 3. Example of wear track

The justification for using methods like this comes from the expense of creating the data for a workable data set. For example, corrosion testing on EMCOR bearings is costly, both monetarily and time-wise. One sample takes about a week to generate according to the ASTM method. Thankfully Nye has a system that can test up to 4 samples at a time. Even with that testing capability, without TL and the NEU data set, it would take about 450 weeks to generate a data set of a similar size (1800 images). The cost to make one of those samples is around 60 dollars per bearing. If Nye wanted to generate 1800 images of bearings to train a model, the cost would be approximately 108,000 dollars. This cost savings is significant when considering some of the other implications, such as a technician's time to operate the tests and perform the analyses. The images would need to be labeled accurately and modified if necessary.

The incorporation of TL means Nye was able to utilize some historical samples to generate data. These samples were already available before this work began, which is why introducing TL was an option. Given that there were 103 samples of bearings, the cost to implement this system only required the time to label the generated images and perform the cropping and isolation of corrosion examples. If we count those bearing costs, the amount of material needed only costs Nye about 6,180 dollars. While this is the cost for the number of samples in the BCA data set, since the testing for these bearing samples occurred before this work, the actual cost considered is just the time for model development and sample labeling. Utilizing historical data like this shows that even though there is a lower accuracy than normal deployment levels, the cost of this model is much more feasible for smaller companies pursuing these advanced methods.

AI and Machine Learning are powerful tools for modern research. Most smaller companies lack the expertise or the data to implement methods like this, but the impact would be noticeable if they could. This paper serves as an example of one such implementation that successfully analyzed corrosion on bearings. It is encouraging to see the support for methodologies like this out there, but the lubricant industry must leverage these capabilities fully, especially Nye. Overall, implementing these methods takes time to introduce, but work such as this helps pave the way for the data-rich lubricant industry to present other examples.

While this work served as an introduction to the potential of this technology in the lubricant industry, future work builds upon this foundation to improve the model's capabilities. Introducing extra dimensions to the data, such as RGB images over greyscale, may provide different feature extractions for the model. Other options include developing a custom activation function that better fits the bearing corrosion model or adding more integrated layers. Overall, the numerous strategies available provide many possibilities as the model continuously evolves [6, 7].

An analysis is a time-consuming task, and by automating it with machine vision and the CNN/TL system, the analysis becomes more robust and more efficient. The savings in cost, time, and manual power free up employees for innovation in other areas. The goal is to integrate systems, where necessary, to help optimize the workflows for all the technicians, engineers, and scientists involved. Removing human error from repetitive tasks such as this drives more time for research and improves the performance characteristics of products. As the technology progresses, these companies become able to leverage data better and continue to improve models, thereby raising the level of their internal capabilities closer to that of a larger company. Reducing the time for this is essential to keep up with the current market trends and technologies across the lubricant customer base. This work provides the starting point for Nye Lubricants Inc. to start the process showing that companies can benefit from using these computational methodologies with minimal starting resources.

Acknowledgments. Our thanks to Northeastern University for providing the stainless-steel defect data set, Nye Lubricants, Inc. for providing the bearing corrosion data, and the technicians who performed testing that helped generate this data. We want to thank students and faculty members participating in the University of Massachusetts Dartmouth Mathematical and Computational Consulting and Data Science Capstone project Fall 2019 course for fruitful discussion.

References

1. ASTM. Standard Test Method for Determination of Corrosion-preventative Properties of Lubricating Greases Under Dynamic Wet Conditions (EMCOR Test). ASTM D6138-13. D6138(13), pp. 1–15 (2013)
2. ASTM. Standard Test Method for Determination of Corrosion-preventative Properties of Lubricating Greases. ASTM D1743-13. D1743(13), pp. 1–15 (2013)

3. Robbins, H., Monro, S.: A stochastic approximation method. Ann. Math. Stat. **22**, 400–407 (1951)
4. Deng, J., Dong, W., Socher, R., Li, L., Li, K., Fei-Fei, L.: Imagenet: a large-scale hierarchical image database. In: 2009 IEEE Conference on Computer Vision and Pattern Recognition, pp. 248–255 (2009). https://doi.org/10.1109/CVPR.2009.5206848
5. Song, K., Yan, Y.: Neu surface defect database. https://drive.google.com/file/d/0B5OUtBsSxu1Bdjh4dk1SeGYtNFU/view. Accessed Sept. 2019
6. MATLAB and Deep Learning Toolbox Release 2019b, The MathWorks, Inc., Natick, Massachusetts, United States
7. Jaakkola, T., Barzilay, R.: Machine Learning MIT Course Book. MIT, Cambridge (2016)
8. He, Y., Song, K., Mend, Q., Yan, Y.: An end-to-end steel surface defect detection approach via fusing multiple hierarchical features. IEEE Trans. Instrum. Measur. **69**, 1493–1504 (2019)

Combining the Spectral Coherence with Informative Frequency Band Features for Condition Monitoring Under Time-Varying Operating Conditions

Stephan Schmidt[1], P. Stephan Heyns[1], and Konstantinos C. Gryllias[2,3](✉)

[1] Centre for Asset Integrity Management, University of Pretoria, Lynnwood Road, Pretoria 0002, South Africa
{stephan.schmidt,stephan.heyns}@up.ac.za

[2] Department of Mechanical Engineering, KU Leuven, Leuven, Belgium
konstantinos.gryllias@kuleuven.be

[3] Dynamics of Mechanical and Mechatronic Systems, Flanders Make, Celestijnenlaan 300, 3001 Heverlee, Belgium

Abstract. Vibration-based monitoring is very popular for expensive gearboxes found in wind turbines for example. However, the fault information of the damaged components is usually masked by high noise levels and time-varying operating conditions. Informative frequency band identification methods allow the fault information in the signal to be enhanced, which facilitates incipient fault detection. In conventional informative frequency band identification methods, Short-Time Fourier Transform or Wavelet Packet Transform estimators are used to identify the frequency bands of interest. However, in the aforementioned estimators there is a compromise between the time and the frequency resolution of the resulting bandlimited signals. This has been one of the motivations of developing the order-frequency spectral coherence-based IFBI_αgram. This IFBI_αgram is constructed by estimating the Signal-to-Noise Ratio of the predetermined component-of-interest in the frequency bands of a set of order-frequency spectral coherences. Thereafter, the optimal band for detecting the component-of-interest is determined by maximising the IFBI_αgram. The order-frequency spectral coherence simultaneously displays the spectral content and cyclic content of the signal under time-varying operating conditions and enhances weak fault information. Hence, it is very effective for identifying informative frequency bands. However, the suitability of using other features instead of the signal-to-noise ratio measure has not been investigated. Hence, in this work, new informative frequency band feature planes are calculated by combining multiple order-frequency spectral coherences with different features such as the L2/L1-ratio, the negentropy, and the kurtosis. The effectiveness of the diagnostic features is investigated on gearbox data acquired under time-varying operating conditions.

Keywords: Gearbox diagnostics · Informative frequency band identification · Time-varying operating conditions

R. Karim et al. (Eds.): IAI 2021, LNME, pp. 189–201, 2022.
https://doi.org/10.1007/978-3-030-93639-6_16

1 Introduction

Gearboxes found in the wind turbine and mining industries operate inherently under time-varying operating conditions [1]. This impedes the application of conventional condition monitoring methods, since both time-varying operating conditions and damage result in amplitude and frequency modulation. Additionally, the fault information is usually buried in specific frequency bands and can be difficult to be detected [2, 3].

Frequency Band Identification (FBI) methods such as the kurtogram [2] and the info-gram [3] can be used to identify informative frequency bands, whereafter a bandpass filter can be designed to extract the information in the bands. The feature plane, generated in the FBI process, is usually calculated using short-time Fourier transform [3] or wavelet packet transform estimators [2]. The benefits of the aforementioned estimators are that they are fast to calculate. However, the aforementioned estimators adhere to the uncertainty principle, i.e. there is a compromise between the spectral frequency and time axes. This results in the corresponding cyclic spectrum to be only appropriate for limited cyclic orders [4].

The order-frequency spectral coherence is a two-dimensional visualisation of the signal under consideration, with the one axis being the cyclic orders (i.e., the periodicity of the impulses) and the other axis being the spectral frequencies (i.e., the time-invariant carriers). This provides a very effective representation of the impulses induced by damaged machine components operating under time-varying operating conditions [5]. The order-frequency spectral coherence is also able to amplify weak fault information, which makes it suitable for incipient fault detection.

This has been one of the main motivations of using the order-frequency spectral coherence in the development of the IFBI_αgram [6]. The IFBI_αgram calculates a set of order-frequency spectral coherences, whereafter a signal-to-noise ratio feature is calculated from the different frequency bands and spectral frequency resolutions. This has performed well on numerical and experimental datasets [6], however, we desire to compare the performance of different features to determine whether its performance can be improved. Hence, in this work, the suitability of different features for frequency band identification are investigated for performing gearbox diagnostics under time-varying operating conditions. The new feature planes are constructed with the same procedure as the original IFBI_αgram method: A set of order-frequency spectral coherences is calculated (instead of the conventional wavelet packet transform or short-time Fourier transform), whereafter different features are extracted to construct new feature planes.

The layout of the paper is as follows. In the next section, Sect. 2, the IFBI_αgram is discussed and the different features that will be investigated are presented. Thereafter, the suitability of the different features for gearbox diagnostics is investigated on two experimental datasets in Sect. 3. Finally, conclusions are drawn in Sect. 4.

2 Overview of Investigation

In this section, a brief overview is given of the IFBI_αgram, whereafter the considered features are presented. Lastly, the performance metric, which will be used to rank the suitability of the different features for frequency band identification, is discussed.

2.1 IFBI$_\alpha$gram

The IFBI$_\alpha$gram, originally presented in Reference [6], is constructed and used as follows for fault diagnosis:

1. For a specific vibration measurement that is under consideration and a set of predetermined window lengths $\{N_w\}$, a corresponding set of Order-Frequency Spectral Coherences (OFSCoh) is calculated, i.e. an OFSCoh is calculated for each window length in $\{N_w\}$. The frequency resolution Δf of a specific OFSCoh is determined by the window length that is used and the resulting squared magnitude of the OFSCoh is denoted by $\gamma(\alpha, f; \Delta f)$, where α denotes the cyclic order variable and f denotes the spectral frequency variable. The OFSCoh is estimated with the Welch estimator as presented in Ref. [5].
2. The prominence of the component-of-interest, with a corresponding cyclic orders set $\{\alpha_c\}$, in the spectral frequency band $[f - \Delta f/2, f + \Delta f/2]$ of the OFSCoh $\gamma(\alpha, f; \Delta f)$ is estimated with a feature. This feature is referred to as the amplitude-to-median feature in this work and is further discussed in Sect. 2.2.1.
3. The features extracted from each spectral frequency band of $\gamma(\alpha, f; \Delta f)$ is used to build a feature plane $FP(f, \Delta f)$. This feature plane quantifies the prominence of the cyclic orders of interest for different combinations of spectral frequencies f and spectral frequency bands Δf.
4. The frequency f and the frequency resolution Δf that maximise the feature plane $FP(f, \Delta f)$ are found and used to design a bandpass filter. The passband of the bandpass filter is set to $[f - \Delta f/2, f + \Delta f/2]$.
5. The bandpass filtered signal is analysed to infer the condition of the gearbox.

In the next step, a description of the different features that will replace the feature used in Step 2 is given.

2.2 Description of Features

The amplitude-to-median feature used in the original IFBI$_\alpha$gram paper [6] is first presented in this section, whereafter a description is given of the other features that will be considered in this work. The categorisation of the features used in Ref. [7] will be used here as well, i.e. the targeted features refer to features that target specific cyclic orders $\{\alpha_c\}$, while blind features do not require the kinematics of the gearbox under consideration to be known.

2.2.1 Feature 1: Amplitude-to-Median (Targeted)

The amplitude-to-median feature is used in the original IFBI$_\alpha$gram paper [6] to measure the signal-to-noise ratio in a specific frequency band. It aims to estimate the prominence of the signal component-of-interest and its harmonics, denoted $\{\alpha_c\}$, in the frequency band $[f - \Delta f/2, f + \Delta f/2]$ under consideration with

$$FP_1(f, \Delta f) = \sum\nolimits_{\alpha\in\{\alpha_c\}} \frac{\gamma(\alpha, f; \Delta f)}{NMed(\alpha, f; \Delta f)} \tag{1}$$

The numerator of Eq. (7) contains the amplitude of the component-of-interest and NMed(α, f; Δf) denotes an estimate of the noise level in the OFSCoh at the cyclic order of α. The noise level is estimated by calculating the median of the squared magnitude OFSCoh in the cyclic order band [α – 1, α + 1].

2.2.2 Feature 2: Amplitude-to-Mean (Targeted)

The amplitude-to-mean feature is used in the construction of the IESFOgram [8]. It is calculated as follows in this work:

$$FP_2(f, \Delta f) = \sum\nolimits_{\alpha \in \{\alpha_c\}} \frac{\gamma(\alpha, f; \Delta f)}{NMean(\alpha, f; \Delta f)} \tag{2}$$

and is very similar to Eq. (1); i.e. instead of using the median in the dominator, the mean (denoted by NMean) of the OFSCoh in the cyclic order band [α – 1, α + 1] is calculated.

2.2.3 Feature 3: Amplitude (Targeted)

The amplitude feature is calculated with

$$FP_3(f, \Delta f) = \sum\nolimits_{\alpha \in \{\alpha_c\}} \gamma(\alpha, f; \Delta f) \tag{3}$$

It calculates the sum of the components-of-interest, but in contrast to Eqs. (1) and (2), it does not account for the noise level in the frequency band under consideration. By comparing the performance of feature 3 with feature 1 and 2, it would be possible to determine whether it is necessary to perform the normalisation step in the frequency band identification problem.

2.2.4 Feature 4: Maximum Amplitude (Blind)

The maximum amplitude feature

$$FP_4(f, \Delta f) = max\{\gamma(\alpha, f; \Delta f)\}_\alpha \tag{4}$$

calculates the maximum of the squared-magnitude OFSCoh over the cyclic order variable α and is therefore considered a blind feature.

2.2.5 Feature 5: Kurtosis (Blind)

The kurtosis is used to construct the kurtogram [2]. It can be used to identify frequency bands that display non-Gaussian characteristics in a signal, i.e. the kurtosis increases as the data becomes more leptokurtic. The kurtosis of the squared magnitude OFSCoh is calculated with

$$FP_5(f, \Delta f) = \frac{\langle(\gamma(\alpha, f; \Delta f) - \langle\gamma(\alpha, f; \Delta f)\rangle)^4\rangle}{\langle(\gamma(\alpha, f; \Delta f) - \langle\gamma(\alpha, f; \Delta f)\rangle)^2\rangle^2} \tag{5}$$

Since it is expected that each frequency band of the squared magnitude OFSCoh would be non-Gaussian, large kurtosis values are expected for all frequency bands. It is however expected that frequency bands with much cyclostationary information would have a higher kurtosis than frequency bands with a flat cyclic spectrum.

2.2.6 Feature 6: Negentropy (Blind)

The spectral negentropy is used in the construction of the infogram [3]. The spectral negentropy of the squared magnitude of the order-frequency spectral coherence is given by

$$FP_6(f, \Delta f) = \langle R(\alpha, f; \Delta f) \cdot log(R(\alpha, f; \Delta f)) \rangle \tag{6}$$

with

$$R(\alpha, f; \Delta f) = \frac{\gamma(\alpha, f; \Delta f)}{\langle \gamma(\alpha, f; \Delta f) \rangle} \tag{7}$$

As the signal becomes more cyclostationary, the negentropy of the cyclic spectrum is expected to increase. Hence, this can be used to detect bands with much cyclostationary content.

2.2.7 Feature 7: L2/L1 Norm (Blind)

The L2/L1 norm is used in the construction of the sparsogram [9] and aims to identify frequency bands that contain much sparseness (i.e. indicative of cyclic components potentially attributed to damage). It is calculated as the ratio of the L2-norm to the L1-norm of the squared magnitude OFSCoh:

$$FP_7(\alpha, f; \Delta f) = \frac{\|\gamma(\alpha, f; \Delta f)\|_2}{\|\gamma(\alpha, f; \Delta f)\|_1} \tag{8}$$

2.2.8 Feature 8: Negative Spectral Flatness (Blind)

Lastly, the Negative Spectral Flatness (NSF) of the squared magnitude OFSCoh

$$FP_8(\alpha, f; \Delta f) = -\frac{GA\{\gamma(\alpha, f; \Delta f)\}}{AA\{\gamma(\alpha, f; \Delta f)\}} \tag{9}$$

is the ratio of the Geometric Average (GA) divided by the Arithmetic Average (AA) and is in the range [−1,0]. The NSF is minimised if the squared magnitude of the OFSCoh is flat (i.e. no cyclic order components) and is maximised if the OFSCoh displays sparse properties in the specific spectral frequency band.

2.2.9 Feature 9: Skewness (Blind)

The skewness

$$FP_9(f, \Delta f) = \frac{\langle (\gamma(\alpha, f; \Delta f) - \langle \gamma(\alpha, f; \Delta f) \rangle)^3 \rangle}{\langle (\gamma(\alpha, f; \Delta f) - \langle \gamma(\alpha, f; \Delta f) \rangle)^2 \rangle^{3/2}} \tag{10}$$

is expected to increase as the cyclostationary content in the signal becomes more prominent with the presence of damage and can therefore be used to detect informative frequency bands.

2.3 Performance Metrics

The Squared Envelope Spectrum (SES) is one of the most popular methods for fault diagnosis. The SES of the filtered signals is firstly calculated, whereafter a performance metric is extracted from the SES. However, the performance metric calculated directly from the SES could be biased for frequency bands with large power spectral densities. Hence, the SES is firstly standardised and then a performance metric is calculated.

In Ref. [10], the authors calculated a standardised spectrum, by estimating the local median and the local median absolute difference of the SES with a moving window. The SES is standardised as follows

$$Z_{SES}(\alpha) = \frac{SES(\alpha) - MED_{SES}(\alpha)}{MAD_{SES}(\alpha)} \tag{11}$$

where Z_{SES} denotes the standardised SES, MED_{SES} denotes the moving median of the SES and MAD_{SES} denotes the moving median absolute difference of the SES. Thereafter, the mean amplitude of the harmonics of the fault component in the standardised SES is calculated and this is used as a performance metric.

3 Results

The different features are compared on experimental data that were acquired under time-varying operating conditions. In the next section, a brief description of the experimental test-rig is given whereafter the results are presented for a gearbox that had localised gear damage. Lastly, the performance of the different features is compared on experimental data from a gearbox that had distributed gear damage.

3.1 Experimental Test-Rig

The experimental test-rig shown in Fig. 1 is situated in the Centre for Asset Integrity Management laboratory at the University of Pretoria.

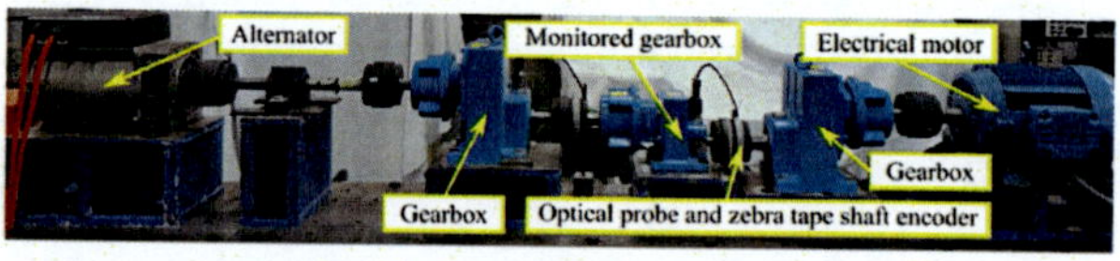

Fig. 1. The experimental setup

The experimental setup consists of three helical gearboxes as well as an alternator and an electrical motor.

The centre helical gearbox (denoted by monitored gearbox in Fig. 1) was damaged for both considered datasets. The axial component of a tri-axial accelerometer, located at the back of the monitored gearbox, is used to obtain vibration condition monitoring measurements. The instantaneous rotational speed of the input shaft of the monitored gearbox is estimated with measurements from an optical probe and a zebra tape shaft encoder.

The alternator and the electric motor are used to exert time-varying operating conditions on the helical gearboxes. The estimated torque and speed are presented in Fig. 2 for the four Operating Condition (OC) classes that are considered in this work.

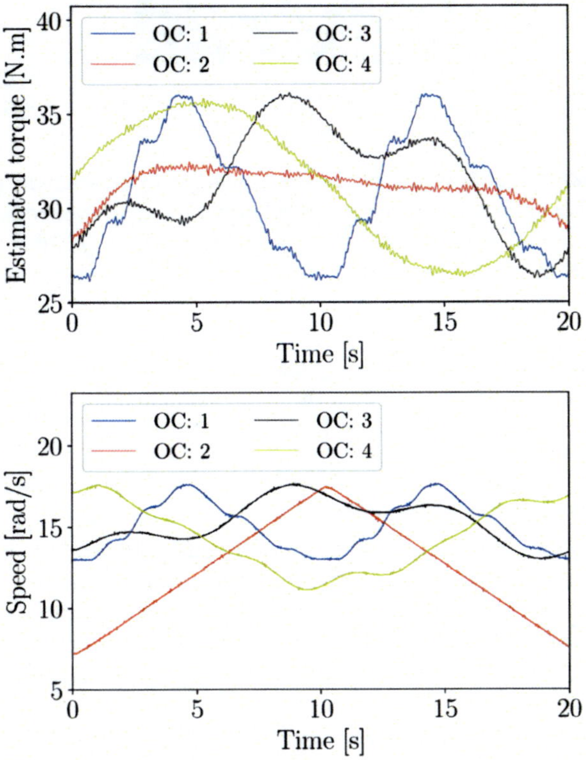

Fig. 2. The operation conditions (torque and speed) that were estimated at the input shaft of the monitored gearbox.

3.2 Localised Gear Damage Results

3.2.1 Overview of the Dataset

In the first experimental dataset, localised gear damage was investigated by seeding a slot in one of the teeth of the gear of the helical gearbox as shown in Fig. 3.

Fig. 3. The gear of the monitored gearbox with localised damage.

This gear was operated for approximately 20 days whereafter the damaged tooth failed. Twenty-five (25) measurements, spaced approximately evenly over the duration of the test and acquired under operating condition 1, are investigated in this work.

The set of OFSCoh was estimated for each of the 25 measurements whereafter the different features, discussed in Sect. 2, were calculated. The set of OFSCoh is estimated for the following window lengths: $\{\log2(N_w)\} = \{4, 5, \ldots, 10\}$.

Since the gear of the gearbox is being monitored for damage and it is connected to the input shaft of the monitored gearbox, the cyclic order set that is used for calculating the targeted features is given by $\{\alpha_c\} = \{1.0, 2.0, 3.0\}$.

3.2.2 Examples of Feature Planes and SES

The resulting feature plane, the SES of the raw signal and the SES of the bandpass filtered signal are presented in this section for one of the considered measurements. Two features, namely, the amplitude-to-median feature (used in the construction of the original IFBI$_\alpha$gram) and the kurtosis (used in the construction of the kurtogram) are presented to emphasise the importance of using the appropriate features for performing frequency band identification.

The feature plane of the amplitude-to-median feature, presented in Fig. 4(a), indicates that a resonance band at 500 Hz was identified. The SES of the raw signal in Fig. 4(c) does not contain any fault information related to the damaged gear. However, the SES of the bandpass filtered signal presented in Fig. 4(c) contains much fault information, because the fundamental component at 1.0 shaft orders and its harmonics are very prominent in the SES of the filtered signal.

In contrast, the kurtosis is maximised by a different frequency band as seen in Fig. 4(b). The corresponding SES of the filtered signal indicates that this frequency band does not contain information related to the condition of the gear and therefore the damage is not detected. It is therefore clear that the amplitude-to-median feature performs much better than the kurtosis feature for the specific measurement under consideration. However, it is important to consider the performance of the different features for all of the 25 measurements considered. This is investigated in the next section.

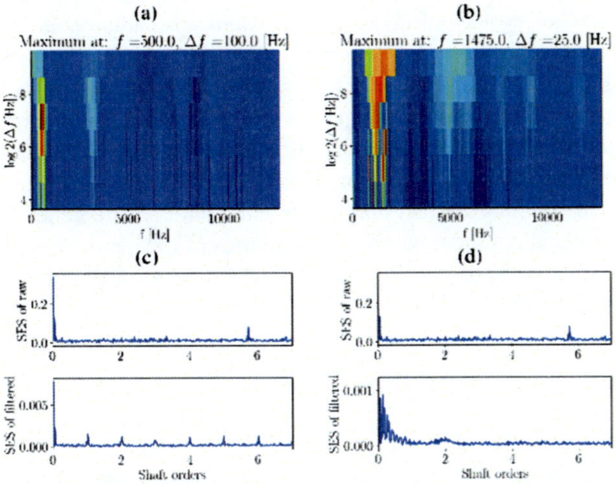

Fig. 4. Feature plane and SES for two features. The results of the Amplitude-to-Median feature are presented in (a) and (c) and the results for the kurtosis feature are presented in (b) and (d).

3.2.3 Performance of Features

The amplitude of the first four harmonics of the component of interest, i.e. {1.0, 2.0, 3.0, 4.0} shaft orders, were extracted from the standardised SES of the filtered signals, whereafter the mean amplitude was calculated. This was performed for the 25 considered measurements.

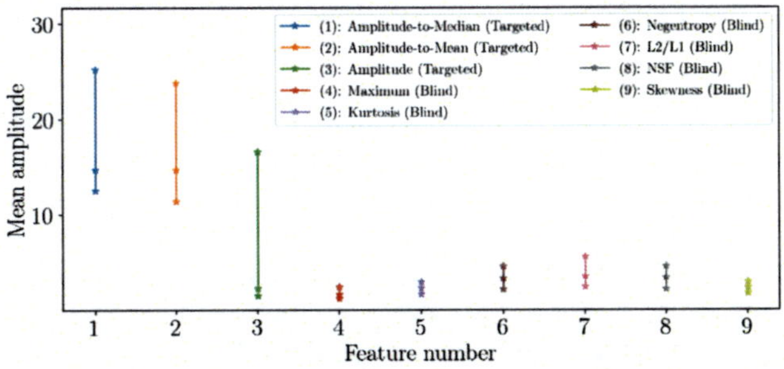

Fig. 5. The performance of the features for localised damage detection. The 10th, 50th, and 90th percentiles are presented for the mean of the amplitudes of the component of interest.

The 10th, 50th and 90th percentiles of the mean of the amplitudes, for the 25 measurements, are presented in Fig. 5 for the localised gear damage case. This makes it possible to visualise the spread of the mean amplitude. Since the gear deteriorated over measurement number, some variation in the amplitudes (and therefore performance) are expected.

According to the results in Fig. 5, the amplitude-to-median and the amplitude-to-mean features performed the best and performed much better than the amplitude feature calculated with Eq. (3). This emphasises that the normalisation of the amplitude by an estimate of the noise as done in Eqs. (1) and (2), could significantly improve the performance of a frequency band identification method. The L2/L1 feature performed the best of the blind features, however, it performed much worse than the targeted features. This corroborates the fact that the targeted features are generally better suited for detecting damaged components if the kinematics of the gearbox is known.

In the next section, the suitability of different features for distributed gear damage detection is investigated.

3.3 Distributed Gear Damage Results

3.3.1 Overview of the Dataset

In the second experiment, distributed gear damage was induced by leaving the gear in a corrosive environment for an extended period. The damaged gear is presented in Fig. 6.

Fig. 6. The gear with distributed gear damage before the experiment started.

The gearbox was operated with the damaged gear for approximately eight days whereafter the experiment was stopped due to excessive vibration. The excessive vibration was attributed to a localised portion of the gear being severely damaged.

Forty-nine (49) measurements, acquired over the test period are used to quantify the performance of the different features. Fifteen (15), 13, 11 and 10 of the 49 measurements were from OC 1, 2, 3 and 4 respectively. The vibration signal is pre-whitened using cepstrum pre-whitening with the procedure described in Ref. [11] to remove some deterministic components in the signal.

3.3.2 Examples of Feature Planes and SES

The feature planes for the amplitude (see Eq. (3)) and amplitude-to-mean feature (see Eq. (2)) are presented in Figs. 7(a) and 7(b) respectively for one of the considered measurements.

The amplitude feature is maximised in a very high-frequency band, while the amplitude-to-mean feature identifies approximately the same frequency band as the localised gear damage case.

The resulting squared envelope spectra for the amplitude and amplitude-to-mean are presented in Figs. 7(a) and 7(c) respectively. It is not possible to identify that the gear is damaged from the SES of the raw signal. The SES of the filtered signal obtained by the amplitude feature contains much noise, which makes it impossible to detect the gear damage.

In contrast, the amplitude-to-mean feature performs much better. The SES of the filtered signal contains prominent signal components at 1.0 shaft orders and 4.0 shaft orders. The fact that it identified the same frequency band as the amplitude-to-median feature for the localised gear damage experiment, further emphasises that this frequency band contains much fault information for this gearbox under consideration.

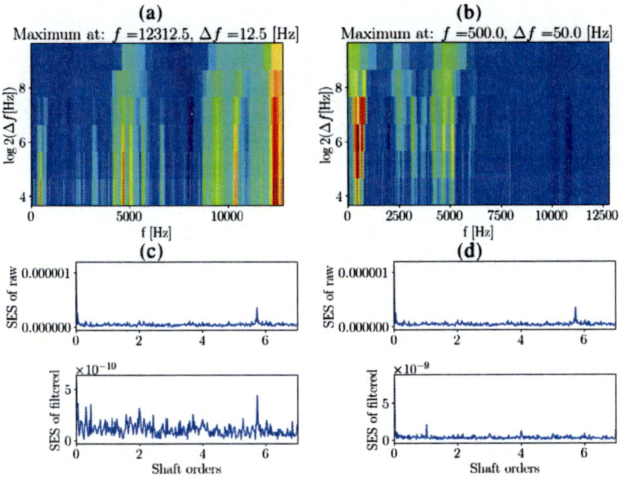

Fig. 7. The feature planes and SES for two features. The results for the amplitude feature are presented in (a) and (c) and the results for the amplitude-to-mean feature are presented in (b) and (d).

3.3.3 Performance of Features

The performance of the features is calculated similarly as for the localised gear damage case. However, only the first and fourth harmonics are prominent in the SES of the filtered signals and therefore they are used in the calculation of the performance measure.

The statistics are summarised in Fig. 8 for the distributed gear damage experiment. The 50th percentile indicates that the targeted methods perform again much better than the blind methods. The amplitude-to-mean performs in this case slightly better than the amplitude-to-median feature. The negentropy, L2/L1 and NSF features perform much better than the other blind features for this case. The maximum feature performs the worst in detecting the informative frequency band, since it can easily be affected by spurious signal components in the spectrum.

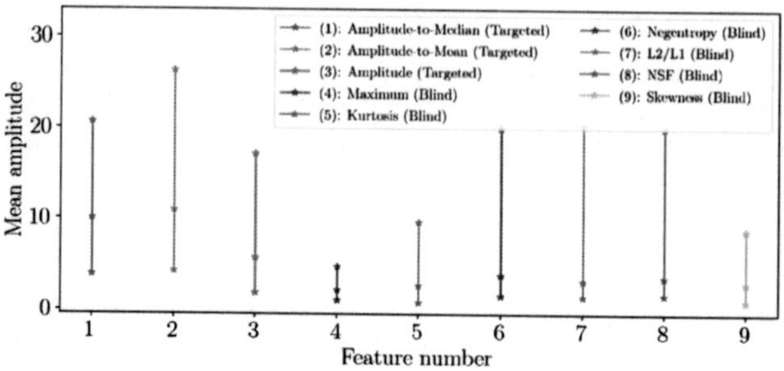

Fig. 8. The performance of the different features for distributed gear damage detection.

4 Conclusions

In this work, the suitability of using different features in the IFBI$_\alpha$gram methodology was investigated on two experimental datasets that were acquired under time-varying operating conditions. The results indicate that the original feature used in the IFBI$_\alpha$gram (referred to as the amplitude-to-median feature in this work) and the amplitude-to-mean feature previously used in the IESFOgram methodology performs the best for localised and distributed gear damage detection. Even though the performance of the two features is very similar for the considered data, it is expected that their performance will be different for datasets where many signal components are present in the cyclic spectrum (e.g. planetary gearboxes). The importance of normalising the features by a measure of the noise floor in the frequency band is also emphasised by the results. Lastly, the targeted methods performed consistently well for both datasets, which is in contrast to the results of the blind methods. This corroborates the fact that targeted methods perform generally much better than blind methods. However, the targeted methods assume that the kinematics of the gearbox and the cyclic orders of the critical components are known. This assumption may not be valid for all considered cases, and therefore better performing blind methods need to be investigated in future work.

References

1. Zimroz, R., Bartelmus, W., Barszcz, T., Urbanek, J.: Diagnostics of bearings in presence of strong operating conditions non-stationarity—a procedure of load-dependent features processing with application to wind turbine bearings. Mech. Syst. Signal Process. **46**(1), 16–27 (2014)
2. Antoni, J.: Fast computation of the kurtogram for the detection of transient faults. Mech. Syst. Signal Process. **21**(1), 108–124 (2007)
3. Antoni, J.: The infogram: Entropic evidence of the signature of repetitive transients. Mech. Syst. Signal Process. **74**, 73–94 (2016)
4. Abboud, D., Antoni, J.: Order-frequency analysis of machine signals. Mech. Syst. Signal Process. **87**, 229–258 (2017)
5. Abboud, D., Baudin, S., Antoni, J., Rémond, D., Eltabach, M., Sauvage, O.: The spectral analysis of cyclo-non-stationary signals. Mech. Syst. Signal Process. **75**, 280–300 (2016)

6. Schmidt, S., Mauricio, A., Heyns, P.S., Gryllias, K.C.: A methodology for identifying information rich frequency bands for diagnostics of mechanical components-of-interest under time-varying operating conditions. Mech. Syst. Signal Process. **142**, 106739 (2020)
7. Smith, W.A., Borghesani, P., Ni, Q., Wang, K., Peng, Z.: Optimal demodulation-band selection for envelope-based diagnostics: a comparative study of traditional and novel tools. Mech. Syst. Signal Process. **134**, 106303 (2019)
8. Mauricio, A., Smith, W.A., Randall, R.B., Antoni, J., Gryllias, K.: Improved envelope spectrum via feature optimisation-gram (IESFOgram): a novel tool for rolling element bearing diagnostics under non-stationary operating conditions. Mech. Syst. Signal Process. **144**, 106891 (2020)
9. Peter, W.T., Wang, D.: The design of a new sparsogram for fast bearing fault diagnosis: Part 1 of the two related manuscripts that have a joint title as "Two automatic vibration-based fault diagnostic methods using the novel sparsity measurement–Parts 1 and 2." Mech. Syst. Signal Process. **40**(2), 499–519 (2013)
10. Kass, S., Raad, A., Antoni, J.: Self-running bearing diagnosis based on scalar indicator using fast order frequency spectral coherence. Measurement **138**, 467–484 (2019)
11. Borghesani, P., Pennacchi, P., Randall, R.B., Sawalhi, N., Ricci, R.: Application of cepstrum pre-whitening for the diagnosis of bearing faults under variable speed conditions. Mech. Syst. Signal Process. **36**(2), 370–384 (2013)

Blockchain Technology for Information Sharing and Coordination to Mitigate Bullwhip Effect in Service Supply Chains

Muthana Al-Sukhni(✉) and Athanasios Migdalas

Luleå University of Technology, Luleå, Sweden
{Muthana.al-sukhni,Athanasios.migdalas}@ltu.se

Abstract. Supply chain management experiences inefficiencies due to several reasons, such as the lack of information sharing and coordination among supply chain participants. Moreover, these participants may not share their information with other parties in the supply chain due to trust issues and considering such information as sensitive asset. This behavior may hinder the supply chain efficiency and motivate the occurrence of bullwhip effect (BWE). This phenomenon has been intensively investigated in manufacturing industries. In addition, information sharing has been considered the main remedy to eliminate BWE in literature. However, few attempts examined the effect of information sharing in reducing this phenomenon in service supply chain (SSC). Digitalization and computerization have the potential to convert and reshape the supply chains in all kind of businesses and improve the coordination among supply chain partners by exchanging real-time information. Blockchain as a disruptive technology has many distinct features like disintermediation and decentralization that provide integrity, visibility and security for the supply chain. To bridge this gap, this paper aims at proposing a blockchain architecture to mitigate BWE in SSC by improving end-to-end visibility among supply chain partners through sharing backlog information. The proposed supply chain-based blockchain enables SSC partners to share securely and transparent backlog information mitigating thus BWE.

Keywords: Digitalization · Blockchain technology · Information sharing · Bullwhip effect (BWE) · Demand amplification · Service supply chain (SSC)

1 Introduction

Supply chain management is developing dynamically over time. This is attributed to the growing of businesses and the pressure of the market and customers. Therefore, improvement of supply chain becomes an imperative need for businesses to keep a competitive advantage [1]. A supply chain includes a network of organizations that shares information and collaborates to produce and deliver products and services to the ultimate consumer [2]. Information transparency and coordination among supply chain partners are required to achieve this goal [3]. Hence, information sharing attracts a great attention in literature as a key tool giving supply chain collaborators the ability to plan, forecast,

R. Karim et al. (Eds.): IAI 2021, LNME, pp. 202–211, 2022.
https://doi.org/10.1007/978-3-030-93639-6_17

and produce [4]. Upstream parties, suppliers, conduct their demand forecasting relying on the available information that passed to them through the preceded party [5]. Supply chain parties, however, may not share their information with other parties, because they consider such information company sensitive asset. In addition, downstream parties, retailers, may share distorted information, particularly distorted demand information, in order to get advantages and thus affect the receiver's decisions [5]. These information distortions cause several problems in supply chain [6], mainly demand amplification. Jay Forrester [7] in 1961 firstly observed this amplification. This phenomenon has become known as bullwhip effect (BWE) [7], which refers to increasing fluctuations in demand as one moves up in the supply chain causing excessive inventory levels [3]. Plenty of studies investigating BWE state that to overcome this amplification, distorted information needs to be avoided and partners need to share demand information in a better way [8, 9]. Blockchain and other disruptive technologies; for instance, Internet of things (IoT) and artificial intelligence (AI) have the potential to transform supply chain [10] and logistics [11] and provide remedies for information asymmetric. Blockchain, which has been initially implemented in the context of cryptocurrency, Bitcoin, by Nakatomo in 2008 [12], is a shared database on a peer-to-peer network that stores and manages data. In addition, it has the ability to share different kinds of data instead of only financial transactions [13]. The implementation of blockchain technology has expanded into different industries, such as healthcare, market monitoring copyright protection, and supply chain management [14–18]. Recent literature on blockchain technology in the context of supply chain management shows an increasing interest to adopt and implement it [19]. Blockchain technology has been applied to improve supply chain performance and partnership efficiency [16] and used as a tool to transform supply chain management [20]. More importantly, it has been implemented for traceability [21], information sharing [22], green supply chain [23], and BWE mitigation in manufacturing supply chain [9]. Most of the works investigating BWE focused on manufacturing supply chains [6, 8, 24–26]. Nonetheless, Service Supply Chains (SSC) should also attract attention due to the increasing significance of services industry. To the authors' knowledge, no study has thus far considered blockchain technology to mitigate BWE in SSCs. Therefore, this paper aims at proposing a blockchain architecture to mitigate BWE in SSC by supporting visibility and transparency of the information along the whole SSC stages.

This paper consists of additionally four sections. Section 2 presents the introduction of this paper. Related work on BWE and the implementation of blockchain technology are presented in Subsects. 2.1 and 2.2. Section 3 introduces conceptual framework for the specific application. Section 4 demonstrate the blockchain architectural realization. Conclusion and future research are given in Sect. 5.

2 Literature Review

2.1 Bullwhip Effect and Information Sharing

BWE is one of the most investigated problem in supply chain management [27]; it is also considered as a forecast driven phenomenon [28], and it is a problem of information asymmetry [27]. BWE first was noticed in 1961 [7]; it refers to the gradual amplification of customers demand far from the actual one as it moves further across the supply chain echelons [29]. BWE occurrence in manufacturing supply chain attributes to numerous causes: price fluctuations, order batching, single demand forecasting, shortage gaming [6], and the lack of information sharing and coordination among supply chain partners [30]. However, due to the nature of SSC, namely perishability, intangibility, simultaneous management of demand and capacity, and the customer-supplier co-production [31], the main cause of BWE in SSC is summarized by the fluctuations of backlog level, its impact on workload level, and the adjustment of capacity [32]. To clarify, BWE existence in SSC appears as a backlog amplification, increases in workload and changes in current capacity due to the delays in noticing, making, and implementing decisions regarding backlog information swings [32–36]. The fact is that BWE causes rely on one key root, which is the poor coordination and collaboration among supply chain partners [30]. In literature, information sharing is noted as a significant remedy to eliminate BWE [29, 37–39]. The table below illustrates the types of information utilized to mitigate BWE in literature (Table 1).

Table 1. Types of information sharing used to mitigate BWE

Authors	The shared information type
[24, 43–45]	Demand information
[40]	Sales information, point of sales (POS)
[41, 48, 51, 55, 56]	Inventory information
[42]	Demand and order information
[25, 49]	Order and inventory information
[46]	Order information
[47]	Demand forecast information
[50]	Ordered quantity information
[52]	Demand and warehouse lead-time information
[53]	Demand forecasting and lead-time information
[54]	Demand and lead-time information

2.2 Blockchain Technology

Blockchain technology can be described as a shared digital database of transactions, records, and events distributed throughout the network and which gives access to such data to authorized participants [57, 58]. Blockchain includes a chain of blocks that are linked together with an encrypted hash (code); each subsequent block holds the hash of the previous block which makes the chain secure, immutable and transparent. All participants (nodes) in blockchain network possess the same copy of the digital ledger [13]. There is no central authority controlling the network. In contrast, all the nodes can access and review the data in the real time [59]. In other words, blockchain technology can make the data visible and transparent for all parties that are allowed to enter the network [60], and it supports real time communications among them [20].

At first, blockchain technology was implemented in financial applications (i.e., cryptocurrencies) [12]. Unlike other information technology, blockchain distinct features expand the utilization of such technology to different businesses: E-government [61], healthcare [62, 63], energy [64], banking [65], and supply chain management [13, 23, 66]. Blockchain applications in supply chain are growing considerably [60]. Plenty of studies focused on blockchain applications in sustainable supply chain [20, 58, 67, 68], products origins traceability [69], enhancement of supply chain resilience [70], collaboration and coordination in maritime supply chain [71], trust sharing [72], information sharing [73, 74], transparency [75], product origins monitoring [76], improvement supply chain partnership efficiency and performance [77], and BWE mitigation in manufacturing supply chain [9]. Few studies have been dedicated to track BWE in SSC [30, 32–35]. Additionally, only one single study has used blockchain technology to mitigate BWE in manufacturing supply chain [9]. As a result, this paper aims to propose an architecture applying blockchain technology to eliminate BWE in SSC.

3 Conceptual Framework

In this paper, we propose a blockchain architecture that focuses on sharing the accumulative number of unfulfilled orders (backlog data) among SSC partners to mitigate BWE [36]. The SSC structure is based on the model by Ellram et al. [78]. Also, our study considers the key stages of SSC service producer, service provider, customer, and other service partners. The focus of the proposed architecture is in supporting the visibility and transparency of the whole SSC by sharing the backlog data among service partners in order to maintain appropriate level of backlog and deliver the right service to the ultimate customer in the right time. Therefore, we consider the service provider as a vital stage that collects and monitors the backlog data in order to eliminate BWE. Sharing real-time backlog data with the upstream stages helps to avoid the occurrence of any management delays and other related consequences. Moreover, we use the hybrid blockchain system (consortium), which applies both permissionless (public) and permissioned (private) ledgers to secure certain information and limit participating in such network by central company (service provider) and share service information with the end customers.

4 Blockchain Architecture

Nakamoto [12] defines 6 steps for running a blockchain network. Some of these steps can be neglected based on the type of transactions are shared: distributed ledger, transaction, block creation, nodes, consensus algorithm, and adding new block [70]. Blockchain network gives access to all nodes, namely, all participants have the same copy of real-time backlog information, which is stored in a chain of blocks. Once the backlog information is added to a new block, the nodes (SSC parties) ensure that the added information is authenticated throughout a consensus algorithm [12]. Then, the new block is distributed

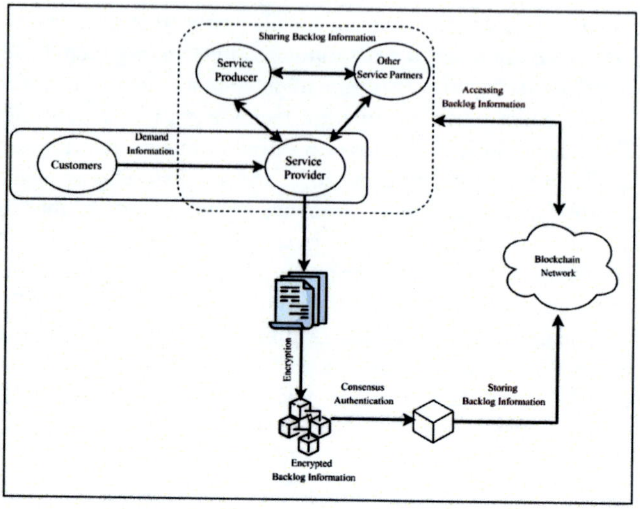

Fig. 1. Blockchain architecture to mitigate BWE via sharing backlog data.

through the whole network and linked with the previous blocks after the validation process [12].

Figure 1 illustrates the proposed framework to mitigate BWE through blockchain technology. The Proposed architecture is built based on the work of Akkhmans and Voss [36], who investigated the occurrence of BWE in two SSC's: the installation of customer broadband services and the installation of glass fibre network services. They found that BWE manifests itself in SSC in the form of variabilities in backlog information across SSC partners. In addition, they found that sudden increases of sales rate happen in customer's calls and complains. This in turn rises the workload for processing such backlogs, which leads to more burden on the companies' resources because it must be fulfilled by people not computers [36]. Additionally, they proposed five root causes of the occurrence of service bullwhip effect (SBWE).

As a result, this study sheds the lights upon the first root cause that concerns about delays and work backlogs, that represented in three main kinds of delays: 1) delays or failure in processing information, 2) delays in making further decisions regarding the sudden actions, 3) delays or failure in implementing adjustment decisions [36]. Therefore, we utilize blockchain technology as a remedy to mitigate the delays and failure of processing backlog information by improving the transparency among SSC parties in terms of sharing real time backlog information. The proposed architecture enables all SSC parties to access the same backlog information database (the database of unfulfilled orders). Additionally, it improves visibility among SSC parties by noticing the fluctuations of backlogs and taking further decisions regarding adjusting their resources in order to process these unfulfilled orders; for instance, adjusting the number of current employees or the current capacity and dividing the overload among SSC parties through outsourcing backlogs in order not to lose sales and maintain the customer satisfaction.

Here is a brief interpretation of the architecture flow: the service provider receives the demand information from customers. Unexpected demand fluctuations affects the service supplier's resources. Therefore, service providers face an accumulation of unfilled orders, called Backlogs. Blockchain technology can play a vital role to clear such unfulfilled orders and mitigate any amplification may occur to the service provider resources such hiring and firing employees. In this endeavour, service provider stores the unfulfilled orders information (backlog information) into blockchain network with a hash key. All parties are involved in singing the backlog information. Then, the backlog information is distributed throughout the network. The service provider adds the backlog information to a block and the block is added to the chain. The new block having the backlog information is distributed throughout blockchain network and the SSC parties decrypt the backlog information and acquire the public key. The other parties in SSC check if the backlog information is signed by the authorized SSC parties. After that, the SSC parties get access to real-time backlog information following a validation process based on specific consensus algorithm and proceed to process the unfulfilled orders. This may mitigate the delays in processing the shared backlog information, which causes inefficiencies for the whole SSC parties.

5 Conclusion and Future Research

Blockchain technology has the potential to revolutionize SSC domain and introduce intelligent and efficient means to optimize its operations. Based on the work of Akkhmans and Voss [36], in this paper, we propose an architecture that takes advantage of such technology to mitigate the well-known BWE phenomenon in SSC. The architecture enables the SSC participants to share real time, secure and transparent backlog information. Backlog information sharing is the key pillar of the architecture as it eliminates the swings in service demand if it is stored and shared in real time. one the other hand, such architecture allows managers to notice the backlog variability in case sudden fluctuations hit service demand and enable them to take instant decisions towards any needed adjustments in order to fullfil the backlog levels before any amplification. Ultimately, the proposed architecture has the potential to improve the process of sharing timely backlog information through reducing delays and failures in SSC. Regarding future research, the proposed framework can be improved by applying the smart contracts technology among SSC participants. Further, the proposed architecture could be tested and validated by conducting an empirical study on service companies that would bring additional insights for researchers and practitioners.

References

1. Helo, P., Hao, H.: Blockchains in operations and supply chains: a model and reference implementation. Comput. Ind. Eng. **136**, 242–251 (2019)
2. Sharma, A., Kumar, R.: Sharing of information in supply chain management. Int. J. Sci. Eng. Technol. Res. (IJSETR) **6**(1), 2278–7798 (2017)
3. Wang, X., Disney, S.M.: The bullwhip effect: progress, trends and directions. Eur. J. Oper. Res. **250**, 691–701 (2016)

4. Trapero, J.R., Kourentzes, N., Fildes, R.: Impact of information exchange on supplier forecasting performance. Omega **40**(6), 738–747 (2012)
5. Deghedi, G.A.: Information sharing as a collaboration mechanism in supply chains. Inf. Knowl. Manag. **4**(4), 82–95 (2014)
6. Lee, H., Padmanabhan, P., Whang, S.: Information distortion in a supply chain: the bullwhip effect. Manage. Sci. **43**(4), 546–558 (1997)
7. Forrester, J.W.: Industrial dynamics - a major breakthrough for decision makers. Harv. Bus. Rev. **36**(4), 37–66 (1961)
8. Bray, R.L., Mendelson, H.: Information transmission and the bullwhip effect: an empirical investigation. Manage. Sci. **58**, 860–875 (2012)
9. van Engelenburg, S., Janssen, M., Klievink, B.: A blockchain architecture for reducing the bullwhip effect. In: Shishkov, B. (ed.) BMSD 2018. LNBIP, vol. 319, pp. 69–82. Springer, Cham (2018). https://doi.org/10.1007/978-3-319-94214-8_5
10. Treiblmaier, H.: The impact of the blockchain on the supply chain: a theory-based research framework and a call for action. Supply Chain Manag. Int. J. **23**(6), 545–559 (2018)
11. Winkelhaus, S., Grosse, E.H.: Logistics 4.0: a systematic review towards a new logistics system. Int. J. Prod. Res. **58**, 18–43 (2020)
12. Nakamoto, S.: Bitcoin: a peer-to-peer electronic cash system (2008)
13. Pournader, M., Shi, Y., Seuring, S., Koh, L.S.C.: Blockchain applications in supply chains, transport and logistics: a systematic review of the literature. Int. J. Prod. Res. **58**, 2063–2081 (2020)
14. Khan, M.A., Salah, K.: IoT security: review, blockchain solutions, and open challenges. Futur. Gener. Comput. Syst. **82**, 395–411 (2018)
15. Scott, B., Loonam, J., Kumar, V.: Exploring the rise of blockchain technology: towards distributed collaborative organizations. Strateg. Chang. **26**(5), 423–428 (2017)
16. Kim, H.M., Laskowski, M.: Toward an ontology-driven blockchain design for supply-chain provenance. Intell. Syst. Account. Financ. Manag. **25**(1), 18–27 (2018)
17. Savelyev, A.: Copyright in the blockchain era: promises and challenges. Comput. Law Secur. Rev. **34**(3), 550–561 (2018)
18. Queiroz, M.M., Telles, R., Bonilla, S.H.: Blockchain and supply chain management integration: a systematic review of the literature. Supply Chain Manag. Int. J. **25**(2), 241–254 (2018)
19. Pawczuk, L., Massey, R., Schatsky, D.: Breaking blockchain open: Deloitte's 2018 global blockchain survey. Deloitte (2018). https://www2.deloitte.com/content/dam/Deloitte/us/Documents/financial-services/us-fsi-2018-global-blockchain-survey-report.pdf
20. Saberi, S., Kouhizadeh, M., Sarkis, J.: Blockchain technology: a panacea or pariah for resources conservation and recycling? Resour. Conserv. Recycl. **130**, 80–81 (2018)
21. Kshetri, N.: Blockchain's roles in strengthening cybersecurity and protecting privacy. Telecommun. Policy **41**(10), 1027–1038 (2017)
22. Van Engelenburg, S., Janssen, M., Klievink, B.: Design of a software architecture supporting business-to-government information sharing to improve public safety and security. J. Intell. Inf. Syst. **52**(3), 595–628 (2019)
23. Kouhizadeh, M., Sarkis, J.: Blockchain practices, potentials, and perspectives in greening supply chains. Sustainability **10**(10), 36–52 (2018)
24. Chen, F., Drezner, Z., Ryan, J.K.: Quantifying the bullwhip effect in a simple supply chain: the impact of forecasting, lead times and information. Manage. Sci. **46**(3), 436–443 (2000)
25. Ouyang, Y.: The effect of information sharing on supply chain stability and the bullwhip effect. Eur. J. Oper. Res. **182**(3), 1107–1121 (2007)
26. Li, C.: Controlling the bullwhip effect in a supply chain system with constrained information flows. Appl. Math. Model. **37**(4), 1897–1909 (2013)

27. Ma, Y., Wang, N., Che, A., Huang, Y., Xu, J.: The bullwhip effect under different information-sharing settings: a perspective on price sensitive demand that incorporates price dynamics. Int. J. Prod. Res. **51**(10), 3085–3116 (2013)
28. Rahman, H., Ali, N., Munna, M., Rafiquzzaman, Md.: A review on the investigation of the causes of bullwhip effect in supply chain management. In: International Conference on Mechanical, Industrial and Energy Engineering. ICMIEE-PI-140125 (2014)
29. Wang, X., Disney, S.M.: The bullwhip effect: progress, trends and directions. Eur. J. Oper. Res. **250**(3), 691–770 (2016)
30. Ross, F.D.: Distribution Planning and Control: Managing in the Era of Supply Chain Management, 3rd edn. Springer, New York (2015)
31. Shahin, A.: SSCM: service supply chain management. Int. J. Logist. Syst. Manag. **6**(1), 1–15 (2010)
32. Akkermans, H., Vos, B.: Amplification in service supply chains: an exploratory case study from the telecom industry. Prod. Oper. Manag. **12**, 204–223 (2003)
33. Anderson, E.G., Jr., Morrice, D.J.: A simulation game for teaching services-oriented supply chain. Prod. Oper. Manag. **9**(1), 40–55 (2000)
34. Anderson, E.G., Jr., Morrice, D.J., Lundeen, G.: The "physics" of capacity and backlog management in service and custom manufacturing supply chains. Syst. Dyn. Rev. **21**, 217–247 (2005)
35. Haughton, M.A.: Distortional bullwhip effects on carriers. Transp. Res. Part E **45**, 172–185 (2009)
36. Akkermans, H., Voss, C.: The service bullwhip effect. Int. J. Oper. Prod. Manag. **33**(6), 765–788 (2013)
37. Chen, L., Lee, H.L.: Information sharing and order variability control under a generalized demand model. Manage. Sci. **55**(5), 781–797 (2009)
38. Bray, R., Mendelson, H.: Information transmission and the bullwhip effect: an empirical investigation. Manage. Sci. **58**(5), 860–875 (2012)
39. Syntetos, A., Babai, Z., Boylan, J.E., Kolassa, S., Nikolopoulos, K.: Supply chain forecasting: theory, practice, their gap and the future. Eur. J. Oper. Res. **252**, 1–26 (2016)
40. Croson, R., Donohue, K.: Impact of POS data sharing on supply chain management: an experimental study. Prod. Oper. Manag. **12**(1), 1–11 (2003)
41. Croson, R., Donohue, K.: Upstream versus downstream information and its impact on the bullwhip effect. Syst. Dyn. Rev. **21**(3), 249–260 (2005)
42. Yu, Z., Yan, H., Cheng, T.C.E.: Benefits of information sharing with supply chain partnerships. Ind. Manag. Data Syst. **101**, 114–121 (2001)
43. Argilaguet Montarelo, L., Glardon, R., Zufferey, N.: A global simulation-optimization approach for inventory management in a decentralized supply chain. Supply Chain Forum Int. J. **18**(2), 112–119 (2017)
44. Asgari, N., Nikbakhsh, E., Hill, A., Farahani, R.Z.: Supply chain management 1982–2015: a review. IMA J. Manag. Math. **27**(3), 353–379 (2016)
45. Chatfield, D.C., Kim, J.G., Harrison, T.P., Hayya, J.C.: The bullwhip effect-impact of stochastic lead time, information quality, and information sharing: a simulation study. Prod. Oper. Manag. **13**(4), 340–353 (2004)
46. Li, S., Chen, J., Liao, Y., Shi, Y.: The impact of information sharing and risk pooling on bullwhip effect avoiding in container shipping markets. Int. J. Shipp. Transp. Logist. **8**(4), 406–424 (2016)
47. Jeong, K., Hong, J.-D.: The impact of information sharing on bullwhip effect reduction in a supply chain. J. Intell. Manuf. **30**(4), 1739–1751 (2017). https://doi.org/10.1007/s10845-017-1354-y
48. Dai, H., Li, J., Yan, N., Zhou, W.: Bullwhip effect and supply chain costs with low and high-quality information on inventory shrinkage. Eur. J. Oper. Res. **250**(2), 457–469 (2015)

49. Agrawal, S., Sengupta, R.N., Shanker, K.: Impact of information sharing and lead time on bullwhip effect and on-hand inventory. Eur. J. Oper. Res. **192**(2), 576–593 (2009)
50. Ding, H., Guo, B., Liu, Z.: Information sharing and profit allotment based on supply chain cooperation. Int. J. Prod. Econ. **133**(1), 70–79 (2011)
51. Hassanzadeh, A., Jafarian, A., Amiri, M.: Modeling and analysis of the causes of bullwhip effect in centralized and decentralized supply chain using response surface method. Appl. Math. Model. **38**, 2353–2365 (2014)
52. Rached, M., Bahroun, Z., Campagne, J.-P.: Decentralised decision-making with information sharing vs. centralised decision-making in supply chains. Int. J. Prod. Res. **54**(24), 7274–7295 (2016)
53. Jiang, Q., Ke, G.: Information sharing and bullwhip effect in smart destination network system. Ad Hoc Netw. **87**, 17–25 (2018)
54. Ojha, D., Sahin, F., Shockley, J., Sridharan, S.V.: Is there a performance tradeoff in managing order fulfillment and the bullwhip effect in supply chains? The role of information sharing and information type. Int. J. Prod. Econ. **208**, 529–543 (2019)
55. Dominguez, R., Cannella, S., Framinan, J.M.: On bullwhip-limiting strategies in divergent supply chain networks. Comput. Ind. Eng. **73**(1), 85–95 (2014)
56. Wang, N., Lu, J., Feng, G., Ma, Y., Liang, H.: The bullwhip effect on inventory under different information sharing settings based on price-sensitive demand. Int. J. Prod. Res. **54**(13), 4043–4064 (2016)
57. Crosby, M., Pattanayak, P., Verma, S., Kalyanaraman, V.: Blockchain technology: beyond bitcoin. Appl. Innov. Rev. **2**, 6–9 (2016)
58. Manupati, V.K., Schoenherr, T., Ramkumar, M., Wagner, S.M., Pabba, S.K., Inder Raj Singh, R.: A blockchain-based approach for a multi-echelon sustainable supply chain. Int. J. Prod. Res. **58**, 2222–2241 (2020)
59. Gupta, M.: Blockchain for Dummies, 2nd edn. Wiley, Hoboken (2018)
60. Bai, C., Sarkis, J.: A supply chain transparency and sustainability technology appraisal model for blockchain technology. Int. J. Prod. Res. **58**, 2142–2162 (2020)
61. Bhardwaj, S., Kaushik, M.: Blockchain—technology to drive the future. In: Satapathy, S., Bhateja, V., Das, S. (eds.) Smart Computing and Informatics. Smart Innovation, Systems and Technologies, vol. 78, pp. 263–271 (2017). https://doi.org/10.1007/978-981-10-5547-8_28
62. Bocek, T., Rodrigues, B.B., Strasser, T., Stiller, B.: Blockchains everywhere - a use-case of blockchains in the pharma supply-chain. In: IFIP/IEEE Symposium on Integrated Network and Service Management (IM) (2017)
63. Mettler, M.: Blockchain technology in healthcare: the revolution starts here. In: IEEE 18th International Conference on e-Health Networking, Applications and Services (2016)
64. Munsing, E., Mather, J., Moura, S.: Blockchains for decentralized optimization of energy resources in microgrid networks. In: IEEE Conference on Control Technology and Applications (CCTA) (2017)
65. Guo, Y., Liang, C.: Blockchain application and outlook in the banking industry. Financ. Innov. **2**(1), 1–12 (2016). https://doi.org/10.1186/s40854-016-0034-9
66. Kshetri, N.: Blockchain's roles in meeting key supply chain management objectives. Int. J. Inf. Manage. **39**, 80–89 (2018)
67. Bai, C., Sarkis, J.: A supply chain transparency and sustainability technology appraisal model for blockchain technology. Int. J. Prod. Res. **58**(7), 2142–2162 (2020). https://doi.org/10.1080/00207543.2019.1708989
68. Saberi, S., Kouhizadeh, M., Sarkis, J., Shen, L.: Blockchain technology and its relationships to sustainable supply chain management. Int. J. Prod. Res. **57**(7), 2117–2135 (2019)
69. Lu, Q., Xu, X.: Adaptable blockchain-based systems: a case study for product traceability. IEEE Softw. **34**(6), 21–27 (2017)

70. Min, H.: Blockchain technology for enhancing supply chain resilience. Bus. Horiz. **62**, 1–35 (2019)
71. Philipp, R., Prause, G., Gerlitz, L.: Blockchain and smart contracts for entrepreneurial collaboration in maritime supply chains. Transp. Telecommun. Inst. **20**(4), 365–378 (2019)
72. Wang, L., Guo, S.: Blockchain based data trust sharing mechanism in the supply chain. In: Yang, C.-N., Peng, S.-L., Jain, L.C. (eds.) SICBS 2018. AISC, vol. 895, pp. 43–53. Springer, Cham (2020). https://doi.org/10.1007/978-3-030-16946-6_4
73. Longo, F., Nicoletti, L., Padovano, A., d'Atri, G., Forte, M.: Blockchain-enabled supply chain: an experimental study. Comput. Ind. Eng. **136**, 57–69 (2019)
74. Zheng, K., Zhang, J.Z., Chen, Y., Wu, J.: Blockchain adoption for information sharing: risk decision-making in spacecraft supply chain. Enterp. Inf. Syst. **15**, 1070–1091 (2019)
75. Francisco, K., Swanson, D.: The supply chain has no clothes: technology adoption of blockchain for supply chain transparency. Logistics **2**(1), 1–13 (2018)
76. Casado-Vara, R., Prieto, J., la Prieta, F.D., Corchado, J.M.: How blockchain improves the supply chain: case study alimentary supply chain. Procedia Comput. Sci. **134**, 393–398 (2018)
77. Kim, J.-S., Shin, N.: The impact of blockchain technology application on supply chain partnership and performance. Sustainability **11**(21), 61–81 (2019)
78. Ellram, L.M., Tate, W.L., Billington, C.: Understanding and managing the services supply chain. J. Supply Chain Manag. **40**(4), 17–32 (2004)

Check for updates

Data Driven Maintenance: A Promising Way of Action for Future Industrial Services Management

Mirka Kans[1(✉)], Anders Ingwald[1], Ann-Brith Strömberg[2], Michael Patriksson[2], Jan Ekman[3], Anders Holst[3], and Åsa Rudström[3]

[1] Linnaeus University, 351 95 Växjö, Sweden
{mirka.kans,anders.ingwald}@lnu.se
[2] Chalmers University of Technology, 412 96 Gothenburg, Sweden
{anstr,mipat}@chalmers.se
[3] Rise ICT, Kistagången 16, 16 440 Kista, Sweden
{jan.ekman,anders.holst,asa.rudstrom}@ri.se

Abstract. Maintenance and services of products as well as processes are pivotal for achieving high availability and avoiding catastrophic and costly failures. At the same time, maintenance is routinely performed more frequently than necessary, replacing possibly functional components, which has negative economic impact on the maintenance. New processes and products need to fulfil increased environmental demands, while customers put increasing demands on customization and coordination. Hence, improved maintenance processes possess very high potentials, economically as well as environmentally. The shifting demands on product development and production processes have led to the emergency of new digital solutions as well as new business models, such as integrated product-service offerings. Still, the general maintenance problem of how to perform the right service at the right time, taking available information and given limitations is valid.

The project Future Industrial Services Management (FUSE) project was a step in a long-term effort for catalysing the evolution of maintenance and production in the current digital era. In this paper, several aspects of the general maintenance problem are discussed from a data driven perspective, spanning from technology solutions and organizational requirements to new business opportunities and how to create optimal maintenance plans. One of the main results of the project, in the form of a simulation tool for strategy selection, is also described.

Keywords: Data driven maintenance · Service-related business models · Maintenance planning · Simulation tool

1 Introduction

The industry demands flexible, safe, environmentally friendly, and available production processes. Industrial digitalisation provides technological solutions to the needs of modern manufacturing—manifested as highly automated, flexible, and just-in-time production—with focus on waste reduction [1]. Large amounts of data created during the

R. Karim et al. (Eds.): IAI 2021, LNME, pp. 212–223, 2022.
https://doi.org/10.1007/978-3-030-93639-6_18

production are utilized for data driven planning and optimization of the operations. This new situation poses increased demands on the flexibility and automation in the maintenance management as well. At the same time, advancements in technologies for data acquisition, data fusion, data analysis, diagnostics, and prognosis provide possibilities to organise the work in new ways, resulting in an increased effectiveness as well as a better adaptation of maintenance to customer needs [2]. Data and information at the right place and at the right time are crucial factors for effective maintenance management. Knowledge sharing and networking enable all personnel involved in maintenance sharing experiences and knowledge, and assistant systems allow for easy interpretation of a vast amount of data for direct use, for instance in assisting the maintenance technicians in their daily work. The technological development has forced the industry to change their business models as well, and therefore the importance of after-sales services has increased [3]. For these technological innovations to succeed, they must be coordinated with, and reflected in, the strategic business models. The traditional view on operations and maintenance services as a necessary cost is shifting to an activity that could potentially lower the total life-cycle cost of the product by the means of performance-based contracts and long-term relationships.

It is a complex task to reform the way maintenance is performed today—to better adapt to the new technological advances, the available data from the processes and systems, and the demand for a flexible and dynamic production. The last decades have seen huge efforts to move towards condition-based and predictive maintenance, in contrast to fixed schedules. In spite of this, the changes have not been as fast and smooth as envisioned. One reason for this is that the change affects several aspects on different levels, all the way from the organizational level down to the technical equipment, which must all be considered. Furthermore, there is a preparedness in organizations for the fact that this change of maintenance strategies is absolutely necessary. Still there is a considerable uncertainty on how to introduce the change, what options there are, how large the potential is, and what risks are present.

Thus, the challenges span from developing new technologies, ways to utilize data, as well as changing the way business is done. The project Future Industrial Services Management (FUSE), a collaboration between Siemens Industrial Turbomachinery, Euromaint Rail, GKN Aerospace, Scania CV, TXG Development, Eduro, Chalmers University of Technology, Linnaeus University, and RISE (formerly SICS Swedish ICT), was launched during 2014–2016 for approaching the challenges from a holistic and interdisciplinary perspective. In the project, researchers from several subject areas collaborated for defining the challenges as well as for providing solutions. This paper reports on the challenges and prerequisites that were defined, and describes some main results achieved in the project.

2 Overview of the General Maintenance Problem

Designing the overall maintenance strategy for a system or a plant, is to determine what specific maintenance strategies are useful for each subsystem, component, or even fault mode. There are several strategies to choose from, but no single one will be suitable to use throughout a whole process or industry. The following two non-condition-based maintenance strategies have traditionally been the most common ones [4]:

1. Scheduled maintenance: service is performed at some predetermined intervals, measured in some unit of usage—often operating hours (or in the case of vehicles, kilometres);
2. Corrective maintenance: service is performed only when a component has failed or is not functioning properly.

The former may of course lead to unnecessary maintenance, and the latter to costly failures and production stops. The idea behind condition-based and predictive maintenance is that—with some information about the condition of the components, and a more dynamic maintenance planning in response to this information—maintenance can to a higher degree be performed when it is actually needed, avoiding both unnecessary maintenance and failures [4–6]. Nevertheless, not all components lend themselves to condition-based maintenance, and there will always be failure modes that cannot be predicted in advance, so there will still be components for which these two traditional strategies constitute the most rational choices.

A general maintenance strategy can be divided into two parts that need to be determined separately: (i) how information about the condition or the remaining life of the components is acquired; (ii) how the maintenance plan can be adjusted in response to such information. Moreover, the maintenance strategy has to generate value to its clients and customers. Two perspectives could be applied; the internal where maintenance is supporting the key processes within a company, and external where maintenance is sold as a service to a customer.

2.1 Acquiring Information

An obvious solution, regarding condition-based maintenance, is to use various sensors to continuously measure some signal related to the component's condition, and use this measure to estimate or predict the component's future condition. The main difference between *continuous condition monitoring with sensors* and *inspections*, is that the inspections occur much more seldom, have associated costs or labor requirements, and typically have to be scheduled in advance, just like any maintenance. Another difference is that at an inspection the condition is often directly measured, while a continuous condition monitoring most often results in indirect measurements of quantities that (only indirectly) affect the (inspected) condition. In some situations and for some components, using sensors to measure the condition may work well and may be the most effective solution. One example is when detecting noise or vibrations in rotating machinery. This can be done by placing a microphone or accelerometer somewhere on the machine, without interfering with the design or function of the machine, and in a location where it is easily available for service.

However, the use of sensors to measure conditions is also connected with several complicating factors that make it less useful in many cases; hence, sensors should be used with care and only when this can be motivated by a technical risk analysis. For example, adding a new sensor means adding a new component in the design as well as a new service point in the maintenance plan, since the sensor itself may need service. So, it is associated with a cost and leads to a more complicated design (and associated maintenance planning) [5]. Further, a measurement results in an estimation of the condition, and the estimate may be rather weakly correlated to the actual condition, and thus of limited value for estimating the service need [5]. If a component has several fault modes and the measurements of a suggested sensor only relates to some of them, adding such a sensor may only have a limited impact on the resulting maintenance. This situation is further aggravated in cases when no dedicated sensor is used, but already existing sensors—installed for completely different purposes—are reused for condition monitoring. There is a risk that such data tells very little about the actual service needs.

Furthermore, often it is not sufficient to know the current condition of the component, but also to be able to predict whether the component will make it until the next service occasion; such information is an outcome of knowledge about at what speed the performance of the component (or the system) decreases due to a degrading condition of the component. These circumstances require the measured signals to indicate well in advance when a failure is approaching [6]. Some fault modes are not easily detected a long time ahead, whereas others may be more predictable.

If a sensor solution is used, one has to make sure that the sensors actually measure properties that are closely related to the component's condition, failure risk, and maintenance requirements, that the most common and/or important fault modes are considered, and that the sensor itself and its communication link are robust and easily maintained. An alternative to using measured signals to estimate the component's actual condition is to estimate the wear of the component from its usage. If the life of a component has a low variance, then the prediction of the remaining life may be quite good [4]. On the other hand, if the variance is high predictions are less reliable.

Advantages of estimation of wear are that (i) it can be used when the component's actual condition is hard to measure, and that (ii) in most systems, the usage (measured in various ways) is already closely monitored, so no new sensors need to be added and maintained. Disadvantages are that (iii) a threshold for an acceptable usage, or (iv) a wear model relating the usage to a probability of failure, must be known.

2.2 Creating Optimized Plans

Today an industrial system is typically composed of a large number of components, having complicated relationships between costs of repair and/or replacement, as well as failure mechanisms. In order to be able to comprehend such complicated systems—that are to be maintained—and decide on which are their (functionally and/or economically) important ingredients, it is fruitful to describe such systems using logical and/or mathematical frameworks. A further complicating fact is that the available information about the (possible) degradations of components in such systems may evolve over time, which calls for a dynamic re-planning of maintenance activities. Maintenance strategies can be tailored to the industrial system as a whole, as well as to the individual components' reliability and failure characteristics, thereby enabling a balance between the objectives to minimize the risk of unplanned system downtime, and to minimize the costs of preventive maintenance of components; this trade-off is naturally treated through a multi-objective optimization. The inherent properties of the systems, combined with the desired properties of the planning solutions—as described in the previous paragraph—illustrate the need for automated decision tools [7]. Moreover, the fact that these systems are characterized by complex relationships among failure mechanisms and the costs involved in the maintenance activities themselves, make it difficult—or even impossible—to create (near-)optimal maintenance plans by hand.

Maintenance could be planned for a single component, as well as for multi-component systems [8]. Advantages of planning for single-component systems include the possibility to regard more properties of the component at hand (e.g., its failure mechanisms, and how condition monitoring may cause component degradation, or even failure), as well as the particularities of how its condition can be monitored. Advantages of planning for multi-component systems include possibilities to handle economic, stochastic, and structural dependencies between components, as well as between components, and the operation of the system. Still, such dependencies are often necessarily modelled in a simplified manner, in order to enable the creation and execution of planning models. The planning can be performed through the use of heuristics or mathematical optimization models. Heuristics (of varying complexity) often comprise parameters related to the system and/or components, and the values of which may in turn be optimized. Mathematical optimization models differ from heuristics in that they aim to mimic the most essential parts of the real systems and usually possess significantly more degrees of freedom. Unsurprisingly, planning using optimization methods tends to result in significant planning improvements over simple heuristics, while heuristics almost always are computationally faster. Since all models developed and used are simplifications of the real systems studied, there is definitely a potential for improvements of both models and corresponding optimal maintenance plans for most types of industrial systems. In reality there may be many relevant objectives, such as to minimize monetary costs, to maximize system availability and reliability, to minimize risks of failure or accidents, and to maximize the quality of service towards customers. In optimization models, it is common to (transform and) combine some of these objectives into a single objective function, monetary costs being the most obvious and common quantity then used. While multiple objectives are viable within continuous, convex optimization, it is however an order of magnitude more complex to find a well-spread collection of near-Pareto optimal solutions in discrete-variable contexts, which is often the case for maintenance planning [9]. A further complication of the situation is that the actual cost of a maintenance decision is not deterministic, since it can not be known with certainty whether a component will last until the next service occasion. This calls for methods for planning under uncertainty, which is even more difficult and an active research field, see e.g. [10–12]. Nevertheless, industrial maintenance decisions would benefit largely from exploring and comparing maintenance plans that are optimized towards different (combinations of) objectives. The effective construction and utilization of relevant maintenance models of course rely heavily on the availability of sufficiently current, reliable, and clearly defined information, the lack of which is a common challenge.

Examples of optimization of the maintenance plan with respect to different degrees of freedom are

- to find the best points in time for certain maintenance activities,
- to find the most important activities to perform at fixed time points,
- to find a dynamic packaging of maintenance activities, and

- to perform a simultaneous planning of maintenance and operation.

2.3 Creating Internal and External Value

Often the maintenance planning problem is phrased in terms of minimizing the costs for maintenance. This could work if the full cost for *lack of maintenance*, such as cost for the lost production in case of a failure, is considered [13]. However, it is generally better to consider maintenance as a *prerequisite for production*, and not as a cost [12, 14]. Then, an important factor is instead the *system uptime* (resulting from the performance of maintenance). Another is *product quality* and *customer satisfaction*. For a complete picture, the *net production profit* is the objective that should be optimized, connecting maintenance with competitive business factors [15]. It is convenient to translate all factors into economic terms, but this is not always possible. For example, it can be hard to estimate the economic value of customer satisfaction, such that a bad product quality or unreliable services may affect the value of the brand in the long run. The *risk of failure* is an important factor [14]. Sometimes a failure will only result in a longer production stop and/or more expensive maintenance. In this case the risk can be translated into economic terms. However, in other cases a failure may cause a major damage, and even risk of health and environment, and may become a long-lasting damage of the brand. In such cases it is more reasonable not to mix the risk with the economic factors, but instead optimize the economy under the constraint that the risk of failure is kept sufficiently low. Another approach to include risk in the planning process is to use *multi-objective optimization*, in which two or more objectives are optimized, resulting in several Pareto optimal solutions. A decision maker may then choose between these solutions and make the compromise between low costs and low risk of failure (and possibly also other objectives) [16].

Changes and innovations in production have led to the emergency of new business models, such as integrated product-service offerings. There are opportunities for both producing companies and service providers to gain benefits by using integrated business models, where products are sold together with services [3]. For achieving these benefits, the view of the business has to change. The focus should be on the values created rather than on what and when to conduct maintenance [17]. The maintenance organisation possesses important resources for realising maintenance in the form of personnel, equipment, knowledge, and data. There resources constitute the foundation for value creation. To sell the value of maintenance opens up for a more effective maintenance, since it will be directly connected to the need of the customer [17]. However, this will also lead to more responsibility and a higher risk for the company delivering the maintenance [18]. To handle this risk, access to more reliable data is essential, and the development in digital technology, optimisation, and simulation enables the transition from simple customer offers to more flexible and effective offers providing utility for the customers.

Most types of maintenance related agreements are regulated in contracts. Resource-based contracts are usually cost-based (the customer pays for the direct resources that are used). In resource-based maintenance contracts, the supplier pays for actual expenses, such as labour costs and spare parts, as well as a profit margin. Performance-based contracts are regulated based on a pre-defined performance qualification. The supplier guarantees a certain level of performance and the customer pays for direct and indirect expenses as well as the increased risk taken by the supplier. The ability to assess the condition of systems and components becomes important for the supplier of performance-based contracts in order to manage the technical risk of downtime and for effective maintenance planning. Advanced forms of performance-based holistic contracts guarantee not only a product's or system's uptime, but also the customer's total operation. In these utility-based contracts the supplier takes a greater risk, in that it ensures the business and not a specific system's or product's operation, and this risk must be carefully priced and regulated in the contract [18].

Factors to consider for optimization of the maintenance business model are [19, 20]: key resources and key partners, availability of data, type of relationship with the customer, net profit of the company, and, business risks and technical risks.

3 Results

3.1 A Simulator to Show the Potential of Strategies

In the context of the present digitalisation of society in general, the envisioned methods and tools should be thought of as realizers of a maintenance upgrade, which is achieved by seizing the digitalisation opportunities. This is put in effect by (i) demonstrating the benefits of novel digitalised and data driven condition-based maintenance strategies, (ii) comparing these novel strategies with the ones used at present, and (iii) providing decision support tools, which as clearly and correctly as possible show the differences between strategies, thereby serving as a foundation that "make the best possible decisions". During the project, a model for evaluation of maintenance strategies was developed. The overall goal is to enable an evaluation and a comparison of maintenance strategies, in terms of the total cost and the total risk taken. The concept of *a general maintenance problem* serves as a basis for development of general methods and tools for analysing maintenance strategies, including business models, maintenance planning, and statistical models for various kinds of mechanical degradation.

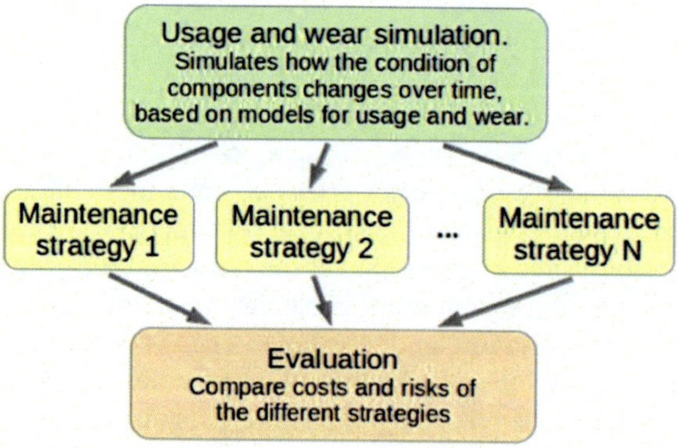

Fig. 1. The maintenance strategy evaluation tool.

The strategy evaluation tool consists of a number of components, as illustrated in Fig. 1. First, there is a simulator to mimic the usage and condition of the system to maintain. It represents all components of the system along with their fault modes. It also keeps track of the usage and wear of the system over time, using wear models and load distributions, which can either be provided by experts or estimated from historical data from the real system. This is followed by the simulation of a number of maintenance strategies to compare. Typically, there is one base strategy, mimicking the way maintenance is performed today. This can then be compared with one or more strategies that are more advanced, such as using more condition information and/or advanced maintenance planning. Finally, the evaluation part computes the cost for the maintenance to be performed according to the different strategies and the risk taken by them, and presents this in a number of plots and tables.

The tool has been used in the case studies with Euromaint and Siemens. For Euromaint (Fig. 2) both maintenance of doors and of the compressor on the studied set of trains can, according to the simulator, be reduced by almost 40% without significantly increasing the risk of failures. In the figure, blue represents the cost (in terms of working hours) of the current maintenance strategy, and green the cost with the suggested maintenance strategy. For Siemens (Fig. 3) a reduction of around 20% seems to be possible without an increased risk of failures (using synthetic but realistic data, so the real reduction can differ from this).

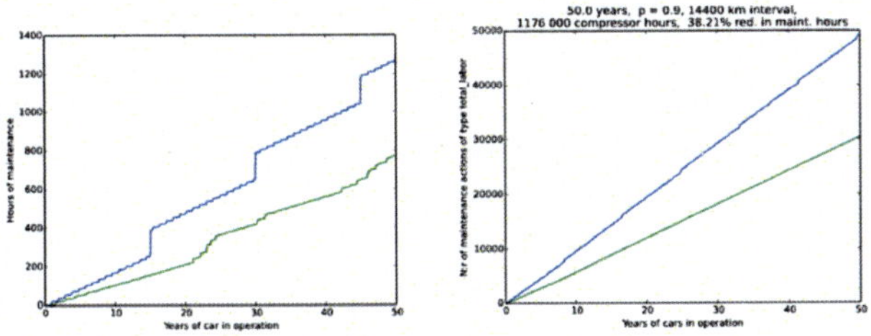

Fig. 2. Euromaint case studies results. *Left:* maintenance of doors. *Right:* maintenance of compressors.

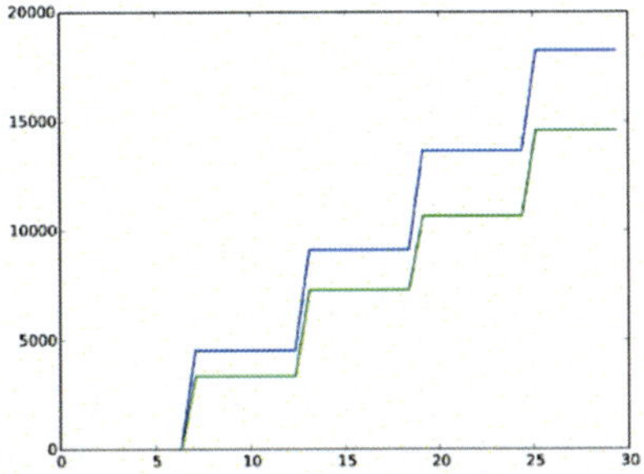

Fig. 3. Siemens case study results of a turbine (with synthetic but realistic costs and wear models).

Thus, for both case studies, the potential reduction of maintenance needed is shown by the simulator to be significant. It also seems that the efforts needed to take the novel, more efficient, maintenance strategies into use should not be a substantial hindrance. The importance of swiftly taking actions to put the proposed strategies into practice cannot be more clearly expressed than shown by the case studies in the project. Of course, it cannot be generally guaranteed that the gain will be as large as in these cases. However, apart from the results concerning reduction of maintenance in these case studies, a prime result of these case studies is the simulator itself, which makes it possible to explore the potential gains and risks of novel maintenance strategies before they are implemented in the organization.

3.2 A Framework to Support the Mental Journey from Product Focus to Value Creation

For supporting the mental shift from viewing maintenance as a cost factor to something that creates value, a framework for business model development was developed; see Table 1. This framework can be viewed as the logical development of the service business model from a narrow technical perspective to a holistic product-service perspective; it can also reflect the technological developments and the service needs from a business modelling perspective, by converting maintenance and other services from being a technical product into a value-creating activity. This shift is visible between levels two and three in Table 1. The approach is applicable for producing enterprises who intend to shift their focus from delivering products to delivering integrated product-services, but also for service providers who want to go from traditional service contracts to long-term, performance-based contracts. It could also be used by a company for positioning themselves in terms of service management maturity. The framework addresses the core business logic and the fundamentals of the value creating process. Without a thorough understanding of the basics, the transition to a more advanced business model is hard. Making a description of the underlying business logic will increase the understanding of necessary changes and support the transformation. In the framework, this is expressed in seven factors that are to be considered when making the shift; four factors connected to the value proposition (Type of offering, Density, Quality dimensions, Business development strategy) and three factors connected to business strategy (Strategic perspective, View on profitability, View on value creation) [20].

The dynamics in the business development strategy is reflected in the modular based maintenance offerings; see Fig. 4 [18]. This means that different value offerings aimed at different customers are extensions of a base offer.

A more advanced offering is utilising the same basic resources, but is packaged so that the customer perceives it as a completely new offer, compared to the base offer. The basic value proposition is maintenance as a resource, for example, based on the supplier's maintenance skills, planning skills, and spare parts' management. The next level considers maintenance from a performance perspective, i.e., the service offering is connected to the system or product functionality. The most advanced offer focuses on the benefits that a maintenance offering provides to the client or customer, and is thus linked to the customer's results (providing utility to the customer). The highest levels are connected with the ability to differentiate the offering depending on, for instance, product characteristics. These offerings are connected with a larger business risk and requires access to more information: Both technical information regarding performance

Table 1. Service management 4.0 framework.

	← Product perspective		Utility perspective →	
Dimension	**Level 1**	**Level 2**	**Level 3**	**Level 4**
Type of offering	Technology	Mainly product	Product and service	Bundled offerings
Density	Low	High	Dynamic	
Quality dimensions	Mainly product		Combination of product and service	
Business development strategy	Technology-driven	Customer-driven	Utility-driven	Dynamic
Strategic perspective	Inside-out		Outside-in	Dynamic
View on profitability	Productivity	Customer satisfaction	Customer satisfaction/ relationships	Relationships (more or less formal)
View on value creation	The own business in focus	Value chain	Value star/network	Ecology

and operation but also regarding use and utility, and in addition other ways of analysing the data.

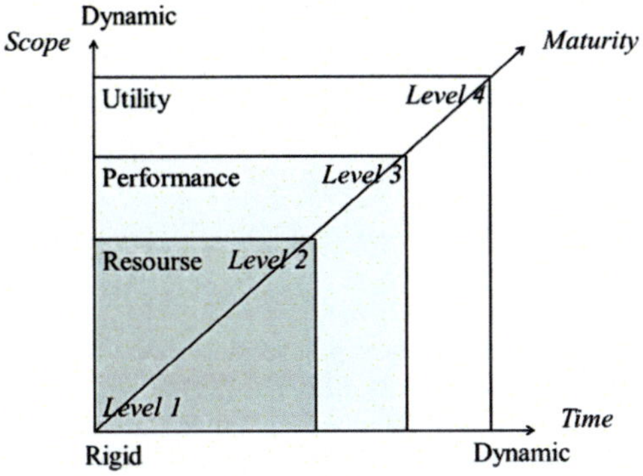

Fig. 4. Modular maintenance offerings.

4 Conclusions

The project Future Industrial Service Management has considered the general maintenance problem from several aspects:

- how to utilize the available condition information in order to create the best maintenance plans;
- how to demonstrate the value of designing such maintenance plans;
- how a better utilization of the available condition information supports the development of performance-based offerings and more advanced business models.

The main idea for the simulator developed in the project is to base the maintenance decisions on those metrics—directly measured or computed—that best reveal the true need for maintenance. A concrete goal for the further development of the simulator is to continue to use counters—measuring component usage rather than momentary physical conditions—for determining maintenance needs, as counters in general imply a swifter way towards the digitalisation of maintenance practice.

Future work includes using these new solutions to better support the decision-making process of selecting the alternatives that result in the best improvement of maintenance practice. Moreover, they support negotiations in the process of finding maintenance solutions that are the best with respect to the needs of different parties, and the most beneficial production solution. That is, the aim is to support decisions concerning complete value chains—from the decision to use new sensors or otherwise available data to business opportunities and strategies. It is also a goal to create trustworthy and realistic models of maintenance strategies suggested by the participating enterprises, including the creation of realistic models of the physical reality in order to predict future maintenance needs. This reduces the element of risk in maintenance decisions and support the effective use of business models for maintenance, where the focus is on the value created rather than on the time and resources used for performing the maintenance tasks. By this, maintenance comes closer to the end user in the value chain, and thus probably will be valued higher by the end user.

Acknowledgments. This research forms a part of the project Future Industrial Services Management funded by the Swedish Innovation Agency Vinnova.

References

1. Sanders, A., Elangeswaran, C., Wulfsberg, J.: Industry 4.0 functions as enablers for lean manufacturing. J. Ind. Eng. Manag. **9**(3), 811–833 (2016)
2. Kans, M., Galar, D., Thaduri, A.: Maintenance 4.0 in railway transportation industry. In: Koskinen, K.T., et al. (eds.) Proceedings of the 10th World Congress on Engineering Asset Management (WCEAM 2015). LNME, pp. 317–331. Springer, Cham (2016). https://doi.org/10.1007/978-3-319-27064-7_30
3. Visnjic Kastalli, I., Van Looy, B.: Servitization: disentangling the impact of service business model innovation on manufacturing firm performance. J. Oper. Manag. **31**(4), 169–180 (2013)
4. Ahmad, R., Kamaruddin, S.: An overview of time-based and condition-based maintenance in industrial application. Comput. Ind. Eng. **63**(1), 135–149 (2012)
5. Prajapati, A., Bechtel, J., Ganesan, S.: Condition based maintenance: a survey. J. Qual. Maint. Eng. **18**(4), 384–400 (2012)
6. Peng, Y., Dong, M., Zuo, J.: Current status of machine prognostics in condition-based maintenance: a review. Int. J. Adv. Manuf. Technol. **50**(1–4), 297–313 (2010)
7. Alaswad, S., Xiang, Y.: A review on condition-based maintenance optimization models for stochastically deteriorating system. Reliab. Eng. Syst. Saf. **157**, 54–63 (2017)
8. Zhu, Z., Xiang, Y., Zeng, B.: Multi-component maintenance optimization: a stochastic programming approach (2019). arXiv.org, 2 July 2019
9. Ehrgott, M.: Multicriteria Optimization. Lecture Notes in Economics and Mathematical Systems, vol. 491. Springer, New York (2005). https://doi.org/10.1007/3-540-27659-9
10. Patriksson, M., Strömberg, A., Wojciechowski, A.: The stochastic opportunistic replacement problem, part I: models incorporating individual component lives. Ann. Oper. Res. **224**(1), 25–50 (2015)

11. Patriksson, M., Strömberg, A., Wojciechowski, A.: The stochastic opportunistic replacement problem, part II: a two-stage solution approach. Ann. Oper. Res. **224**(1), 51–75 (2015)
12. Laksman, E., Strömberg, A., Patriksson, M.: The stochastic opportunistic replacement problem, part III: improved bounding procedures. Ann. Oper. Res. 1–23 (2019). https://doi-org.proxy.lnu.se/10.1007/s10479-019-03278-z
13. Al-Najjar, B.: The lack of maintenance and not maintenance which costs: a model to describe and quantify the impact of vibration-based maintenance on company's business. Int. J. Prod. Econ. **107**(1), 260–273 (2007)
14. Barberá, L., Crespo, A., Viveros, P., Stegmaier, R.: Advanced model for maintenance management in a continuous improvement cycle: integration into the business strategy. Int. J. Syst. Assur. Eng. Manag. **3**(1), 47–63 (2012)
15. Rishel, T., Canel, C.: Using a maintenance contribution model to predict the impact of maintenance on profitability. J. Inf. Optim. Sci. **27**(1), 21–34 (2006)
16. Gustavsson, E., Patriksson, M., Strömberg, A., Wojciechowski, A., Önnheim, M.: Preventive maintenance scheduling of multi-component systems with interval costs. Comput. Ind. Eng. **76**(C), 390–400 (2014)
17. Kans, M., Ingwald, A.: A framework for business model development for reaching service management 4.0. J. Maint. Eng. **1**, 398–407 (2016)
18. Kans, M., Ingwald, A.: Modular-based framework of key performance indicators regulating maintenance contracts. In: Mathew, J., Lim, C.W., Ma, L., Sands, D., Cholette, M.E., Borghesani, P. (eds.) Asset Intelligence through Integration and Interoperability and Contemporary Vibration Engineering Technologies. LNME, pp. 301–310. Springer, Cham (2019). https://doi.org/10.1007/978-3-319-95711-1_30
19. Kujala, S., Kujala, J., Turkulainen, V., Artto, K., Aaltonen, P., Wikström, K.: Factors influencing the choice of solution-specific business models. Int. J. Project Manage. **29**(8), 960–970 (2011)
20. Kans, M., Ingwald, A.: Business model development towards service management 4.0. Procedia CIRP **47**, 489–494 (2016)

Rail View, Sky View and Maintenance Go – Digitalisation Within Railway Maintenance

Rikard Granström(✉), Peter Söderholm, and Stefan Eriksson

Trafikverket, Box 809, 971 25 Luleå, Sweden
{rikard.granstrom,peter.soderholm,
stefan.eriksson}@trafikverket.se

ABSTRACT. The work presented in this paper is performed in connection to the research project "Reality Lab Digital Railway" (VDJ). The paper contains a description of three parts of the project, i.e. "Rail View", "Sky View", and "Maintenance Go". Infrastructure data was collected through railway vehicle and helicopter, while user needs was collected through interviews, workshop, observations, and document studies. The analysis of qualitative material is mainly based on international dependability standards, e.g. the IEC 60300-series. The result encompass a description of some use cases as well as some specific user needs in relation to a generic maintenance process. Finally, the paper concludes with a discussion of some results and their application, but also their extension with Artificial Intelligence (AI) as part of digitalised asset management. A tentative analysis model is also described to support the positioning of different digitalisation initiatives within asset management and support their implementation.

Keywords: Railway · Digitalisation · Laser scanning · 360 pictures · Reality Lab · Augmented Reality (AR) · Asset management · Planning · Sweden

1 Introduction

As all parts of society, the railway is experience a paradigm shift due to digitalisation. The railway infrastructure is digitalised through new systems such as the European Rail Traffic Management System (ERTMS), but also Internet of Things (IoT), where traditional systems such as Switches & Crossings (S&C) are becoming connected. Hence, asset management of the railway system also have to change due to new possibilities and challenges to manage the system throughout its lifecycle. In fact, Trafikverket (the Swedish Transport Administration), as all other authorities in Sweden, is required to be managed efficiently, take care of the state's resources, obey present laws and obligations, and present its performance in a reliable and fair manner; see SFS (2007:515) at [1]. Hence, to fulfil these requirements, Trafikverket as national railway infrastructure manager has to adapt to the digitalisation. However, even though the technology is changing very fast, a successful implementation requires changes also of more slow changing areas such as individuals and organisation, but also of regulations (see, e.g., Martec's Law).

R. Karim et al. (Eds.): IAI 2021, LNME, pp. 224–239, 2022.
https://doi.org/10.1007/978-3-030-93639-6_19

The work presented in this paper has been carried out in connection to the research project "Reality Lab Digital Railway" (VDJ) [2]. Within VDJ, research and development is conducted to promote the digitalisation of railway maintenance. One of the project's goals is to find new areas of application for existing technologies, while simultaneously considering existing organisation and regulations. For example, data about the railway infrastructure has been collected through railway vehicles, drones, helicopters and satellites, see Fig. 1. This data has in turn been analysed by a wide range of approaches spanning from manual analysis to advanced statistical approaches and AI. This paper describes three parts of the project, i.e. "Rail View", "Sky View" and "Maintenance Go".

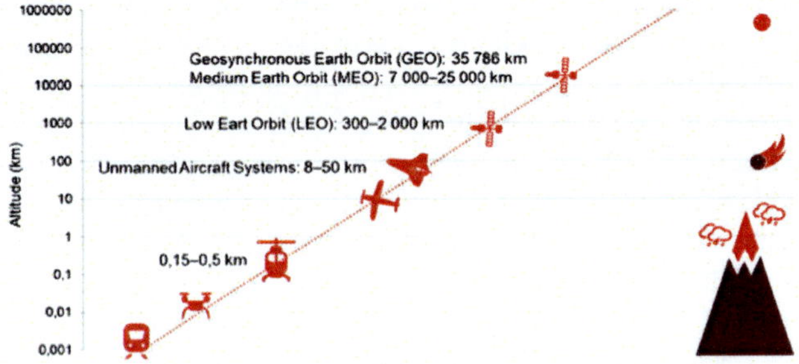

Fig. 1. Examples of vehicles used to monitor the railway infrastructure. From [2].

"Rail View" is an application of 360° images of railway infrastructure. A similar technology is used by, for example, Google in their service "Street View". Images are positioned using GPS. In parallel with the photography, the railway infrastructure has also been laser scanned. Hence, the results can be used to, for example, measure the position, size and distance of infrastructure objects and surrounding objects in relation to each other.

"Sky View", is an application of helicopter 3D photography and laser scanning of the railway infrastructure. This corresponds to technology currently used by Trafikverket for inspection of infrastructure for non-linear power. The applied technology has similar possibilities for measuring objects as in "Rail View".

"Maintenance Go" is an application of Augmented Reality (AR) in the management of railway infrastructure assets and is an extension of "Rail View" and "Sky View".

2 Method and Material

Based on systematic selection criteria [3] (i.e. type of research question, no required control over behavioural events, and focus on contemporary events, but also criticality and extremeness of the case) a single case study of the Swedish Iron ore line was chosen as an appropriate research strategy. The choice was also supported by accessibility and available resources, since the Iron ore line is part of the physical assets of the project "Reality Lab Digital Railway". The choice also enabled the use of action research (see discussion in, e.g. [4] and [5]).

Qualitative data was collected through interviews, a workshop, document studies, and observations. Information about the railway infrastructure that constitute the basis for Rail View and Sky View was collected by railway vehicle and helicopter respectively. The analysis of qualitative data is mainly based on theories from the dependability area, e.g., international dependability standards within the IEC 60300-series. Finally, the paper has been reviewed by key informants and roles to verify its content.

2.1 Overall Case Study – The Iron Ore Line

The Iron ore line is about 500 km long and runs between Narvik in Norway in the north-west and Luleå in Sweden in the south-east. The largest ore deposits and refining plants are located in-between, at Kiruna, Gällivare/Malmberget and Svappavaara. The iron ore is transported around the clock throughout the year in an extreme subarctic climate. Large temperature differences and weather changes are demanding both for the infrastructure and the rolling stock. The Iron ore line allows a train weight of 8,600 metric tonnes and an axle load of 32.5 metric tonnes. This are the heaviest train transports in Europe, and within some years, the aim is to increase the axle load to 40 metric tonnes. Hence, asset management of the Iron ore line is crucial and continuous dependability improvements are necessary to meet ever increasing operational requirements.

2.2 Collection of Infrastructure Data

The vehicles and equipment used for collection of data used in Rail View and Sky View can be seen in Figs. 2 and 3 respectively. Collected data for Rail View was stored, analysed and provided in the application "Orbit 3DM Publisher". To manage data for Sky View in a similar way, the application "DPM 3D Inspection" was used. To enable a qualitative inspection of the infrastructure, it was desirable to collect data after the snow had melted and before the leaves on trees start to grow, but also to have good weather and lighting conditions.

The original assignment for Rail View included laser scanning and photography of 360° images, with the purpose to aid projecting of ERTMS at the Iron ore line. The requirement for data collection was that it should be carried out under lighting and weather conditions so that sign texts, cabinet designations etc. could be identified. Furthermore, additional requirements were [6]:

- The data positioning accuracy of ±0.05 m in plane and height.
- At least two scanners should be used, rotating 135° relative to the driving direction, to minimize shadow effects.
- The spot density for laser data would be at least 1,000 points per square meter.
- The images should have a 360° coverage and no part of the image should be covered by equipment and vehicles.
- The images should have a resolution of at least 30 Megapixels and a colour depth of at least 12 bits.
- Distance between images should be 5–10 m.

However, in order to be able to evaluate whether the opportunities for inspection could be improved by changing the requirements, which Trafikverket traditionally uses in procurement of mobile data collection, supplementary data were used where the density of laser data and resolution in downward images were significantly improved [6].

Complementary data consisted of comprehensive images photographed with downward-facing cameras with a ground resolution of about 1 mm. As well as directed line scanners with an approximate point distance in the line of 0.5–1 mm, each rail was covered and with a distance between each line of about 10 mm [6].

Fig. 2. Picture of locomotive with measuring equipment used for 360 photography and laser scanning of the Southern iron ore line.

Sky View data was collected through a Hughes 500: a helicopter. The helicopter is turbine-powered and the system for image collection and laser scanning is EASA certified to be installed on the model. The applied measurement system consists of:

- GPS/GLONASS L1L2
- Inertial navigation sensor
- Two 3D-cameras that create a 3D-image of the power line and its surroundings.
- Two cameras that take pictures for documentation and possibly inspections, one is aimed forward and one backward to get pictures of, e.g., posts, on both sides.
- Laser scanners that generate between 30–80 points per square meter.
- Multiple echoes
- Camera for orthophoto.

2.3 Collection of User Needs

To identify various applications of Rail View, Sky View and Maintenance Go in the management of Trafikverket's railway infrastructure a number of activities have been

Fig. 3. Picture of helicopter with measuring equipment used for 3D photography and laser scanning of the Iron ore line.

conducted. The approach has been both inductive (starting from user needs and identifying contribution of available solutions) and deductive (matching available solution to user needs).

A deductive approach was to perform individual demonstrations of Rail View and Sky View for various stakeholders within Trafikverket and maintenance entrepreneurs. Examples of internal stakeholders are: project managers railway maintenance; maintenance engineers; project engineers; railway system representatives; asset data managers; and claims manager. Included stakeholders from maintenance entrepreneurs are site managers, supervisors, and technicians.

Another deductive approach was to conduct a workshop focusing on Rail View and Sky View solutions. At the workshop, representatives from Trafikverket, maintenance entrepreneurs and the suppliers of respective solution discussed their functionality and usability. Examples of participant stakeholders are maintenance districts, measuring units, asset data, railway systems, business management, technology and environment, inspectors, and maintenance contractors

The inductive approach was based on three smaller case studies. These case studies were initiated by needs in the ongoing operations, where employees at Trafikverket had to study the railway infrastructure and its surroundings, and Sky View or Rail View was judged to give valuable support.

In addition to the above activities, a meeting with SL (Stockholm local traffic) was conducted since they are an infrastructure manager that has their own application of Rail View. This meeting was used as a validation of the findings of the case studies at Trafikverket.

3 Results

The results of the study can be divided in two parts depending on the applied approach for collection of user needs, and if it was inductive or deductive.

The results related to the inductive approach is described in the context of the three subcases where needs emerged in the organisation and valuable support was judged to be available through Rail View and Sky View. The results based on the deductive approach, where Rail View and Sky View were presented for potential stakeholders to judge its usefulness, is related to some of the phases of a generic maintenance process.

3.1 Results from Subcases

Roles related to the management of Trafikverket's road maintenance contracts, make frequent use of services such as Google Maps and Street View. These are valuable tools for road project managers who may be responsible for assets that can be up to 2,000 km long. Hence, a person cannot possibly have a total local knowledge about these assets. The tools are used to quickly get information about the area and its surroundings (e.g., geographical positioning and an overview of how it looks in the immediate nearby area), which can save Trafikverket and the taxpayers considerable money. For example, regarding working time, vehicles and fuel when field visits can be carried out at the office instead of out in the field. The tools can support a person to quickly get a picture of the asset and its surroundings, e.g., regarding fences, railings, water sources, ground conditions, and slope conditions. In different types of consultations with stakeholders, the tools are also valuable. For example, in contact with the public the responsible person can quickly get an idea of what it looks like in and around the road area that is adjacent to the property owner concerned, e.g., the school or the sports ground.

In a similar manner, project roles related to the management of railway maintenance contracts from time to time use Google Earth, and to some extent Street View (in cases where the railway is visible). Common applications are, e.g., control of intersections between road and rail (level crossings, bridges, and tunnels), planning of field visits, and the identification and localisation of connecting roads. However, as indicated above, the usability of these commercial solutions for rail purposes are very limited compared to the road applications.

3.1.1 Timber Terminal

One case was consultation related to modifications of the timber terminal in Murjek. This consultation involved two external stakeholders, i.e., railway operator and paper manufacture (SCA). Internal stakeholders at Trafikverket that participated were representatives from the unit responsible for strategical and tactical management of the railway infrastructure, the unit responsible for long-term planning, and the unit responsible for operational management of the railway infrastructure and the maintenance contract.

The paper company SCA wanted to improve their logistic solution by using longer timber trains and thereby reduce the number of trains from three to two per day. However, to achieve this, the timber terminal has to be modified. Trafikverket's contact with SCA goes through the unit of Planning (who is responsible for external contacts), which prepared a consultation meeting with the primary external stakeholders (SCA and the railway operating company Hector-Rail) together with the unit responsible for the operational infrastructure maintenance. The consultancy meeting was performed via Skype and used Google Earth and Rail View as main information sources. In addition to the internal and external stakeholders mentioned above, the unit at Trafikverket responsible for strategic and tactic maintenance of the infrastructure asset also participated.

The physical constraints of the infrastructure and its surroundings could easily be identified. The information was even better than would be obtained through a field visit since the consultancy meeting was performed during winter and snow was covering the area. Two different scenarios could be evaluated through the available information, i.e., extension of the terminal in the far end (limited by water) or loading of train in two parts from two sides of a crossing road, followed by a coupling of the train.

The Skype meeting based on information from Google Earth and Rail View took about one hour. A field visit would take roughly eight hours. The distance from Luleå to Murjek is about 300 km. the labour cost of the five participating persons are about SEK 24,000 (SEK 650 per person and hour) for a field visit compared to about SEK 3,000 for a Skype meeting. In addition, each participant can use seven hours for other activities, i.e., 35 h in total saved for all participants.

3.1.2 Snow Protection and Avalanche Warning

Another case was the investigation of measures in avalanche areas, where Sky View was used to investigate local conditions in the immediate area where avalanches have emerged. The purpose of this investment project was to build avalanche protection and to investigate the location of avalanche warning devices. From a work environment point of view, it would be beneficial to avoid building the systems to far into the terrain, as it in many cases is very hilly with steep slopes. Figure 4 illustrates an avalanche area, which is identified through an area with fallen trees in one direction. In Fig. 5, the user can move around in the laser point cloud, and identify the topology of an area, e.g. in this case steep slopes at the position of a cross section (white line in Fig. 5). Figure 6 illustrates the cross section at the white line in Fig. 5. This is a valuable tool in this case to assess steepness. In other cases it can be used for mass calculations and other assessments of areas where work is to be executed.

By using Sky View it was possible assess the characteristics of the terrain, to identify, position, and measure, e.g., buildings, roads, walking trails, water, existing avalanche systems, and areas of earlier avalanches.

The investigation of the area took about one hour by the use of Sky View. A field visit would take about one week to achieve the same result. The quality of information from Sky View was judged to be of much higher quality than services available from, e.g., Google, Eniro or Favy (a GIS system within Trafikverket).

3.1.3 Accident Investigation

When an accident occurs, it may be possible to use Rail View or Sky View to gather some of the necessary information. One example of this is the third case, which was an accident investigation, where visibility conditions at a level crossing was studied in relation to an accident with one fatality. Since the information in Rail View was gathered shortly before the accident, it was possible to establish that the line of sight was according to the requirements at the time of the accident. Hence, vegetation clearance was performed as it should and some other contributing factors to the accident had to be identified. In addition, it was possible to verify that the project manager and the entrepreneur

Fig. 4. Fallen trees used to identify an avalanche area.

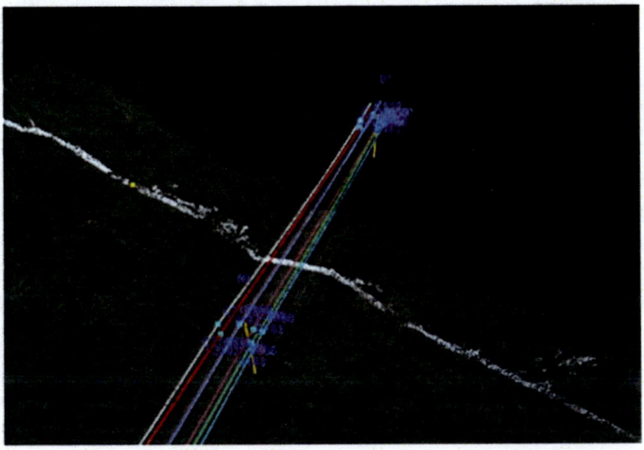

Fig. 5. Laser point cloud, including position of cross section (white line).

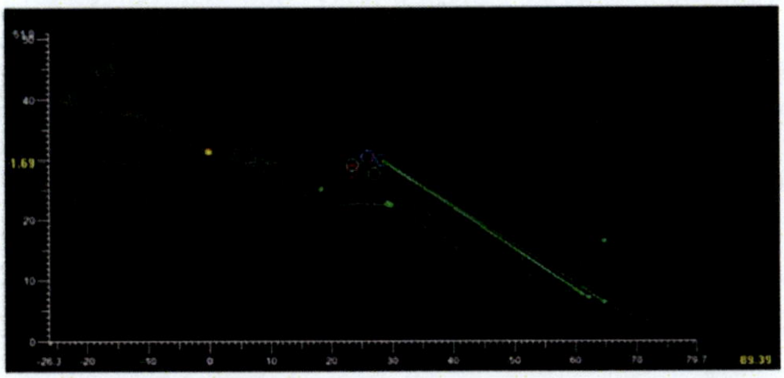

Fig. 6. Cross section of terrain, correlating to white line in Fig. 5.

had fulfilled their responsibilities even though their internal communication had been informal and not documented.

3.2 Results Related to User Needs in the Maintenance Process

The generic maintenance process is an extension of the improvement cycle and adapted for maintenance practice. However, the findings are also relevant for other asset management activities such as operation and investment.

3.2.1 Asset Information Management

One central part of asset management is the asset register. The reason being that it is at the core of Trafikverket's systems for management of safety and dependability. For example, the asset register gives input to the system "Inspection plan" that manages the status of both safety and maintenance inspections. Hence, the system makes it possible to follow up that the right objects are inspected in a correct way regarding type, frequency, time and place. The asset register also gives input to the inspection protocol, which is used by the inspection personnel to ensure that the right inspection points are included for individual items in the infrastructure depending on its unique features such as configuration and model. Besides the inspection systems, the fault reporting system is also dependent on the asset register to connect faults to individual parts of the infrastructure. The data from the inspection and fault reporting systems, are in turn used to follow up the safety and dependability performance of the asset or the related organisation and make decisions about appropriate actions within, e.g. maintenance, reinvestment or investment.

Today, errors regarding asset register data is manually reported by personnel using the data in their work, e.g. for inspections. The follow up of these error reports are time consuming and the quality of the reports may in turn be rather low adding to the burden. Field visits to measure and position objects take time, and the use of alternative sources of information such as film from measurement wagons or commercial map services have limited usability. The measurement wagons only provide film as a secondary service, which makes the quality heavily depending on daylight, weather, season, and direction of travel. The commercially available information services have limited usability with regard to evaluate individual objects within the railway infrastructure.

Hence, information about the asset from Sky View and Rail View can be used to update and improve the quality of information in the asset register, e.g., missing, extra or erroneous objects as well as their position. The inventory of existing items of a specific type (e.g. bird spikes) in the asset may also be supported by the use of Sky View and Rail View. This management of information can be performed through a system interface instead of field visits, which improves the efficiency tremendously.

3.2.2 Planning of Actions in the Infrastructure

Rail View and Sky View can be used to support the planning of actions in the infrastructure in a strategical, tactical and operational perspective, e.g. regarding maintenance, reinvestment and investment. By providing information about the infrastructure and its surroundings, it is possible to:

- Plan safety measures.
- Plan the use of machinery, e.g. which type, access paths, and working sites.
- Plan material logistics, e.g., which material shall be used, where should it be stored and loaded, and how to manage disposed material.
- Combined planning of multiple actions that are beneficial to coordinate at a specific site and time.
- Amount of work that is necessary depending on the assets condition and the characteristics of the surroundings.

From a maintenance perspective, the actions can be single major actions such as felling of risk trees, vegetation clearing, and drainage or fencing works. However, also the planning of additional work, e.g. when replacing component in Switches & Crossings (S&C), it is possible to investigate the surroundings. Hence, it might be possible to coordinate the component replacement with replacement of broken or missing equipment in other asset types (e.g. ducting caps) or removal of disposal that are on the site.

Another example is coordination of multiple actions that requires railway vehicles and are to be performed in the infrastructure at about the same time and place. Two examples of this type of actions are the changing of sleepers and mechanical vegetation clearance.

3.2.3 Inspection

Inspection is a central part of condition-based maintenance in railway. Rail View can be used for many visual inspections where the perspective is similar to a train driver. Hence, Sky View is a good complement to visual inspections with the view of a helicopter pilot. To perform inspections through an interface to Rail View and Sky View has a number of benefits compared to inspections in the field, e.g.:

- Time savings, the time for traveling to and in the infrastructure changes from hours to minutes or seconds.
- Improved work environment, it is more safe and comfortable for the personnel to perform the inspections at an office than out in the field.
- Improved capacity for traffic, since the time in field for maintenance activities is reduce there are more time for traffic.
- Improved effectiveness and efficiency of maintenance, the number of inspectors may be reduced and they can focus on yards instead of on lines.
- Assessment of performed work, depending on the relation in time between different maintenance actions and gathering of information as input to Rail View and Sky View.

With regard to inspection, data from Rail View was evaluated with regard to Trafikverket's regulation for safety inspection (TDOK 2014:0240). The possibilities for off-site inspection were calculated for each asset type. The greatest opportunities for inspection was related to track, where approximately 85% of the inspection was judged possible. Automation were most developed in some cases where it is possible to see changes between images from different data collections, for example missing or damaged fasteners and sleepers, damaged signals and signs, but also cracking on various construction objects [6].

Other areas where automation of inspection proved to be possible based on analysis of Rail View laser data were, for example, wear on rail heads, deviating track geometry, control of ballast levels, erosion and damage to the embankments, free space, and control of catenary system [6].

Much of the inspection points according to TDOK 2014: 0240 was carried out by use of Rail View. The application showed that the completeness could be improved by adding high-resolution laser data and track images. Inspection of tracks, switches, bank, intersections, platform, signals, boards, and fences showed a special suitability with completeness of 50–100% at inspection [6].

4 Discussion

To achieve the potential benefits with Rail View and Sky View, the geographical positioning of data and integration with the asset register is fundamental. Hence, this is an integration with the spatial and the system domains. In addition, it would be beneficial to have an integration with the time domain, where past, present and future actions or asset condition can be visualised. Hence, the time domain also supports a life cycle perspective where maintenance, reinvestment and investment actions can be considered simultaneously. By a combination of spatial, system and temporal domains, both coordination and optimisation of actions in the infrastructure from a line (instead of subsystem) perspective is enabled. However, the need for actions may very well be initiated from a subsystem perspective, e.g., track, catenary, or signalling systems. However, from a practical point of view, the bottleneck is the physical asset, where all these actions have to be performed in coordination with each other and with the railway traffic.

Virtual Reality (VR) can be used as an enhanced interface in order to move around in the infrastructure while being off site. In this way it may be possible to perform all the actions described in this paper, but in a more vivid way than through a screen. In addition, education and training may also be supported in a safe, effective and efficient way. However, with or without VR, both Rail View and Sky View provides the possibility to gain local knowledge about the infrastructure and its surroundings. This local knowledge is fundamental for performing actions in the infrastructure both effectively and efficiently.

There are technologies to combine information in Rail View and Sky View with other information to enable Augmented Reality (AR) when being out in the field. The most famous application of AR is probably Pokémon Go (see Fig. 7), where a similar application for maintenance purposes could be named Maintenance Go. For example, inspection remarks generated off site by inspection personnel using Rail View or Sky View can be visualised through goggles for the personnel in field when they are acting upon the remarks. Hence, the remarks can be superimposed on the asset type concerned, which facilitates identification of the right action to perform on the right object. However, the time constraints of this study limited the demonstration of any solution for Maintenance Go.

The data in Rail View and Sky View can be used as parts of a digital twin. For example, in Sky View, the catenary system have been digitalised and classification of laser data has been performed of, e.g., land, buildings, road, water, and vegetation. In Rail

Fig. 7. Example of Pokémon in the track area.

View, multiple asset types have been digitalised. When digitalisation of objects are done, Artificial Intelligence (AI) can be used to automatize many analyses, e.g., identification of risk trees, vegetation in the track area, estimation of line of sight, volume of ballast, volume of vegetation, and tilted or missing objects (e.g., fasteners, sleepers, and poles). Additional benefits with AI in combination with machine measurement are increased objectivity compared to manual inspection. The digital twin can also be used to predict the future condition of the asset (e.g. degradation of items) or its surroundings (e.g. when individual trees becomes a risk) based on different scenarios. However, this prediction requires consecutive measurement occasions [6, 7].

Hence, the frequency of measurement related to Rail View and Sky View becomes central. Based on the study, the recommendation is to update Rail View twice a year, once in the fall (after defoliation) and once in the spring (after snow melting and before leaves budding). Besides the conditions for performing measurements, it is good to have a view of the asset after winter and summer respectively. In the spring, it becomes possible to judge if there are any damages in the infrastructure due to winter actions by comparing with the autumn images. In the autumn, it is possible to evaluate some of the work performed during summer by comparing with the images from spring. Regarding Sky View, the measurement for update could be carried out every other year, for example to harmonize with the catenary system maintenance carried out every second year.

The application of commercial system platforms facilitates integration of different systems. For example, the platform for Rail View, Orbit 3DM Publisher, has been successfully integrated with asset management systems, such as IBM Maximo, by other infrastructure managers. Sky View is also provided to users in Trafikverket and externally by an agreement, which gives access to DPM 3D Inspection through "My page" at Trafikverket's web page (see Fig. 8).

It is necessary to review specifications when purchasing mobile data collection if the purpose is inspection instead of projecting. Mainly with regard to high-resolution data of the track, both so that the outside and inside of the rail are visible. As well as ensuring that the visibility is good in panoramic images both forward and backward, to avoid that

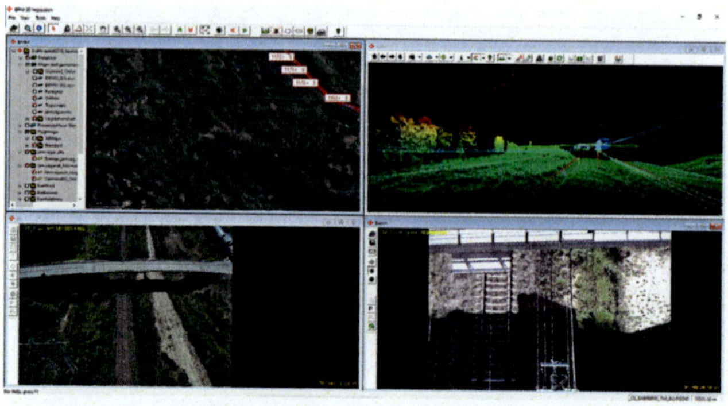

Fig. 8. Example of interface to digital twin of the Iron ore line based on "Sky View". Upper left corner - town photo with distance and km-marking, upper right corner - laser point clouds with digitized catenary system, lower left corner - overview photo, lower right corner - detailed photo. From [2]

the image is covered by the locomotive in one direction and that there is backlight in the other [6].

As indicated above, due to requirements on image quality, collection of data for Rail View and Sky View is performed during nice conditions. Existing collection approaches (e.g. traditional measurement wagons) and applications (e.g. projecting instead of inspection) have lower requirements and hence lower quality.

There are many similarities between Rail View and Sky View. However, there are also some differences. One obvious difference is the perspective. Rail View provides the perspective of a train driver, while Sky View provides the perspective of a pilot. Hence, they provide complementary perspectives on the infrastructure and its surroundings. Sky View enables a wider view of the infrastructure's surroundings and the possibility to view assets from above (e.g. roofs on buildings, top of poles, wirings and cables). Another difference is that collection of data for Rail View requires time in track, while Sky View does not. Other differences of the present applications depends mainly on the sensors installed on the measurement vehicles, which in turn depend on the present use and related requirements. The functionality of Rail View is mainly based on requirements related to projecting, with some additional functionality for inspection purposes. The functionality of Sky View is optimised for inspection of non-linear power infrastructure (e.g., high resolution and overview cameras, identification of warm objects by infrared camera and dead or dry trees by near-infrared camera). However, if inspection of railway infrastructure is becoming a common practice, sensors and functionality will be adapted for both Rail View and Sky View.

Inspection by use of Rail View and Sky View can currently be seen as a complement to ordinary manual inspection activities, where it can be used to plan and optimize existing resources. Furthermore, where it is possible to apply automation, an increased objectivity and an opportunity to register remarks at an earlier stage are achieved without increasing the effort. However, one limitation with Rail View and Sky View is that it is not possible to perform inspections points that include functional tests where the inspector, for example, should feel that screws and bolts are tightened. On the other hand, you can get an indication of a risk for inspection remarks, which is not captured by the statistics, e.g., by an indication of a loose fastener if there are, e.g., large cracks in a sleeper, the base plate is oblique, or the bolt is tilted.

When working with digitalisation of railway asset management it is possible to use different models to support implementation. One way is to relate different levels of analysis to the model of organisational learning and its three loops of learning, see [8]. One first level of analysis may be seen as related to "doing things the right way" (efficiency), or "following the rules" (single-loop learning). Hence, independent of what kind of maintenance that is applied (i.e., corrective, preventive or improvement), it can be enhanced by the use of digitalisation. However, the development is performed within single boxes of Fig. 9. This practice might be exemplified with the comprehensive work related to Rail View, where the new technology was evaluated with regard to each inspection point (about 600 in total) in the existing regulation (see [6]). Even though new technology is used to enhance the maintenance, the regulation itself is not questioned to any large extent. Here it is worth noticing, that even though the focus is on the application of AI within condition-based maintenance, digitalisation can be applied to increase the efficiency also of corrective maintenance (see Fig. 9).

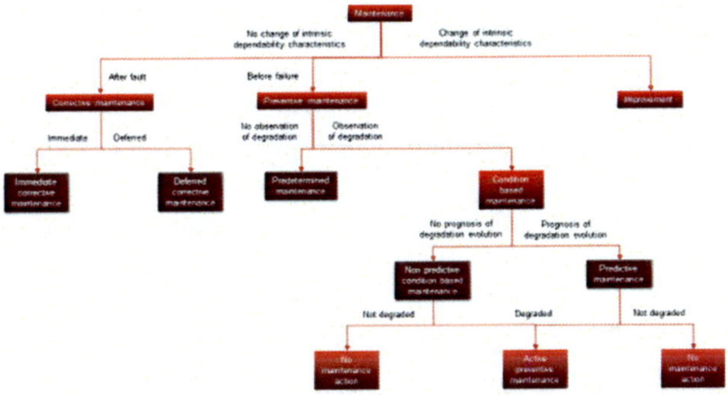

Fig. 9. Different types of maintenance. Inspired by EN 13306:2017 (Maintenance – Maintenance terminology).

Another level of analysis may be seen as related to "doing the right things" (effectiveness), or "changing the rules" (double-loop learning). Here, the regulation itself is questioned and an analysis may be performed to evaluate new technologies and change the current practice. In this way, it is possible to achieve a more dynamic, or "living", regulation. One example of this is to apply an infrared camera to inspect the function of heating systems in S&C (i.e. obtain a certain temperature in the S&C to melt snow and ice), instead of using snow or water to indicate the temperature in combination with measuring the resistance in cables powering the heating elements (see, e.g., Granström et al. 2019). On this level of analysis, it is also possible to change between different types of maintenance (i.e., which box to select in Fig. 9), where digitalisation often tends to be used as a means to strive towards predictive maintenance. This may be due to new technologies that makes monitoring (cf. Fig. 1) and analysis more accessible to detect degradation or faults that earlier were hidden or performed manually (cf., first level of analysis). However, as mentioned earlier, digitalisation can be used to improve any type of maintenance (see Fig. 9).

A third level of analysis is to reflect upon the methodology used to decide upon which the right things are (i.e., how to select the right box of Fig. 9) by a systematic use of best practise, historical data and expert judgement. This level of analysis might be related to "learning about learning" (triple-loop learning). Examples of this third level is to apply Failure, Modes, Effects & Criticality Analysis (FMECA) or Reliability-Centred Maintenance (RCM) in combination with barrier-analysis to achieve a more dynamic maintenance regulation, see, e.g. [9] and [10] respectively. Design of Experiments (DoE) is another useful methodology to evaluate what the right things to do are. See, e.g., [11] for a description of DoE and [12] and [13] for applications within a railway context. As mentioned earlier, corrective maintenance might be the right (effective) choice and digitalisation might be one way to make it more efficient.

On a more aggregated level, the third level of analysis may be illustrated by the "Four-step principle", which is an approach that Trafikverket applies to identify the most efficient and sustainable solutions in the transport system (cf. the dependability standard SS-EN 441 05 05 – Dependability terminology, and its description of relationship between stakeholder requirements, operation, maintenance, and modification). The first step is to rethink, e.g., if it is possible to transport things in a different way, or to use another mode of transport (cf. stakeholder requirements). If the rethinking does not result in any valuable solution, the next step is to see if it is possible to make more effective use of existing infrastructure (cf. operation and maintenance). For example, if it is possible to control the traffic or adjust the installations that currently control the traffic. In the third step, it is evaluated if it is possible to resolve the issue by making some minor alterations, e.g., add a track, redesign a level crossing (cf. maintenance and modification). It is not until the fourth and final step that the most expensive solution is considered, i.e., carrying out a major rebuild or extension, or investing in new infrastructure (cf. modification).

The performed study and related experiences indicate that the technology is available and develops very fast. However, to support an implementation it is necessary to at least consider the related regulations, but also organisation and roles. The regulations may be considered on any of the described three levels of analysis. However, only providing a technical solution without relating it to user needs and linked regulations is only one of the very first steps towards an implementation. The supplier should many times take a larger responsibility to provide mature user-centred solutions instead of requiring the user to adapt technical solutions with low or no integration in practice. The three levels of analysis related to digitalisation of asset management can be compared to the four business requirements that all Swedish authorities should fulfil; see SFS (2007:515) at [1]. Hence, digitalisation initiatives related to asset management should fulfil the following requirements:

- objectives that are aligned with and support the mission (effectiveness, i.e., to do the right things);
- operations with an efficient use of resources (efficiency, i.e., to do things the right way);
- reliable operational and financial reporting (i.e., the data and information included in the technical solution, e.g. Rail View and Sky View, but also the management of a

more dynamic maintenance regulation through the use of FMECA, RCM and barrier analysis);

- compliance with applicable laws and regulations (e.g., the maintenance regulations as part of the safety management system).

Acknowledgments. The Reality lab digital railway project was made possible thanks to funding from Vinnova (grant number: 2017-02139), Trafikverket (FUDid: 6538) and Luleå University of Technology.

References

1. Riksdagen (The Swedish Parliament): Svensk författningssamling (2020). http://www.riksdagen.se/webbnav/index.aspx?nid=3910. Accessed 17 Feb 2020. (In Swedish)
2. Söderholm, P., et al.: Verklighetslabb digital järnväg: Förmåga för ökad digitalisering och hållbarhet (Reality lab digital railway: capability for increased digitalisation and sustainability). Report, Trafikverket, Luleå (2019). (In Swedish)
3. Yin, R.K.: Case Study Research: Design and Methods, 3rd edn. Sage, Thousand Oaks (2003)
4. Patel, R., Tibelius, U.: Om osäkerhet vid insamlandet av information, i Grundbok i forskningsmetodik (Uncertainty in collection of information, in fundamentals of research methodology). Studentlitteratur, Lund (1987). (In Swedish)
5. Gummesson, E.: Qualitative Methods in Management Research. Sage, Thousand Oaks (2000)
6. Östrand, P.: Utvärdering av mobil datainsamling för besiktning av järnvägsanläggningar (Evaluation of mobile data collection for inspection of railway assets). Report, eP119-244-2019. Luleå University of Technology, Luleå (2020). (In Swedish)
7. Granström, R.: Rail View, Sky View och Maintenance Go – Tillämpningar inom Trafikverket – Ett projekt inom Verklighetslabb Digital Järnväg (Rail view, sky view and maintenance go – applications within Trafikverket – a project within reality lab digital railway). Report, Trafikverket, Luleå (2020). (In Swedish)
8. Argyris, C., Schön, D.A.: Organizational Learning: A Theory of Action Perspective. Addison-Wesley, Boston (1978)
9. Granström, R.: Tillförlitlighetsbaserat underhåll inom Trafikverket - Demonstrator med växelvärme (Dependability centred maintenance within Trafikverket – demonstrator with heating of switches and crossings). Report, Trafikverket, Luleå (2019). (In Swedish)
10. Söderholm, P., Nilsen, T.: Systematic risk-analysis to support a living maintenance programme for railway infrastructure. J. Qual. Maint. Eng. **23**(3), 326–340 (2017)
11. Montgomery, D.C.: Design and Analysis of Experiments, 10th edn. Wiley, Hoboken (2019)
12. Patra, A.P., Söderholm, P., Kumar, U.: Uncertainty estimation in railway track life-cycle cost: a case study from Swedish National Rail Administration. Proc. Inst. Mech. Eng. Part F J. Rail Rapid Transit **223**(3), 285–293 (2009)
13. Lundkvist, P.: Stresstest i samband med besiktning av växelvärme - Tillämpning av arbetssätt inom ramen för Verklighetslabb digital järnväg (Stress-test at inspection of heating system for switches & crossings – application of methodology within reality lab digital railway). Report, Licab AB, Luleå (2018). (In Swedish)

Reality Lab Digital Railway – Digitalisation for Sustainable Development

Peter Söderholm[1], Veronica Jägare[2(✉)], and Ramin Karim[2]

[1] Trafikverket, Box 809, 971 25 Luleå, Sweden
peter.soderholm@trafikverket.se
[2] Luleå University of Technology, 971 87 Luleå, Sweden
{veronica.jagare,ramin.karim}@ltu.se

Abstract. This paper summarizes the project "Reality Lab Digital Railway" (VDJ). The goal of the project is to open up Trafikverket's core business within railway for the development and demonstration of digital information solutions that are tested by primary actors and end users. The project is mainly financed by Vinnova and Trafikverket.

The implementation takes place within the framework of Trafikverket's main process "Research and develop innovation" and in accordance with Trafikverket's project model (XLPM). The project is carried out in close collaboration with the project "ePilot" at Luleå Railway Research Centre (JVTC). A key project strategy is to address prioritised stakeholder needs through various initiatives to increase Trafikverket's ability to contribute to the project goal.

The results of the project can be summarized in regulations, organization, roles and technologies that contribute to Trafikverket's ability to achieve the project goals. Examples of results are methodologies and tools for establishing and actively managing a more dynamic regulatory framework that clarifies the need for innovation as well as the specification of the application of digital technologies to strengthen the industry's ability. To support industry collaboration during development and demonstration projects, physical and digital assets as well as IT-solutions have also been established. The project has also produced proposals for agreements regarding information management between Trafikverket and external parties.

The future organization of VDJ will mainly be based on informal and formal competence networks, both within Trafikverket and externally. The approach is based on existing processes related to research and innovation, testing activities and the development of systems and components. In order to enable ongoing operations within VDJ at Trafikverket, the following functions and activities are central:

- A platform for receiving, storing, analysing and providing large amounts of industrial information. This platform should be able to be used during all stages of development, testing and ongoing operations.
- Standardized agreements with operators and system suppliers regarding management of asset-related information. This is to support demonstration, testing and application of vehicle-based solutions for monitoring the infrastructure.

R. Karim et al. (Eds.): IAI 2021, LNME, pp. 240–255, 2022.
https://doi.org/10.1007/978-3-030-93639-6_20

- Adaptation of maintenance contracts to support development and demonstration in railway operations. This is in addition to today's opportunity for research and innovation projects to use amendments, additions and outgoing work in the maintenance contracts.
- Further development of implemented methodologies and tools to identify functions to monitor so that they also meet the requirements for traffic safety within railway. This is to support a more active management of dynamic maintenance programs and strategies, which in turn support the identification and establishment of need-based solutions for monitoring of the railway infrastructure.

Keywords: Railway · Digitalisation · Reality Lab · Innovation · Information logistics · eMaintenance · Sweden

1 Introduction

In 2016, the Swedish Government launched the initiative "Testbed Sweden" to create environments where ideas and prototypes can be tested and developed with the aim of strengthening the possibilities for conducting globally competitive and qualified business in Sweden. The Government considers that the greatest development needs, as well as the greatest potential, exist within the category of real-world environments that are developed and operated together with need owners, within real-life conditions. Test beds in real environments are also an area where Sweden, with its well-functioning public sector and many developed system solutions, has the best possibilities for creating something unique. This type of environment provides good opportunities to test and develop services and products, as well as regulations, organizational forms and policies. Vinnova's announcement "Reality Lab in Public Business" is part of this ambition [1].

The purpose of Vinnova's announcement is to facilitate publicly funded players to open up their core business for experimental innovation work in collaboration with business, academia and the idea-driven sector in real-life conditions. This contributes to solutions that are designed to a greater extent to meet real needs, become cheaper to develop and faster to use. It also challenges and makes visible shortcomings in current policies and creates the conditions for better policy development. In the longer term, it provides the organisations with an improved and faster ability to adapt their business to [1]:

- changed conditions
- specific target groups, and
- increased individualization of welfare services.

A closer experimental collaboration between public actors, business, academia and the idea-driven sector creates increased demand for solutions based on real needs. This increases the competitiveness of Swedish companies, increases the relevance of the research to societal challenges, and streamlines the commitment from the idea-driven sector [1].

In the long term, the initial investment in reality labs supports a society that better and more efficiently responds to the needs and expectations of the inhabitants and a more attractive climate for Swedish and international entrepreneurs, investors and innovators [1].

By "reality lab" Vinnova refers to the development of a public enterprise that enables testing and demonstration of new solutions in the core business. Like regular labs, reality labs have equipment, plant, organization, processes and methodologies for conducting limited risk-managed experiments. Unlike ordinary labs, reality labs are integrated with the core business. This means that premises, equipment, employees and customers are included in testing and demonstration while the ordinary core business is ongoing [1].

The goal of Trafikverket's project "Reality Lab Digital Railway" (VDJ) is to open up Trafikverket's core business in railway for the development and demonstration of digital information solutions that are tested by performers and end users. This includes [2]:

- Ability, organization and digital and physical assets to, in collaboration with others, carry out rapid, smaller tests in real railway operations, learn from them and carry out further corrected tests.
- Established processes and procedures well integrated with regulations and agreements to support collaboration between primary stakeholders and thus short lead times from idea to application.

Expected effects and results from the project "Reality Lab Digital Railway" are [2]:

- A more socio-economical sustainable and digitized railway service.
- Improved equality in the railways with an increased integration with the IT-industry.
- Improved capacity utilization and punctuality, as well as more efficient and more coordinated and condition-based operation and maintenance of the railway system based on information logistics solutions.
- Adapted regulations, agreements, contracts and business models for increased digitalisation in railways.
- Internationally recognized and sought-after test and demonstrator operations within digital railway.

2 Method and Material

The project is carried out in accordance with Trafikverket's project model XLPM and the main process "Research and develop innovation".

In order to develop Trafikverket's ability in the activities covered by VDJ, a number of needs have been addressed within the project. These needs have been initiated on the basis of internal and external stakeholders and assessed to contribute to the project's goals by the project and steering groups.

The areas that are mainly covered within the framework of VDJ are the management of data related to the monitoring of the infrastructure via vehicle-based or integrated solutions, as well as infrastructure-based monitoring of vehicles that can affect the infrastructure. Monitoring of vehicles that are based on solutions integrated into the vehicle itself is not directly included in VDJ's operations because it is outside the responsibility of Trafikverket. However, in cases where information from integrated vehicle monitoring is used to assess the condition of the infrastructure, initiatives have been supported by VDJ (e.g. [3–5]).

In order to expose VDJ to relevant needs initiated by its stakeholders, communication efforts have been an important part of the project. This is to create awareness of the existence of VDJ among its stakeholders and to build up a demand by showing the possibility of managing their needs within the framework of VDJ.

The management has been adapted based on the scope and complexity of the need and the relevant stakeholder. In this management, the overall ability of the project has been applied and, if necessary, supplemented with additional capabilities of each project partner. In some cases, new capabilities have also been developed to complement the existing ones.

To complement the project's ability, ongoing collaboration has been carried out with other projects and programs, e.g. the national project "ePilot" and the EU-program "Shift2Rail".

A basic prerequisite for the project is also the physical assets within "Testbed railway" (i.e. the Iron ore line and the Haparanda line with test sites at Sävast and Sunderbyn), the traffic control centre in Boden (digital traffic management and simulator), Trafikverket's regional office in Luleå (cutting-edge railway knowledge), research and educational resources at Luleå Railway Research Center (JVTC), and innovative information logistics solutions at eMaintenance LAB. The design of the basic contract for railway maintenance on the Southern Iron ore line and the Haparanda line also supports development and demonstrator activities [6, 7].

In order to support the implementation of project results, relevant administrations and operations have been involved on a continuous basis. Examples of this are parts of Trafikverket's main process "Research and develop innovation", the support process "Managing systems and component development", specific asset types (e.g. heating system of switches and crossings) as well as maintenance districts at Trafikverket and external maintenance entrepreneurs. The responsibility of members of steering and reference groups, as well as communication efforts, also support the implementation of project results.

3 Results and Deliverables

The results from VDJ can be described based on deliverables in the four areas: regulations; organisation; roles; and technology. See Fig. 1.

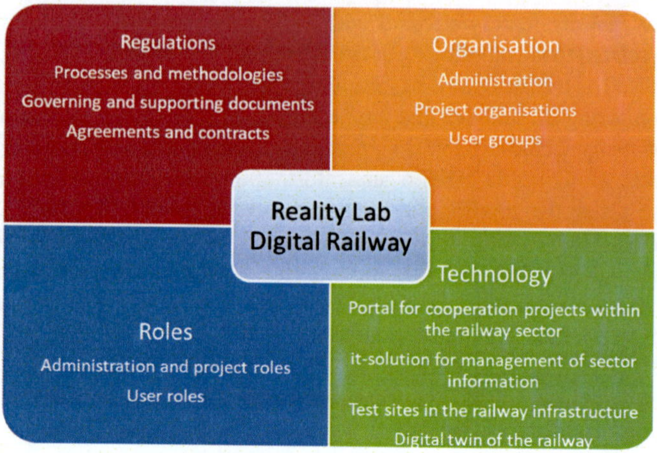

Fig. 1. Some of the results from VDJ.

3.1 Regulations

Some of the project results related to regulations are shortly discussed below. However, they have been implemented to varying degrees, which is indicated in the description.

One result is methodologies and tools for conducting Failure, Modes, Effects & Criticality Analysis (FMECA). A template for FMECA has been implemented in the support process "Manage system and component development" [7–10]. The FMECA is used as a central support to other reliability, safety and security analyses, and thereby also a more dynamic maintenance program that can adapt to new and innovative digital solutions (see Fig. 2).

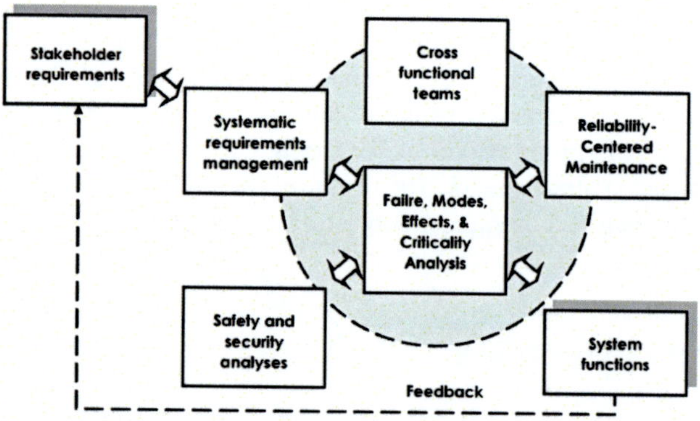

Fig. 2. Failure, Modes, Effects & Criticality Analysis (FMECA) as a central support to other reliability, safety and security analyses.

Another result is a proposal for guidance on extending FMECA for the preparation of analysis for Reliability Centred Maintenance (RCM). See [11]. Trafikverket is increasingly using FMECA, but also parts of RCM; in the development of maintenance strategies and programs in activities related to the implementation of Asset management according to ISO 55001.

In addition, the project has delivered proposals for methodologies and tools for the application of Design of Experiment (DoE). Supporting documents and template are submitted to the support process for continued management. See [9, 10, 12–14]. This result is further developed in a research project related to the establishment of cause-effect relationships for asset degradation and maintenance actions.

Other results from the project are examples of information sharing agreements between Trafikverket and the operator regarding the management of data from vehicles [15]. This agreement has been extended further to include additional vehicle data that support Trafikverket's operation.

Another result is the example of a Non-Disclosure Agreement (NDA) with a system supplier regarding the management of condition information about the rail infrastructure collected via vehicles in regular traffic [15].

The project also delivered examples of requirements specification for procurement of analytical services based on a demonstration of integration of heterogeneous data sources in the management of switches and crossings (S&C) at strategic, tactical and operational level [16]. This has resulted in a combined procurement of sensors and analytical services for monitoring of 1,000 S&C.

The project also resulted in a number of R&D agreements (Research & Development) with external parties to provide information from Trafikverket's system for the purpose of developing information services related to the condition of the railway infrastructure [15]. A further development of this is Trafikverket's innovation procurement that focus on information solutions that supports improved condition-based maintenance of track and catenary systems.

The project also demonstrated the application of an adapted basic contract in railway maintenance to support development and demonstration in the railway infrastructure. Today, Trafikverket are performing a number of procurements related to maintenance contracts based on extended collaboration. This type of maintenance contracts supports primarily innovation and development initiatives within the contract itself, but facilitates collaboration with other initiatives as well. Hence, this type of maintenance contract is valuable to use in combination with the innovation procurement mentioned above.

One result that was developed and applied within the framework of a research centre prioritised by Trafikverket (JVTC) is collaboration agreements for the implementation of research and development projects concerning digitalized railways [17]. Another result within the context of JVTC are the development and application of governing and supporting documents for the implementation and management of collaboration projects within the railway industry [17]. These results are further developed and applied within the research project "Artificial Intelligence Factory – Railway (AIF/R)" financed by Vinnova.

The project also provided proposals for designing and establishing new information services (proactive trend alarms instead of traditional safety-related alarms) linked to detectors for monitoring of railway vehicles, based on the needs of Trafikverket and operators. See [18] and [19].

In addition, the project performed a number of information security analyses regarding test site for measuring wheel geometry [20] and infrared measurement of the heating system for switches and crossings [21]. The analyses are found in six compiled documents for each application: summary; external environment and information analysis; business requirements analysis and risk identification; action plan; support for continuity planning; and status regarding requirements in accordance with the information security regulations.

3.2 Organisation and Roles

Organization and roles for VDJ are mainly based on Trafikverket's regular organization with some additional capabilities. The main focus is to let the project organization for VDJ (see [2]) change to ongoing operations. See Fig. 3 for some of the capabilities within VDJ.

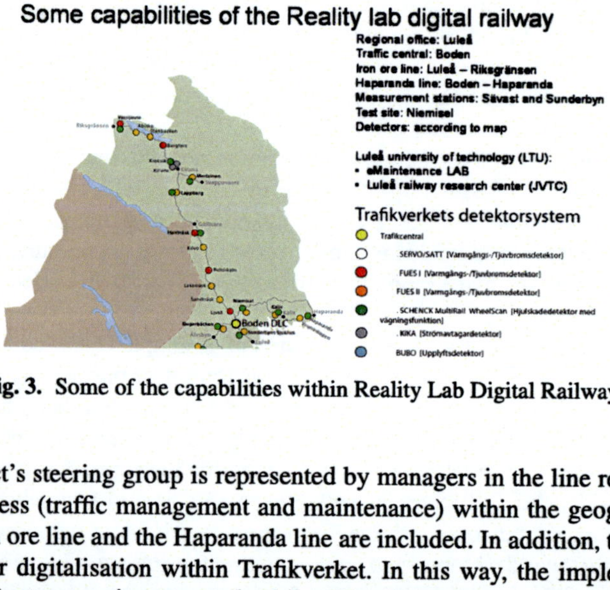

Fig. 3. Some of the capabilities within Reality Lab Digital Railway.

The project's steering group is represented by managers in the line responsible for the core business (traffic management and maintenance) within the geographical area where the Iron ore line and the Haparanda line are included. In addition, there are roles responsible for digitalisation within Trafikverket. In this way, the implementation of tasks in the infrastructure is supported, while taking into account strategic digitalisation initiatives. After project completion, the steering group will transform into an informal steering network where the line manager in charge, if necessary, makes decisions on development and demonstration with the support of the others. Mainly, the task will be linked to physical infrastructure within the framework of "Testbed Railway" and the establishment of test sites.

The project group is mainly represented by people who participate in research and business development at European and national levels, both at Trafikverket and at the competence centre (JVTC) prioritized by Trafikverket regarding digitalised railway maintenance. The project team is expected to transform to an informal competence network, where the project manager as future manager of VDJ through his participation in the development management at the business area of maintenance is convening. Activities within VDJ that are run in the form of business development or research projects are managed within the framework of the main process "Research and develop innovation" by resources in the development management within the business area of maintenance.

In addition to managers, an administrative resource should support the lab. Since VDJ's operations are largely aimed at stakeholders outside of Trafikverket, it is important that a prepared communication plan also is managed after the project end. Therefore, resources from the central function of communication must also be involved.

The external reference group consists primarily of a steering group for the project ePilot and members of JVTC. In the future, ePilot will focus on becoming an innovation engine and providing an industry-wide platform (eMaintenance LAB) for collecting, storing, processing and providing data from industry players. In this way, relevant development and research projects are jointly prepared by the industry within ePilot, which is then addressed to Trafikverket's R&D portfolios. Granted projects can then use eMaintenance LAB in their implementation, while overall project management is handled by Trafikverket. In this way, ePilot contributes to an external preparation and assessment of relevance regarding project proposals. Today, ePilot activities are driven within JVTC and the project AIF/R.

Parts of Trafikverket's formal network of competence for monitoring of infrastructure also function as a preparation and reference group. The competence network focus and collaborates on current issues for monitoring of infrastructure for roads, railways and IT. The purpose of the competence network is to promote increased competence in technologies and methodologies for the future monitoring of Trafikverket's infrastructure. It provides an arena for exchange of experience and presentations of both internal and external initiatives. The competence network thus contributes to an internal preparation and assessment of relevance regarding project proposals.

The internal reference group for development or research projects consists, on a general level, mainly of the steering group for the R&D portfolio Asset management, which means that activities that are designed as development or research projects are anchored in the business. In cases where research projects are conducted within the framework of other R&D portfolios, each steering group constitutes the project's internal reference group at the overall level.

3.3 Technologies

There are a number of technical solutions that support the business or are offered as services within the framework of VDJ.

One solution is a Webpage on Trafikverket's website under "Research and innovation" with introductory information and contact information [22]. Another solution is "My Page" on Trafikverket's website [23], to which established agreements provide access to:

- Project portal for administration of collaborative projects in the industry [24].
- External workspace for sharing data between participants in collaborative projects within the industry [25].

- Systems with data from Trafikverket's operations [26], including DPM 3.50 with "Sky view" as part of established digital twin of the Iron ore line.

Another capability are physical assets in the form of the Iron ore line and the Haparanda line, the Traffic Control Centre in Boden and test sites in Sävast, Sunderbyn and Niemisel [2]. It is also possible to connect up to 157 nodes with power and communication via existing remote monitoring of heating systems for switches and crossings [10].

Another capability are laboratories at LTU, primarily eMaintenance LAB with the ability to receive, store, analyse and provide large amounts of data [27].

The project also performed an adaptation of the inspection system (Bessy) to record test and demonstrator data collected in the infrastructure [1].

The project also applies the possibility to publish results via Trafikverket (e.g. FUDInfo and diariet) or Luleå University of Technology's websites.

Another result of the project is part of a digital twin of the railway system. Two parts that have been established are "Rail view" and "Sky view", see Fig. 4. These parts are based on data in the form of images and laser point clouds collected via railroad vehicles and helicopters. Demonstrator of augmented reality (Augumented Reality, AR) for maintenance purposes based on "Sky view" has also been developed under the name "Maintenance Go" (cf. "Pokémon Go"), see [10]. Even more parts of a digital twin have been set up in collaboration with the European program Shift2Rail, see, e.g., [27].

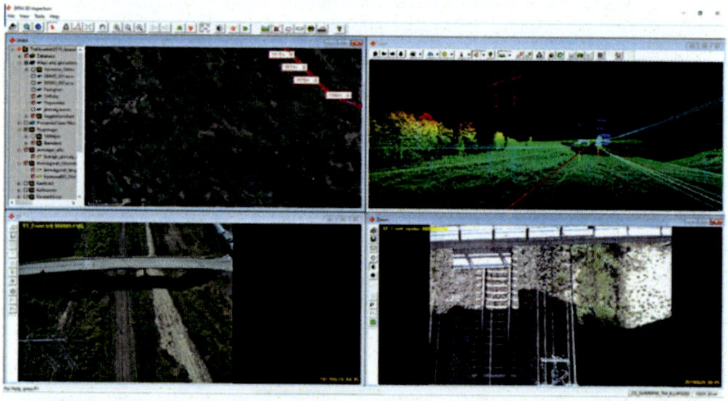

Fig. 4. Example of interface to the part of the digital twin of the Swedish Iron ore line named "Sky view". In the upper right quadrant is the catenary system digitalised in a laser cloud. The other quadrants display a map with track section numbers, overview and detailed photos respectively.

A platform for receiving, storing, analysing and delivering railway related data is a central capability. During the project period, two different platforms have been used to varying degrees. One platform is eMaintenance LAB at Luleå University of Technology, which can be used to manage data from several different players within the railway sector (LTU, 2019). Another platform is "pilotSmartFlow", which was established within the framework of the project "Strategy and basis for monitoring the infrastructure" and is primarily used to combine different types of data for which Trafikverket is responsible [28]. Trafikverket is currently working to establish solutions for this type of activities.

4 Discussion and Conclusions

Reality Lab Digital Railway offers a number of services. The first service that an external stakeholder encounters is a unified entrance to Trafikverket regarding monitoring of the railway system. This entrance has also been called the "Green door". At present, the door corresponds to a webpage on Trafikverket's website, where contact information is available to the door manager, who at this stage has the role of a gatekeeper.

Once the door is opened, guidance is offered and the manager assumes the role of a guide. Depending on the need, there are a number of possible solutions that can be offered.

- The need is not related to Trafikverket's mission, which is communicated to the stakeholder.
- The need is not considered to be related to monitoring of the railway system, but is in line with Trafikverket's mission. In this case, reference is made to alternative contacts within Trafikverket.
- The need can be solved using open data. In this case, reference is made to "Lastkajen" on Trafikverket's website.
- The need can be met with access to systems at Trafikverket that require permission. In this case, VDJ can help to establish agreements and provide access.
- The need requires a work effort that is financed and is within the scope of Trafikverket's mission. In this case, the following options are available:

  - Project grant (non-profit association). Trafikverket awards project grants to non-profit organizations. The project grant is submitted to projects aimed at achieving the transport policy goals of accessibility, safety, environment and health. The projects can be carried out at local, regional or national level.
  - Degree projects and summer jobs. Trafikverket has the opportunity to offer students degree work and summer jobs. Students are given the opportunity to get to know the organization and to carry out qualified work in a number of areas. Student work can also be conducted in collaboration with any external organization. Within the framework of VDJ, a degree project and three summer jobs have been completed, see [29–32].
  - Projects at one of Trafikverket's three prioritised research centres within railway or one industry program. Within VDJ, the focus has been on collaboration with the project ePilot, which is run by JVTC at Luleå University of Technology (LTU) on behalf of Trafikverket. The reason is that, in addition to the regional link to designated railway lines, there is the ability to work in research-oriented collaboration projects in digitalisation, for example via eMaintenance LAB. Examples of collaboration projects with ePilot that have also involved other parts of Trafikverket and the industry are models for collaboration, innovation, implementation, agreements and business models [31], implementation of digital collaboration [32], application of vehicle-based monitoring of the infrastructure [33], data sharing laws and

regulations [34], and implementation of a national measuring station [35]. Future focus on research in the field at JVTC is also described in an external analysis [36], which can be seen as a complement to VDJ's activities at Trafikverket.
- Research and innovation. Trafikverket procures research and innovation according to its mission in the transport area.
- Business development within Trafikverket. Business development is carried out in close connection with the ongoing operations. Business development can be the step after research and innovation, namely to adapt and utilize the results for specific operations.
- Continuous improvement within Trafikverket. Continuous improvement are continuous and incremental, often minor, improvements to existing processes and working methodologies. They can constitute the step after business development, namely implementation.

For grants, student work, research and innovation, business development and continuous improvement, there are established processes with supportive methodologies and tools that are applied.

Regarding research, innovation or business development, methodologies in development management are applied in the maintenance area or methodologies in collaboration with the project ePilot. In both cases, Trafikverket's project model XLPM is applied with a number of standardized decision points and established documents from project idea to final report. Which of the two options that are selected depends on the conditions for the project. Within ePilot, the project, for example, must include at least three project partners and there is a requirement for co-financing. Even in cases where data from several players in the industry are to be combined, ePilot has been applied as this functionality has not been fully available within Trafikverket. However, during the project period, VDJ also had some access to the "pilotSmartFlow" platform to receive, store, analyse and deliver large amounts of data. This platform has been established within the framework of the internal Trafikverket project "Strategy and basis for monitoring the infrastructure". Today, there are initiatives to establish such a platform for running operations within Trafikverket.

In cases where the research and innovation project is conducted in collaboration with the industry, but not within the framework of ePilot, VDJ offers a project portal for project management. An external workspace is also offered to share data between project partners. Access to these two services requires agreements and permissions that provide access via "My Page" on Trafikverket's website. The corresponding functionality is also available external to Trafikverket through ePilot and managed by JVTC.

There are also examples of collaboration with the industry conducted on the basis of agreements outside of project activities, e.g. with operator and system supplier to evaluate technology for monitoring the infrastructure. In cases where the initiatives were related to VDJ's activities, the lab has contributed to the coordination of necessary resources for the conclusion of agreements, provision of data or evaluation. During the project period, the main actors involved have been organizations that develop information solutions aimed at other organizations, often provided through some type of portal, see, e.g., [5, 37–40]. One reason for this is that Trafikverket's suppliers and customers on the railway side are organizations where the relationship is regulated by agreements and not by individuals. There are also examples of operators who have contacted VDJ for guidance on designing information services for travellers or establishing test sites for evaluation of measurement systems for monitoring railway vehicles, where resources and capabilities at Trafikverket have been identified and involved.

If the need requires a physical test site in the infrastructure, this functionality is also offered to some extent. The infrastructure that is mainly covered is the Iron ore line (including measurement sites at Sävast and Sunderbyn) and the Haparanda line (including measurement site at Niemisel). Along the lines, it is also possible to use the remote monitoring of the heating system for switches and crossings, which provides both power and communication possibilities. In all cases, the support process "Test maintenance" offers methodologies and tools that can be adapted to the purpose, e.g. technically approved railway material or research and innovation.

There are a number of critical capabilities that should be ensured in order for VDJ to remain successful after transitioning from project to normal operation. Four of these capabilities are described in some more detail in the following paragraphs.

One capability is related to all railway maintenance contracts affecting the Northern maintenance district, which should be designed in a similar way to the basic railway maintenance contract. This is to support the possibility of conducting experiments and tests in the infrastructure. The other eleven types of maintenance contracts affecting the district (and all other maintenance districts) besides the basic contract are: chemical vegetation control on line; chemical vegetation control on yards; detectors; traffic information; NDT measurements (Non-Destructive Testing); milling of rails; grinding on line; grinding of switches and crossings; track and catenary measurement (measuring cars); tree fencing and clearing; as well as FOMUL measurement (i.e. the fixed objects in the vicinity of tracks being measured). However, even if the different types of maintenance contracts are not adapted, it is possible to use them for testing activities, e.g. by purchasing needs from R&D projects. However, there are two decisions that must be made. For each specific case, it is necessary to determine if the need in question is possible to include as an amendment, supplement or outgoing work in the intended maintenance contract. On the part of the R&D project, it depends on what the agreement looks like. If the design of the R&D agreement makes it a matter of direct procurement, this must be justified and approved via a motive attachment for direct procurement. See [41].

Another central capability is that standardized information sharing agreements with operators and system suppliers should be established. The project has delivered proposals that can be used as a starting point. This is to support the management of vehicle-based solutions for monitoring of the infrastructure.

A third capability is a platform for receiving, storing, analysing and providing large amounts of data, which must be established to support increased digitalisation [42]. This platform should be usable during development, testing and normal operations. During the project period, eMaintenance LAB has been applied externally and pilotSmartFlow internally Trafikverket. There are initiatives in the area at Trafikverket, but it is unclear how long it will take before these results in useful solutions for VDJ's operations (see [43]).

A fourth and fundamental capability is that the developed FMECA should be supplemented to support a more complete RCM-analysis in accordance with the standard "Management of reliability - Part 3–11: Guidelines for functional safety-oriented maintenance" (EN 60300-3-11: 2010) [44]. This is necessary to enable the establishment of effective, efficient and dynamic maintenance programs for different subsystems of the infrastructure. These, in turn, form the basis for assessing the benefits of new digital solutions for surveillance and monitoring. In cases where supervised functions are critical from a traffic safety perspective, the methodologies must also be supplemented by a formal risk analysis, e.g. barrier analysis in accordance with the description of [45] to fulfil requirements according to Common Safety Methods (CSM) [46]. Requirements for monitoring systems should be based on established FMECA with a supplement from the standard "Maintenance - Part 5: Testability and diagnostic testing" (EN 60706-5: 2009) [47], e.g. requirements regarding undetected faults and false alarms (see, e.g. [48]).

The project Reality Lab Digital Railway (VDJ) can also be valued based on the long-term effect that Vinnova intends to achieve by meeting the program's performance goals. These are a Swedish public enterprise that [1]:

- is internationally renowned for its ability to experiment with external actors
- has rapid adaptability to changing demands and opportunities in the surroundings as well as to different segments and individuals
- offers open environments that lead to more companies, entrepreneurs and researchers choosing to place their operations and production in Sweden, which strengthens the Swedish global competitiveness.

As far as internationally known ability is concerned, Reality lab digital railway can take advantage of the good reputation that has already been established in the region. This is largely thanks to the development and R&D activities conducted in relation to the Iron ore line, at Luleå University of Technology (primarily Luleå Railway Research Center and eMaintenance LAB) and EU programs such as Shift2Rail. This work has been further deepened and developed during the project period. Examples of collaborative activities at European level are the digitalisation of bridges, systems for monitoring used fatigue consumption and photographic methods for monitoring stresses in real constructions, see In2Track [49]. The problems with transition zones, which have been addressed with link plates at VDJ's infrastructure assets, are described in In2Track [50]. Another example is fully automated equipment for finding missing fasteners, which was presented at the final conference of the project In2Smart [51].

As far as adaptability and open environments are concerned, the reality lab has, through concrete sub-projects, contributed to building an improved ability. However, a number of activities remain to improve both speed and flexibility.

Acknowledgements. The Reality lab digital railway project was made possible thanks to funding from Vinnova (grant number: 2017-02139), Trafikverket (FUDid: 6538) and Luleå University of Technology.

References

1. Vinnova: Verklighetslabb inom offentlig verksamhet (Reality Lab in Public Sectors). Announcement. Number: 2016-04792, Vinnova, Stockholm (2017). (in Swedish)
2. Trafikverket: Projektspecifikation för Verklighetslabb digital järnväg (Project specification - Reality Lab Digital Railway). Project specification, Trafikverket, Luleå (2017). (in Swedish)
3. Vinnova: UPPSAMT 2.0 - UPPkopplade & SAMverkande järnvägar och medarbeTare: punktlighet, rapporterande tåg & driftledning (2019). https://www.vinnova.se/p/uppsamt-2.0---uppkopplade--samverkande-jarnvagar-och-medarbetare-punktlighet-rapporterande-tag--driftledning/. Accessed 5 Nov 2019. (in Swedish)
4. Melander, P.: Slutrapport till VDJ – Baserad på godkänd slutrapport till Vinnova, m.fl. Report, Outflight AB, Åkersberga (2019). (in Swedish)
5. Enplore: Enplore Data Analytics Platform (2019). https://bombardier.admin.enplore.io/login. Accessed 5 Sept 2019
6. Trafikverket: Implementering av resultat från ePilot i baskontrakt järnväg (Implementation of results from ePilot in basic maintenance contracts in railway). Working material, Trafikverket, Luleå (2016). (in Swedish)
7. Lundkvist, P., Söderholm, P.: Handledning FMECA (User Manual FMECA). User Manual, Trafikverket, Luleå (2018). (in Swedish)
8. Trafikverket: Administrativa föreskrifter – Utförandentreprenad - För utförande av basunderhåll av järnvägsanläggningar – Luleå-Murjek samt Haparandabanan. Ärendenummer: 136550, Trafikverket, Luleå (2016). (in Swedish)
9. Trafikverket: TMALL 0967 – FMECA. Template, Trafikverket, Luleå (2018). (in Swedish)
10. Granström, R.: Tillförlitlighetsbaserat underhåll inom Trafikverket - Demonstrator med växelvärme (Dependability-based maintenance within Trafikverket – Demonstrator with heating system in S&C). Report, Trafikverket, Luleå (2019). (in Swedish)
11. Lundkvist, P.: Förslag på nytt TDOK (Suggestion of new regulation). Report. Licab AB, Luleå (2018). (in Swedish)
12. Granström, R.: Rail View, Sky View och Maintenance Go – Tillämpningar inom Trafikverket – Ett projekt inom Verklighetslabb Digital Järnväg. Report, Trafikverket, Luleå (2019). (in Swedish)
13. Lundkvist, P., Söderholm, P.: Handledning RCM (User Manual RCM). User Manual, Trafikverket, Luleå (2018). (in Swedish)
14. Trafikverket: TMALL 0979 – Statistisk försöksplanering (Template for Design of Experiments). Template, Trafikverket, Luleå (2018). (in Swedish)
15. Lundkvist, P.: Stresstest i samband med besiktning av växelvärme - Tillämpning av arbetssätt inom ramen för Verklighetslabb digital järnväg (Stress test during inspection of heating systems in S&C – Application of Methodology within Reality Lab Digital Railway). Report, Licab AB, Luleå (2018). (in Swedish)
16. Lundkvist, P., Söderholm, P.: Handledning statistisk försöksplanering (User manual - Design of Experiments). User Manual, Trafikverket, Luleå (2018). (in Swedish)
17. Trafikverket: Verklighetslabb digital järnväg. FUD-ärende till Portfölj 2: Vidmakthålla - Ärendebenämning: Verklighetslabb digital järnväg id 6538. Diariet, Number: TRV 2017/67785, Trafikverket, Borlänge (2017). (in Swedish)
18. Nilsson, A.: Analytiska tjänster – Slutrapport (Analytical Services – Final Report). Report, eMaintenance 365, Luleå (2019). (in Swedish)
19. LTU: ePilot avtal. (Number: LTU 843-2017). Agreement, Luleå university of Technology, Luleå (2017). (in Swedish)
20. Eriksson, C., Eriksson, P.: Lösningsidé WPM-Detektorer – Underlag inför IT-samråd (Idea of Solution for WPM Detectors – Material for IT-Consultation). Report, Trafikverket, Borlänge (2019). (in Swedish)

21. Sammeli, C.: Tjänsteidé hjulgeometri-/hjulprofilmätare (Idea of Solution for Wheel Geometry Detectors). Decision support service idea, Trafikverket, Borlänge (2019). (in Swedish)
22. Trafikverket: Sammanfattning informationssäkerhetsanalys Hjulunderhåll Provplats öst. Protokoll (Ärendenummer TRV2018/101620), Information security analysis, Trafikverket, Borlänge (2018). (in Swedish)
23. Trafikverket: Sammanfattning informationssäkerhetsanalys Värmekamera för växelbesiktning. Protokoll (Ärendenummer TRV 2018/16883), Information security anlysis, Trafikverket, Borlänge (2018). (in Swedish)
24. Trafikverket: VDJ – Verklighetslabb digital järnväg, 11 October 2019 (2019). https://www.trafikverket.se/resa-och-trafik/forskning-och-innovation/aktuell-forskning/transport-pa-jarnvag/vdj-verklighetslabb-digital-jarnvag/. Accessed 21 Oct 2019. (in Swedish)
25. Trafikverket: Min sida (My Page) (2019). https://authweb.trafikverket.se/authweb/. Accessed 21 Oct 2019
26. Trafikverket: Projektportalen (Project Portal) (2019). https://www.trafikverket.se/tjanster/system-och-verktyg/projekthantering/projektportal/. Accessed 21 Oct 2019. (in Swedish)
27. Trafikverket: Externt arbetsrum Verklighetslabb digital järnväg (External Working Space Reality Lab Digital Railway) (2019). https://earum.sp.trafikverket.se/sites/20181011130320/home/Sidor/start.aspx. Accessed 21 Oct 2019. (in Swedish)
28. Trafikverket: Alla tjänster från A-Ö (2019). https://www.trafikverket.se/tjanster/alla-fran-ao/. Accessed 21 Oct 2019. (in Swedish)
29. In2Track: Improvement of tunnels and bridges. In: Hermosilla, C. (ed.) Deliverable D4.2, European Project: Research into Enhanced Tracks, Switches and Structures, 135 p. (2019). https://projects.shift2rail.org/mwg-internal/de5fs23hu73ds/progress?id=0aKHNMYzxyN4ia9qZFLvxCXL2Vy4yo1ll5rCH58d26A,&dl. Accessed 25 Nov 2019
30. LTU: eMaintenance LAB, 21 October 2019 (2019). https://www.ltu.se/research/subjects/Drift-och-underhall/Laboratorium-och-utrustning/eMaintenance-LAB-1.78754. Accessed 28 Nov 2019
31. Trafikverket: Big Data och kvalificerad analys/AI i tillgångsförvaltningen - en rapport från projektet Strategi och grund för övervakning av anläggning i Trafikverket 2019. Report, Trafikverket, Borlänge (2019). (in Swedish)
32. Josefsson, H.: Verklighetslabb digital järnväg – Inledande studie (Reality Lab Digital Railway – Initial Study). Report, Trafikverket, Luleå (2017). (in Swedish)
33. Wahlstedt, F., Nilsson, J.: Tillämpning av drönare för tillståndsbedömning av järnvägsterräng (Application of Drones for Condition-based Inspection of Railway Surrondings). Master thesis, Luleå University of Technology, Luleå (2018). (in Swedish)
34. Juntti, U., Jägare, V., Lindgren, M., Olofsson, B., Johansson, N., Norrbin, P.: EP200 - Modeller för samverkan, innovation, implementering, avtal och affärsmodeller (SIIAA). Project report eP200-201-2017, ePilot, Luleå Railway Research Centre (JVTC), Luleå University of Technology, Luleå (2019). (in Swedish)
35. Markgren, J.: Riskanalys ERTMS (Risk Analysis ERTMS). Report, Trafikverket, Luleå (2018). (in Swedish)
36. Kratz, S., et al.: Implementering av digital samverkan för hjul-räl på sträckan Uppsala-Sundsvall-Umeå. ePilot, Luleå Railway Research Centre (JVTC), Luleå University of Technology, Luleå (2019). (in Swedish)
37. Strömbom, J.: Nyttjandestudie av TrackLogger – ett mobilt mätsystem för tillståndskontroll i ordinarie tågtrafik. Report, Trafikverket, Luleå (2019). (in Swedish)
38. Juntti, U., et al.: EP214 - Implementering av tillståndsövervakning av infrastruktur. Projektrapport eP20-214-ImpINFRA. ePilot, Luleå Railway Research Centre (JVTC), Luleå University of Technology, Luleå (2019). (in Swedish)

39. Juntti, U., Jägare, V.: EP219 - Juridik för datadelning. Project report eP20-219-2018-ImpD, ePilot, Luleå Railway Research Centre (JVTC), Luleå University of Technology, Luleå (2019). (in Swedish)
40. Juntti, U., Jägare, V., Karim, R.: EP220 - Implementering av 'Mätstation Sverige' (ImpMSE) – Handlingsplan. Project report eP20-220-2018-ImpMSE. ePilot, Luleå Railway Research Centre (JVTC), Luleå University of Technology, Luleå (2018). (in Swedish)
41. JVTC: JVTC research topics towards 2030. Appendix 1 – Appendix to JVTC's business plan 2019–2021. Luleå Railway Research Centre (JVTC), Luleå University of Technology, Luleå (2019)
42. Jones, M.: Perpetuum Track Monitoring on the Swedish Rail Network. Report document number: 01041477. Perpetuum Ltd., England (2019)
43. Jones, M.: Perpetuum Track Health Website User Manual. User manual, document number: 01040827. Perpetuum Ltd., England (2019)
44. Perpetuum: Perpetuum Track Portal (2019). https://www.perpetuumtrackhealth.com/Account/Login. Accessed 5 Nov 2019
45. D-Rail: D-Rail portal (2019). https://portal.d-rail.se/. Accessed 5 Nov 2019
46. Laine, J.: Sammanställning underhållsavtal 2019-10-28. Excel-document, Trafikverket, Luleå (2019). (in Swedish)
47. Fredriksson, A., Rendalen, T., Söderholm, P.: Mottagande av informationstjänst från leverantör med avseende på data om anläggningens tillstånd – En studie inom Verklighetslabb digital järnväg (VDJ). PM, Trafikverket, Borlänge (2019). (in Swedish)
48. Söderholm, P.: Inventering av initiativ inom Trafikverket som är relaterade till övervakning av anläggningen. PM (number: 2016/32601), Trafikverket, Luleå (2019). (in Swedish)
49. SS-EN 60300-3-11:2010: Ledning av tillförlitlighet - Del 3-11: Riktlinjer för funktionssäkerhetsinriktat underhåll. Svensk elstandard (SEK), Stockholm (2010). (in Swedish)
50. Söderholm, P., Nilsen, T.: Systematic risk-analysis to support a living maintenance programme for railway infrastructure. J. Qual. Maint. Eng. **23**(3), 326–340 (2017)
51. EU: Commission implementing regulation (EU) 2015/1136 of 13 July amending Implementing Regulation (EU) No. 402/2013 on the common safety method for risk evaluation and assessment. EU, Brussels (2015)
52. SS-EN 60706-5:2009: Underhållsmässighet - Del 5: Provbarhet och diagnostisk provning. Svensk elstandard (SEK), Stockholm (2009). (in Swedish)
53. Söderholm, P.: Maintenance and continuous improvement of complex systems: linking stakeholder requirements to the use of built-in test systems. Ph.D. thesis, Luleå University of Technology, Luleå (2005)
54. In2Track: Inspection and monitoring techniques for tunnels and bridges. In: Aleksieva, N. (ed.) Deliverable D4.1, European Project: Research into Enhanced Tracks, Switches and Structures, 521 p. (2019). https://projects.shift2rail.org/download.aspx?id=e2e528fd-6990-46f7-bda2-b76ab2ffc415. Accessed 25 Nov 2019
55. In2Smart: Final event, 10 October 2019 (2019). https://projects.shift2rail.org/download.aspx?id=0da0e7cc-d85f-4af9-8b2f-0d60dba9ad4f. Accessed 25 Nov 2019

An Integrated Procedure for Dynamic Generation of the CRC in View of Safe Wireless Communication

Larissa Gaus, Michael Schwarz(✉), and Josef Börcsök

University of Kassel, Kassel, Germany
{lgaus,m.schwarz}@uni-kassel.de

Abstract. Due to the rapid technological development in recent decades, increasingly complex technical systems are finding their way into all areas of human activity. The growing complexity and scope of technical systems result in higher requirements on their safety.

Due to the technological possibilities, in modern industrial applications, more and more important data are exchanged between several industrial machines, monitoring or control devices. Therefore, safe and secure communication is one of the most important aspects besides safe hardware and software. The use of innovative technologies makes it necessary to develop new concepts and methods to ensure safety.

A very common method for detecting transmission errors in the digital communication process is a Cyclic Redundancy Check (CRC). Due to the increasing use of wireless technologies in the field of digital communication, this method offers no longer sufficient protection.

This paper presents an approach for a novel dynamic procedure for CRC generation in a digital communication channel. The procedure is based on the conventional CRC method. However, the procedure allows the use of several specified generator polynomials, dynamically. Thus, using the procedure in a communication process obstructs unauthorized access and falsification of the transmission data.

In this paper, an initial concept for the procedure is introduced, the detailed description of some developed algorithms and aspects is explained and the selection of the eligible generator polynomials is discussed.

Keywords: Wireless communication · Safety · Security · Linear code · CRC · Dynamic CRC

1 Introduction

Constantly increasing flexibility and complexity of the industrial machinery and systems lead to higher requirements on their safety. The avoidance of process accidents and the maintenance of the safe operation are the core requirements in the field of functional safety. Not only operating devices has to comply with the safety requirements, but also these rules shall apply to the networking system.

R. Karim et al. (Eds.): IAI 2021, LNME, pp. 256–265, 2022.
https://doi.org/10.1007/978-3-030-93639-6_21

There are several norms and standards that exclusively or partially address the requirements for safe digital communication. IEC 61508 [1], EN 50159 [2] or IEC 61784 [3] can be mentioned as examples.

Generally, errors, repetitions, deletion, insertion, re-sequencing, corruption, delay and masquerading can damage the proper communication. To perform the safe communication one of the available safety application protocols can be used. Such protocols implement the measures for avoiding, detecting and where appropriate correcting the faulty transmission. Those measures can be: cyclic exchange of messages, watchdog timer on each communication member and codename for each communication relationship [3]. Also, one of the required measures by IEC 61784-3 is proving integrity of the message by the cyclic redundancy check.

Data integrity is one of the most important aspects of the transmission process in terms of safety. Here, it is necessary to ensure that the data received at the actuators side are the same that have been sent by the sensors at the opposite side and has not been falsified or manipulated. The incorrect data can lead to wrong adjustments of the actuators. By the manipulation of data, the process can be affected so that it results in an unsafe state of the plant or even a dangerous situation. For example, even the slightest adjustment of a valve in a chemical plant could lead to a wrong mixture of chemicals and may cause an explosion.

Industry 4.0 and the associated concepts of smart factory and Internet of Things (IoT), which demand intelligent and flexible connectivity, substantial affect the tendency to apply the wireless solutions for the manufacturing systems [4, 5]. But, using wireless concepts instead of established wired technologies is fraught with challenges. As an example, in [6] the major technical challenges for the use of industrial wireless sensor networks are outlined as follows: limited resource constraints, malfunctions due to dynamic topologies and harsh environmental conditions quality-of-service (QoS) requirements, data redundancy, large-scale deployment and ad-hoc architecture, packet errors and variable-link capacity, security violation as well as integration with the Internet and other networks.

In this paper an approach for the integrated procedure for the dynamic generation of the CRC checksum is presented, which addresses the last three stated technical challenges. To introduce that approach, initially the background and the relevant theory issues are provided in Sect. 2. Following that, in Sect. 3, the theoretical proposal of methodology is outlined. Section 4 gives an example for the possible application within the Ethernet protocol. Finally, the paper is concluded in Sect. 5.

2 Background

2.1 Wireless vs. Wired Communication

The transmission system is a crucial component of the technical application besides the sensors, actuators and logical elements. Damages in the transmission system can have a negative effect of the entire system and subsequently lead to the hazardous impact. Electrical, mechanical, environmental and application-specific stresses are the factors which can defeat the cables and consequently comprise the reliability, signal integrity and life performance of the transmission system [7]. These factors can be avoided by using the wireless technologies for transmission tasks.

A decisive asset of wireless networks is the possibility of using them in sectors of application, where the wired communication is not appropriable or too expensive, for example in harsh environments. But, the nature of the wireless communication techniques carries drawbacks. Because of their open way of transmission the wireless networks are more vulnerable to malicious attacks, such as eavesdropping or jamming. Additionally, due to the dynamic topology and mobility of wireless networks the implementation and managing of security measures are complicated [8].

Regardless of wired or wireless technology is used, the communication part follows the OSI reference model. In both cases, the protocol architecture consists of the physical layer, the link layer, the network layer, the transport layer, the session layer, the presentation layer and the application layer. The processing of the data in the five upper layers is performed identically with both, wired and wireless, technologies. The two lower layers, the physical layer and the data link layer, are depending on the network technology and different by wired vs. wireless networks [9].

Protocols are implemented in each OSI (Open System Interconnection) layer, which can lead to measures to avoid safety and security failures. While these measures in the upper layers are the same for both types of technology, wireless communication in the lower two layers requires different measures than wired solutions.

Due to the same protocols, wired and wireless networks in the upper layers have the same security weaknesses.

2.2 Wireless Security Attacks

Different OSI layers fulfill different tasks, which are realized by different protocols. This variety leads to the fact that each layer has its own security gaps and problems and therefore has to fulfill its own security requirements. For this reason, each layer implements its own appropriate measures to defend against the relevant security attacks.

In general, attacks on wireless networks can be divided into passive and active attacks [10]. Table 1 lists the most common attack strategies in wireless networks.

Table 1. Types of security attacks in wireless networks [10]

Passive attacks	Active attacks
• Traffic analysis • Eavesdropping	• Denial of Service (DoS) • Resource consumption • Masquerade • Replay attack • Information disclosure • Message modification

Especially critical attacks in terms of safety are active attacks: DoS, resource consumption, masquerade, information disclosure and message modification.

2.3 Security Requirements for Wireless Networks

In the context of security, wireless networks have to satisfy requirements listed in the Table 2.

Table 2. Wireless network security requirements [10]

Security requirements	Requirement objective
Authentication	Only the authorised users or processes are allowed access to the resources of the network
Non-Repudiation	Non-repudiation is the guarantee that the sender cannot deny that the message was sent and the recipient cannot deny that the message was received
Confidentiality, Access Control	Confidentiality is the ability to protect the transmitted data from access by, or disclosure to, unauthorized users. The sent data should be accessible only for provided receiver node
Integrity	Integrity is the guarantee that exactly the same data, without falsification, has reached the recipient as was sent
Availability	Availability guaranties the access to the transmission whenever a legitimate user requires them
Resistance to Jamming	Resistance to jamming is the ability to protect each wireless transmission from intentionally external interference
Resistance to Eavesdropping	Resistance to eavesdropping is the ability to ensure that the eavesdropper cannot gain any information send from the transmitter to the receiver

2.4 CRC

Firstly described in 1961 by W.W: Peterson and D.T. Brown [11], the CRC is generally accepted and a widely used method for error detection in data transmission and storage processes.

The CRC is based on the fact that each sequence of binary symbols can be represented as an polynomial in the *Binary Galois Field* $GF(2) = \{0, 1\}$ in the form

$$a_{n-1}x^{n-1} + a_{n-2}x^{n-2} + \ldots + a_1x^1 + a_0 \tag{1}$$

The computation of the CRC value is based on polynomial division and the error detection process proceeds as follows:

- Depending on the application, a generator polynomial $G(x)$ with the degree r is selected.

- Before sending, the transmitter extends the message by r zero bits at the end and then divides this extended message by the selected $G(x)$. Then the reminder of the division, named CRC checksum, is appended to the original message. The resulting message is divisible by $G(x)$ and in this form is sent to the receiver.
- The receiver divides the obtained message by the same selected $G(x)$. If the message is divisible by $G(x)$ without reminder, then the message was received correctly. Otherwise, the message was corrupted by the transmission.

The selection of the generator polynomial depends on the type of application. Here, for example, the length of the messages plays a major role. But also, the Hamming Distance (HD) is an important metric for the choice of $G(x)$. Mathematically speaking HD is the measure of the deference of two code words in a block code. From a practical point of view, HD can be described as "the smallest number of bit errors that are undetectable by the error code" [12].

3 Methodology

The idea of the described approach is a procedure, which allows supplementary dynamically generating of the CRC checksum at runtime using different generator polynomials within the link layer.

A transmission channel with two communicating devices is assumed. Both transmission members, the transmitter and the receiver, include the same list of the defined generator polynomials. Firstly, a single transmission cycle is considered. Bevor sending the transmitter builds an additionally CRC checksum of the payload field. Thereby, instead of the fixed generator polynomial the polynomial from the integrated list is used. The choice of the generator polynomial occurs at run-time dynamically depending on the actually set operating mode (see Sect. 3.2).

The reorganization of the table entries is performed at regular intervals. For this purpose, synchronization messages are exchanged between the master and the slave. A synchronization message is not sent separately, but is integrated into a regular message and transmitted in parallel with process data, in order not to increase the communication effort unnecessarily (for details see Sect. 3.3).

3.1 Selection of the Generator Polynomials

The CRC generator polynomials shall be selected based on the data word length, the admissible CRC checksum length and the provided HD. Thereby, the aim is to keep the probability of undetected error as low as possible.

For the purpose of the current approach, an existing listing of the CRC generator polynomial created by Philip Koopman [13] was taken as a basis. Due to the background of this research, the highest priority has been given to the safety related aspects.

In the safety-critical applications the CRC polynomials with the HD = 6 are commonly used [12]. On that account, the list of the generator polynomials for the described approach was in the first step restricted to the polynomials with HD = 6.

In the second step, the code word length served as a factor by the composition of the list. The code word length is composed of the data worth length and the checksum length. The minimum feasible code word length has to correspond to the conceivable frame length of the used protocol. In this approach, the PROFIsafe® message format was chosen as a reference. The maximum allowed length of a PROFIsafe® message amounts 128 bytes. Thus, all polynomials, which cover this length, can be used for the outlined approach.

On this basis a table of 12 generator polynomials was compiled, as shown in Table 3.

Table 3. Generator polynomials

Degree	Polynomial	Data word length
32	0x9960034c	32738
31	0x74f9e7cb	32738
30	0x2ad4a56a	16356
29	0x1cf492f3	16356
28	0xd120245	8166
27	0x6c3ff0d	8166
26	0x2186c30	4072
25	0x1b9189d	4072
24	0xbd80de	2026
23	0x6bc0f5	2026
22	0x395b53	1004
21	0x1edfb7	1004

3.2 Operating Modes

There are two operating modes available: sequential mode and mixed mode.

In the sequential mode, the list of the generator polynomials is processed sequentially. For every transmission transaction the generator polynomial with the next index is used. If in the previous step the last entry in the table has been reached, the pointer is located on the first position of the list and the table is processed again from the start.

In the mixed mode the choice of the generator polynomial is handled in the assorted way. It means the transceiver uses the generator polynomial with randomly selected index. The index of the used polynomial is notified in the same message to the receiver. For that the corresponding GPI (Generator Polynomial Index) field, containing 5 bits, is allocated in the message. After receiving the message the receiver can find the index in the assigned field and then use the same generator polynomial for decryption.

Note: in the current phase 4 bits are sufficient to announce the index. However, since future extensions of the GP table are planned, the GPI field has been designed to be larger in order to be prepared for later extensions. Thus, the table can contain up to 32 generator polynomials.

3.3 Algorithm for the Table Reorganization

Two bits are reserved in the message for the table reorganization (TR). These bits are not set (value “00”) as long as regular message exchange takes place. If the master wants to order a table reorganization, it sends a message with the set 1st bit of the TR field (value “10”). If the recipient has received the request for table reorganization, it carries out the reorganization and sends a confirmation message, whereby the 2nd bit is set (value “01”). The table reorganization takes place with the same algorithm for both stations, so that the tables are the same for both stations after the reorganization. Table 4 provides an overview of the values of the TR field.

Table 4. TR-Field

TR-Value	Description
00	No table reorganization requested
10	Request to the reorganization of the table
01	Confirmation of the performed table reorganization
11	Not allowed

The procedure of the table reorganization is as follows:

1. The master sets the first TR bit in the next message being transmitted (TR-Value = “10”), starts the transmission process and waiting for a confirmation message.
2. After receiving the message by the slave, in the first step the process data is handled. Then the slave checks the TR-Field. If the value is “10”, the slave fulfills the requested table reorganization. If the reorganization was successfully performed, the TR-field in the following message is set to “01” and the message is send to the master. Otherwise, the value of the TR-field in the confirmation message is set to “10”.
3. After receiving the confirmation message, the master modifies his table and closes the reorganization process.
4. If the confirmation message was not received, the table remains unchanged and the master starts the new reorganization request.

4 Case Study Ethernet

This section gives an example for the possible application of the described approach.

For now, several wireless transmission technologies exist. One of them, the Wireless Local Area Network (WLAN), is specified in the IEEE 802.11 standard. One of the MAC-frame types described in this Standard is the Data Frame. Data Frames consists of three sections: a MAC header, Frame Body and Frame Check Sequence (FCS). The Data Frame contains frames of the overlying layer, e.g. Ethernet-Frames, in the section Frame Body. For this reason, the Ethernet frame was used as an example for the present work.

Until now, the Ethernet described in the standard IEEE 802.3 is an established protocol in the data link layer. The length of the payload in an Ethernet frame can amount up to 1500 Byte. Thus, the Ethernet frame offers sufficient capacity for an integrated dynamic checksum (IDC) protocol data unit (PDU).

The IDC PDU contains three sections: the coordinating byte, the data field and the IDC field, as shown in Fig. 1.

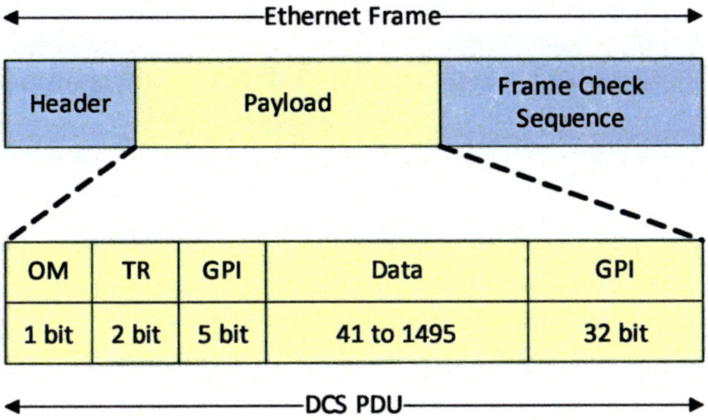

Fig. 1. Ethernet frame with the integrated DCS PDU

The coordinating byte consists of one OM bit, two TR bits and five bits reserved for the GPI.

The last four bytes in the IDC PDU are assigned for the dynamic CRC checksum.

The allocation of five bytes for the integrated method effects in the limitation of the length of the network packets, the PDU of the network layer, to 1495 bytes.

5 Conclusion

In this paper, the approach for the integrated procedure for dynamic generation of the CRC checksum was presented. For this purpose, a general concept of the procedure was developed, the operational modes were described, the table of the CRC generator polynomials was determined and the algorithm for the table reorganization was defined.

The introduced approach operates solely with the payload of the interchanged messages. The other parts of the frame, header end metadata, remain untouched.

The described approach does not affect the behaviour and effects of the other methods for security protection. For this reason the approach can be used in parallel to all existing defence mechanisms.

This is an initial approach. Therefore, the selection of the generator polynomials was made based on the available sources. In the later work, the polynomials to be used have to be validated beforehand with regard to the use cases.

Also, the table was created for worst-case scenarios. If the length of the messages for a concrete application is less than 128 bytes, additional polynomials can be used accordingly. For later, extended work, generator polynomials with different minimum distances would be conceivable. Among other things, dynamic reaction to the length of the message and thus the extension of the GP table is possible in the future.

The possibility to adjust the CRC to the length of the message has the advantage that the communication effort can be optimized by sending messages only as long as necessary. This allows the sparing use of resources, such as decreasing the average transmission energy per message.

Another advantage is that the generator polynomial is selected anew in each cycle and this is only known to the two communication participants at each cycle. This makes the intentional modification of the messages on the communication channel more difficult. Security attacks, such as information disclosure and message modification, can thus be made more difficult, which in the sense of safety-critical applications means less probability of unnoticed processing of incorrect process data and can thus increase the safety of the system. In general, by using the described procedure in a wireless network the integrity of data is supported.

HESSEN **Acknowledgments** This paper is part of the project MEPHYsto which is funded by the German federal state Hessen through the programme Distr@l. We thank Distra@l for the financial support, with the project-nr.: Distra@l 493 20_0021_1.

References

1. IEC 61508: Functional safety of electrical/ electronic/ programmable electronic safety-related systems. International Electrotechnical Commission (2000)
2. DIN EN 50159: Railway applications- Communication, signalling and processing systems-Safety-related communication in transmission systems. European Committee for Electrotechnical Standardization CENELEC (2011)
3. IEC61784-3: Industrial communication networks – Profiles – Part 3: Funktional Safety fieldbuses – General rules and profile definitions International Electrotechnical Commission (2016)
4. Yoshigoe, K.: Data-driven transmission mechanism for wireless sensor networks in harsh communication mironment. In: Proceedings of Globecom Workshop on Torwards Smart Communications and Network Technologies applied on Autonomous Systems. IEEE (2010)
5. Zhu, J., et al.: Foundation study on wireless big data: concept, mining, learning and practices (2008). Invited Paper 15: 1–5. China Comun
6. Gungor, C.V., Hancke, G.P.: Industrial wireless sensor networks: chalenges, design principles, and technical approaches. In: Proceedings of IEEE Transactions on industrial electronics, Vol. 56, No. 10 (2009)
7. GORE: Improving cable Performance in harsh environments. White Paper (2013). https://www.gore.com. Accessed 11 Feb 2020
8. Lin, Y., Chang, J.: Improving wireless network security based on radio fingerprinting. In: Proceedings of the 19th International Conference on Software Quality, Reliability and Security Companion (QRS-C). IEEE (2019)
9. Zou, Y., et al.: A survey on wireless security: technical challenges, recent advances and future trends. In: Proceedings of the IEEE, vol. 104, no. 9 (2016)

10. Shiu, Y.-S., et al.: Physical layer security in wireless networks: a tutorial. In: Proceedings of IEEE Wireless Communications (2011)
11. Peterson, W.W., Brown, D.T.: Cyclic codes for error detection. In: Proceedings of the IRE, vol. 49 (1961)
12. Koopman, P., Driscoll, K., Hall, B.: Selection of Cyclic Redundancy Code and Checksum Algorithms to Ensure Critical Data Integrity. Final Report DOT/FAA/TC-14/49 (2015). https://users.ece.cmu.edu/~koopman/pubs/faa15_tc-14-49.pdf. Accessed 18 Feb 2020
13. Koopman, P.: Best CRC Polynomials. https://users.ece.cmu.edu/~koopman/crc/. Accessed 18 Feb 2020

Operational Security in the Railway - The Challenge

Ravdeep Kour(✉), Adithya Thaduri, and Ramin Karim

Division of Operation and Maintenance Engineering, Luleå Tekniska Universitet, 97187 Lulea, Sweden
{ravdeep.kour,adithya.thaduri,ramin.karim}@ltu.se

Abstract. Information and Communication Technologies (ICT) in the railway has replaced the legacy systems with modern IP communication networks designed for open-market sectors. This adoption and convergence of Information Technology (IT) and Operational Technology (OT) in railway has brought significant benefits in reliability, operational efficiency, capacity as well as improvements in passenger experience but also increases the vulnerability towards cyber-attacks from individuals, organizations, and governments. This paper proposes a methodology on how to deals with OT security in the railway signalling using failure mode, effects and criticality analysis (FMECA) and ISA/IEC 62443 security risk assessment methodologies.

Keywords: Operational security · ISA/IEC 62443 · FMECA · Railway · Cyber threat · Risk assessment

1 Introduction

Information and communication technology (ICT or commonly IT) has benefited railway greatly in reliability, operational efficiency, capacity as well as improvements in passenger experience. Previously, railway transportation has been generally considered safe in the context of cybersecurity because it rely on almost exclusive proprietary technology. But because of the adoption of new ICT technologies they are replacing their existing legacy systems with modern IP communication networks which are designed for more open-market sectors. This digital transformation of railway has benefited technological systems and methods of operation such as the European System Control System (ETCS), Automatic Train Operations (ATO) and Operations Control Centre (OCC). But this digital transformation has also increased the risk of cyber-attacks from individuals, organizations, and governments. The safety and well-being of passengers, employees, and public in general, including nearby traffic and pedestrians along with transporting cargo will be the priority of rail operators. However, this safety is on risk due to cyber-attacks, which are increasing over the last years. Hackers' already targeted rail companies in Belgium, China, Denmark, Germany, Russia, South Korea, Sweden, Switzerland, the UK, and the US. The impacts of these cyber-attacks were derailment of a tram, data

R. Karim et al. (Eds.): IAI 2021, LNME, pp. 266–277, 2022.
https://doi.org/10.1007/978-3-030-93639-6_22

unavailability, paralyzed rail operations including ticketing systems and the communication infrastructure, leakage of the private data about every vehicle in the country, and disrupted railway signals [1–5].

Safety is the main value of the railway system and with increase in the cybersecurity instances in railway, the issue of OT security has received more attention particular for safety-related technical systems.

Railways are considered critical infrastructure, which means that they must continue operations even under unusual circumstances like cyber-attacks. However, it may be a safe option to shut down the rail system when an OT security breach is suspected, but this is not possible and not unacceptable. To meet the need for continued safe operations in the field of OT security, a comprehensive view of operational processes regarding security and security issues is required.

As cyber-attacks become increasingly automated and sophisticated, control systems become more vulnerable. The move away from standalone control systems to those that are connected with other computer systems and networks also increases their exposure to cyber-attacks. The objective of this paper is to identify various cyber security threats in railway signalling system and propose a methodology to calculate Risk Priority Number (RPN) by using Failure Mode, Effects & Criticality Analysis (FMECA) and ISA/IEC 62443 security risk assessment methodologies. Morant et al. [6] discussed various different signalling subsystems and their interfaces as presented in Fig. 1.

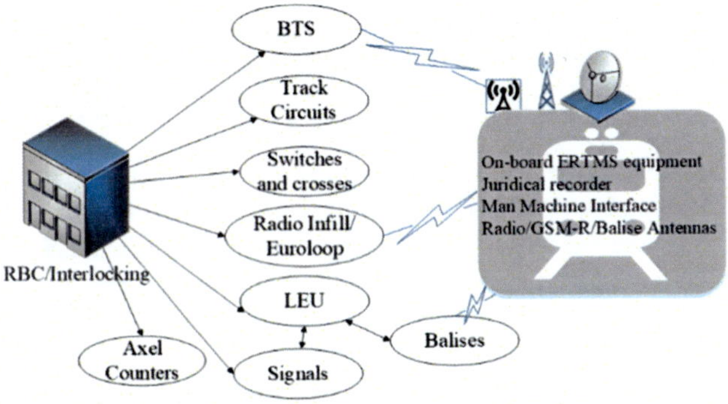

Fig. 1. Subsystems of railway signalling system [6].

2 Cybersecurity and ISA/IEC 62443 Standard

Cybersecurity is the preservation of confidentiality, integrity and availability (CIA) of information in the cyberspace [7]. The operational goals of Information Technology (IT) security are confidentiality, integrity, and availability (CIA) and the operational goals of Operational Technology (OT) security are safety, reliability, and availability (SRA) [8]. Any compromise to these security goals impact railway services. Railway signalling consists of operational technology and, therefore, it is very critical that the signal's availability and integrity must be maintained. On the other hand, IEC-62443 standard provides guidance to reduce the risk of compromising availability, integrity and confidentiality of components or systems used for industrial automation and control, thus it enables the implementation of secure Industrial automation and control systems (IACS) [9]. Therefore, this paper proposes a methodology for cybersecurity risk assessment in railway signalling system by using ISA/IEC 62443 standard (Fig. 2) and propose a methodology to calculate Risk Priority Number (RPN) by using FMECA methodology.

3 Failure Mode, Effects and Criticality Analysis (FMECA)

FMECA is a risk management method [10] that can be applied to cybersecurity to identify various cyber threats within railway signalling assets, their effects and criticality. Criticality is the combined impact of the probability that cyber threat will occur and the severity of its effect i.e. Criticality = occurrence X severity.

FMECA helps in identifying weaknesses or vulnerabilities in the system [10]. The list of possible threats and vulnerabilities specific to the railway signalling are:

- Interlocking devices in the switches [11].
- Risks in Communications-based train control (CBTC) [12].
- Vulnerabilities in the wireless technology in the signalling [13].
- Confidentiality issues in wireless technologies from the third parties [13].
- Jamming, spoofing and flooding of replay attacks on ERTMS Communications [14].
- Breaches of train movement safety, set conflicting routes and other breaches of functional safety and reliability that indirectly affect railway safety and operation [15].

By evaluating the impact of failures due to particular cyber threat on system's performance and safety using FMECA, helps in considering potential timely countermeasures. Some of the cybersecurity countermeasures in the form of a matrix called Railway Defender Kill Chain (RDKC) has been provided in previous work by Kour et al. [16]. Table 1, Table 2, and Table 3 shows the scales for severity, occurrence, & detection of cyber threats respectively.

Severity of the cyber threat is calculated as sum of operation loss, human loss, and economy loss [17].

The detection ranking may be used based on the capabilities and accuracy of defenses and detection mechanisms, the expertise of personnel, and the sophistication of the threat [18].

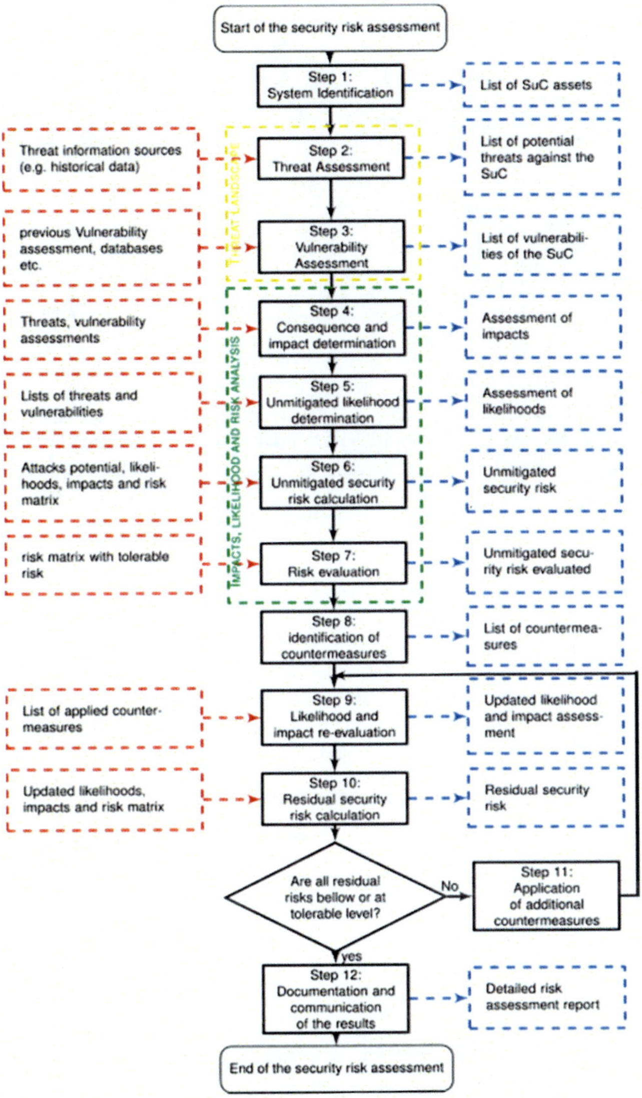

Fig. 2. ISA/IEC 62443 security risk assessment methodology [9].

The occurrence or likelihood of a cyber threat is the combination of needed time to perform a cyber-attack, necessary expertise to perform the attacks, needed knowledge of the scenario, opportunity and needed equipment [19]. Thus, occurrence of cyber threat = Time + Expertise + Knowledge + Opportunity + Equipment.

Based on the values of severity, occurrence, & detection, RPN value is calculated which is a critical indicator for each cyber threat.

Table 1. Severity ranking [17].

Severity					
Operation Loss (OL)	Rank	Economy Loss (EL)	Rank	Human Loss (HL)	Rank
No relevant effect, i.e. at most, an unimportant function is affected and the vehicle can be used without restrictions	0	<5% of annual sales	0	No injuries	0
Comfort affected, i.e. a function is affected. The vehicle can be used with some restrictions	1	5%–20% of annual sales	10	Light and moderate injuries	100
Service required, i.e. an important function is affected. The vehicle can be used only with massive restrictions	10	20%–30% of annual sales	100	Severe and life-threatening injuries (Survival probable)	1000
Vehicle unusable, i.e. one or more fundamental functions are affected	100	>30% of annual sales	1000	Life-threatening injuries (Survival uncertain), fatal injuries	10000

Table 2. Occurrence ranking [19].

Factor	Range	Rank
Time (T) (elapsed time)	≤1 day	0
	≤1 week	1
	≤2 weeks	2
	≤1 month	4
	≤2 months	7
	≤3 months	10
	≤4 months	13
	≤5 months	15
	≤6 months	17
	>6 months (see note 1)	19
Expertise (Ex)	Layman	0
	Proficient	3
	Expert	6

(*continued*)

Table 2. (*continued*)

Factor	Range	Rank
	Multiple experts	8
Knowledge (K)	Public	0
	Restricted	3
	Sensitive	7
	Critical	11
Opportunity (O)	Unnecessary/unlimited access	0
	Easy	1
	Moderate	4
	Difficult	10
	None (see note 2)	999
Equipment (Eq)	Standard	0
	Specialized (see note 3)	4
	Bespoke	7
	Multiple bespoke	9

NOTE 1: A successful attack requires in excess of 6 months.
NOTE 2: None means that the window of opportunity is not sufficient to perform the attack.
NOTE 3: If clearly different groups of specialized equipment are required for distinct steps of an attack, this should be rated as bespoke.

Table 3. Detection ranking [18] and [20].

Detection based on defensive controls	Rank
Nearly certain to detect ($p = 0$)	1
Extremely high probability of detection ($p \geq 0.01$)	2
High probability of detection ($0.01 < p \geq 0.05$)	3
Likely to be detected ($0.05 < p \geq 0.20$)	4
Possibly detected ($0.20 < p \geq 0.50$)	5
Unlikely to be detecteds ($0.05 < p \geq 0.70$)	6
Highly unlikely to be detected ($0.70 < p \geq 0.90$)	7
Poor chance of detection ($0.90 < p \geq 0.95$)	8
Extremely poor chance of detection ($0.95 < p \geq 0.99$)	9
Nearly certain it will not be detected ($p = 1$)	10

4 Cybersecurity Threats in Railway Signalling System

The convergence of IT and OT technology in the railway has brought significant benefits but at the same time has made it vulnerable to cyber threats. This vulnerability also depends upon the maturity of the integration of IT with OT; e.g., ERTMS (European Rail Traffic Management System) level 3, which is fully digital, is more vulnerable to cyber threats. OT security generally deals with industrial control systems (ICS) like SCADA systems. The rationale of this paper is to propose a methodology for cybersecurity risk assessment in railway signalling system by using ISA/IEC 62443 standard and propose a methodology to calculate Risk Priority Number (RPN) by using FMECA methodology. Table 4 lists cyber threats cyber threat with their effects on railway signalling system.

Table 4. Cyber threats and their effects on railway signalling system.

Railway signalling assets [17]	Cyber Threats	Effect
Axel Counter (trackside equipment)	An attacker could damage axel counter (Vandalism)	Control room will not receive axel counter and direction information that could lead to stoppage of several trains and in extreme case an accident
Axel Counter (trackside equipment)	An attacker could eavesdrop /tap the data	Operational information could be released
Axel Counter (trackside equipment)	An attacker could change the axel counter value and direction to send falsified information to the control room	This could lead to train collisions
Track circuit	An attacker could damage track circuit (Vandalism)	Signaller could't detect the presence or absence of train on track and this could lead to train accident
Signal (ERTMS level 1 and 0)	An attacker could damage the signal (Vandalism)	This could lead to stoppage of several trains
Point/switch Machine	An attacker could damage the point machine (Vandalism)	This could lead to train collisions and affect the train movement severely
Balise	An attacker could damage the Balises (Vandalism)	This may cause operational delays to trains
Balise	Tampering attack to modify or rewrite telegram on a balise to inject false data	This could lead to train collisions
Balise	Cloning attack to copy a valid telegram from one balise onto another to compromise multiple balises	This could lead to train collisions

(*continued*)

Table 4. (*continued*)

Railway signalling assets [17]	Cyber Threats	Effect
BTS (Base Transceiver Station)	Man in the middle/session hijacking attack could occur because GSM-R is built on the top of GSM, which uses weak encryption algorithm	This may cause operational delays to trains
BTS (Base Transceiver Station)	An attacker could insert messages into the communications channel	These inserted messages could cause a train to stop for an unpredictable amount of time
BTS (Base Transceiver Station)	An attacker could break the availability of the network by jamming, blocking, or interfering with wireless communications	This may cause operational delays to trains
RBC	Man in the middle/session hijacking attacks	Since session establishment process does not use time-stamps and, therefore, messages could be replicated. In this case, once the session is established, the train does not verify the identity of the RBC anymore [21]
RBC	An attacker could eavesdrop /tap the information from optical fiber	Operational information could be released
Local ERTMS control	An attacker could eavesdrop /tap the commands send to RBC	Operational information could be released
Local Maintenance Aid System ERTMS	An attacker acting as a maintenance engineer requests physical and logical access to the local control centre using malware. The threat agent installs remotely accessible malware allowing remote maintenance command and control of the network access from any available Internet connection	This could lead to the deletion/modification/unavailability of historic
Juridical recorder	An attacker could damage Juridical recorder (Vandalism)	This could lead to the unavailability of system events
Temporary Speed restriction Manager	An attacker acting as a maintenance engineer could report falsified temporary speed restriction to the network controller	This may cause operational delays to trains
Temporary Speed restriction Manager	An insider as an attacker intentionally could not report temporary speed restriction to the network controller	This may lead to derailment

(*continued*)

Table 4. (*continued*)

Railway signalling assets [17]	Cyber Threats	Effect
Key Management Center	An attacker could launch brute-force attacks beacause EuroRadio still relies on DES and 3DES ciphers [22]	This may cause operational delays to trains and accidents
Interlocking	An attacker could damage interlocking (Vandalism)	This may cause operational delays to trains
Interlocking	Man in the middle/session hijacking attacks	This could lead to train accident
Interface control equipment ERTMS	An attacker could damage the gateway (Vandalism)	This could block the communication between RBC and GSM-R network
Communications Front-end	An attacker could send falsified information through communication channels to the control center	This could cause a train to stop for an unpredictable amount of time
Centralized Maintenance Aid (MAS) system ERTMS	An attacker acting as maintenance engineer could infiltrate into interlocking and RBC systems in order to check failures and review the historic	This could lead to the deletion/modification/unavailability of historic
Data server ERTMS	An insider as an attacker could acquire data server ERTMS authentication credentials to hack this system	This could lead to malfunctioning of the ERTMS or stopping it completely
Graphics interface server ERTMS	An insider as an attacker could damage the graphics interface server ERTMS (Vandalism)	Controllers could not visualize the data and this could cause the operational delays to trains
External interface system ERTMS	A threat agent acting as a railway staff could request access to the railway enterprise network using malware	The threat agent installs remotely accessible malware allowing remote command and control of the ERTMS from any available Internet connection
Operation and Management workstation	An attacker conducts man in the middle/session hijacking attack to send falsified information to the control centre	This could lead to trains accident
Operation and Management workstation	Denial of service (DoS) attack to overload the control unit	This could lead to the unavailability of ERTMS system and could cause trains to stop for an unpredictable amount of time
On-board ERTMS equipment	An attacker could install virus in the on-board ERTMS equipment by connected it to the on-board network/Internet	This could lead to malfunctioning of the on-board ERTMS equipment or stopping it completely

(*continued*)

Table 4. (*continued*)

Railway signalling assets [17]	Cyber Threats	Effect
On-board ERTMS equipment	An attacker conducts man in the middle/session hijacking attack to sniff the data and command traffic exchanged between wayside equipment	Operational information could be released
On-board ERTMS equipment	An attacker could conduct man in the middle/session hijacking attack to send falsified information to the train driver	Train driver could take wrong decision that could lead to an accident

5 Cybersecurity FMECA Worksheet for Calculating Risk Priority Number

Table 5 presents a worksheet which needs to be filled by railway infrastructure managers along with IT/OT team. Based upon the values of severity, occurrence, & detection of cyber threat, corresponding RPN values will be calculated. More the RPN value, more is the risk due to that cyber threat.

Table 5. Cybersecurity FMECA Worksheet for calculating Risk Priority Number (RPN) value for each cyber threat.

Railway Signalling Assets	Cyber Threats	Effect	$S = OL + EL + HL$	$O = T + Ex + K + O + E$	D	RPN (SxOxD)

S = the rank of the severity of the cyber threat; **O** = the rank of the occurrence of the cyber threat; **D** = the rank of the likelihood the cyber threat will be detected.

6 Conclusions and Future Work

This paper has proposed a methodology to assess cybersecurity risks in railway signalling system by using FMECA and ISA/IEC 62443 risk assessment methodologies.

The research is in constant touch with European Railway to get actual values of severity, occurrence, & detection of cyber threats to calculate RPN value (Table 5) which is the critical indicator for each cyber threat.

Acknowledgments. The authors would like to thanks Luleå Railway Research Center (JVTC) for sponsoring research work.

References

1. Baker, G.: Schoolboy hacks into city's tram system. The Telegraph (2008)
2. The Local 2017: Swedish transport agencies targeted in cyber attack. https://www.thelocal.se/20171012/swedish-transport-agencies-targeted-in-cyber-attack. Accessed 01 Feb 2020
3. BBC 2018: Great Western Railway accounts breached. https://www.bbc.com/news/technology-43725640. Accessed 01 Feb 2020
4. Whittaker, Z.: Rail Europe had a three-month long credit card breach (2018). https://www.zdnet.com/article/rail-europe-had-a-three-month-long-credit-card-breach/. Accessed 01 Feb 2020
5. Paganini, P.: Massive DDoS attack hit the Danish state rail operator DSB (2018). https://securityaffairs.co/wordpress/72530/hacking/rail-operator-dsb-ddos.html. Accessed 01 Feb 2020
6. Morant, A., Galar, D., Tamarit, J.: Cloud computing for maintenance of railway signalling systems. In: International Conference on Condition Monitoring and Machinery Failure Prevention Technologies, vol. 1, pp. 551–559 (2012)
7. ISO/IEC 27032: 2012, Information technology–Security techniques–Guidelines for cybersecurity
8. Force CIT 2013: Operational levels of cyber intelligence
9. ISA-62443 2016: Security for Industrial Automation and Control Systems, Standard, International Society of Automaton (ISA)
10. Lipol, L.S., Haq, J.: Risk analysis method: FMEA/FMECA in the organizations. Int. J. Basic Appl. Sci. **11**(5), 74–82 (2011)
11. Marrone, S., Rodríguez, R.J., Nardone, R., Flammini, F., Vittorini, V.: On synergies of cyber and physical security modelling in vulnerability assessment of railway systems. Comput. Electr. Eng. **47**, 275–285 (2015)
12. Chen, B. et al.: Security analysis of Urban railway systems: the need for a cyber-physical perspective. In: Koornneef, F., van Gulijk, C. (eds.) Computer Safety, Reliability, and Security. SAFECOMP 2014. LNCS, vol. 9338, pp. 277–290. Springer, Cham (2015). https://doi.org/10.1007/978-3-319-24249-1_24
13. Hartong, M., Goel, R., Wijesekera, D.: Communications security concerns in communications based train control. WIT Trans. Built Environ. **88** (2006)
14. Lopez, I., Aguado, M.: Cyber security analysis of the European train control system. IEEE Commun. Mag. **53**(10), 110–116 (2015)
15. Gapanovich, V., Rozenberg, E., Gordeychik, S.: Signalling cyber security: the need for a mission-centric approach. Int. Railw. J. **56**(7) (2016)
16. Kour, R., Thaduri, A., Karim, R.: Railway defender kill chain to predict and detect cyber-attacks. J. Cyber Secur. Mobil. **9**(1), 47–90 (2010)
17. Cyrail Report. 2017. CYbersecurity in the RAILway sector. D2.1 – Safety and Security requirements of Rail transport system in multi-stakeholder environments. https://ec.europa.eu/research/participants/documents/downloadPublic?documentIds=080166e5b678c2dc&appId=PPGMS. Accessed 01 Feb 2020
18. Flashpoint-intel. 2018. Using FMEA to Measure Relative Risk. https://www.flashpoint-intel.com/blog/using-fmea-to-measure-relative-risk/. Accessed 23 Dec 2019
19. ETSI T. 102 165-1: CYBER Methods and Protocols. Part 1: Method and Pro Forma for Threat, Vulnerability. Risk Analysis (TVRA). -2017. Technical Specification. European Telecommunications Standards Institute
20. Chang, K.H.: Evaluate the orderings of risk for failure problems using a more general RPN methodology. Microelectron. Reliab. **49**(12), 1586–96 (2009). 1

21. Arsuaga, I., Toledo, N., Lopez, I., Aguado, M.: A Framework for Vulnerability Detection in European Train Control Railway Communications. Security and Communication Networks (2018)
22. Chothia, T., Ordean, M., De Ruiter, J., Thomas, R.J.: An attack against message authentication in the ERTMS train to trackside communication protocols. In: Proceedings of the 2017 ACM on Asia Conference on Computer and Communications Security, pp. 743–756. ACM (2017)

Design for Vibration-Based Fault Diagnosis Model by Integrating AI and IIoT

Natalia F. Espinoza-Sepulveda and Jyoti K. Sinha(✉)

The University of Manchester, Oxford Road, Manchester M13 9PL, UK
{natalia.espinozasepulveda,jyoti.sinha}@manchester.ac.uk

Abstract. An intelligent fault diagnosis model of identical machines with different operating conditions has been developed earlier. The model is based on the experimental vibration data with several rotor related faults in the experimental rig and their faults diagnoses through artificial neural network (ANN). This method is further validated through the finite element model of the experimental rig by simulating the different rotor faults. The concept needs now to be integrated together to realise a centralised vibration-based condition monitoring (CVCM) system by putting all identical machines in a pool. The CVCM system can then perform the data collection from machines, data storage and data processing leading to the machine diagnosis. The design concept of the CVCM system is proposed in this paper mainly using artificial intelligence (AI), cloud computing, machine identifier using the global positioning system (GPS) location and Industry 4.0 internet of things (IIoT). The paper also highlights the requirements and challenges to meet and implement the proposed CVCM in practice.

1 Introduction

Condition monitoring through mechanical vibrations in rotating machinery nowadays has an important role in industrial maintenance. Having the ability of delivery an opportune and accurate diagnosis on the machine condition is fundamental in terms of safety, production and costs for any industry. Currently, an expert who analyses the collected vibration data and provides the possible diagnoses based on their knowledge and experience. The possibility to remove this dependence on the experts as well to provide standardised criteria to assess the faults identification and diagnosis is feasible by sensible and physics-based application of the artificial intelligence (AI) based machine learning approach.

There are many such studies related to mechanical systems can be found in literature with the use of a wide range of techniques, from traditional machine learning methods (ML) to deep learning techniques (DL) [1]. In general, the results are promising. However, most of the models are developed for a machine with a specific fault and/or specific operational conditions, limited to only laboratory produced data or mathematically simulated data. Therefore, in order to overcome some of these limitations, a smart model using the AI-based machine learning approach has already been developed, which includes several fault types and two operational speeds on experimental data form a rotating rig [2, 3]. The model has further been validated through mathematically simulated

R. Karim et al. (Eds.): IAI 2021, LNME, pp. 278–285, 2022.
https://doi.org/10.1007/978-3-030-93639-6_23

data in a finite element (FE) model of the experimental rig [4]. This smart model can identify whether machine is healthy or faulty when applied blindly to another identical machines or same machine with different operating conditions [2, 3]. It is now important to define the next steps required to integrate and adapt this model to the current industrial requirements.

Prototypes and pilots of maintenance assessment through the concept of Industry 4.0 have been mainly developed for their application to manufacturing equipment, such as machine tools and machine centres [5–9]. A common notion among literature is the required "integration" of multiple technologies and concepts.

Industry 4.0 is characterised for the implementation of smart systems, the ubiquity regarding the internet of things (IoT), the cyber-physical systems (CPS) and the clouded information. These concepts are integrated in this paper to propose a design for the earlier developed AI-based machine learning model for the rotating machines into a centralised vibration-based condition monitoring (CVCM) system by putting all identical machines in a pool.

2 Earlier Fault Diagnosis Model

An earlier smart model has been developed for the fault diagnosis in a rotating machine [2, 3] using the available experimental vibration data from a rig [10]. The experimental rig is shown in Fig. 1. The rig represents a typical rotating machine with 4 bearing supported on the flexible foundations. The smart model was developed using the available vibration data at two operating speeds, 1800 RPM (30 Hz) and 2400 RPM (40 Hz) with 5 different rotor conditions [1]. The available vibration data were simultaneously collected at the 4 bearing housings using 4 accelerometers (one accelerometer per bearing). These vibration data were processed to extract 4 time domain features per bearing. These features are the root mean square (RMS), variance (V), skewness (S) and kurtosis (K). These parameters were used to create the input vectors to develop the smart vibration-based fault diagnosis (SVFD) model for the fault diagnosis. The artificial neural network (ANN) model in the AI and machine learning approaches was used to develop for this smart model. Details can be found in the papers by Espinoza-Sepulveda and Sinha [2, 3].

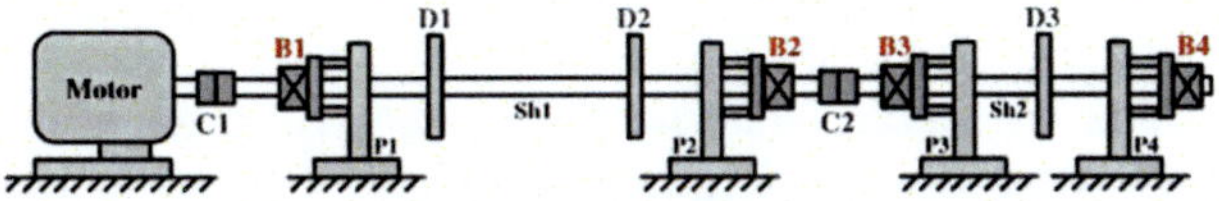

Fig. 1. Experimental rig.

Initially the SVFD model was developed and tested at the machine speed of 1800 RPM. The performance of the SVFD model at the speed of 1800 RPM is listed in Table 1.

The developed SVFD model at 1800RPM was the blindly tested at the machine speed of 2400 RPM. The model has successfully done diagnosis of machine conditions

Table 1. Performance of SVFD model in the faults classification at 1800 RPM.

Diagnosis	Actual rotor condition				
	Healthy	Misalignment	Bow	Looseness	Rub
Healthy	100%	–	–	–	–
Misalignment	–	100%	–	–	–
Bow	–	–	100%	–	–
Looseness	–	–	–	100%	–
Rub	–	–	–	–	100%

i.e., 100% of accuracy in the separation of healthy from faulty machine status. These results are listed in Table 2. The finite element (FE) simulation of the rig also validate and support the experimental model [4].

Table 2. Performance of the developed SVFD model at 1800 RPM when applied blindly to the machine operating at 2400 RPM

Diagnosis	Actual rotor condition	
	Healthy	Faulty
Healthy	100%	–
Faulty	–	100%

This developed SVFD model for a rotating machine is nothing but the model based on the AI and machine learning approaches, which reliably perform the main task of accurately identifying the machine faults even for different machine operating condition. Therefor the model has potential to extend to further into the CVCM system by integrating with other technologies relates to the Industry 4.0 internet of things (IIoT).

3 Design Integration of AI and IIoT

The proposed design considers the fault detection/diagnosis of multiple identical rotating machines operating under slightly different conditions. They can be identified several stages through which the information data flow occurs. It physically starts from the data acquisition at the machine, moving forward until their cyber storage in the cloud, as detailed in the next sections.

3.1 Individual Machine Setup

The model design includes a series of components locally installed in the rotating machine, which allow to conduct the condition monitoring with a real-time access to the physical asset. Here, vibration data are collected and locally prepared at the machine for the later analysis.

Knowledge-based models rely on the quality of the processed data. These should be consistent and capable to represent or map the condition of the entire machine. Thus, a multiple sensor approach is taken into account for the data collection.

The main components considered to be installed at the individual machine, are:

- *Accelerometers*, permanently mounted at each of the bearing cases along the rotor. The quality of the acquire data will be highly dependent on the mounting technique used to set up the sensors.
- *DAQ card*, which acquires simultaneously the vibration data from all the sensors installed along the rotor at a defined sampling frequency and time length. The data samples are transferred to the local controller.
- *Controller*, present at each machine, it allows to process and prepare the samples within the requirements of the smart model by extracted the defined parameters (RMS, V, S, K). Hence, the samples are concentrated into 4 parameter vectors per bearing. By reducing locally the size of the sample, the information transference to further stages of the process becomes easier and less expensive. As well, it allows an optimisation of the resources when storing the information in the cloud.

The configuration for the individual machine is presented in Fig. 2. Same installation is considered to be repeated for all the identical machines subject of monitoring.

It is essential to unify the process across all identical machines connected to the CVCM system. This means that the instrumentation including sensor positons, data collection and their processing to estimate the required number of parameter vectors per bearing should be same. It is better to have a compact instrumentation and data processing unit that can be fitted to the machine.

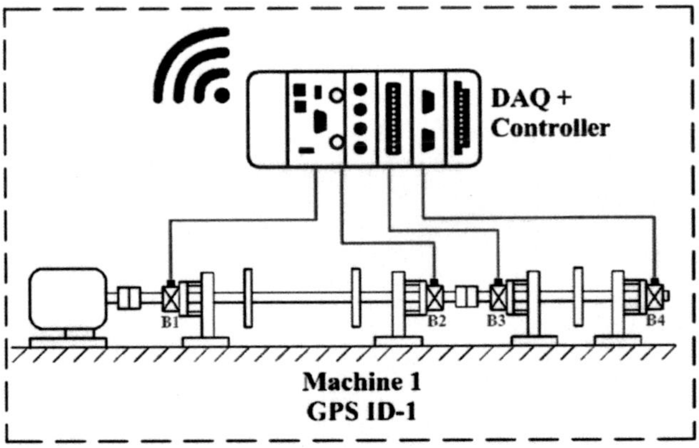

Fig. 2. Individual machine setting.

3.2 Identical Machines Identification

Each machine needs to have a specific ID designated with their GPS position. This information must be transferred along with the features extracted from each machine. Data coming from the same machine will be clustered together in the cloud for trending purposes if required.

The machine location will allow to understand any particular circumstances regarding their operational conditions and environmental factors that could potentially affect the analysed responses. On site inspections could be triggered. This may be related with the network issue, quality of data transferred or problem with installation of sensors, etc.

3.3 Data Transference

Transferring information among the system should be efficient, effective and secure. Different channels of communications and technologies are involved in the process in order to assure these conditions. The general information flow design can be observed in Fig. 3, where the technologies involved are highlighted at each of the processes or stages.

From the machine's controller, the extracted information is transferred through a wireless connection to the local network. Once data reach the local network, the information coming from the machine, goes to the local server, were the analysis processes are conducted.

At this point, some of the information, such as the results provided by the AI model, remains in the intranet and are extended to the area of maintenance that is responsible for the decisions over the assets. Other information, such as output data, are transported through Internet to the Cloud, where they are stored.

Information at the Cloud stage can transit *to* and *from* it. Being possible by this to update the training inputs in the ML model by feeding the ANN every certain time with new information.

The advantage of using a wireless connection in the first information transference, is the possibility to have physical access and simpler connection within the machine in operation. Reducing the amount of cables and physical connections makes it easier and smart. The information is transferred into the local network, which provides safety and it becomes faster and always available. Finally, the transference through Internet allows to get together more pieces of data coming from different geographical locations, i.e. different plant.

3.4 Defect Identification

The assessment of the defect or fault identification is the core function of the CVCM model proposed. The AI applied to this task should provide an accurate diagnosis with a higher reliability rate without concerning the experience of the experts involved in the diagnosis.

At the second main stage of the information flow, the data reaches the local server and then the analyses are conducted. The measurements come to this stage as an arrangement

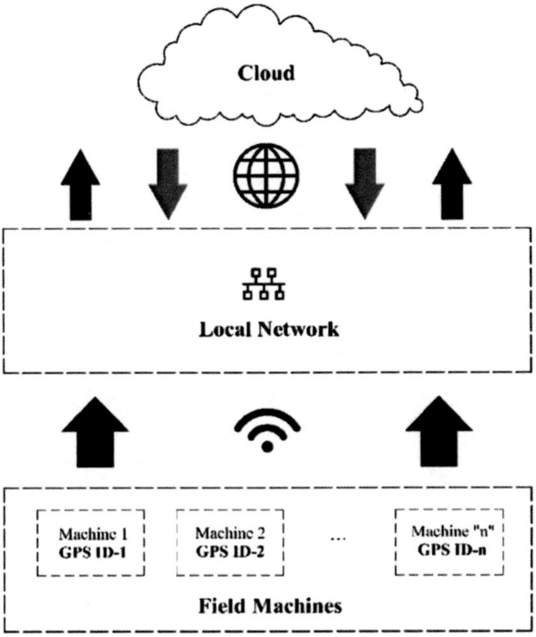

Fig. 3. General data transferring design.

of selected features that characterise the original signals. Here the data sample is processed through the CVCM system, providing a diagnosis regarding the actual machine condition. This will also define a target for the analysed fault, which is communicated along with the other values to the cloud for their storage, as represented in Fig. 4.

When a rotor condition is diagnosed and verified, the information is directly transferred and stored in the cloud. The new data are periodically used to update the knowledge of the CVCM model through the process of deep machine learning approach.

Assuming there is a limited number of fault types initially included in the model, but there is a possibility to find samples with another faulty condition which is not currently included in the model. Since the healthy pattern is already introduced in the model, an unknown fault should be misclassified as another fault. Identifying this case can be done through inspection on site.

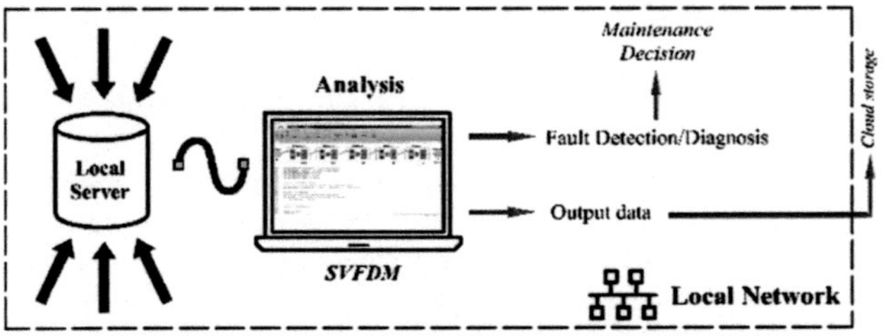

Fig. 4. Data analysis at local network.

If a new faulty condition is identified, a new class has to be created in the model using the deep and self-machine learning approach.

The fault size assessment implies several challenges on this proposed smart model after detection of any defects. Few prominent points are listed here:

- Determining accurate lead time to maintenance (LTTM) if a defect is identified in a machine.
- Providing automated maintenance decisions and prognosis approach.

3.5 Cloud Storage

In this proposed design, all the data is considered to be stored in a Cloud. The information is transferred from the local network through Internet, which allows to save together data coming from different geographical locations. For instance, a company may have many identical machines in their plants around the globe. These identical machines can be added to this proposed CVCM system and share the data in a common platform which may further unify the process of machine fault diagnosis across the different plants. This is the biggest limitation in the current industrial conditions because many vital knowledge is either sitting with experts and within individual plant.

4 Conclusions

A design for the integration into the IIoT of an earlier developed SVFD model in rotating machines, has been outlined in this paper.

The existent technologies allow to address the challenging application of AI in the industrial maintenance field. The proposed CVCM model seems to be feasible option for the unified industrial approach in the machine condition monitoring across the globe. However, further studies need to be carried out regarding the reliability of the smart CVCM model to be used for the fault detection in real industrial environments.

Acknowledgments. This study was supported by CONICYT (Comisión Nacional de Investigación Científica y Tecnológica/Chilean National Commission for Scientific and Technological Research) "Becas Chile" Doctorate's Fellowship programme; Grant No. 72190062 to Natalia Fernanda Espinoza Sepúlveda.

References

1. Liu, R., Yang, B., Zio, E., Chen, X.: Artificial intelligence for fault diagnosis of rotating machinery: a review. Mech. Syst. Signal Process. **108**, 33–47 (2018)
2. Espinoza Sepulveda, N.F., Sinha, J.K.: Comparison of machine learning models based on time domain and frequency domain features for faults diagnosis in rotating machines. In: Proceedings of VETOMAC XIV. MATEC Web Conferences, vol. 211, p. 17009 (2018)
3. Espinoza Sepulveda, N.F., Sinha, J.K.: Non-dimensional pattern recognition approach for the faults classification in rotating machines. In: Proceedings of the 3rd International Conference on Maintenance Engineering IncoME-III, pp. 211–219 (2018)

4. Espinoza Sepulveda, N.F., Sinha, J.K.: Theoretical validation of experimental rotor fault detection model previously developed. In: Zhen, D., et al. (eds.) IncoME-V 2020. MMS, vol. 105, pp. 169–177. Springer, Cham (2021). https://doi.org/10.1007/978-3-030-75793-9_17
5. Li, Z., Wang, K., He, Y.: Industry 4.0-potentials for predictive maintenance. In: Advances in Economics, Business and Management Research, pp. 42–46 (2016)
6. Ferreira, L.L., et al.: A pilot for proactive maintenance in industry 4.0. In: 13th International Workshop on Factory Communication Systems (WFCS), pp. 1–9. IEEE (2017)
7. Ferreiro, S., Konde, E., Fernández, S., Prado, A.: Industry 4.0 : predictive intelligent maintenance for production equipment. In: European Conference of the Prognosis and Health Management Society, pp. 1–8 (2016)
8. Li, Z., Wang, Y., Wang, K.-S.: Intelligent predictive maintenance for fault diagnosis and prognosis in machine centers: industry 4.0 scenario. Adv. Manuf. **5**(4), 377–387 (2017). https://doi.org/10.1007/s40436-017-0203-8
9. Yan, J., Meng, Y., Lu, L., Li, L.: Industrial big data in an industry 4.0 environment: challenges, schemes, and applications for predictive maintenance. IEEE Access **5**, 23484–23491 (2017)
10. Nembhard, A.D., Sinha, J.K.: Comparison of experimental observations in rotating machines with simple mathematical simulations. Measurement **89**, 120–136 (2016)

Statistical Representation of Railway Track Irregularities Using Wavelets

Mariana A. Costa(✉), António R. Andrade, and João N. Costa

IDMEC, Instituto Superior Técnico, University of Lisbon, Av. Rovisco Pais 1, 1049-001 Lisboa, Portugal
{mariana.a.costa,antonio.ramos.andrade, joao.n.costa}@tecnico.ulisboa.pt

Abstract. Railway track deteriorates through time, and a good maintenance strategy should keep its condition at an acceptable level. Detection of railway track geometry defects places an additional challenge to the maintenance team, as these irregularities come in different shapes, types, and sizes, and may potentially cause service disruption or, worse, serious safety hazards. A good inspection/monitoring policy can lead to more accurate and reliable fault detection. However, one obstacle is how to deal with the data provided by the monitoring equipment or inspection vehicle, as well as the choice of suitable methods to translate the data into useful information for maintenance planning and prioritization. Recent research has indicated that, though observed irregularities and its quality indicators may still lie within limit values established by current standards, without the consideration of wavelength content along with dynamic wheel-rail forces, it might be difficult to detect geometric defects that may potentially influence vehicle safety. This study explores the use of Wavelet Analysis (WA) in the statistical modelling of railway track irregularities, namely (1) longitudinal level, (2) alignment, (3) gauge, (4) cross-level. After reconstruction of signals using WA, the impact on Y/Q through vehicle dynamics simulations is measured and probabilities of derailment based on Nadal's criterion are computed through Importance Sampling. Observed results suggest that Y/Q values are sensitive to the scale/frequencies of the track defects, hence, investigating the effects of varying wavelength defects in the force transmissions between vehicle and rail may be benefitial for better condition monitoring and track maintenance prioritization.

Keywords: Wavelet analysis · Railway track irregularities · Vehicle dynamics simulation · Railway maintenance · Inspection

1 Introduction

Railways play a fundamental role in transportation systems worldwide. To guarantee the safety and proper risk assessment of operations, they must comply with existing rules and standards. In terms of railway track, lack of reliability could potentially cause a simple train speed restriction to a total infeasibility of circulation, under risk of a safety hazard. A poor maintained track is also associated with increased wear rates in rolling stocks

R. Karim et al. (Eds.): IAI 2021, LNME, pp. 286–299, 2022.
https://doi.org/10.1007/978-3-030-93639-6_24

and higher maintenance costs [1, 2]. As track deteriorates over time, inspections are essential to gather knowledge about its current condition and monitor various parameters, guaranteeing that they lie within the tolerance limits established in the standards. In a recent work, Andrade and Teixeira [3] have explored different maintenance strategies for track geometry degradation considering these tolerance limits.

Railway standards, such as EN 13848, are long-established and widely used, as they focus on promoting interoperability and harmonisation across large and complex railway systems existing in different countries and operating under different conditions. In terms of track geometry, which is defined by its layout and by track irregularities [4], fault detection and diagnosis is a challenging problem since the track is subject to fluctuating load and stress conditions, all involving randomness. Most railway infrastructure managers assess the track quality either by comparing isolated track defects with recommended statistical indicators, commonly referred to as track quality index (TQI), which usually depend on standard deviations of the isolated track irregularities [5]. Although these approaches have been widely used, they have some drawbacks. The first one is the averaging effect caused by considering long maintenance sections (MAINS), as highlighted by Esveld [6]. Long maintenance sections, usually 200 m, are often employed to justify the assumption of independence among the various sections since the autocorrelation approaches zero as the length of the track increases. Second, such point-wise comparisons may be flawed since they ignore track information up to that point. Li et al. [5] highlight the importance of considering dynamic responses at the wheel-rail interface at different train speeds and loads. They state that short-wavelength defects, such as rail welds, dipped joints and hanging sleepers are difficult to detect by evaluating only the measured track data and show that short wavelength geometric irregularities below the limits specified in the standards still generate high dynamic wheel-rail forces. Andrade and Teixeira highlight some other limitations of using TQIs in railway track geometry degradation [7].

Recent literature also highlights a research gap related to the challenge of finding the statistical correlations between track geometry quality and vehicle dynamics. The current models give varying relationships even on areas of identical track quality [8]. The work by Costa et al. [9] is a contribution to this area and is based on using ARMA models to create synthetic irregularities to study the vehicle response. Some studies [5, 10–13] have pointed out the importance of considering wavelength content in the analysis of dynamic wheel-rail forces to develop models that identify track sections that are likely to produce high track forces and unsafe vehicle responses.

In this paper, wavelet analysis is used to study and reconstruct different track geometry irregularity signals. The data, transformed into a set of coefficients in the wavelet domain, is reconstructed based on pre-defined decomposition levels of a chosen mother wavelet, and its coefficients increased in magnitude. The aim is to to study how the coefficients in each wavelength affect the force transmissions between vehicle and rail, i.e., by reconstructing different irregularities signals and studying their impact on Y/Q through vehicle dynamics simulations, it is possible to relate the frequency of signals associated with a higher risk of derailment.

2 Wavelets

Wavelets are mathematical objects which have broad application in "time-scale" types of problems. The term is usually associated with a function $\psi \in L_2(\mathbb{R})$ such that the translations and dyadic dilations of ψ, $\psi_{jk}(x) = 2^j\psi\left(2^jx - k\right), j, k \in \mathbb{Z}$, constitute an orthonormal basis of $L_2(\mathbb{R})$ [14]. In comparison to Fourier transformations (FT), although they behave similarly, wavelet transformations have the advantage of being able to describe the evolution of the spectral features of a signal as it evolves in time or space [15]. Because wavelets come in different shapes and are limited in time and frequency, decomposing a signal using wavelets can give a much better resolution than using FT, where the signal is decomposed into sinusoidal waves that are infinitely long. Therefore, the Fourier Transform extracts details from the signal frequency, but all information about the location of a particular frequency within the signal is lost [14]. The best that can be done with FT is to sample a range of time or space and find a range of frequencies existing over that amount of time or space, meaning that one has to evaluate the trade-off between knowing precisely the frequency or time of the signal. On the other hand, decomposing a signal by adding up wavelets which are finite, can "slide" along time domain and be compressed together (high frequencies) or stretched out (low frequencies) is an alternative to cope with the time/location loss of the FT by combining both time/location and frequency domains.

Wavelet transforms can be either continuous or discrete. Let $\psi_{a,b}(x), a \in \mathbb{R}\backslash\{0\}, b \in \mathbb{R}$ be a family of functions defined as translations and re-scales of a single function $\psi(x) \in L_2(\mathbb{R})$, also called the wavelet function or "mother" wavelet:

$$\psi_{a,b}(x) = \frac{1}{\sqrt{|a|}}\psi\left(\frac{x-b}{a}\right)$$

The Continuous Wavelet Transform (CWT) is obtained by convolving a signal with an infinite number of functions, generated by translating and re-scaling a certain mother wavelet $\psi(x)$. The CWT for a signal $f(x)$ is defined as a function of two variables, as follows:

$$CWT_f(a,b) = f, \psi_{a,b} = \int f(x)\overline{\psi_{a,b}(x)}dx$$

In the resulting transform the dilation and translation parameters, a, and b, respectively, vary continuously over $\mathbb{R}\backslash\{0\} \times \mathbb{R}$. The signal $f(x)$ is a function of one parameter and its CWT is a function of two. To make the transformation "less redundant" one can select discrete values of a and b and still have a transformation that is invertible: $a = 2^{-j}, b = k2^{-j}, j, k \in \mathbb{Z}$ will produce a minimal basis, so that any coarser sampling will not give a unique inverse transformation; i.e. the original function will not be uniquely recoverable [14]. Such sampling, under some mild conditions, produces the aforementioned orthogonal basis $\left\{\psi_{jk}(x) = 2^j\psi\left(2^jx - k\right), j, k \in \mathbb{Z}\right\}$. The "mother" wavelet can be rewritten accordingly:

$$\psi_{j,k}(x) = \frac{1}{\sqrt{2^j}}\psi\left(\frac{x - 2^j.k}{2^j}\right)$$

In CWT the output comes in the form of smoothly varying local frequency and scales. However, a more efficient way to extract the signal's interesting features and reconstruct the signal by using only a few important coefficients is to consider Discrete Wavelet Transforms (DWT). The DWT uses the concept of multiresolution, which can be roughly understood as a decomposition of a signal on a grid of time and scale (frequency). In practice, only a limited number of scales are used, and the rest (of the information) is stored by the scaling function ϕ, also called "father" wavelet, which is defined as:

$$\phi_{j,k}(x) = \frac{1}{\sqrt{2^j}}\phi\left(\frac{x - 2^j.k}{2^j}\right)$$

The scaling function is the complement of the wavelet, and just like the mother wavelet, the father wavelet can be translated and dilated over the signal. Following the above notation, and considering a signal $f(x)$, where x is taken to be a certain location of the track, $x = 0, 1, 2, \ldots, N - 1$, and with $N = 2^J$, where N represents the total track length. Then, $j = 1, 2, 3, \ldots, J_0$ indexes the scale $S_j = 2^j\left(S_1 = 2, S_2 = 4, \ldots, S_{J_0} = 2^{J_0}\right)$ to which the wavelet has been dilated ($J_0 < J$ defines the maximum scale of the analysis) and $k = 1, 2, 3, \ldots, K_j$ indexes the location in the space to which it has been translated, where $K_j = \frac{N}{2^j}$ implying $K_1 = \frac{N}{2^1}, K_2 = \frac{N}{2^2}, \ldots, K_{J_0} = \frac{N}{2^{J_0}}$. The wavelet coefficients $d_{j,k}$, associated with the mother wavelet, and the scaling (expansion) coefficients $a_{j,k}$, associated with the father wavelet, are defined in a linear signal decomposition as:

$$f(x) = \sum_{k=1}^{2^{J-J_0}} a_{J_0,k}\phi_{J_0,k}(x) + \sum_{j=1}^{J_0}\sum_{k=1}^{2^{J-j}} d_{j,k}\psi_{j,k}(x)$$

The above equation shows how a signal $f(x)$ may be decomposed as the sum of an approximate signal (expanded in terms of scaling functions), and of several detail signals expanded in terms of wavelets. The scale corresponding to J_0 represents the biggest scale of the wavelet. Calculation of the expansion coefficients may be performed relying on a two-channel filter bank downsampled by a factor 2. Hence, the DWT works by iteratively filtering the signal with a low-pass filter (father wavelet) and a high-pass filter (mother wavelet), downsampling the results by a factor of 2, so that the remaining coefficients are of the same dimension of the original signal and repeat these steps on the smaller low-passed signal a number of times (according to the level of decomposition chosen). When several two-channel filter banks are connected repeatedly at the output of a low-pass filter a tree dyadic structure is obtained: this is the conventional way to build up the DWT [15].

The DWT coefficients will have the same dimension as the original signal. These coefficients represent the energy level of the decomposed signal in each frequency band, with the first iteration having the high-frequency coefficients or fine-detailed features, moving to iterations linked to low frequency or coarse features.

Some authors have used Wavelet Analysis for fault diagnosis in the railway track. For example, Xu et al. [16] build a model for predicting track portions with deteriorated wheel/rail forces by combining wavelet transform and Wigner-Ville distribution for characterizing time-frequency characteristics of track irregularities and a three-dimensional nonlinear model for describing vehicle-track interaction. Caprioli et al. [15] show the promising aspects of the wavelet approach (both continuous and discrete) in comparison to Fourier Analysis for fault detection in rails from measurements carried out on axle box accelerations running on the line. They propose the creation of an ad-hoc wavelet, suitable to get out a particular defect "stamp" and by using wavelet packets (a natural extension of DWT) they show the method's ability to detect short pitch corrugation, changes in track sub-structure and local defects. Toliyat et al. [17] also use wavelet packets as the main approach for the detection of defects in rail, and they show the effectiveness of the method by comparing the deviation of wavelet coefficients in the "healthy" rail from the rail containing defects. In a related application, Andrade et al. [18] use the wavestrapping technique [19] to reconstruct signals with same statistical properties of an initial signal and to estimate extreme values in train aerodynamics, comparing the novel approach to the standard.

3 Application

3.1 Track Irregularities

EN 13848 divides track geometry measures into five classes: (1) longitudinal level, (2) alignment, (3) cross-level, (4) gauge, and (5) twist. Following Soleimanmeigouni et al. (2018), the longitudinal level is the track geometry of track centreline projected onto a longitudinal vertical plane. Alignment is the track geometry of track centreline projected onto a longitudinal horizontal plane. Cant (cross-level) is the difference in height of the adjacent running tables computed from the angle between the running surface and a horizontal reference plane. Gauge is the distance between the gauge faces of two adjacent rails at a given location below the running surface. The twist is the algebraic difference between two cross-levels taken at a defined distance apart, usually expressed as a gradient between the two points of measurement [20].

Track geometry deteriorates under the influence of dynamic track loads. These loads cause stresses and elastic displacement and, depending on the total stress level, also permanent deformations [6]. The deterioration is typically quantified as track irregularities of the longitudinal track profile [21]. Two main reasons for this choice are (1) defects in the vertical direction usually grow faster than defects in the horizontal direction and, (2) the defects in the horizontal and cross-level directions would be automatically recovered by track maintenance activities [20]. In this study, however, the option is to look at four classes of measures (1) longitudinal level, (2) alignment, (3) gauge, (4) cross-level. As observed by Mohammadzadeh et al. [22], the random nature of these geometric irregularities, wear on the rail profile, variations in track stiffness, and track structural issues in addition to deterioration of systems used in the railway fleet are sources of the indeterminate nature of track–rail interaction. A model that takes into consideration how the irregularities are correlated with each other and study the track behavior in different configurations of these irregularities is preferred.

3.2 Y/Q Criterion

Derailment occurs when a vehicle runs off its rails. It may be caused by a number of factors, having as an immediate consequence the temporary disruption of the train operations, and may potentially involve serious safety hazards. Following Mohammadzadeh et al. [22], there are two types of derailment: (i) sudden derailment, caused by the wheelset jumping the rails and (ii) flange climb derailment, caused by a wheel gradually climbing to the top of the railhead and then running over the rail. The latter is usually related to forces caused by track irregularities [22], and it will be the focus of this study.

In terms of derailment, many criteria (see [22]) have been developed to provide railway companies with safe, and yet most times conservative, limits for operation. In this paper, Nadal's criterion [23], which was developed in 1908 but it is still widely used in derailment researches [22], especially those related to flange climbing, is considered. The criterion is based on a single wheel lateral to vertical force ratio (Y/Q), and the European Standard EN 13463 defines the limits for safety operations against derailments, based on a track with curve radius of $R \geq 250$ m and Y/Q per wheel previously filtered with a simple sliding mean over 2 m of track, to be: $Y/Q < 0.8$. If this inequality holds, derailment due to the flange climbing will not occur.

3.3 Wavelet Analysis

Wavelet transforms, as mentioned earlier, are limited in time and frequency, have a much better resolution than FT, making them an attractive alternative for fault diagnosis, as they can give better information about (spatial) location of occurrence of the fault.

The track portion considered in this study is a 2048 m line (from km 109.3575 to km 111.4055), consisting of UIC 60 kg/m rail profiles laid out in Iberian gauge (1.668 m), with curvature profile:

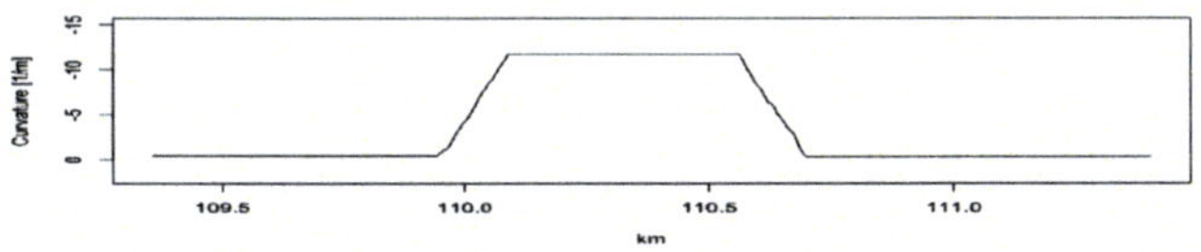

Fig. 1. Curvature profile of track.

Figure 2 displays the four signals corresponding to the measured track geometry components of longitudinal level, alignment, cross-level and gauge in the track section studied:

Each of the four measured signals in Fig. 2 is a vector with length equal to $8192 = 2^{13}$, meaning that data (the 2048 m track segment) was sampled at 0.25 m intervals.

The first step of the wavelet analysis is to choose a mother wavelet. As mentioned earlier, the DWT coefficients represent the energy level of the decomposed signal in each frequency band, with the first iteration having the high-frequency coefficients or fine-detailed features, moving to iterations linked to low frequency or coarse features. Three different wavelet families were tested: daubechies, symlet and coiflet. Various

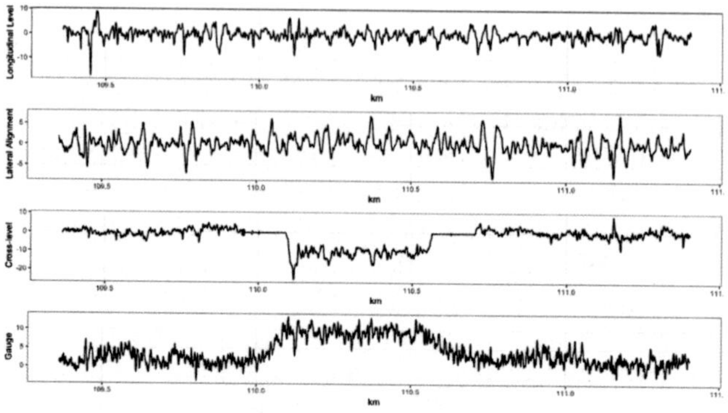

Fig. 2. Track irregularities.

vanishing moments were tested for each of them: daubechies (1:45), symlet (1:45) and coiflet (1:5). Wavelets were decomposed down to decomposition level 8, encompassing all the relevant frequencies.

The daubechies wavelets family was the one chosen to represent all signals, with the following vanishing moments: 18 for longitudinal level and lateral alignment and one (equivalent to "haar" wavelet) for cross-level and gauge. This was the mother wavelet that, with the least number of resolution levels and coefficients, had the most percentage of energy associated. For example, the longitudinal level and alignment D(18) wavelets had over 95% of the energy associated with less than 5% of the maximum amplitude wavelet coefficients and the highest % of the energy accumulated at decomposition level 8. Table 1 summarizes some statistics for the chosen wavelets.

Table 1. Summary of wavelet analysis

Measure	Wavelet	% Energy Accum. level 8	% Energy (5% of coeffs.)	Detail, level 8 coeff. Mean (m/m, SD (m/m)	Scaling, level 8 coeff. Mean (m/m), SD (m/m)
Level	D(18)	99.90%	96.64%	2.45, 12.08	0.04, 1.29
Align	D(18)	92.97%	98.47%	1.19, 12.22	4.68, 7.45
CL	D(1)	16.78%	99.26%	−3.89, 24.14	−34.49, 73.94
Gauge	D(1)	8.04%	98.61%	0.27, 8.83	65.20, 51.58

Following the analysis, the signals were reconstructed based on the first eight decomposition level of the chosen wavelet. The highest wavelet coefficients observed for each level (after decomposition) were multiplied by 2 to increase their amplitudes, and then the reconstruction was made based on replication of these coefficients according to the level's frequency. For example, considering the highest coefficients in the decomposition level 4 for longitudinal level (multiplied by 2), reconstruction of the signal considering a replicated sequence of daubechies wavelets in the frequency band of decomposition level 4 is shown in Fig. 3:

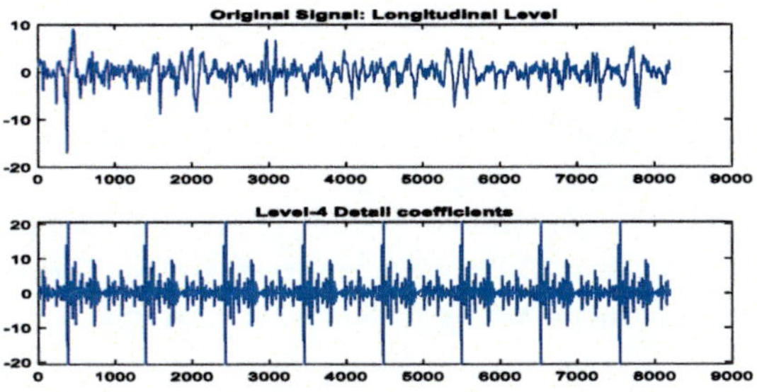

Fig. 3. Wavelet reconstruction on level 4.

Starting with level 1 of the finest details, if a signal is reconstructed based only on each decomposition level frequency band, different signals will have different frequencies and the effects of those in the Y/Q can be further investigated through simulations.

3.4 Simulations

In vehicle dynamics, a train running on a track is usually modelled with detailed multi-body models and suitable wheel-rail contact representations [24]. The multibody simulation package used is a commercial program named Vampire®. The inputs are the track irregularities and track layout.

A set of 17 multibody simulations were run. The output comes in the form of eight columns displaying the Y/Q for each wheel of the vehicle (four wheelsets or eight wheels - left and right - total). To evaluate the results according to the criterion established in EN 13463, the Y/Q values per wheel were filtered with a simple sliding mean over 2 m of track. The maximum value of Y/Q across the eight wheels is taken at each 1 m. Simulations 1 to 8 are made considering the reconstructed signals at decomposition levels 1 to 8 for all four irregularity signals. The reconstruction at each level was made as described in Sect. 3.3. Then, simulations 9 to 16 considered a shift of the reconstructed signal, by assuming that all irregularities were zero up to the half-point of the simulated track., with the other half signal being identical to the same half in simulations 1 to 8. The goal of this last simulation experiments was to check for possible accumulated Y/Q effects by comparing the half-track irregularities responses to the full-track ones. The last simulation (number 17) consisted of the original track points (irregularities measured), and it is the base scenario for comparison.

4 Results and Discussion

The multibody simulations output comes in the form of $\max(Y/Q)$ per 1 m segment of the simulated track. Figure 4 displays the results in terms of $\max(Y/Q)$ in the y-axis and

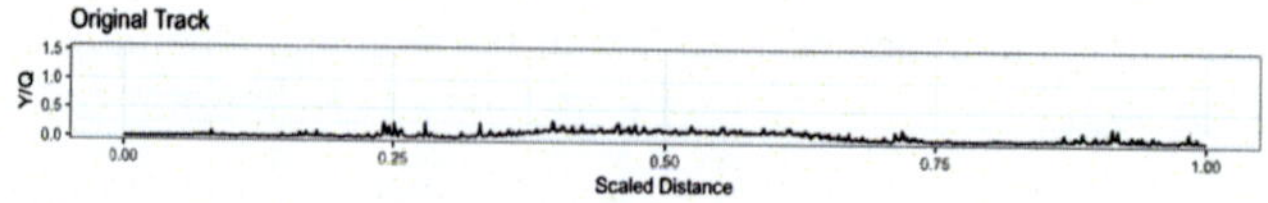

Fig. 4. Original track Y/Q.

the scaled distance (km 109.3575 to km 111.4055 scaled to interval [0, 1]) in the x-axis obtained for the original track measurements.

As it should be evident from visual inspection of Fig. 4, the $\max(Y/Q)$ are all under the standard limit of 0.8. Figure 5 displays the results for simulations 1 to 8 (left side) and 9 to 16 (right side). Same levels are compared in each row.

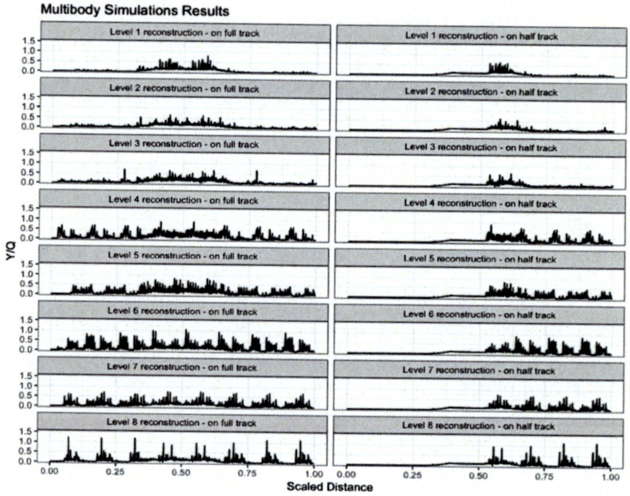

Fig. 5. Multibody simulations results (Y/Q).

Table 2 summarizes the statistics for each of the simulation runs.

In Table 2, the runs marked with an asterisk were the ones where the Y/Q exceeded the limit value of 0.8, i.e. the 'Max' column value is higher than 0.8. This happens for the runs corresponding to full track reconstructed at levels 1, 4, 5, 6 and 8 and for half track reconstructed at the same levels, except for level 1.

Many interesting aspects arise from inspection of Fig. 5 and Table 3. The first, and probably the most important one, is that the half reconstructed signals are not identical to the same half in the fully reconstructed signal, although they are very similar. Especially at high frequencies this makes a difference as, for example, at level 1, the Y/Q maximum was never beyond the limit value in the half reconstructed signal, but it surpassed the limit in the fully reconstructed one.

The second aspect is related to the response on the transition-curve-transition segment (see Fig. 1). It seems that, the higher the frequency, the higher the curve effect on Y/Q, as it can be inferred by noticing that the signal oscillates more in the segment

Table 2. Multibody simulations Y/Q statistics

Run	Min	Mean	Max	SD
1 – Level 1 full*	0.000	0.075	0.805	0.103
2 – Level 2 full	0.000	0.085	0.637	0.094
3 – Level 3 full	0.000	0.100	0.694	0.104
4 – Level 4 full*	0.000	0.166	0.892	0.160
5 – Level 5 full*	0.001	0.148	0.838	0.144
6 – Level 6 full*	0.001	0.190	1.080	0.185
7 – Level 7 full	0.000	0.127	0.751	0.126
8 – Level 8 full*	0.000	0.114	1.230	0.147
9 – Level 1 half	0.000	0.053	0.648	0.080
10 – Level 2 half	0.000	0.056	0.617	0.074
11 – Level 3 half	0.001	0.063	0.669	0.080
12 – Level 4 half*	0.000	0.097	0.899	0.132
13 – Level 5 half*	0.001	0.091	0.834	0.116
14 – Level 6 half*	0.001	0.109	0.905	0.149
15 – Level 7 half	0.001	0.081	0.750	0.106
16 – Level 8 half*	0.000	0.076	1.160	0.118
17 – base (measured)	0.001	0.066	0.303	0.064

corresponding to the transition-curve-transition section of the track. At level 8, it is hard to notice that the signal changes when on that same curved track section.

For the third and last aspect, a closer look at level 3 reconstructed signals is displayed in Fig. 6, as follows:

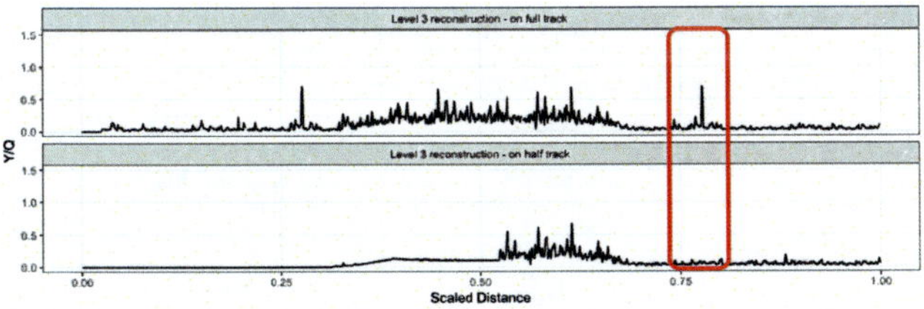

Fig. 6. Multibody simulations results (Y/Q) at level 3.

The half reconstructed signal in Fig. 6 had a maximum Y/Q of 0.669 (located in the transition section after the curve), while the full one had 0.694 (the peak inside the highlighted region in Fig. 6, part of the straight track). This comparison suggests that there may be a cumulative effect of the irregularities signals that causes peaks and/or excessive vibration in the response, although further investigation is needed to understand all possible triggers associated with that peak value.

To relate the above results to actual probabilities of derailment, the Importance Sampling method is used. Details of this statistical method can be found at [25]. Each of the Y/Q reconstructed signals has an empirical density function associated. From this empirical density, it is desired to know the probability associated with values of Y/Q greater than the limit of 0.8, i.e. $P(Y/Q > 0.8)$ for each one of the simulation run empirical distributions. Importance sampling is particularly useful in this case because values exceeding 0.8 are rare, and hence, this method arises as a good alternative to simulate "tail" events. It is further assumed that an exponential distribution with rate λ estimated through maximum likelihood fitting is a good candidate (or "auxiliar") density, from where samples will be drawn. Let $g(x)$ define the empirical density of the signal (simulation run), a candidate density truncated at 0.8 $h(x) \sim Exp((x-0.8).\lambda)$ and a uniform majorizing function $f(x)$, an estimation of the integral $\int_{0.8}^{\infty} g(x)dx$ by generating a random sampling of size n from another density $f(x)$, i.e. $\int_{0.8}^{\infty} g(x)f(x)dx$, can be obtained by:

$$\int_{0.8}^{\infty} g(x)\frac{f(x)}{h(x)}h(x)dx \approx \frac{1}{n}\sum_{i=1}^{n} g(y_i)\frac{f(y_i)}{h(y_i)}$$

where y_i, for $i = 1, \ldots, n$, corresponds to samples from the candidate (exponential) density without truncation. The results for the rates and probabilities can be found on Table 3, which puts into evidence some of the remarks made earlier, where the runs with higher and more frequent peaks are associated with a higher probability of having Y/Q greater than the limit of 0.8, validating statistically the previous discussion. These probabilities represent the likelihood of derailment according to Nadal's criterion. However, as emphasized in the EN 13643 standard, it is unlikely that a single peak of Y/Q would cause a derailment. Instead, this usually requires having high amplitudes of Y/Q along a significant portion of the track.

Table 3. Importance sampling results

Run	Rate λ	P(Y/Q > 0.8)
1 – Level 1 full	13.23	0.03%
2 – Level 2 full	11.79	**
3 – Level 3 full	9.99	**
4 – Level 4 full	5.99	4.65%
5 – Level 5 full	6.74	2.81%
6 – Level 6 full	5.25	7.43%
7 – Level 7 full	7.87	**

(continued)

Table 3. (*continued*)

Run	Rate λ	P(Y/Q > 0.8)
8 – Level 8 full	8.73	0.75%
9 – Level 1 half	18.93	**
10 – Level 2 half	17.84	**
11 – Level 3 half	15.98	**
12 – Level 4 half	10.23	0.24%
13 – Level 5 half	11.04	0.13%
14 – Level 6 half	9.15	0.50%
15 – Level 7 half	12.40	**
16 – Level 8 half	13.15	0.03%
17 – base (measured)	15.05	**

**Not computed since no data points exceeded the value of 0.8, making the computation of probabilities from the empirical distribution unreliable.

5 Conclusions

This study explored the use of Wavelet Analysis in statistical modelling of railway track irregularities and, by reconstructing the irregularities signals into different frequency levels, it related the levels/frequencies more closely associated with increased values of Y/Q, especially those exceeding the 0.8 limit established by the European Standard EN 13463 for safety operations.

Observed results suggest that the Y/Q metric is sensitive to the scale/frequencies of the track defects, and hence, relying solely in a point-wise comparison of the maximum amplitude of Y/Q with the alert limit established in current standards may be insufficient for proper condition assessment. A statistical validation of the results observed from the simulations was obtained by computing the probabilities associated with values of Y/Q greater than the limit of 0.8, showing that certain frequencies are able to generate higher and more frequent peaks of the response than others.

Many opportunities arise from this investigation. A next step would be to create a Design of Computer Experiments to allow the simulation of different combinations of irregularities signals, and to enable answering the question: how can wavelet analysis improve how maintenance data is analysed and interpreted?

Acknowledgments. The authors gratefully acknowledge the financial support from the European Union's Horizon 2020 research and innovation programme under the Grant Agreement No. 881805 (LOCATE research projects under the Shift2Rail Joint Undertaking). This work was also supported by FCT, through IDMEC, under LAETA, project UIDB/50022/2020. The first author acknowledges the support of the Brazilian National Council for Scientific and Technological Development (CNPq) under grant 203130/ 2015-4. The second author thanks the FCT financial support through SFRH/BSAB/150396/2019. The third author expresses his gratitude to the Portuguese Foundation for Science and Technology (Fundação para a Ciência e a Tecnologia) and the

Luso-American Development Foundation (Fundação Luso-Americana para o Desenvolvimento) through the grants PD/BD/128138/2016 and project – 140/2019, respectively.

References

1. De Almeida Costa, M., De Azevedo Peixoto Braga, J.P., Ramos Andrade, A.: A data-driven maintenance policy for railway wheelset based on survival analysis and Markov decision process. Qual. Reliab. Eng. Int. **37**(1), 176–198 (2021)
2. Sancho, L.C., Braga, J.A., Andrade, A.R.: Optimizing maintenance decision in rails: a markov decision process approach. ASCE-ASME J. Risk Uncertain. Eng. Syst. Part A Civ. Eng. **7**(1), 04020051 (2021)
3. Andrade, A.R., Teixeira, P.F.: Exploring different alert limit strategies in the maintenance of railway track geometry. J. Transp. Eng. **142**(9), 04016037 (2016)
4. Pombo, J., Ambrósio, J.: An alternative method to include track irregularities in railway vehicle dynamic analyses. Nonlinear Dyn. **68**(1–2), 161–176 (2012)
5. Li, M.X.D., Berggren, E.G., Berg, M.: Assessment of vertical track geometry quality based on simulations of dynamic track-vehicle interaction. Proc. Inst. Mech. Eng. Part F J. Rail Rapid Transit **223**(2), 131–139 (2009)
6. Esveld, C.: Modern Railway Track (1989)
7. Andrade, A.R., Teixeira, P.F.: Statistical modelling of railway track geometry degradation using Hierarchical Bayesian models. Reliab. Eng. Syst. Saf. **142**(October), 169–183 (2015)
8. Higgins, C., Liu, X.: Modeling of track geometry degradation and decisions on safety and maintenance: a literature review and possible future research directions. Proc. Inst. Mech. Eng. Part F J. Rail Rapid Transit **232**(5), 1385–1397 (2018)
9. Costa, J.N., Ambrósio, J., Frey, D., Andrade, A.R.: A multivariate statistical representation of railway track irregularities using ARMA models. Veh. Syst. Dyn., 1–17 (2021)
10. Balouchi, F., Bevan, A., Formston, R.: Development of railway track condition monitoring from multi-train in-service vehicles. Veh. Syst. Dyn. **59**(9), 1397–1417 (2021)
11. di Scalea, F.L., McNamara, J.: Wavelet transform for characterizing longitudinal and lateral transient vibrations of railroad tracks. Res. Nondestruct. Eval. **15**(2), 87–98 (2004)
12. Zhang, X., Feng, N., Wang, Y., Shen, Y.: Acoustic emission detection of rail defect based on wavelet transform and Shannon entropy. J. Sound Vib. **339**, 419–432 (2015)
13. Zeng, Z., Liang, F., Zhang, Y.: Wavelet analysis of track profile irregularity for Beijing-Tianjin intercity high speed railway on bridge. In: 2010 International Conference on Intelligent Computation Technology and Automation ICICTA, vol. 3, pp. 1155–1158 (2010)
14. Vidakovic, B.: Statistical Modeling by Wavelets, 1st edn. Wiley-Interscience, Hoboken (1999)
15. Caprioli, A., Cigada, A., Raveglia, D.: Rail inspection in track maintenance: a benchmark between the wavelet approach and the more conventional fourier analysis. Mech. Syst. Signal Process. **21**(2), 631–652 (2007)
16. Xu, L., Zhai, W., Chen, Z.: On use of characteristic wavelengths of track irregularities to predict track portions with deteriorated wheel/rail forces. Mech. Syst. Signal Process. **104**, 264–278 (2018)
17. Toliyat, H.A., Member, S., Abbaszadeh, K., Rahimian, M.M., Olson, L.E.: Packet Decomposition. IEEE Trans. Ind. Appl. **39**(5), 1454–1461 (2003)
18. Andrade, A.R., Johnson, T., Stow, J.: Application of wavestrapping statistical technique to estimate an extreme value in train aerodynamics. J. Wind Eng. Ind. Aerodyn. **175**(August 2017), 419–427 (2018)
19. Percival, D.B., Sardy, S., Davison, A.C.: Wavestrapping time series: adaptive wavelet-based bootstrapping. Nonlinear Nonstationary Signal Process., 442–471 (2000)

20. Soleimanmeigouni, I., Ahmadi, A., Kumar, U.: Track geometry degradation and maintenance modelling: a review. Proc. Inst. Mech. Eng. Part F J. Rail Rapid Transit **232**(1), 73–102 (2018)
21. Gustavsson, E.: Scheduling tamping operations on railway tracks using mixed integer linear programming. EURO J. Transp. Logist. **4**(1), 97–112 (2014). https://doi.org/10.1007/s13676-014-0067-z
22. Mohammadzadeh, S., Sangtarashha, M., Molatefi, H.: A novel method to estimate derailment probability due to track geometric irregularities using reliability techniques and advanced simulation methods. Arch. Appl. Mech. **81**(11), 1621–1637 (2011)
23. Nadal, J.: Locomotives a Vapeur, Collection Encyclopedie Scientifique, Biblioteque de Mecanique Appliquee et Genie. Paris (1908)
24. Pagaimo, J., Magalhães, H., Costa, J.N., Ambrosio, J.: Derailment study of railway cargo vehicles using a response surface methodology. Veh. Syst. Dyn., 1–26 (2020)
25. Christian, R., Casella, G.: Monte Carlo Statistical Methods. Springer Science & Business Media, Heidelberg (2013)

Use of Artificial Intelligence in Mining: An Indian Overview

Pragya Shrivastava(✉) and G. K. Pradhan(✉)

AKSU, Satna, India

Abstract. In Mining in India contributes over 2% to its GDP. The growing population had a direct impact on the need for more electricity, steel, cement, fertilizers etc. Most of these consumables are mining products. Thus expansion and modernization of mining activities had always gone for introducing new initiatives, innovations, safe and cost cutting areas. In order to make operations safe and economical an attempt has been made to introduce Artificial Intelligence in most mining activities. Operations attached to mining areas like transportation, shovelling, beneficiation etc. are manpower intensive and unsafe having considerable health related issues. By adopting Artificial Intelligence most complex operations, computations, analysis becomes easier and accurate. Although, global mining companies have made significant advances in AI applications, in India it has been in nascent stage. The areas of AI applications has been identified and this paper highlights the roadmap of the future.

Keywords: Artificial intelligence · Machine learning · Virtual reality · Neural networks

1 Introduction

In an industry such as mining where improving efficiency and productivity is crucial to profitability, even small improvements in yields, speed and efficiency can make an extraordinary impact. Mining companies basically produce interchangeable commodities. The mining industry employs a modest amount of individuals - just 670,000 Americans are employed in the quarrying, mining and extraction sector - but it indirectly impacts nearly every other industry since it provides the raw materials for virtually every other aspect of the economy. When it comes to India, the mining sector creates 13 times more employment than agriculture and six times more than manufacturing. One estimate indicates, by 2025 the Indian mining sector has the potential to provide employment opportunities to about 5×10^6 people directly and create overall employment opportunities for about 5×10^7 people.

2 Indian Mining Scenario

Mining contributes to around 2% of the GDP of Indian economy and is one of the largest employer. With burgeoning demand for energy (over 50% coal based thermal

R. Karim et al. (Eds.): IAI 2021, LNME, pp. 300–306, 2022.
https://doi.org/10.1007/978-3-030-93639-6_25

and lignite0, mining had an important position in India. The global competition as well as rising consumable process had forced the industry to adopt automation and innovation.

India is home to 1,531 operating mines and produces 95 minerals – 4 fuel-related minerals, 10 metallic minerals, 23 non-metallic minerals, 3 atomic minerals and 55 minor minerals. India is the 2nd largest producer of coal. Coal production grew at CAGR 4.6% over FY14-FY19 (to 730.35 MT) and is expected to grow 6–7% Y-o-Y over FY20 as miners focus on the surface mining of coal. Coal's share in India's primary energy consumption is expected to be 48% in 2040. India is the 2nd largest crude steel producer in the world, generating an output of 106.5 MT in 2018, a growth of 3.7% Y-o-Y. India's steel consumption rose 7.5% Y-o-Y and 7.9% Y-o-Y over the last 2 years, outpacing a 2.1% to 5% growth globally.

Over 2030–31, crude steel demand/production forecasted to reach 255 MT, and per capita finished steel consumption also expected to rise to 158 kg by 2030–31 (from 74.1 kg in 2018).

•

3 Indian National Strategy for Artificial Intelligence

In June 2018, the Indian government defined a national policy on AI in a working paper titled, "National Strategy for Artificial Intelligence #AIforAll." The NITI Aayog (highest Planning body of the Central Government) paper identifies five focus areas where AI development could enable both growth and greater inclusion: healthcare, agriculture, education, urban-/smart-city infrastructure, and transportation and mobility. The paper also discusses five barriers to be addressed: lack of research expertise, absence of enabling data ecosystems, high resource cost and low awareness for adoption, lack of regulations around privacy and security, and absence of a collaborative approach to adoption and applications.

The paper proposes a two-tiered framework for promoting AI research organizationally. This includes the creation of Centres of Research Excellence in AI (COREs), which will be academic research hubs, and International Centres for Transformational Artificial Intelligence, which will be industry-led. According to the report, "#AIforAll will focus on harnessing collaborations and partnerships, and aspires to ensure prosperity for all. Thus, #AIforAll means technology leadership in AI for achieving the greater good" (Sinha et al. 2016).

4 Mining Industry

Presently, the mining industry is all set to adopt automation and men less operations. It's already been 10 years since the British/Australian mining company Rio Tinto began to use fully autonomous haul trucks, but they haven't stopped there. Here are just a few ways Rio Tinto and other mining companies are preparing for the 4th industrial revolutions by creating intelligent mining operations.

Rio Tinto's operations include 16 mines, 1,500 km of rail, three ports and more, and it creates 2.4 terabytes of data every minute from all of its mobile equipment and sensors that collect and transmit data in real-time to help monitor equipment. When it comes to underground operations like LKAB's Kiirunavaara mine is the largest iron ore mine in the world with a production of 27,3 million tonnes of crude ore, the mine had always taken several innovative steps to encourage safe and sustainable mining. Figure 1, presents the development and automation in UG mining. Figure 2, further illustrated the present day status of innovation in this sector in Operation and maintenance (Kumar 2019).

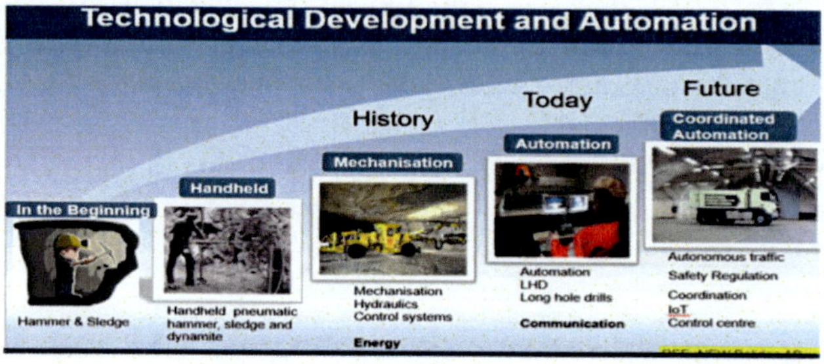

Fig. 1. Development and automation in mining

Fig. 2. Status of Innovation in operation and maintenance

BHP Mitsubishi Alliance (BMA) plans to introduce autonomous haulage at the Goonyella Riverside coal mine in Queensland in 2020 without any forced redundancies. Goonyella Riverside, which will be the first BMA site to implement autonomous haulage, will undergo a staged conversion to a fleet of up to 86 Komatsu trucks over the next two years. BMA plans to deliver more than 40,000 h of training to help prepare for Goonyella Riverside's autonomous future. BHP has committed to the autonomous project soon after Anglo American shied away from introducing the technology at one of its Queensland coal operations.

Areas of innovation vis-a-vis use of AI, Machine Learning and Robots have been briefly outlined in this paper.

5 Mineral Exploration

Artificial intelligence and machine learning can help mining companies find minerals to extract, a critical component of any smart mining operation. Although this is a fairly new application of AI and machine learning, many mining companies are excited about the prospect. Goldspot Discoveries Inc. is a company that aims to make finding gold more of a science than art by using machine learning. Similarly, Goldcorp and IBM Watson are collaborating to use artificial intelligence to review all the geological info available to find better drilling locations for gold in Canada. These efforts to be more precise when finding areas to mine by using machine learning can help the mining industry to be more profitable.

6 Autonomous Vehicles and Drillers

While many of us have been focused on the progress Uber, Google and Tesla have made with autonomous vehicles many people don't realize that Rio Tinto had already been using autonomous haul trucks that can carry 350 tons and operate totally independently since 2008. These trucks have impacted the company's bottom line by reducing fuel use by 13% and are safer to operate. While arguably the challenges of autonomous driving in a quarry aren't as daunting—the trucks move slow, they don't have to worry about pedestrians—it's still a notable accomplishment. This year, the company's long-haul autonomous rail system will go live and is the next step in developing the Mine of the Future. With 244 cars, the autonomous train has been in development for five years, but will make its debut by the end of the year after some software and communication glitches have been worked out.

In addition, Rio Tinto has used autonomous loaders and drilling systems for several years. Just as with other autonomous applications, the company asserts the innovation has improved productivity by 10%.

7 Sorting Minerals

In the majority of mining operations, a much larger volume of materials needs to be removed to find the valuable materials they are mining for. Inevitably, separating the useless rocks and debris to get to what you're mining for tends to be an expensive endeavor. Some companies have begun to use smart sorting machines that can sort the mined material based on whatever criteria a company wants. This work can lead to savings in fuel and energy during processing.

8 Digital Twinning

As part of making the pit-to-port operations as intelligent as possible, companies like LKAB of Sweden and Rio Tinto have plans to have an intelligent mine. There are more than 100 innovations available, but one initiative called digital twinning, first created by NASA, is now being adopted by many in the industrial sector. By creating a virtual model that is fed real-time data from the field, scenarios can be quickly tested, and operations and production can be optimized. This ability to test out decisions before they are implemented in a replica system leads to better outcomes and savings.

9 Safety and Maintenance

Thanks to Internet of Things technology and sensors, mining equipment can be monitored and maintained before breakdowns occur. Sensors can monitor temperature, speed, and vibration on machines to take action transforming preventative maintenance into predictive maintenance. By assessing real-time data and analytics, mining operations can be safer for all involved.

This adoption of this new tech requires re-skilling the mine workers, and many mines have already taken steps. Collectively, they will spend $10 million to up-skill potential and existing workers to handle tasks in analytics, IT and robotics.

10 Intelligent Mine

Several mining companies have drawn ambitious plans to create "intelligent" mine, where all assets are networked together and capable of making decisions themselves "in a microsecond".

Rio Tinto disclosed plans to have the first intelligent mine at Koodaideri, which will deliver its first tonnes of ore in 2021, assuming it meets regulatory approvals.

11 Other Areas

Apart from the above mentioned systems in surface mines, today all out effort is being made to switch over to eco-friendly techniques. Most mines have identified the mined out area, old dumps, vacant land area to install Solar Panel system to generate electricity.

Unmanned slope monitoring systems, surveying by use of scanners, various fire detection systems, using paste fill technology for mine reject disposal and storage, etc. are gaining more application in Indian mines. Working in 3-D -.University of Technology Sydney (UTS) and its research partners, Downer's Mineral Technologies and the Innovative Manufacturing CDC (IMCRC), have reached an important milestone: one year of advancing their efforts to create a bespoke 3-D printer for the production of mineral separation and mining equipment.

Drones have been launched by Ministry of Mines and Indian Bureau of Mines (IBM) in association with mining companies (ex: Tata Steel at Sukinda and Noamundi). In the coming days, drones will help mine management to prepare accurate plans and sections of the mine and neighbouring areas so as to understand the changes in the land use. Use of DGPS is another milestone achieved by the regulators to ensure quick survey of the mine area, movement of ores and minerals etc.

12 Skilling Mine Workers

The rapid technological transformation of the Australian mining industry to the 'digital mine' has companies concerned that Australia may not be able to fill its growing and future workforce needed. It has prompted mining giant Rio Tinto to join forces with the Western Australian Government and vocational training provider South Metropolitan

College of TAFE, to develop a new curriculum to fill the gap. Together, they will spend $2 million developing courses that will up-skill potential and existing workers to undertake tasks in analytics, robotics, IT, and other technology tasks.

Rio Tinto already operates a large autonomous truck fleet at its Pilbara iron ore mines and increasingly autonomous trains, from 1,200 kms away in the West Australian capital Perth.

13 Application of Artificial Intelligence Techniques in Blasting Operation

ANN has been popularly adopted in design-assessment of blasting operations both in India and abroad. One key areas involving safety is 'flyrocks'. Its prediction plays an important role in the minimization of related hazards. In past years, various empirical methods were developed for the prediction of flyrock distance using statistical analysis techniques, which have very low predictive capacity. Artificial intelligence (AI) techniques are now being used as alternate statistical techniques. Two predictive models were developed by using AI techniques to predict flyrock distance in Sungun copper mine of Iran. One of the models employed artificial neural network (ANN), and another, fuzzy logic. The results showed that both models were useful and efficient whereas the fuzzy model exhibited high performance than ANN model for predicting flyrock distance. The performance of the models showed that the AI is a good tool for minimizing the uncertainties in the blasting operations. Pradhan (2020) extensively dealt with use of drones and computer assisted blast design.

ANN and fuzzy logic are two branches of AI that are considered as the most intelligent tools for simulating complex problems. Within recent years, an increase of the ANN and fuzzy logic applications in the field of mining, rock mechanics, and geological engineering has been observed.

14 Conclusion

The depleting ore reserves, uncertainities in commodity prices, environmental regulations on processing/beneficiation units had made the mining industry to adopt new technologies and also go for men less systems and a combination of AI, Machine Learning and IoT. Although Indian mining industry has initiated several steps in this direction, efforts are in nascent stage awaiting cooperation from global manufacturers of mining machineries and IT experts.

Acknowledgement. Thanks are due to the management of AKS University Satna India .Special thanks are due to professor Uday kumar of LTU, Sweden for his guidance and encouragement. Our thanks to sponsors of IAI2020 Conference for their intellectual and financial support.

References

https://daciandata.tech/artificial-intelligence-is-changing-the-mining-industry-examples-of-successful-applications

https://blog.prototypr.io/mining-companies-using-ai-machine-learning-and-robots-e6dcdebaccc3

Kumar, U.: New Technologies Empowered Transformations in Mining Issues and Challenges, Key Note Presentation at the International Conference on Mining – Present and Future Investments, Issues and Challenges, 23–25 Oct 2019, Hyderabad, pp. 1–2 (2019)

Pradhan, G.K.: Explosives & Blasting Techniques, 4th edn. Mintech Publications, Bhubaneswar, India (2020)

Shrivastava, P.: Use of AI, machine learning and robots. In: Mining, Proceedings of the 2nd International Conference on Opencast Mining Technology & Sustainability (ICOMS-2019), 13–14 December 2019, Singrauli, India (2019)

Sinha, A., et al.: AI in India: A Policy Agenda, The Centre for Internet & Society, 5 September 2018

Application of artificial intelligence techniques in developing for predicting the flyrock distance caused by blasting operation. Arabian J. Geosci. **7**(1), 193–202 (2014)

Symbolic Representation of Knowledge for the Development of Industrial Fault Detection Systems

Andrew Young(✉), Graeme West, Blair Brown, Bruce Stephen, Craig Michie, and Stephen McArthur

University of Strathclyde, 204 George Street, Glasgow G1 1XW, UK
{andrew.young.101,graeme.west,blair.brown,bruce.stephen, c.michie,s.mcarthur}@strath.ac.uk

Abstract. In critical infrastructure, such as nuclear power generation, constituent assets are continually monitored to ensure reliable service delivery through preempting operational abnormalities. Currently, engineers analyse this condition monitoring data manually using a predefined diagnostic process, however, rules used by the engineers to perform this analysis are often subjective and therefore it can be difficult to implement these in a rule-based diagnostic system. Knowledge elicitation is a crucial component in the transfer of the engineer's expert knowledge into a format suitable to be encoded into a knowledge-based system. Methods currently used to perform this include structured interviews, observation of the domain expert, and questionnaires. However, these are extremely time-consuming approaches, therefore a significant amount of research has been undertaken in an attempt to reduce this. This paper presents an approach to reduce the time associated with the knowledge elicitation process for the development of industrial fault diagnostic systems. Symbolic representation of the engineer's knowledge is used to create a common language that can easily be communicated with the domain experts but also be formalised as the rules for a rule-based diagnostic system. This approach is then applied to a case study based on rotating plant fault diagnosis, specifically boiler feed pumps for a nuclear power station. The results show that using this approach it is possible to quickly develop a system that can accurately detect various types of faults in boiler feed pumps.

Keywords: Condition monitoring · Nuclear power plants · Expert systems · Knowledge-based systems · Automation

1 Introduction

Fault detection and diagnostics is an active research area, especially in the nuclear industry for rotating machinery [1,6,10,13]. There are two main approaches that can be adopted for the development of systems for fault detection or diagnostics. These are either data-driven approaches, e.g. machine learning, or knowledge-based approached, e.g. expert systems. While both of these techniques have similar aims and can provide similar results they differ quite significantly in their implementation.

R. Karim et al. (Eds.): IAI 2021, LNME, pp. 307–318, 2022.
https://doi.org/10.1007/978-3-030-93639-6_26

The basis for data-driven approaches is centred around statistical models of the problem data. The individual parameters of the model are learned through a process called training where a large volume of data is input into the model and the model attempts to produce the correct output for the majority of cases. It should also be noted that related to many data-driven approaches a balanced dataset is required, i.e. there is an equal number of samples for both normal data and fault states. Due to the nature of these models and the lack of explicability for many data-driven approached; for critical assets (especially in the nuclear industry) that can present an issue. This is because supporting evidence is often required when making decisions on these assets as there is a significant amount of cost involved in the repair, replacement or downtime of these assets, another consideration is that "black box" techniques cannot support the writing of new safety cases.

Knowledge-based approaches are the second technique that can be used to solve this problem, they attempt to solve (or support the resolution of) complex problems where a significant amount of human expertise or expert knowledge is required. This knowledge is acquired from the engineers or domain experts through a process called knowledge elicitation, this is then formalised into a format that is compatible with the technique, e.g. as the rules for a rule-based expert system. The one main advantage of this type of approach over data-driven approaches is the ability to not only solve a problem but also to explain and justify the reasoning behind why a decision was made. However, this comes with the disadvantage that a significant time cost is associated with the capturing of the knowledge, then formalising this into a knowledge-based system. Because of this disadvantage, there his been a significant amount of research undertaken across numerous fields in order to streamline the knowledge elicitation process [9, 12], as this is the most time consuming part of the development of an expert system.

The next section of this paper provides background information into rule-based expert systems, a type of knowledge-based approaches. Section three proposes a new methodology for knowledge elicitation through the use of symbolic representation of the expert's knowledge and the parametrization of this knowledge. While section four presents a case study of this methodology applied to boiler feed pumps from an advanced gas-cooled reactor in the UK. Finally, the conclusions and future work are presented in section five.

2 Rule-Based Expert Systems

Knowledge-based systems can be used for a variety of applications to provide not only accurate decisions but also the explanation and reasoning behind these decisions. One example of these is rule-based expert systems [5], they store the knowledge captured from the domain expert as a set of rules. These rules are formalised in a way that mimics the domain experts reasoning process and are mainly applied to knowledge or time-intensive problems.

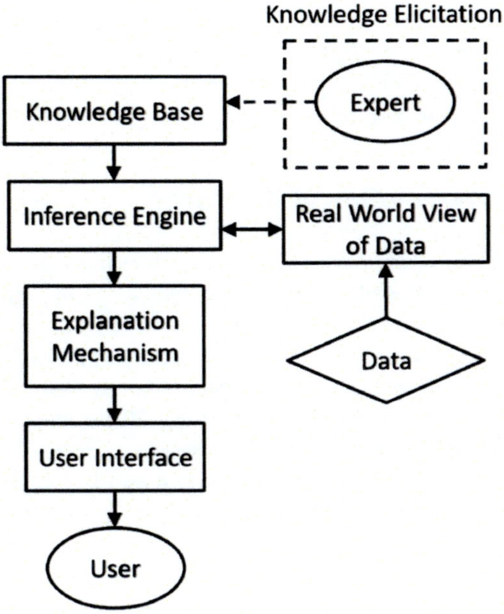

Fig. 1. Typical rule-based expert system architecture

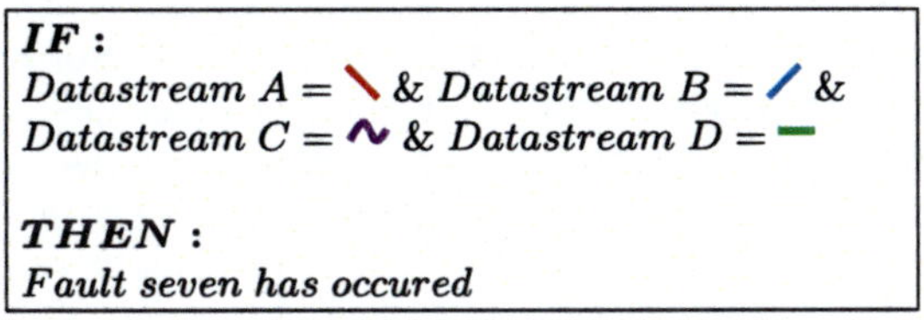

Fig. 2. Structure of rules stored in knowledge base

A typical rule-based expert system contains five main components, see Fig. 1. The first of these is the **Knowledge Base** and this contains all the domain-specific captured knowledge from the experts. Figure 2 shows an example of how the rules are expressed in an expert system as a set of IF-THEN rules. This can be considered as a fixed set of data, therefore it remains the same throughout the decision making process. The **Real World View of Data** is the next component, this contains all the data, and facts relating to the asset under analysis. This can be considered the current state of the machine and hence is fluid and constantly changing. The facts relating to the asset are then compared with the IF condition in the knowledge-based to determine intermediate facts, which can then be stored in the Real World View of Data or a diagnostic conclusion. The third component is the **Inference Engine**, this performs the analysis by comparing the rules in the knowledge base to the facts stored in the real world view of data. The **Explanation Mechanism** provides justifications and an explanation as to why the inference engine has decided on a conclusion. This component is crucial for the system to be accepted by the user or by industry. Finally, a **User Interface** allows for communication between the user and the system, whether this is for the input of new facts relating to the data or the output of the diagnostic conclusions, this information can also be passed to external programs or systems.

3 Symbolic Representation of Knowledge

For many industrial applications fault diagnosis involves the engineers following a predefined diagnosis process. Therefore, the expert knowledge has already been acquired to some extent, although this is not always complete enough to be formalised into a set of rules for a rule-based expert system. There is often a significant amount of subjectivity involved when the engineers asses the problem, due to their own experiences with the asset, rules of thumb, or different formal training. However, at a high level, they are often looking for standard data trends such as increases or decreases in specific data, or an increased noise or fluctuation. There is often no prescribed quantitative information relating to these trends that they analyse, such as how much increase or decrease relates to a specific rise or fall, or how much increase in fluctuation relates to a signal moving from stable to fluctuating, as these values will change based upon multiple factors, such as the type of machine, the age of the machine and the operational profile of the machine. Therefore, before the knowledge can be formalised into a rule-based expert system this additional knowledge must be acquired from the domain experts through the knowledge elicitation process. There are several different approaches for performing this knowledge elicitation [2], some of these include: structured and unstructured interviews: observation through active participation or focused observation; and task or decision analysis. For complex problems this is an extremely time-consuming task, this bottleneck in the development of a knowledge-based system has long been recognised and has hence been called the "knowledge elicitation bottleneck" [4].

The rest of this paper focuses on a new methodology to streamline this knowledge elicitation process by simplifying the knowledge into a set of symbols, or common language, that can be easily communicated between domain experts and data engineer. The proposed methodology is a three staged process that involves a minimum of two structured interviews.

3.1 Definition of Symbols

These symbols were selected as low-level predicates that could be used to broadly describe a time series at any instant. The trends that were selected are shown in Fig. 3, these symbols are a stable symbol which relates to normal behaviour, a rise and fall symbol for an increase or decrease over a specific time with a specified limit, and a fluctuating symbol for an increase in noise present in the signal. These symbols were selected as they are the most basic trends that can be present in time series data, and any complex trend can be constructed from these primitives. This allows for the domain experts to easily communicate the diagnostic process they follow using a common language.

(a) Stable (b) Rising (c) Falling (d) Fluctuating

Fig. 3. Four selected symbols/trends

3.2 Definition of Rule Base

The next stage of the process is to set up a structured interview with the domain experts to agree on a definition of the rule base. This requires a definition of the individual faults that are being analysed, the specific datastreams necessary to determine those faults, and the associated trends for each of those datastreams. For each fault, a table can be produced that contains all the information discussed above, the example format of this table is shown in Table 1. Additionally, any comments that the engineers can provide at this stage will also prove to be extremely useful, this develops a rationale behind each piece of knowledge and for example could be: the physical reasoning behind the associated trends: or clarification on a subset of faults where a full data set or other operational influence is unavailable to fully diagnose a specific problem.

Following the meeting, each of the tables for the individual faults are combined to produce an overall rule-base for the asset being analysed, an example of this is shown in Table 2. Regarding system development, it is now possible to construct a prototype rule-based expert system using placeholder values for the quantitative parameters relating the each of the individual trends, which will be set in the next stage.

Table 1. Example format for individual fault diagnostic rules.

Datastream	Trend	Comments
Datastream A	/	
Datastream B	\	
Datastream C	∿	
Datastream D	—	
Datastream E	—	

Table 2. Example format for asset specific rule-base.

Cause	Datastream A	Datastream B	Datastream C
Fault 1	\	—	\
Fault 2	—	/	—
Fault 3	∿	∿	\
Fault 4	\	\	—
Fault 5	/	∿	\

3.3 Definition of Parameters

Having defined the necessary symbols to accurately interpret related data streams; agreed with the domain experts the individual faults and the associated trends used to asses these faults: the next stage is for all this information to be tabulated and parametrised. Subsequently, a second structured interview is arranged to determine the individual magnitudes for each specific trend associated with each specific rule. The previously mentioned symbols that are now shown in Fig. 3 are regarded as the most basic trends that are present in the data. The expert knowledge that is required to qualify the diagnostic rules shown in Table 2 is the subtle differences in the trends in Fig. 3. For each symbol various parameters must be assigned to them that accurately describe the possible variations in the symbols for different rules, this is shown in more detail in Fig. 4 and Table 4.

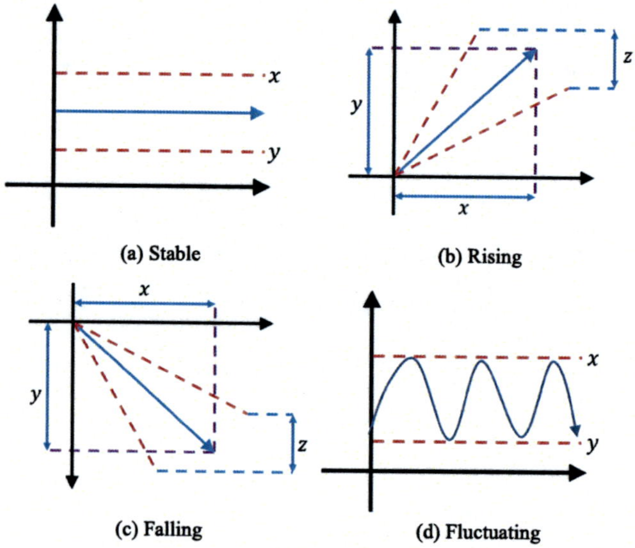

Fig. 4. Definition of parameters for subtle difference in symbols/trends

This information and the corresponding parameters can be tabulated and presented to domain experts in a structured interview knowledge elicitation session. An example of this structure is shown in Table 3. The parametrisation of the knowledge allows for efficient and accurate capture of the domain specific knowledge by focusing the domain experts on a simplified version of the problem. This also facilitates the ease of formalising this knowledge into the rules for a rule-based expert system, without the need to listen to hours of audio recording or to interpret the engineer's answers to specific questions.

Table 3. Example structure for rule specific table to be completed during knowledge elicitation session

Rule	Parameters
Datastream A - Rising	$x =$, $y =$, $z =$
Datastream B - Falling	$x =$, $y =$, $z =$
Datastream C - Fluctuating	$x =$, $y =$
Datastream D - Stable	$x =$, $y =$

Table 4. Description of parameters for quantifying the subtle difference in the trends

Trend	Parameter	Description
Stable	x	The upper limit in variation for a signal to be consider stable
	y	The lower limit in variation for a signal to be consider stable
Rising	x	The time period for the rise to occur over
	y	The minimum change in the measurement
	z	Two values relating to the spread of the x and y parameters
Falling	x	The time period for the fall to occur over
	y	The minimum change in the measurement
	z	Two values relating to the spread of the x and y parameters
Fluctuating	x	The upper limit for the transition between stable to fluctuating
	y	The lower limit for the transition between stable to fluctuating

3.4 Implementation

After gathering all the expert knowledge from the knowledge elicitation meetings, the methodology proposed to evaluate the diagnostic rules on time series data is to first segment the data into specific time regions, see Fig. 5. Splitting up the time series data into various timesteps, or segments, based on the information provided by the expert, each data stream/channel timestep is assigned a symbol which is either rising, falling, fluctuating or stable.

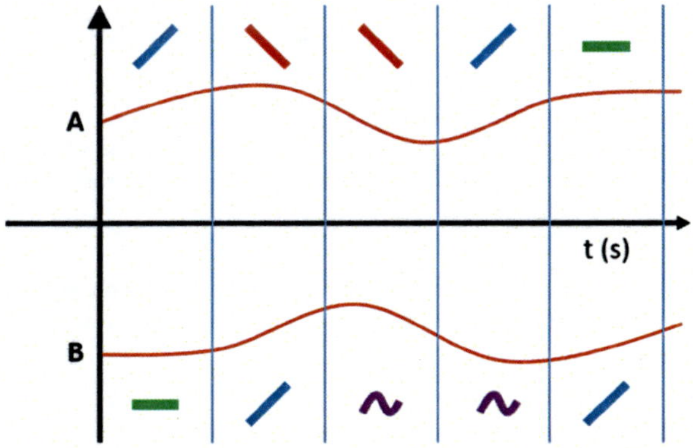

Fig. 5. Example of signal to symbol transformation for two time series data sources.

The assigning of the symbols is performed using a technique based on Signal to Symbol transformation [8] which has been successfully used for rotating plant in the nuclear industry previously [3]. For this application, the symbols are assessed by first calculating the average of the first and last 10% of the timestep, a comparison is then performed to determine which of the following four categories best describes the timestep. The categories are defined as: **Stable** less than 50% of the data is out with the thresholds set by x and y, **Fluctuating** more than 50% of the data is out with the thresholds set by x and y, **Rising** the mean value for the first 10% of the data is greater than y times the mean value of the last 10%, and **Falling** the mean value for the last 10% of the data is greater than y times the mean value of the first 10%, where x, y and z are defined in Fig. 4. Algorithm 1 shows the pseudocode for this calculation and an example is shown in Fig. 6 for two generic pressure and temperature datastreams.

Algorithm 1: **Signal to symbol transformation. Where x, y and z are defined in Figure 4**

```
if 50% of data (< x*mean(data) or > y*mean(data)) then
  | Result: Stable
else if 50% of data (> x*mean(data) or < y*mean(data)) then
  | Result: Fluctuating
else
  | Calculate average of first and last 10% of data for x period of time;
  | if First < y*Last then
  |   | Result: Rising
  | else if First > y*Last then
  |   | Result: Falling
```

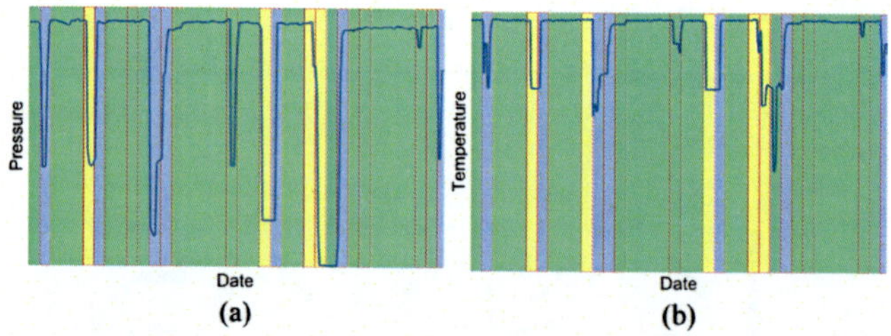

Fig. 6. Example of signal to symbol transformation for (a) pressure and (b) temperature. (Green - Stable, Blue - Rising, Yellow - Falling, and Red border - Fluctuating)

Having formalised the rules and implemented the Signal to Symbol Transformation as described above, it is possible to detect faults in near-realtime across multiple data sources. As new timesteps are input into the system each datastream can be assigned a symbol. When all datastreams have been assigned a symbol the expert system can then compare the symbols with the rule base to determine if any fault has occurred. If a positive correlation occurs this timestep is marked with the corresponding fault type.

Over time it is possible to build up a history of any faults that have occurred historically in the asset, an example of this over a small time period is shown in Table 5.

Table 5. Example processing of 4 datastreams for 5 timesteps.

Datastream	T1	T2	T3	T4	T5
A	∿	—	\	—	∿
B	—	/	—	/	\
C	—	∿	—	—	/
D	∿	—	\	\	—
Fault	N/A	7	6	N/A	N/A

4 Case Study: AGR Boiler Feed Pumps

Following the proposed methodology, a case study was performed for data gathered from boiler feed pumps of an advanced gas-cooled (AGR) reactor in the UK. This case study was selected as these assets are critical to the continued operation and electrical generation of an AGR power station, therefore, it is imperative that the pumps are monitored for any abnormal behaviour that may contribute to accelerated plant degradation or to tripping the plant which would result in reduced or zero power generation.

The diagnostics rules for the asset were supplied by the domain experts at the beginning of the project. This determined each data stream necessary to diagnosis a given predefined list of faults, These rules were represented by a set of trends, i.e. stable, fluctuating, rising or falling. The data contained 37 faults and the associated trends for 10 specific data streams covering pressure, temperature, speed, vibration, and flow. The additional data required to formalise this knowledge into the rules for a rule-based expert system were acquired through knowledge elicitation meetings following the proposed methodology.

Having captured and formalised the domain experts knowledge it was possible to develop a prototype demonstrator for quickly and accurately identifying faults in the boiler feed pump data in realtime. To implement the knowledge base, all the knowledge that was acquired from the knowledge elicitation meetings is stored in a Microsoft Excel spreadsheet. It was stored in this format so that any engineers using the system would easily be able to view all the captured knowledge and therefore provides greater acceptance of the system and also that the captured knowledge is correct. If the analyst wishes to amend a specific rule or add a new fault type, this can be done by editing the spreadsheet directly. Any updates that are made to the rule-based are automatically detected by the system, and displayed to the analyst in the "Changes to Rule Base" panel in formatted text, see Fig. 7. Currently, any amendments made to this file will only be saved for the same session, however, the functionality to load the rule base from historical sessions can be added in the future. This functionality will also require for validation of any new, or amended rules using historical data to ensure that the quality of the knowledge base is maintained. When the analyst is satisfied with the knowledge stored in the rule base the analyst can begin to perform the analysis. The average analysis for the

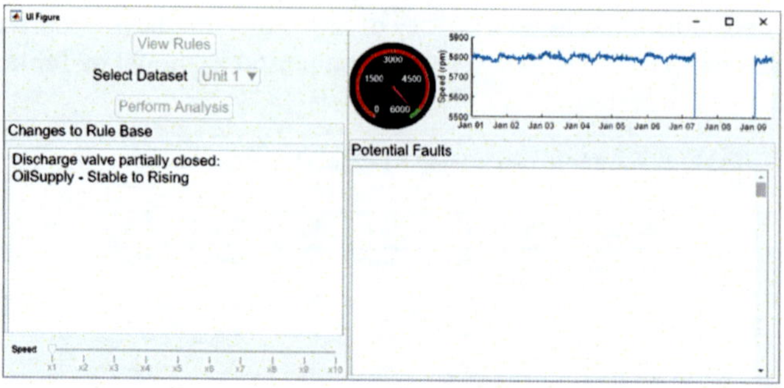

Fig. 7. Main GUI for automated boiler feed pump diagnostics

current rule base (37 faults and 10 datastreams) takes less than 0.5 s to complete one timestep of the analysis. When potential faults are detected, they are displayed to the analyst in the "Potential Faults" panel, see Fig. 7. The date of any fault detected is displayed and the analyst can select this to open up a new window (Fig. 8). This presents the analyst with a drop down menu that contains all faults detected and the associated data streams used in the analysis. The analyst can then view each of these datastreams to display the data covering the time in question on the axis to the right of this window and confirm that the correct fault has been identified.

This methodology has allowed for the rapid development of a rule-based expert system for the purpose of fault detection in boiler feed pumps. Due to the novel approach adopted for the knowledge elicitation process, it was possible to minimise the amount of time required from the domain experts but still accurately elicit all the knowledge necessary to develop the system.

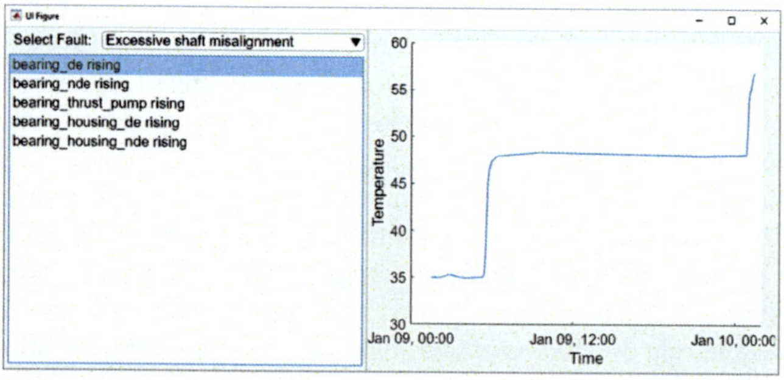

Fig. 8. Fault justification window for boiler feed pump diagnostics

5 Conclusions and Future Work

This paper has proposed a new approach to knowledge elicitation for the development of a knowledge-based fault detection system, specifically a rule-based expert system.

The benefits of knowledge-based systems over data-driven approaches are the increased explicability, however, the increased cost in the development time has been highlighted as a disadvantage. The methodology discussed in this paper attempts to reduce the burden placed on the domain experts by streamlining the knowledge elicitation process, the most time consuming part of developing an expert system. Through the use of symbolic representation of knowledge and the parametrisation of these symbols, it was possible to set out a framework to follow for these streamlined knowledge elicitation sessions.

Using this framework it was possible to develop a rule-based expert system for boiler feed pumps from an AGR power station in the UK. Having further developed the expert system beyond the knowledge elicitation process it has been possible to implement all 37 faults that occur on the boiler feed pumps for the corresponding 10 datastreams. The resulting system can detect faults in the data in real-time due to the segmentation of timesteps into symbols and the efficient inference engine deployed in CLIPS (C Language Integrated Production System) [11].

Future work will involve the development of a human in the loop automated system to improve the elicited knowledge during the system operation. By initially using the methodology discussed in this paper to set the initial parameters for the knowledge and the formalisation of the rule base it should be possible to develop an active learning system [7] to query the analyst to determine any false positives. These labelled false positives will then be used to amend the current parameters to improve the overall system performance. Due to the change in the system knowledge a further piece of work will also explore how to verify and validate this new knowledge using historical data without the need for the domain experts to manually label all historical occurrences.

Acknowledgements. This work was funded by the Engineering and Physical Sciences Research Council under grant EP/R004889/1.

References

1. Ayodeji, A., Liu, Y.K., Xia, H.: Knowledge base operator support system for nuclear power plant fault diagnosis. Prog. Nucl. Energy **105**, 42–50 (2018)
2. Cooke, N.J.: Varieties of knowledge elicitation techniques. Int. J. Hum.-Comput. Stud. **41**(6), 801–849 (1994)
3. Costello, J.J.A., West, G.M., McArthur, S.D.J., Campbell, G.: Self-tuning routine alarm analysis of vibration signals in steam turbine generators. IEEE Trans. Reliab. **61**(3), 731–740 (2012)
4. Cullen, J., Bryman, A.: The knowledge acquisition bottleneck: time for reassessment? Expert. Syst. **5**(3), 216–225 (1988)
5. Grosan, C., Abraham, A.: Rule-based expert systems. In: Grosan, C., Abraham, A. (eds.) Intelligent Systems. ISRL, vol. 17, pp. 149–185. Springer, Heidelberg (2011). https://doi.org/10.1007/978-3-642-21004-4_7
6. Guan, X., He, J.: Life time extension of turbine rotating components under risk constraints: a state-of-the-art review and case study. Int. J. Fatigue **129**, 104799 (2019)
7. Holzinger, A.: Interactive machine learning for health informatics: when do we need the human-in-the-loop? Brain Inform. **3**(2), 119–131 (2016). https://doi.org/10.1007/s40708-016-0042-6

8. Nii, H.P., Feigenbaum, E.A., Anton, J.J.: Signal-to-symbol transformation: HASP/SIAP case study. AI Mag. **3**(2), 23 (1982)
9. O'Hagan, A.: Expert knowledge elicitation: subjective but scientific. Am. Stat. **73**(sup1), 69–81 (2019)
10. Tang, S., Yuan, S., Zhu, Y.: Deep learning-based intelligent fault diagnosis methods toward rotating machinery. IEEE Access **8**, 9335–9346 (2020)
11. Wygant, R.M.: CLIPS - a powerful development and delivery expert system tool. Comput. Ind. Eng. **17**(1), 546–549 (1989)
12. Xiao, C., Jin, Y., Liu, J., Zeng, B., Huang, S.: Optimal expert knowledge elicitation for Bayesian network structure identification. IEEE Trans. Autom. Sci. Eng. **15**(3), 1163–1177 (2018)
13. Young, A., West, G., Brown, B., Stephen, B., McArthur, S.: Improved explicability for pump diagnostics in nuclear power plants. In: 2019 ANS Winter Meeting and Nuclear Technology Expo, 17–21 November 2019 (2019)

A Novel Intelligent Compound Fault Diagnosis Method for Piston Engine Valves Using Improved Deep Convolutional Neural Network

Yufeng Guan[1], GuanZhou Qin[1], Jinjie Zhang[2], and Zhiwei Mao[2](✉)

[1] Jiangsu Nuclear Power Corporation, Lianyungang, China
{guanyf,qingz}@cnnp.com.cn

[2] Key Lab of Engine Health Monitoring-Control and Networking of Ministry of Education, Beijing University of Chemical Technology, Beijing, China

Abstract. In recent years, intelligent data-driven deep learning-based fault diagnosis methods play a vital role in the reliability and security for modern industrial machinery. However, in real industrial application, the issues of large amounts of data and environmental noises make it difficult to detect the compound fault for piston engine based on traditional intelligent fault diagnosis method. To break through the problem and further enhance efficacy, one-dimension hierarchical joint convolutional neural network (1-DHJCNN), an improved deep convolutional neural network is proposed to extract directly useful information for vibration signal and then give diagnosis results in this paper. The effectiveness of the proposed method is verified through the dataset of valve fault for piston engine, which is composed of single valve fault and compound valve fault with different fault degrees. Benefiting from this novel network structure, the experimental results show that the proposed method can not only outperform the traditional diagnosis methods such as integral deep convolutional neural network, but also present a small amount of model parameter calculation. In the end, feature visualization is adopted as a promising tool to analyse the mechanism behind the good diagnosis result of the proposed model.

Keywords: Compound fault diagnosis · Piston engine · Deep convolutional neural network

1 Introduction

In recent years, with the rapid development of mechanical fault diagnosis technology, the methods of research are changing rapidly, especially in the field of rotating machinery, which has achieved good application results. However, for piston engine, a reciprocating machinery, its practical application has not achieved satisfactory results because of its complex structure, unknown excitation sources and unstable operation, although scholars at home and abroad have done a lot of research work in this area. Fault mechanism model of reciprocating machinery is difficult to establish due to lack of enough cognitive and expert diagnostic knowledge such as fault tree and fuzzy logic.

R. Karim et al. (Eds.): IAI 2021, LNME, pp. 319–329, 2022.
https://doi.org/10.1007/978-3-030-93639-6_27

Sun R et al. [1] combined with bootstrap and genetic algorithm to study the optimal composite feature search method and apply it to engine fault diagnosis. The best compound feature found by genetic algorithm could discriminate among the four kinds of commonly operating statuses of the machine. Li et al. [2] developed a novel vibration-based fault diagnostic method to identify the vital components of a piston engine that have abnormal clearance. The nominal baseline was obtained via theoretical modeling and the abnormal clearance was then determined by inspecting the timing of impacts created by the components that had abnormal clearance during operation. Li [3] proposed a fuzzy diagnosis method which used the concepts of membership function and fuzzy relation matrix theory to solve the uncertain relationship between fault and symptom.

Diagnosis method based on signal processing and machine learning can obtain a variety of feature vectors in the time and frequency domain, by which can recognize the location of the fault source through the relationship between feature vectors and the system fault source. The basic idea is to extract fault features from the collected data, and then classify the existing fault data. For example, Li et al. [4] firstly presented the combination of EMD and KICA to estimate IAS signals from a single-channel IAS sensor and applied also the KICA to select distinguished features extracted by Wigner bispectrum. Liu et al. [5] proposed a multi-core SVM fault diagnosis method for piston engine based on dimension measurement, the method improved the fault diagnosis effect of piston engine. However, it seems to be out of hand for traditional diagnosis method when a compound fault for piston engine occurs and simultaneously exists the issues of large amount of data and environmental noises.

With the further development of neural network deep learning technology, it is gradually applied to solve various problems in fault diagnosis [6]. Neural network technology is widely used in automatic control, signal processing, auxiliary decision-making and other aspects because of its own nonlinear, non-limited, non-convex and other outstanding characteristics [7]. Neural network is composed of a large number of nodes connected with each other. Each node has different signal processing algorithms. Therefore, it can integrate and abstract amounts of sensor data, distinguish some characteristic similar faults and fault symptoms [8, 9]. Desbazeille, M et al. [10] optimized mechanical and combustion parameters of the model with the help of actual data. Then proposed an automated diagnosis based on an artificially intelligent system and Neural networks are used for pattern recognition of the angular speed waveforms in normal and faulty conditions. Porteiro, J et al. [11] presented a multi-net fault diagnosis system which designed to provide power estimation and fault identification. They applied the data obtained to train a three-level multi-neural network system designed to estimate the load of the engine, its condition status (between failure and normal performance) and to identify the cause of the failure. However, this paper mainly focuses on the research convolution neural network (CNN). For original CNN, the inputs of network are the raw images or signals without any preprocess and it may lead low prediction accuracy no matter how hyper-parameters change because the useful information of the input is not sufficient for precise classification. Thus, it is meaningful to enhance the model prediction by applying information enhancement technology. There are some common approaches for information enhancement technology applied into 2-D CNN. However, the information enhancement technology hasn't been applied into 1-D CNN when process the piston engine. Therefore, it is necessary to design a self-adaptive information enhancement method for 1-D CNN.

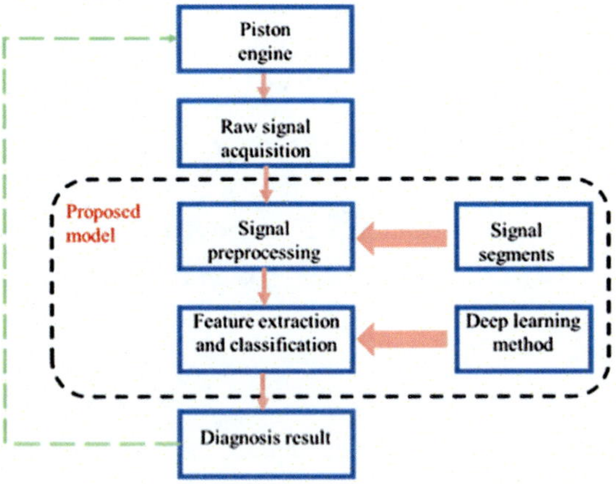

Fig. 1. Flow chart of 1-DHJCNN

In this paper, we propose a model 1-DHJCNN to detect the compound fault for piston engine based on the CNN. The CNN is mainly composed of four sub-CNNs, corresponding to four strokes of piston engine including fire, exhaust, intake and compress. As shown in Fig. 1, the proposed model divides the raw signal into signal segments according to the phase to increase the diagnosis accuracy of piston and is applied to piston vibration datasets under normal and noisy conditions.

2 Proposed of 1-DHJCNN

2.1 Convolutional Layer

All material on each page should fit within a rectangle of 18 × 23.5 cm (7" × 9.25"), centred on the page, beginning 2.54 cm (1") from the top of the page and ending with 2.54 cm (1") from the bottom. The right and left margins should be 1.9 cm (.75").The text should be in two 8.45 cm (3.33") columns with a .83 cm (.33") gutter.

The convolutional layer is an integral part of the 1-DHJCNN architecture and it is responsible for multiplying the input signal (input local regions) with the convolutional kernel (filter kernel) and generating the output features by the activation function. Raw vibration signal is divided into four parts according to the phase position and then sent into the architecture respectively. As a result of it, the input size is 900 which is a quarter of that of CNN. Then in order to generate nonlinear mapping between input data and output data, the input data is convolved in the convolutional layer and then is sent to the activation function to be activated by the activation function. Compared to the saturating nonlinearities such as sigmoid and tangent functions, the employed non-saturating RELUs are capable to deal with the gradient diffusion problem and show better fitting ability with respect to large-scale training datasets. [11]The convolutional layer uses a set of kernels $\boldsymbol{k}^l \in \Re^{D\times J\times H}$ to learn feature vectors, where J indicates the number of the kernels and D × H is the depth and height of a kernel. The *j*th output feature vector $\boldsymbol{x}_j^l$ is obtained by [12]

$$\boldsymbol{x}_j^l = \sigma(\boldsymbol{u}_j^l) \tag{1}$$

$$\boldsymbol{u}_j^l = \boldsymbol{k}_j^l * \boldsymbol{x}^{l-1} + \boldsymbol{b}_j^l = \sum\nolimits_d \boldsymbol{k}_{j,d}^l * \boldsymbol{x}_d^{l-1} + \boldsymbol{b}_j^l \tag{2}$$

where, $\boldsymbol{u}_j^l$ is the nonlinear activation, $\sigma(\cdot)$ is the activation function, $\boldsymbol{x}_d^{l-1}$ is the *d*th feature vector in $d = 1,2,\ldots D$, $j = 1,2,\ldots J$, $\boldsymbol{k}_{j,d}^l$ is an H-dimension vector, and $\boldsymbol{b}_j^l$ is the bias vector.

2.2 Pooling Layer

In order to reduce the feature dimension and reduce the error caused by feature offset to a certain extent, the dataset is sent to the pooling layer after being activated by the RELU activation function. The size of all the pooling layer is 2 × 1. After the pooling layer, in order to solve the problem of slow convergence caused by the data divergence in the training process of each layer, batch normalization layer [13] is applied to the structure of 1-DHJCNN. In the pooling layer l, the max pooling is conducted by [12]

$$\boldsymbol{x}_j^l = down(\boldsymbol{x}_j^{l-1}, s) \tag{3}$$

where, *down*(·) represents the down-sampling function of max pooling, $\boldsymbol{x}_j^l$ is the output feature vector of the pooling layer, $\boldsymbol{x}_j^{l-1}$ is the feature vector in the previous layer *l*-1, and *s* is the pooling size.

2.3 Full Connection Layer

The full connection layer is used to map the distributed feature representation to the sample marker space. The first full connection layer is connected with the second pooling layer directly. Then the four full connection layers are combined as input to the next full connection layer. Dropout layer is added to the full connection layer to prevent over fitting and improve the generalization ability of the model. The output $\boldsymbol{x}^l$ is the lth fully connected layer is obtained by [12].

$$\boldsymbol{x}_j^l = \sigma(\boldsymbol{u}_j^l) \tag{4}$$

$$\boldsymbol{u}_j^l = (\boldsymbol{w}^l)^T * \boldsymbol{x}^{l-1} + \boldsymbol{b}^l \tag{5}$$

where, $\boldsymbol{u}^l$ is the nonlinear activation, $\boldsymbol{x}^{l-1}$ is the output vector of the previous layer $1-1$, $\boldsymbol{w}^l \in \Re^{M\times N}$ is the weight matrix of the fully connected layer, M is the dimension of $\boldsymbol{x}^{l-1}$, N is the dimension of $\boldsymbol{x}^l$, and $\boldsymbol{b}^l \in \Re^N$ is the bias vector.

In the end, the characteristic is sent to the softmax layer which is used to guide learning process as network objective function.

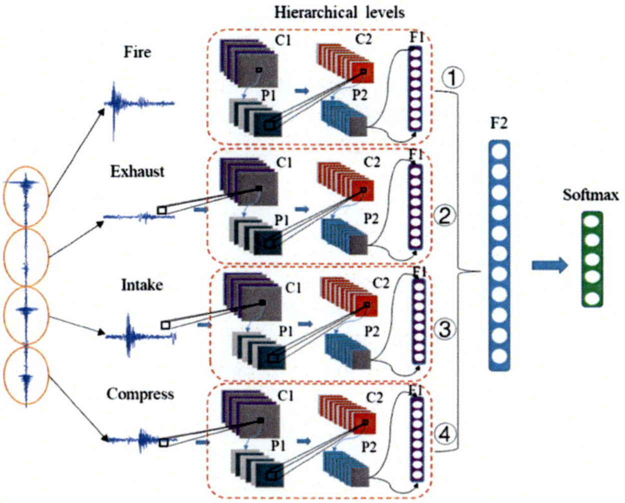

Fig. 2. Architecture of 1-DHJCNN

2.4 Training Strategy of 1-DHJCNN

To show the architecture clearly, the architecture chart of 1-DHJCNN is shown in Fig. 2. The original signal is segmented according to the characteristic phase and then put into the neural network for learning. While the convolution neural network structure layer is too few, it is difficult to learn the characteristics of the signal, too much will make the learning time too long, and even lead to over fitting of the model. Therefore, two convolution layers and two pooling layers are used in the proposed model. The detailed parameters of CNN and 1-DHJCNN are shown in Table 1(a) and Table 1(b) respectively.

Once the optimization object of the propose method is built, it is convenient to train the proposed method by stochastic gradient descent algorithm. Therefore, the parameters w^l is updated as follow,

$$w_j^l = w_j^l - \varepsilon \frac{\partial L}{\partial w_j^l} \tag{6}$$

where, L is the loss function, ε is the learning rate. According to many times of debugging, we found that the learning rate which is set at 0.5 will get the best results.

3 Results of Proposed Model

3.1 Data Description

The cylinder diesel engine provides the experimental data which built by laboratory. The dataset was collected by the acceleration sensor in 25600 sampling frequency and 1500r/min rotation rate. According to the signal characteristics of diesel engine, every two periods are selected as a data sample. Then interpolation preprocessing transforms the raw data to 3600 length samples.

The failure data collected by the testing are mainly valves malfunction. The failure dataset is divided into: normal state, intake valve small clearance fault, intake valve large clearance fault, exhaust valve small clearance fault, exhaust valve large clearance fault, small clearance compound fault of intake and exhaust valve, large clearance compound fault of intake and exhaust valve. From the description, we can conclude that the dataset can be divided into 7 health conditions (named one, two, tr, fr, fi, si, se) under different levels of White Gaussian Noise.

3.2 Comparison with DHJCNN and CNN

According to the data, we compare the 1-DHJCNN model with the traditional fault diagnosis model. The accuracy comparison results are shown in Fig. 3.

Table 1. The detail of models

Layers	Tied parameters	Activation	Output size
Input	/	/	3600
C1	Kernels:20 × 1 × 32, bias:32	RELU	720 × 32
P1	S:2 × 1	/	360 × 32
C2	Kernels:10 × 32 × 64, bias:64	RELU	180 × 64
P2	S:2 × 1	/	90 × 64
F1	Weights:90 × 64 × 1024, bias:1024	RELU	1024
F2	Weights:1024 × 7, bias:7	softmax	7
(a) parameters of CNN			
Layers	Parameters size	Activation	Output size
Input	/	/	900
C1	Kernels:20 × 1 × 8, bias:8	RELU	180 × 8
P1	S:2 × 1	/	90 × 8
C2	Kernels:10 × 32 × 16, bias:16	RELU	45 × 16
P2	S:2 × 1	/	23 × 16
F1	Weights:23 × 16 × 256, bias:256	RELU	256
F2	Weights:1024 × 7, bias:7	softmax	7
(b) parameters of 1-DHJCNN			

Obviously, the accuracy of 1-DHJCNN is higher than the traditional CNN. The average diagnostic accuracy is up to 99.28%. The CNN performs not well in one and two which testing accuracy only are 89% and 90% respectively.

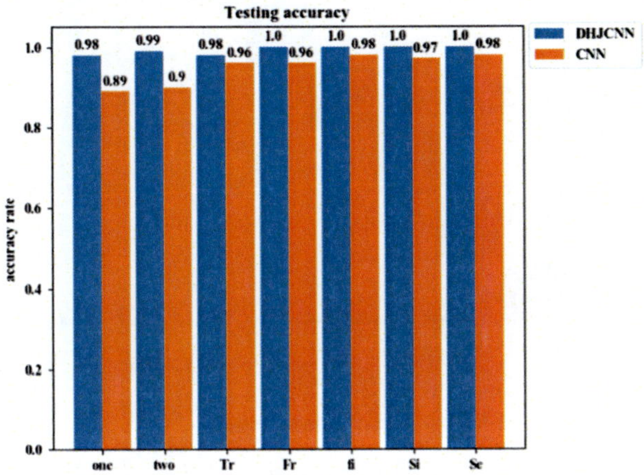

Fig. 3. Testing accuracy of 1-DHJCNN and CNN

In order to illustrate the diagnostic accuracy of the proposed model under different states, confusion matrices without normalization and normalization are shown in Fig. 4. Ordinate presents the true label and Abscissa presents the predicted label. From the image, we can observe the classification result clearly. It can be seen that accuracy of model has achieved 100% under four categories including exhaust valve small clearance fault, exhaust valve large clearance fault, small clearance compound fault of intake and exhaust valve, large clearance compound fault of intake and exhaust valve. However, the performance of the proposed method for the intake clearance faults are relatively poor, which is 98% and 99%.

3.3 Comparison Under Different Level Noise Conditions

Signal Noise Ratio (SNR) is the ratio of the power, which is usually expressed in decibels, of the output signal of the amplifier to the noise power output. The higher the SNR of the device, the less noise it produces. Generally speaking, the larger the signal-to-noise ratio, the smaller the noise mixed in the signal. The SNR is expressed as:

$$\mathrm{SNR} = 10\log_{10}\frac{P_{signal}}{P_{noise}} + 1 \tag{7}$$

To verify the robustness of 1-DHJCNN, the different levels of White Gaussian Noise are added. Figure 5 shows the testing accuracy of the noise corrupted at different SNRs, whose value is respectively 5, 10 and 15. It can be seen that testing accuracy at the most of conditions would be affected with the increased SNRs. However, the testing accuracy is large than 90%, which is suitable to practical application.

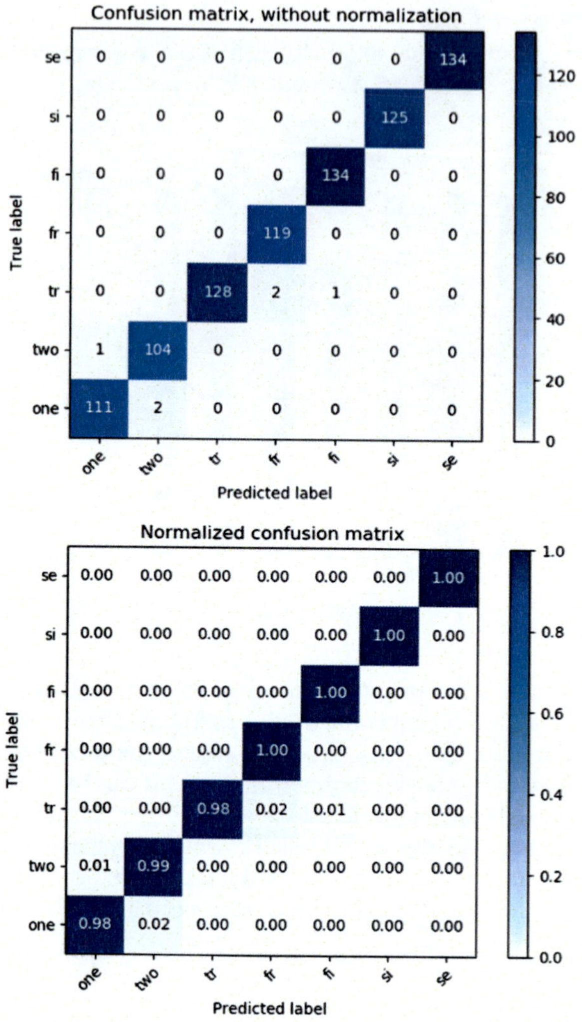

Fig. 4. Testing results through confusion matrix

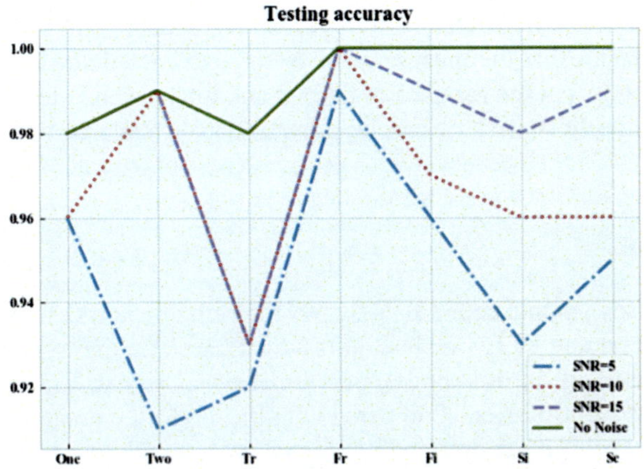

Fig. 5. Testing accuracy at different SNRs.

4 Feature Visualization

To show the extracted feature clearly and demonstrate the model can classify the fault features in the training processing, feature visualization is adopted as a promising tool to visualize the training data behind the results of 1-DHJCNN model. For clarity, we use label 0 to denote features of health condition 'normal' from original dataset, and the others are represented in the same way as shown in Fig. 6. Apparently, the images of features classification of layer 'input' are not enough to recognize each feature. For layer F1, some samples of the same features are far away from each other but different features are always overlapped or gathered very close, which implies the conditional distributions of them are not aligned successfully. As we can see from the layer F2, the result performs well than other layers. samples of the same features are grouped closer

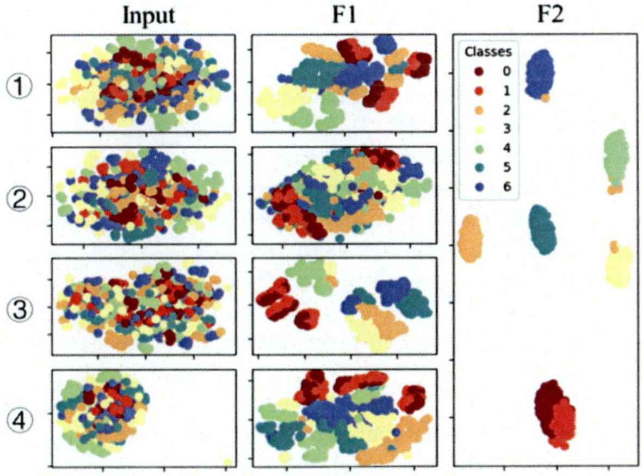

Fig. 6. Feature visualization

together and keep away from other different features except label 0 and label 1. As a result of it, it presents that the most parts of features extracted via 1-DHJCNN can be separated excellently, but the samples of normal and intake small clearance fault aren't separated distinctly. In reverse, it can be proven that the state of two sample is very similar.

5 Conclusion

This paper proposes a model called 1-DHJCNN for fault diagnosis of piston engine. The batch normalization and RELU activation is added based on the traditional intelligent fault diagnosis method. In this model, one-dimensional signals are divided into four branches for fault identification. Compared with the CNN, the diagnostic performance of 1-DHJCNN model is more outstanding. Signal dataset with four conditions under no noise and noisy environments is used to verify the proposed model. The classification results show that 1-DHJCNN has high classification accuracy in most conditions. But the model can't recognize the 'normal' and 'intake valve clearance fault' precisely, so the model still needs to be improved.

Acknowledgments. This work was supported by the Fundamental Research Funds for the Central Universities (JD2107) and the Double First-rate Construction Special Funds (ZD1601).

References

1. Sun, R., Tsung, F., Qu, L.: Combining bootstrap and genetic programming for feature discovery in diesel engine diagnosis. Int. J. Ind. Eng. **11**(3), 273–281 (2004)
2. Li, Y., Tse, P.W., Yang, X.: EMD-based fault diagnosis for abnormal clearance between contacting components in a diesel engine. Mech. Syst. Sig. Process **24**(1), 193–210 (2009)
3. Li, W.: Advance of intelligent fault diagnosis for complex system and its present situation. Comput. Simul. **21**(10), 4–7 (2004)
4. Li, Z., Yan, X., Yuan, C.: Intelligent fault diagnosis method for marine diesel engines using instantaneous angular speed. J. Mech. Sci. Technol. **26**(8), 2413–2423 (2012). https://doi.org/10.1007/s12206-012-0621-2
5. Liu, X., Yang, L., Sun, D.: Multi core SVM fault diagnosis of diesel engine based on dimension measurement. Control Eng. China **26**(12), 2211–2217 (2019)
6. Jia, F., Lei, Y., Guo, L.: A neural network constructed by deep learning technique and its application to intelligent fault diagnosis of machines. Neurocomputing **272**, 619–628 (2018)
7. Wang, Z., Zhao, H.: Fault diagnosis method for diesel engines using local wave neural network. Chin. Int. Combust. Eng. Eng. **6**, 59–62 (2006)
8. Zhu, J.: Diesel engine fault diagnosis based on BP neural network and time series analysis. J. Shanghai Marit. Univ. **4**, 22–27 (2006)
9. Yu, S., Zhu, B., Deng, J.: Fault diagnosis expert system of marine diesel engine based on rule reasoning. Ship Ocean Eng. **2**, 40–42 (2005)
10. Desbazeille, M., Randall, R.B.: Model-based diagnosis of large diesel engines based on angular speed variations of the crankshaft. Mech. Syst. Sig. Process. **24**, 1529–1541 (2010)
11. Porteiro, J., Collazo, J., Patino, D.: Diesel engine condition monitoring using a multi-net neural network system with nonintrusive sensors. Appl. Therm. Eng. **31**(17), 4097–4105 (2011)

12. Yang, W., Jin, L., Tao, D., Xie, Z., Feng, Z.: DropSample: a new training method to enhance deep convolutional neural networks for large-scale unconstrained handwritten Chinese character recognition. Pattern Recogn. **58**, 190–203 (2016)
13. Jia, F., Lei, Y., Lu, N., Xing, S.: Deep normalized convolutional neural network for imbalanced fault classification of machinery and its understanding via visualization. Mech. Syst. Sig. Process. **110**, 349–367 (2018)
14. Zhang, K., Zuo, W., Chen, Y., Meng, D., Zhang, L.: Beyond a Gaussian denoiser: residual learning of deep CNN for image denoising. Trans. Image Process. **26**, 3142–3155 (2017)

Multi-unit Fusion Learning in the Application of Gearbox Fault Intelligent Classification Under Uncertain Speed Condition

Yinghao Li, Lun Lin, Xiaoxi Ding, Liming Wang, and Yimin Shao(✉)

State Key Laboratory of Mechanical Transmission, Chongqing University, Chongqing 400044, People's Republic of China
{dxxu,lmwang,ymshao}@cqu.edu.cn

Abstract. The health monitoring and fault identification has played a crucial role in modern industrial machinery for an intelligent industrial production. Due to complex service conditions, gears often operate under complex multi-speed conditions. When under multi-speed conditions or when the speed information is unknown, the application of traditional fault identification methods based on the analysis of the characteristics of the signal spectrum, waveform and sideband distribution will be limited. To overcome this issue under multi-speed conditions, a novel intelligent identification method, called multi-unit fusion learning (MFL) has been proposed in this study. Different for the traditional network, MFL network firstly builds a series of speed identification units for different speed conditions based on artificial neural network theory, where the fault knowledge under multi-speed conditions would be collected respectively. Meanwhile, a fusion vector unit is also dynamically designed to collect the speed knowledge. Finally, the fault knowledge and the speed knowledge would be fused together, and a fusion feature will be learned for intelligent classification of gearbox under multi-speed conditions. An experiment for gearbox fault classification, with five gearbox fault status, was obtained under various rotating speeds. Compared to the conventional methods, the proposed approach can achieve much higher classification accuracies with a clearer scatter distribution. The experimental results also show the potential of the proposed MFL method in gearbox fault intelligent classification under uncertain speed conditions.

Keywords: Gearbox fault intelligent classification · Uncertain speed condition · Multiunit fusion learning · Neural network · Classification accuracy

1 Introduction

Gear is one of the important components in the mechanical transmission system, which is generally used in the power transmission and speed regulation of the mechanical system [1]. Long-term, variable-speed and variable-load operation can easily lead to gear wear, fatigue spalling, or breakage, causing serious economic losses and safety accidents [2]. Therefore, the study of gear fault identification methods is of great significance.

R. Karim et al. (Eds.): IAI 2021, LNME, pp. 330–341, 2022.
https://doi.org/10.1007/978-3-030-93639-6_28

Traditional mechanical equipment fault identification methods analyse the characteristics of signal waveform distribution, frequency spectrum, and sideband distribution, make judgments on gearbox fault based on expert experience [3]. In recent years, with the development of machine learning, data-driven fault identification methods have provided new ideas for gearbox fault identification [4]. Machine learning methods such as support vector machines (SVM), decision trees, random forests, and Bayesian networks have achieved significant results in gearbox fault identification [5–7].

However, gearboxes often operate under variable speed and variable load conditions. And it can be known that the working speed has a serious impact on the distribution of vibration signal characteristics of gearbox. However, traditional fault identification methods cannot work well in this situation. Due to the ignoring of the effect of working speed on fault samples, traditional machine learning applications will be severely limited on gearbox fault identification. Some researchers have noticed this phenomenon and carried out some research on mechanical equipment fault identification methods under variable speed conditions. For example, Ding proposed a novel energy-fluctuated multiscale feature mining approach based on wavelet packet energy (WPE) image and deep convolutional network for spindle bearing fault diagnosis under various speeds [8]. Wang proposes a reweighted and reconstruction strategy of the decomposed IMFs to obtain the de-noised signal based on the new criterion, a time-frequency representation based demodulation analysis is employed, which guarantees an accurate extraction of the fault feature at the early stage of fault. The results show that the proposed method achieves a better performance in terms of SNR improvement and fault feature detection, it can successfully detect the fault features in the presence of Gaussian and non-Gaussian noises. [9]. Ding uses the wavelet decomposition coefficient energy ratio as the characteristics of the fault signal, and combines it with an improved convolutional neural network model for the fault identification of rolling bearings, which makes the model have high fault identification ability under unknown speed conditions [10]. However, these methods focus on ensuring the accuracy of fault identification through feature selection and feature processing. As the range of gearbox speed increases, the effectiveness of these fault identification methods will decrease. Therefore, gearbox fault identification under multi-speed conditions is a very valuable research topic.

Artificial neural network, as a new pattern identification technology or a new processing method, describes the complex relationship between features and fault types, and shows great application potential in the field of equipment fault identification. When faced with a complex problem, many scholars have proposed new network architectures for specific identification problems, and those networks have achieved good identification results. Yang proposed a rolling bearing fault diagnosis method based on empirical mode decomposition (EMD) and neural network. This method first performs empirical modal decomposition on the original signal, and then selects several IMF components containing the main fault information for further analysis. The energy characteristic parameters are extracted from each IMF component as input parameters of the neural network to identify the fault type of the rolling bearing [11]. Hu proposed a combined fault pattern identification method based on wavelet neural network. A multi-layer wavelet neural network was constructed, and the residual signal was subjected to binary discrete wavelet transform at the input layer, and the detail coefficients at multiple scales were extracted as the fault feature vector, which is put into a neural network classifier for corresponding pattern classification [12]. Jin proposed an imbalanced fault diagnosis method for planetary gearbox samples. This method uses the conditional variational self-encoder's coding network to obtain the distribution of the original fault samples, and then generates a large number of valid fault samples through the decoding network. The network parameters of the generator, discriminator and classifier are continuously trained, and finally the trained model is used for intelligent fault diagnosis of the planetary gearbox [13]. It can be seen that, based on specific problems, a well-designed and reasonable network model is the mainstream method for solving complex fault identification problems, which provides a reference solution for gearbox fault identification under multi-speed conditions.

In this study, gearbox vibration signals are employed to characterize the fault status, where nine time-domain and frequency-domain features are constructed. The nine signal features include four time-domain features (RMS, P2P, Kr, If) and five frequency-domain (F3, F4, FM0, FM4, M6A) [14, 15]. The gearbox fault identification units corresponded to different speed nodes are constructed, which can realize the accurate identification under different speed conditions for gearbox faults. Finally, by constructing a fusion vector unit for fusing the multi fault identification results, gearbox fault identification under multi-speed conditions was achieved. The experimental results obtained by this proposed method and other traditional methods show the feasibility and effectiveness of the proposed method.

The rest of paper is structured as follows: Sect. 2 introduces the basic theory of gearbox fault identification method based on traditional network unit; Sect. 3 proposes gearbox fault identification method based on multi-unit fusion learning under multi-speed conditions; Sect. 4 introduces gearbox fault verification experiment; Sect. 5 introduces the identification results of gearbox faults by different methods, and analyses some valuable phenomena; Sect. 6 summarizes the work and results of this paper.

2 Gearbox Fault Identification Based on Traditional Network Unit

The structure of the gearbox fault identification based on the traditional network unit is shown in Fig. 1. It is a multilayer feedforward neural network with a fully interconnected networks consisting of an input layer, a hidden layer, and an output layer.

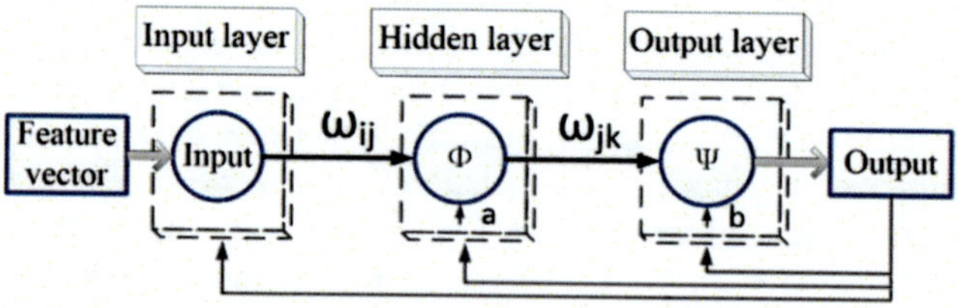

Fig. 1. Structural diagram of gear box fault identification of traditional network unit

The training process of the traditional network unit identification method consists of two parts: the forward propagation of information and the back propagation of error. The input variable is passed from the input layer through the hidden layer to the output layer. If the network cannot get the desired output, the error is propagated back. By repeatedly adjusting the weights and thresholds between different layers, the network output error reaches the desired accuracy [16].

(1) Information forward propagation
During the forward propagation of information, the input variables are passed from the input layer through the hidden layer to the output layer to get the result. The state of the neurons in each layer will only affect the state of the neurons in the next layer.
The output of the node i of the hidden layer is shown as [16]:

$$o_j = \varphi(\sum_{j=1}^{M} \omega_{ij} x_j + \theta_i) \tag{1}$$

The output of the node k of the output layer is calculated as [16]:

$$o_k = \psi[\sum_{i=1}^{q} \omega_{ki} \varphi(\sum_{j=1}^{M} \omega_{ij} x_j + \theta_i) + a_k] \tag{2}$$

where x_j is the input of input layer node j, ω_{ij} is the weight coefficient between hidden layer node i and input layer node j, ω_{ki} is the weight coefficient between hidden layer node i and output layer node k, and θ_i is the offset variable of the hidden layer node i, a_k is the offset variable of the output layer node k, and $\varphi(x)$, $\psi(x)$ are the output functions of the hidden layer and the output layer, respectively.

(2) Error back propagation
Let T be the training target of the network and O be the real output of the network. Generally, there is a certain error between O and P. The error calculation function of the network on the training samples is shown as [16]:

$$Ep = \frac{1}{2} \sum_{p=1}^{P} \sum_{k=1}^{L} (T_k^p - o_k^p)^2 \tag{3}$$

The error back propagation process mainly calculates the weight correction amount $\Delta\omega_{kj}$ of the output layer, the threshold correction amount $\Delta\theta_k$ of the output layer, the weight correction amount $\Delta\omega_{ij}$ of the hidden layer, and the threshold correction amount $\Delta\theta_i$ of the hidden layer, which are shown as follows [16]:

3 Gearbox Fault Identification Based on Multi-unit Fusion Learning

Under the multi-speed conditions, for each specific speed, the MFL network firstly builds a series of speed identification units for different speed conditions based on artificial neural network theory, where the fault knowledge under multi-speed conditions would be collected respectively. Then, using the fusion vector output by the fusion vector unit to fuse the results of different fault identification unit under different speed nodes, considering the influence of speed on the signal characteristics, the fault identification network of multi-unit fusion learning that is not sensitive to speed is constructed, which can achieve accurate identification of gearbox fault under multi-speed conditions.

$$\Delta\omega_{kj} = \eta\sum_{p=1}^{P}\sum_{k=1}^{L}(T_k^p - O_k^p).\psi'(\sum_{i=1}^{q}\omega_{ki}\varphi(\sum_{j=1}^{M}\omega_{ij}x_j + \theta_i) + a_k).y_i \tag{4}$$

$$\Delta\theta_k = \eta\sum_{p=1}^{P}\sum_{k=1}^{L}(T_k^p - O_k^p).\psi'(\sum_{i=1}^{q}\omega_{ki}\varphi(\sum_{j=1}^{M}\omega_{ij}x_j + \theta_i) + a_k) \tag{5}$$

$$\Delta\omega_{ij} = \eta\sum_{p=1}^{P}\sum_{k=1}^{L}(T_k^p - O_k^p).\psi'(\sum_{i=1}^{q}\omega_{ki}\varphi(\sum_{j=1}^{M}\omega_{ij}x_j + \theta_i) + a_k).\omega_{ki}.\varphi'(\sum_{j=1}^{M}\omega_{ij}x_j + \theta_i).x_j \tag{6}$$

$$\Delta\theta_i = \eta\sum_{p=1}^{P}\sum_{k=1}^{L}(T_k^p - O_k^p).\psi'(\sum_{i=1}^{q}\omega_{ki}\varphi(\sum_{j=1}^{M}\omega_{ij}x_j + \theta_i) + a_k).\omega_{ki}.\varphi'(\sum_{j=1}^{M}\omega_{ij}x_j + \theta_i) \tag{7}$$

The structure of gearbox fault identification method based on multi-unit fusion learning is shown in Fig. 2. First, the gearbox fault signals are pre-processed and the signal features are extracted, and they are input into the single fault identification unit that has completed training, and the fault identification results of different units are obtained. Then, the signal feature vector is input into the fusion vector network, and the fusion vector unit generates a fusion vector according to the signal features to correct the results of different units and output the final fault identification result. Finally, by processing the fault identification results, the fault type with the highest allocation probability is taken as the fault type of the input sample.

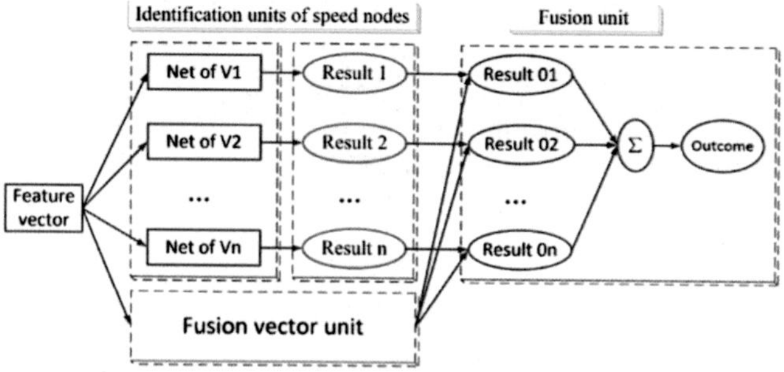

Fig. 2. Structural diagram of gearbox fault identification of multi-unit fusion learning

4 Experiment Platform

In this section, by designing a gearbox fault simulation experimental platform, the typical gearbox fault vibration signals are collected within a certain speed range and speed gradient, and gearbox fault sample data sets under multi-speed conditions are constructed to provide data support for analysis of fault identification capability.

The structure of the gearbox fault experimental platform, as shown in Fig. 3, mainly composed of five parts: a drive motor, a two-stage spur gear reducer, a speed sensor, a magnetic powder brake and a control cabinet.

The gearbox fault experiment platform uses a two-stage spur gear reducer. The fault gear in the experiment is set to the high-speed gear of the second transmission stage, and the rest are normal standard gears. Specially, the schematic diagram of the internal structure of the gearbox, the arrangement of measuring points and the position of fault gear are respectively as shown in Fig. 4. This gearbox contains 2 pairs of meshing gears, the number of teeth is 23/39, 25/53, the modulus is 3, and the total transmission ratio is 3.59.

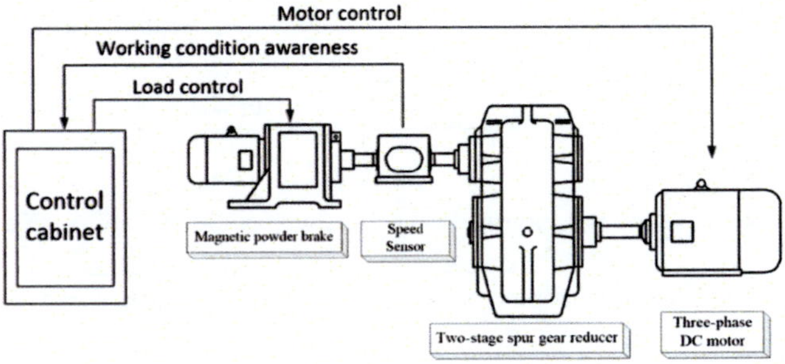

Fig. 3. Gearbox fault experimental platform system

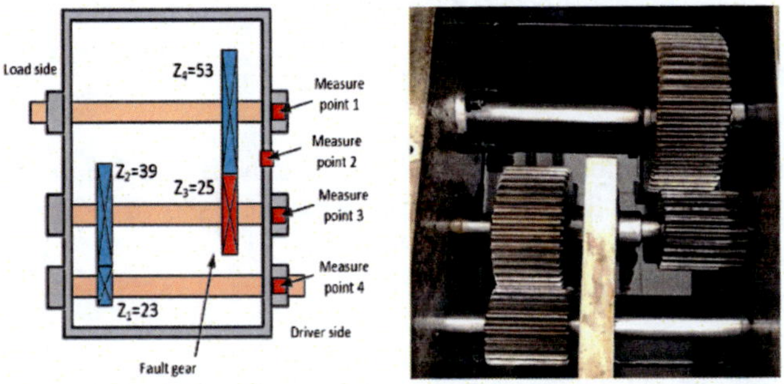

(a) Gearbox structure diagram (b) Gearbox interior physical map

Fig. 4. Schematic and physical diagram of the internal structure of the gearbox

According to the common fault types of gearboxes, four kinds of common gear fault such as tooth surface spalling, root crack, tooth surface pitting, and single tooth 30% broken teeth were selected as the research objects., the experiment included a total of 5 gear states with normal gears. Details of the processing methods, position dimensions of each gear are shown in Table 1.

In order to reflect the distribution of gearbox fault vibration signal characteristics with speed changes, the gearbox fault simulation experiment sets 5 speed gradients in the range of 400rpm to 600rpm, and the interval between two adjacent speeds is 50rpm. In this manner, combining 5 gear states and 5 speed operating conditions, the experiment includes 25 sets of gearbox fault simulation data to verify the effectiveness of multi-unit

Table 1. Fault gear parameter table

Fault type	Fault location	Fault parameter	Processing method
Tooth surface spalling	Gear tooth meshing line	Length 60 mm, width 3 mm, depth 0.5 mm	Electric spark
Tooth root crack	Tooth root	Depth 2 mm, width 60 mm, crack propagation angle 60°	Wire cutting
Tooth surface pitting	gear tooth mesh	radius 1 mm, uniform distribution	Electric spark
Broken-teeth	\	30% tooth width	Milling

Table 2. Gear fault simulation experimental conditions setting table

Experimental groups	Fault type	Input speed (rpm)
1	Normal	400:50:600
2	Tooth surface spalling	400:50:600
3	Tooth root crack	400:50:600
4	Tooth surface pitting	400:50:600
5	Broken-teeth	400:50:600

fusion learning network model under multi-speed conditions. The detailed operating information of the gearbox fault simulation experiment is shown in Table 2.

5 Experiment Analysis

In this paper, the vibration signals are characterized by nine statistical features which are used as the input of the fault identification networks. The nine signal features include four time-domain features (RMS, P2P, Kr, If) and five frequency-domain (F3, F4, FM0, FM4, M6A) [14, 15], For the time domain signal $x = [x_1, x_2 \ldots x_N]$, the calculation formulations of these features are shown as Table 3.

For the time domain signal$x = [x_1, x_2 \ldots x_N]$, the spectrum can be calculated by Fourier transform, as shown in Eq. 8:

$$F(k) = \sum_{n=0}^{N-1} x(n)e^{-2\pi jnk/N} \quad (8)$$

In Table 2, N is the total number of spectral lines of the signal spectrum, μ_x represents the mean of the signal, PP_x represents the maximum peak-to-peak value of the time domain signal, P_h represents the amplitude of the signal frequency multiplier, d_n represents the differential signal, and its calculation is shown in Eq. 9:

$$d_n = x - y^d \quad (9)$$

The differential signal d_n contains only high-order sideband components and system noise components in the original vibration signal. f_{p^1} and f_{p^2} represent the average frequency and frequency second-order moment of the signal spectrum respectively, the calculation formulations are shown as Eq. 10 and Eq. 11:

$$f_{p^1} = \frac{\sum_{k=1}^{N} f_k}{N} \tag{10}$$

$$f_{p^2} = \frac{\sum_{k=1}^{N} \left(f_k - f_{p^1}\right)^2}{N-1} \tag{11}$$

Table 3. The calculation formulations of feature extraction

Feature	Calculation formulation	Feature	Calculation formulation
F3	$F3 = \frac{\sum_{k=1}^{N}\left(f_k - f_{p1}\right)^3}{N\sqrt{f_{p2}}^3}$	*RMS*	$x_{rms} = \sqrt{\frac{1}{N}\sum_{i=1}^{N} x_i^2}$
F4	$F4 = \frac{\sum_{k=1}^{N}\left(f_k - f_{p1}\right)^4}{N{f_{p2}}^2}$	*P2P*	$x_{P2P} = \max(x) - \min(x)$
FM0	$\mathrm{FM0} = \frac{PP_x}{\sum_{h=0}^{H} P_h}$	K_r	$K_r = \frac{\sum_{i=1}^{N} x_i^4}{N x_{rms}^4}$
FM4	$FM4 = \frac{N\sum_{n=1}^{N}(d_n - \mu_d)^4}{\left[\sum_{n=1}^{N}(d_n - \mu_d)^2\right]^2}$	I_f	$I_f = \frac{x_{max}}{\lvert\mu_x\rvert}$
M6A	$M6A = \frac{N^2\sum_{n=1}^{N}(d_n - \mu_d)^4}{\left[\sum_{n=1}^{N}(d_n - \mu_d)^2\right]^3}$		

To verify the superiority of the multi-unit fusion learning network (MFL) in gearbox fault identification under multi-speed conditions, this section randomly selects 60% from the data of 400rpm, 500rpm, and 600rpm as the training set, the remaining 40% and the data of 450rpm and 550rpm are the test set. The k-means clustering method based on the principal component analysis (PCA) and the traditional network unit identification method are as comparison methods, to analyse the ability of different fault identification methods under multi-speed conditions.

The average identification accuracies of these three fault identification methods under multi-speed conditions are shown in Table 4.

From Table 4, it can be seen that the PCA-based k-means clustering method has a low identification accuracy of the gear fault sample set under multi-speed conditions, which is only 59.96%. Meanwhile the traditional network unit method is still low, only 58.15%, and the MFL identification method can reach 99.43%, which is the highest fault identification accuracy among three methods.

Table 4. Average identification accuracies of different methods

Fault identification method	PCA	Traditional network unit	MFL
Average identification accuracy	59.96%	58.15%	99.43%

The clustering plots of the three fault identification methods are shown in Fig. 5. We can find that the clustering plots of the fault identification results of different methods show the similar results to the fault identification accuracy in Table 3. In the clustering plots of the PCA-based k-means clustering method, only the tooth surface spalling faults were accurately separated, and the distribution areas of the remaining three types of fault had serious overlap phenomenon, so the fault identification accuracy was low. In the clustering plot of traditional network unit, more types of fault samples are distinguished, but there are still overlaps between different types of fault sample distribution areas. Therefore, its identification accuracy is still low. In the MFL-based fault identification clustering plot, various types of fault are accurately concentrated in different areas, and there is no overlap between the various distribution areas.

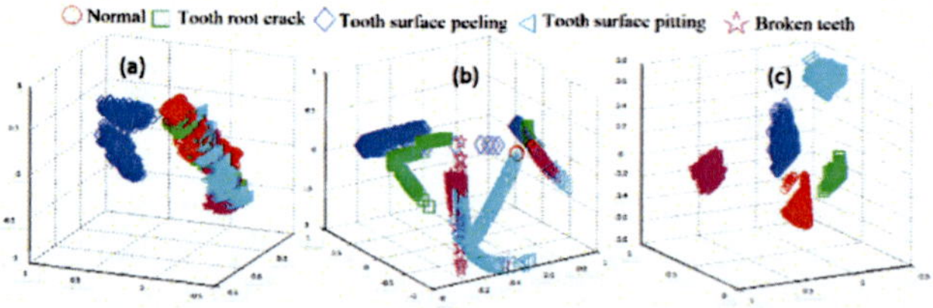

(a) PCA method (b) Traditional network unit (c) MFL method

Fig. 5. Clustering plots of fault identification results of different methods

In order to further analyse the fault identification capability of the gearbox under multi-speed conditions in different methods, the fault identification accuracy of the above three methods under 5 speeds is calculated, as shown in Table 5:

Table 5. Fault identification accuracy of different methods under multi-speed conditions

Operating conditions	PCA	Traditional network unit	MFL
400 rpm	59.8%	60.5%	100%
450 rpm	60%	52.88%	97.33%
500 rpm	60%	59.00%	100%
550 rpm	60%	59.00%	100%
600 rpm	60%	59.38%	99.83%

In Table 5, 60% of the fault samples of 400 rpm, 500 rpm, and 600 rpm are the training set, and the remaining 40% are used as the test set. Therefore, the gear fault sample set of the three speeds belongs to known speed information for the fault identification model. The data sets of the speeds of 450 rpm and 550 rpm are used as test sets, and belong to unknown speed information for the fault identification model.

As listed in Table 5, we can find that the k-means clustering method based on PCA and the traditional network unit method have low identification accuracy for the known speed test set and unknown speed test set. The MFL identification method has high identification accuracy under all speed conditions. The corresponding clustering plots of the fault identification results under 5 speeds are shown in Fig. 6.

The PCA-based k-means clustering method identification result clustering plots, the test samples at different speed are only divided into three categories, which shows the effect of the speed change is significantly greater than the effect of the fault type on the signal characteristic distribution, which makes its fault identification ability under multi-speed conditions extremely low. The traditional network unit identification result clustering plots, the test samples at different speed are divided into 5 categories. However,

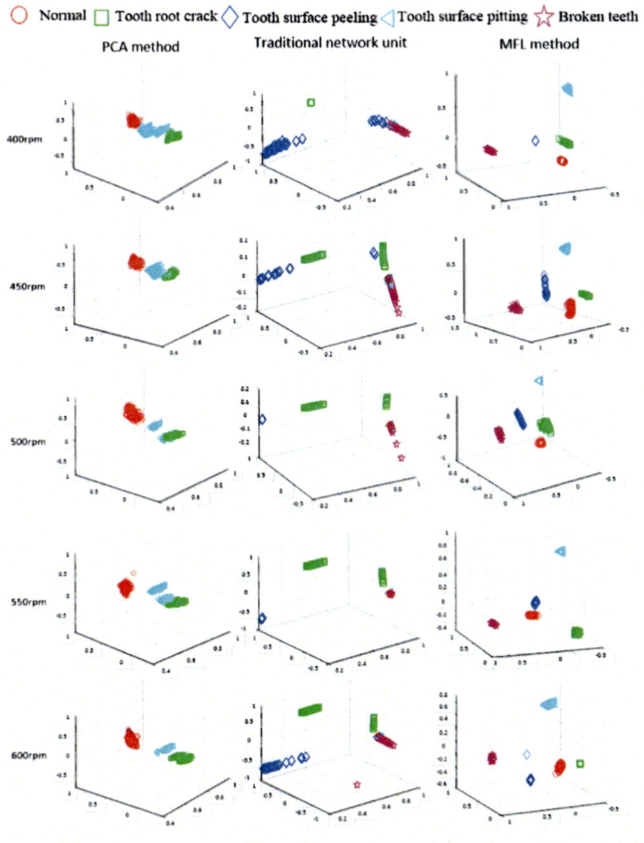

Fig. 6. Clustering plots of the fault identification results of different methods at five speeds

there is serious overlap phenomenon in the distribution areas of the 5 categories of fault samples, which makes the fault classification accuracy very low, only about 60%. It can be seen that the traditional network unit model does not accurately describe the effect of fault types on signal characteristics, which makes its fault identification ability low. The MFL-based identification result clustering plots, the fault samples at different speed are accurately divided into 5 categories. There are obvious boundaries between various types of sample distribution areas and there is no overlap. This shows that the multi-unit fusion learning network model can not only accurately extract the influence of the change of speed conditions on the signal characteristic distribution, but also accurately describe the effect of fault types on the signal characteristics. It is effective, and has accurate and stable fault identification capability at all speeds within the set speed range.

Therefore, it can be found that, compared with the traditional fault identification methods, the proposed multi-unit fusion learning network can not only accurately describe the influence of speed change on the signal characteristic distribution, but also accurately analyse the correlation between fault types and signal characteristics, and this further indicate that MFL has the ability to accurately recognize gearbox fault under uncertain multi-speed conditions.

6 Conclusion

This paper proposes a multi-unit fusion learning method for gearbox fault identification method under different speeds. Different for the traditional network architecture, with the consideration of artificial neural network theory, a series of speed identification unit is first built for different speed operation conditions, where the fault knowledge and signal characteristics under different speed working conditions would be mined in different net units respectively. Meanwhile, a fusion vector unit is also dynamically adjusted to evaluate the features from the multiple speed identification units. Finally, the fusion feature will be learned based on the states of speed knowledge embedded in this multiple net unit fusion architecture. Gearbox fault experiments of five gearbox statues at different speeds were established. Results and comparisons show that this method can achieve higher fault identification accuracy under multi-speed conditions. In future research, this method will be applied to bearing fault identification to achieve accurate fault identification results under multi-speed conditions.

Acknowledgments. This work was supported by the National Natural Science Foundation of China under Contract No. 51475053, No. 51805051 and No. 51905053.

References

1. Wang, L.-M., Shao, Y.-M.: Crack fault classification for planetary gearbox based on feature selection technique and K-means clustering method. Chin. J. Mech. Eng. **31**(01), 242–252 (2018). https://doi.org/10.1186/s10033-018-0202-0
2. Su, Z., Wang, F., Xiao, H., et al.: A fault diagnosis model based on singular value manifold features, optimized SVMs and multi-sensor information fusion. Meas. Sci. Technol. **31**(9), 095002 (2020)

3. Azamfar, M., Singh, J., Li, X., et al.: Cross-domain gearbox diagnostics under variable working conditions with deep convolutional transfer learning. J. Vib. Control **8**, 107754632093379 (2020)
4. Singh, J., Azamfar, M., Ainapure, A., et al.: Deep learning-based cross-domain adaptation for gearbox fault diagnosis under variable speed conditions. Meas. Sci. Technol. **31**(5), 055601 (2020)
5. Shao, Y., Nezu, K.: Prognosis of remaining bearing life using neural networks. Proc. Inst. Mech. Eng. Part I: J. Syst. Control Eng. **214**(3), 217–230 (2000)
6. Li, Y.: Fault diagnosis of gearbox of wind turbine based on decision tree support vector machine. North China Electric Power University, Beijing (2018)
7. Cerrada, M., Zurita, G., Cabrera, D., Sánchez, R.-V., Artés, M., Li, C.: Fault diagnosis in spur gears based on genetic algorithm and random forest. Mech. Syst. Sig. Process. 70–71 (2016)
8. Ding, X., He, Q., Luo, N.: Energy-fluctuated multiscale feature learning with deep ConvNet for intelligent spindle bearing fault diagnosis. J. IEEE T Instrum Meas. **99**, 1–10 (2017)
9. Wang, L., Shao, Y.: Fault feature extraction of rotating machinery using a reweighted complete ensemble empirical mode decomposition with adaptive noise and demodulation analysis. Mech. Syst. Sig. Process. **138**, 106545 (2020)
10. Ding, X., He, Q., Luo, N.: A fusion feature and its improvement based on locality preserving projections for rolling element bearing fault classification. J. J SOUND VIB. **335**, 367–383 (2015)
11. Chen, S., Du, M., Peng, Z., Feng, Z., Zhang, W., et al.:Fault diagnosis of planetary gearbox under variable-speed conditions using an improved adaptive chirp mode decomposition. J. Sound Vib. **468**
12. Zheng, X., Wang, S., Qian, Y.: Fault feature extraction of wind turbine gearbox under variable speed based on improved adaptive variational mode decomposition. Proc. Inst. Mech. Eng. Part A J. Power Energy **234**(6), 095765091988572 (2019)
13. Jin, Q.: Research on Fault Diagnosis Technology of Planetary Gearbox based on Generative Network. Nanjing University of Aeronautics and Astronautics (2019)
14. He, Z., Chen, J., Wang, T.: Theories an Applications of Machninery Fault Diagnostics. Higher Education Press, Beijing 2 (2012)
15. Theodoridis, P., kourtroumbas, K.: Pattern Recognition, Fourth Edition. Academic press, Burlington (2008)
16. Liu, B., Guo, H.: Super Learning Manual of Neural Network. Posts & Telecom Press, Beijing, 8 (2017)

Fault Diagnosis Capability of Shallow vs Deep Neural Networks for Small Modular Reactor

Hanan Ahmed Saeed(✉), Peng Min-jun, and Hang Wang

Harbin Engineering University, Nantong Street 145-1, Harbin 150001, China
heupmj@hrbeu.edu.cn

Abstract. NPPs warrant a reliable and effective fault diagnosis system. Data driven approaches especially neural networks have gained popularity in the past few years in the field of fault diagnosis. However, most of the researches in this field for NPP apply shallow neural networks which do not amply cater the requirements of this application. This paper is an attempt to show advantages of deep network i.e. LSTM over commonly used feed forward network. Moreover, commonly applied network architecture is 1-D which is not in sync with the requirement of NPP 2-D dataset. In manuscript, IP-200 which is a Small Modular Reactor was simulated in RELAP-5 thermal hydraulic code in different conditions. The results show obvious superiority of deep network over shallow networks.

Keywords: Deep learning · Fault diagnosis · Neural Network · Long short-term memory · Small Modular Reactor

1 Introduction

Despite, Nuclear Power being able to accrue obvious advantages like low emissions or low fuel requirements. Public opinion about this field have been filled with suspicion. The major cause for this attitude has been past incidents in Chernobyl, Three miles island and Fukushima NPPs [1]. Therefore, making safe and reliable Nuclear Power Plants has always remained the top priority of nuclear industry and academia. Although, Nuclear Technology and Nuclear Power Plants have come a long way since then in achieving more reliable, safe and economical operation. However, the strive for research in this field is an ongoing process which is to be continued to achieve higher standards, public support and accident free operations. In this regard, Instrumentation and Control (I&C) perhaps plays the most important role. Of which fault diagnosis has remained the prime focus of most researchers [2]. Over the period of time, different approaches have been applied to the problem of Nuclear Power Plant fault diagnosis like data-driven, signal-based or model-based approaches [3].

However, NPPs are complex machines with integrated sub-systems generating a large variety of parameters each effecting the other. Therefore, model based and signal based approaches have generally resulted in not so efficient means of diagnosing a fault [4]. Sometimes, system parameters and alarms could not point out the type and/ or location of the exact fault and operators may not be able to diagnose the fault in a timely

R. Karim et al. (Eds.): IAI 2021, LNME, pp. 342–351, 2022.
https://doi.org/10.1007/978-3-030-93639-6_29

manner. On the other hand, data driven approaches generally work well with non-linear inputs and outputs of a particular system [5] and it is envisaged that such a technique if applied correctly could be able to timely define the state of a NPP and its sub-systems making an efficient control system. Therefore, timely and correct Fault diagnosis and Identification (FDI) makes up the most challenging part of such a system [6].

Recent years have seen a profound interest in data-driven approaches including Artificial Neural Networks (ANNs), Support Vector Machine (SVM) and Fuzzy Inferences. Of these, ANNs have been the most popular and successful one due its capability of auto-association, self-organization, self-learning and non-linear modelling [7]. Normally, Neural networks with 1 or 2 hidden layers are regarded as shallow networks. Example of researches in this field include: Shi et al. applied Genetic Radial Basis Function Neural Network (GRBFNN) for diagnosing faults in condensation and feed water system of nuclear power plant [8]. Hadad et al. used Principle Component Analysis (PCA) for dimensionality reduction and then used neural network for fault identification [9]. Du et al. applied Data fusion in combination with Neural Network for fault analysis of Electric Actuator [10]. Yong-kuo et al. applied a hybrid of fuzzy logic and neural network for the same purpose [3]. Gomes & Medeiros researched the possibility of fault identification using Gaussian Radial Basis Function [11]. Ayodeji et al. applied RBFN and Elman NN to classify between 5 different faults [12].

Abovementioned examples and many others like these clearly point out that in nuclear industry, most fault diagnosis applications are based on shallow ANNs applied through regression mode. However, ANN based methods detailed in literature for nuclear industry has two obvious disadvantages: (1) Commonly apply networks for fault diagnosis have 1-Dimensional structure (shallow ANNs) while data received from any NPP or simulator has 2 dimensions. The importance of this fact lies in the way a neural network works. When any neural network is applied in 1-D architecture, it processes a bunch of parameters as single input with no relation to next or previous inputs (in time). In other words, it processes instantaneous values of any power plant for fault diagnosis while NPP is a complex machine whose state could only be correctly identified by trend analysis of its parameters. This trend analysis can only be achieved by observing these parameters over time which requires 2-Dimensional network models. (2) Secondly, shallow networks have relatively simple architectures and cannot handle big data. Many researchers try to cope this deficiency through dimensionality reduction which basically undermines the capabilities of machine learning. Moreover, such simple architectures limit the capability of machine learning to learn complex non-linear relationships between various intervening variables.

On the other hand, deep learning networks have the ability to overcome these deficiencies which will invariably produce more reliable and accurate intelligent diagnosis methods. Deep learning is a specific class of machine learning which applies non-linear transformations systematically through multiple layers of data processing. In this way, it can approximate complex non-linear functions with minimal error. Since its inception in computer sciences, deep learning has been the focus of researchers in wide variety of fields. Most notably, deep networks have produced record breaking results in the fields of Game of GO [13], Image recognition [14], Speech recognition [15] and particle physics

[16]. Unfortunately, deep learning has not been widely applied for intelligent fault diagnosis research of NPPs despite its obvious advantages. In one case, Peng et al. applied Deep Belief Network for fault diagnosis but applied correlation analysis for dimensionality reduction which may severely reduce deep networks ability. This paper is an effort to clear common perception about deep network applications and its superiority over shallow networks in fault diagnosis of nuclear power plant.

2 Methodology

2.1 IP-200 Nuclear Power Plant

The reactor selected for this research is IP-200 which is an advance Small Modular Reactor (SMR) based on Integral PWR designed at Harbin Engineering University. Recent years have seen a growing interest in small and medium sized reactors (SMR) due to their compactness, ease of assembly (at factory) and remote areas accessibility. Such a reactor is ideal for providing energy to coastal and remote areas. Key parameters of the reactor are stated in Table 1.

Table 1. IP-200 key parameters

System parameters	Rated values
Full core power	220 MW
Reactor core inlet temp	562.15 K
Reactor core outlet temp	594.15 K
Pressurizer pressure	15.5 MPa
Primary coolant mass flow rate	1200 kg/s
Feed water mass flow rate	81.6 kg/s
Feed water temp	373.15 K
Steam pressure	3.0 MPa
Superheat of steam	40.0 K
Number of main pumps	4
Number of OSTGs	12

RELAP5 has been used for reactor thermal hydraulic modelling. RELAP5/MOD 4.0 is a two-fluid system analysis code. Figure 1 details the nodalization of the plant.

2.2 Long Short Term Memory Network

Long Short Term Memory (LSTM) neural networks are an effective solution to sequence regression problems. In many ways, LSTM are far superior to Recurrent neural networks (RNN) or conventional feed forward neural networks. A common network consists of input, output and hidden layers as shown in Fig. 2:

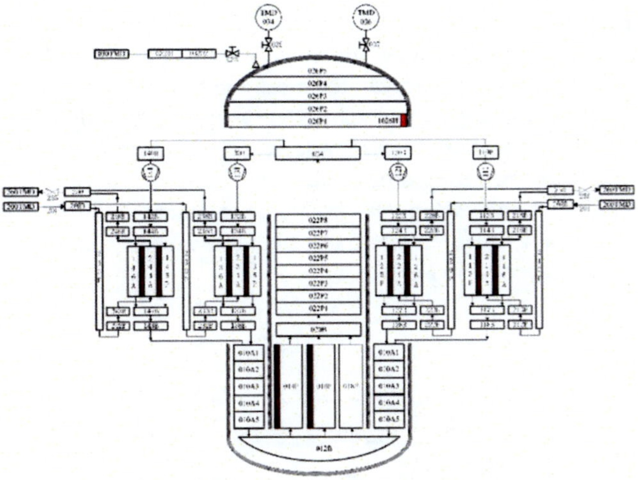

Fig. 1. IP-200 nodalization

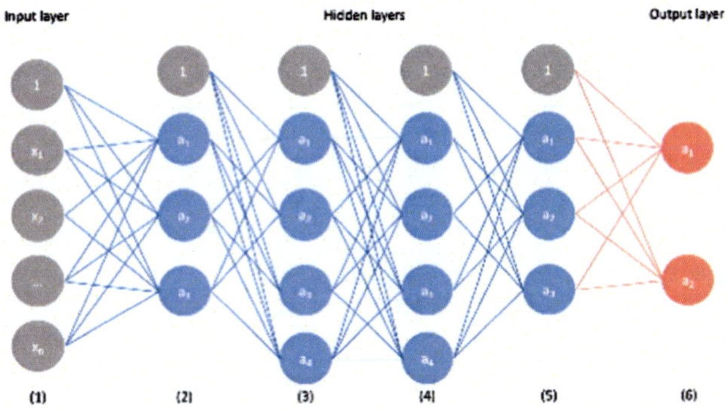

Fig. 2. Common network model

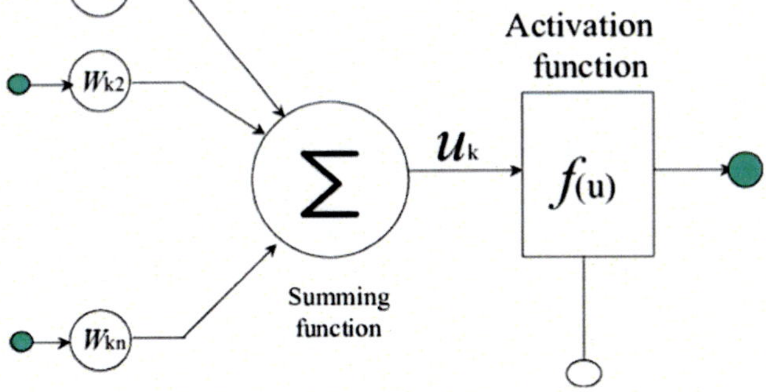

Fig. 3. Structure of a common neuron

Here, a simple cell or neuron can be shown as (Fig. 3).

Hence, output of each cell can be written as $y_j = f(\sum_{i=1}^{n} W_i X_i)$ where W and X are weight and input matrices. However, the most common algorithm used for training such networks uses backpropagation which update weights from output layer all the way up-to input layer. If the values exponentially decrease, this cause vanishing gradient or if increase in a similar manner, exploding gradient is faced. With deep networks, number of hidden layers is enormous making this problem prominent. To solve this problem, an LSTM cell not only has a time step information but also memory information. Each cell of LSTM network is depicted in Fig. 4.

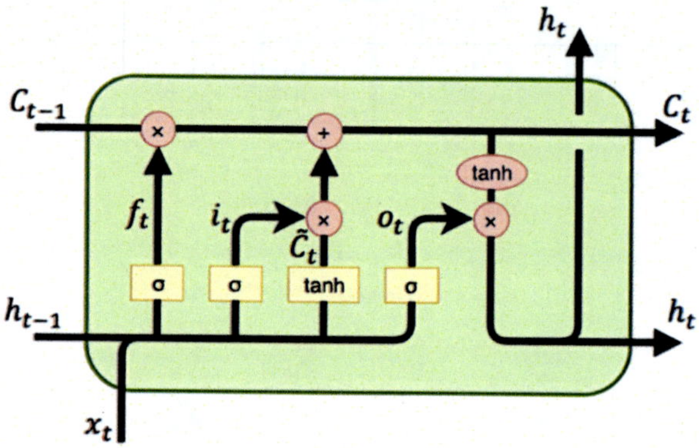

Fig. 4. Structure of an LSTM cell

Here, x_t is input vector, h denotes cell output and C denotes cell memory. For each time step, four vectors $f_t, i_t, \tilde{C}_t$ *and* o_t are calculated by concatenating dot products of previous hidden states and input vectors as given by Eq. 15–18

$$o_t = \sigma(x_t * u_o + h_{t-1} * W_o) \tag{1}$$

$$\tilde{C}_t = \tanh(x_t * u_c + h_{t-1} * W_c) \tag{2}$$

$$i_t = \sigma(x_t * u_i + h_{t-1} * W_i) \tag{3}$$

$$f_t = \sigma(x_t * u_f + h_{t-1} * W_f) \tag{4}$$

Here f_t is forget gate, i_t is input gate, $\tilde{C}_t$ is candidate gate and o_t is output gate. Depending on value of forget gate, the value of next memory gate is calculated as

$$C_t = f_t * C_{t-1} + i_t * \tilde{C}_t \tag{5}$$

This allows for required memory state to be transmitted to next cell unaffected. This way long term dependencies are passed through the whole network unaffected as forget gate is applied to filter only redundant parameters. Finally, output h_t of each cell is calculated as

$$h_t = o_t * \tanh(C_t) \tag{6}$$

The network topology applied is shown in Fig. 5.

Fig. 5. Applied LSTM layers

3 Results and Discussion

In this paper, we investigated the performance of different networks by first modeling IP-200 NPP through use of RELAP5 code. During RELAP5 simulation, one steady state, one transient and four incipient faults were introduced. 43 different operating parameters were recorded for each scenario to obtain corresponding time-series databases. Table 2 list each scenario with its encoding for regression.

Table 2. Simulated plant conditions

Data set	Simulated faults	Encoding
1	Normal operation	0.0
2	Change of reactor power from 100% to 80%	0.15
3	01 x main pump failure	0.3
4	Loss of feed water	0.45
5	Inadvertent opening of pressurizer relief valve	0.6
6	Small break LOCA	1.0

In a Nuclear Power Plant, plant status can be ascertain from instrument readings. When in steady state, the readings are normally consistent. However, during transients like power change or introduction of a defect causes these instrument readings to develop specific time-dependent patterns which if correctly identified would allow a control system to pin point the type of transient or defect. After steady state operation, each fault was injected after 50 s and parameters were recorded for 200 time steps. During next stage, the datasets were pre-processed. The steps included are encoding and normalization. Each fault encoding is shown in Table 2 while normalization is carried out through Newtonian method.

3.1 Network Optimization

Subsequently, data was processed through LSTM Network. However, in order to apply any NN correctly, it is necessary to find optimum values of concerning parameters for that NN according to the type of data and application. Table 3 below lists the results along-with applied parameters for LSTM.

Table 3. Network optimization results

Parameters	Applied values	Accuracy	Final values
Hidden units	30	66%	150
	50	83%	
	100	83%	
	150	99%	
Batch size	1	83%	2
	2	83%	
	3	66%	
LSTM layer type	BiLSTM	83%	BiLSTM
	LSTM	66%	
Max epochs	50	66%	15
	100	83%	
	150	99%	

After application of optimum parameters, each of these networks were trained on training data of 200 instances of these 6 cases which makes a total of 1200 dataset. It is worth noting that each of these instances were varied a bit (only 1–2% noise) by adding white gaussian noise to the pre-processed data. This allowed for realistic training of the neural network. Results of optimized LSTM training is shown in Fig. 6.

Similarly, optimized values of hyper parameters for feed forward network was found and network was trained on these parameters. Figure 7 shows the results of final network training.

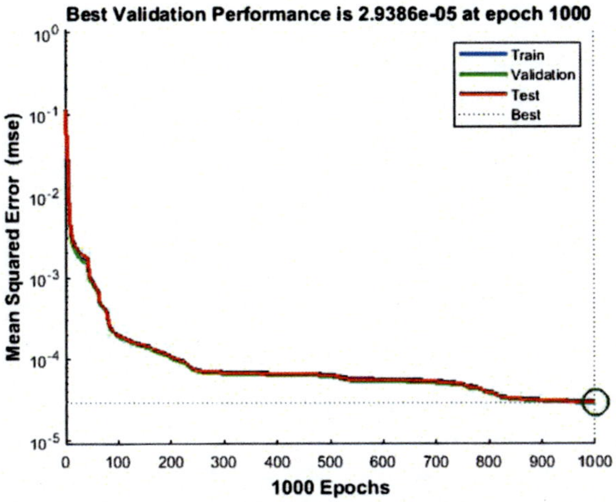

Fig. 6. LSTM training results

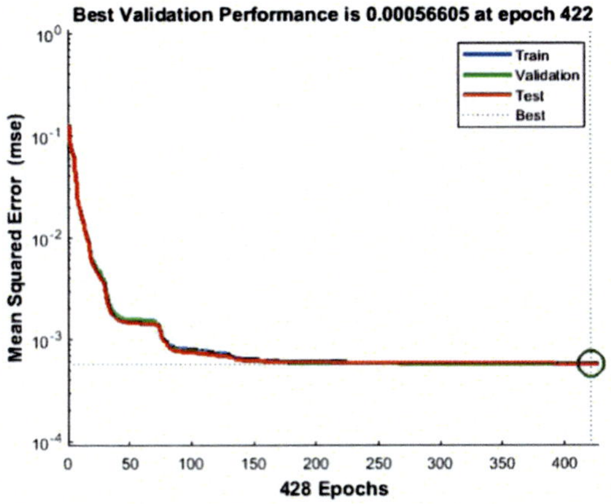

Fig. 7. Feed forward network training results

3.2 Network Comparison

Network performance is generally assessed by the amount of modelling accuracy achieved on test data. In this regard, Mean Square error (MSE) is a valid indication of a network's performance. In a real plant data, observed parameters vary due factors like noise induction, machinery/ sensor degradation and effect of operator actions. A smart neural network must therefore be able to satisfactorily regress this data. In order to make testing more realistic, each test data was created by inducing different level of noise to the pre-processed dataset of total 1800 instances taken from 6 cases, mentioned earlier. To access performance of optimized LSTM network, its performance was compared with most commonly applied shallow network i.e. feed forward network with back propagation algorithm. Like LSTM network, the network was first optimized to achieve best results on sample data. Then the performance of two networks were compared. Table 4 lists the results achieved:

Table 4. Result comparison

	0% noise	10% noise	20% noise	33% noise
Feed forward	0.00056	0.1894	0.2748	0.2987
LSTM	0.0000293	0.0022	0.0556	0.1044

Results clearly shows that Deep Neural Networks performs much better than shallow neural networks specially when faced with realistic data. LSTM's accuracy can be contributed to its 2D neuron structure. Based on this research, it can be inferred that Deep learning shows obvious superiority over conventional neural networks.

4 Conclusion

The researched method of fault diagnosis based on deep learning is highly prospective for industrial application due outstanding diagnosis accuracy. Therefore, future research work on application of this promising technology (Deep Learning) must be focused as opposed to shallow networks. However, in a real plant, not all faults could be anticipated before-hand, making it impossible for a supervised trained neural network to detect each situation. Moreover, multiple faults could also be confronted. Therefore, an effective control strategy must be devised that can tackle every situation. Our future work will also focus on application aspects of latest machine learning techniques for NPP fault diagnosis.

Acknowledgments. This work is funded by the Chinese national research project "Research of Online Monitoring and Operator Support Technology". The authors are also grateful for the support from the Chinese national scholarship council (201706680057).

References

1. Soni, A.: Out of sight, out of mind? investigating the longitudinal impact of the Fukushima nuclear accident on public opinion in the United States. Energy Policy **122**, 169–175 (2018)
2. Lee, D., Seong, P.H., Kim, J.: Autonomous operation algorithm for safety systems of nuclear power plants by using long-short term memory and function-based hierarchical framework. Ann. Nucl. Energy **119**, 287–299 (2018)
3. Yong-kuo, L., et al.: Research and design of distributed fault diagnosis system in nuclear power plant. Prog. Nucl. Energy **68**, 97–110 (2013)
4. Moshkbar-Bakhshayesh, K., Ghofrani, M.B.: Transient identification in nuclear power plants: a review. Prog. Nucl. Energy **67**, 23–32 (2013)
5. Patan, K.: Artificial Neural Networks for the Modelling and Fault Diagnosis of Technical Processes. Springer, Heidelberg (2008). https://doi.org/10.1007/978-3-540-79872-9

6. Ma, J., Jiang, J.: Applications of fault detection and diagnosis methods in nuclear power plants: a review. Prog. Nucl. Energy **53**(3), 255–266 (2011)
7. Liu, Y.-K., et al.: Improvement of fault diagnosis efficiency in nuclear power plants using hybrid intelligence approach. Prog. Nucl. Energy **76**, 122–136 (2014)
8. Shi, X.-C., Xie, C.-L., Wang, Y.-H.: Nuclear power plant fault diagnosis based on genetic-RBF neural network. J. Mar. Sci. Appl. **5**(3), 57–62 (2006). https://doi.org/10.1007/s11804-006-0064-1
9. Hadad, K., et al.: Enhanced neural network based fault detection of a VVER nuclear power plant with the aid of principal component analysis. IEEE Trans. Nucl. Sci. **55**(6), 3611–3619 (2008)
10. Du, H., et al.: Study of fault diagnosis method based on data fusion technology. Procedia Eng. **29**, 2590–2594 (2012)
11. Gomes, C.R., Medeiros, J.A.C.C.: Neural network of Gaussian radial basis functions applied to the problem of identification of nuclear accidents in a PWR nuclear power plant. Ann. Nucl. Energy **77**, 285–293 (2015)
12. Ayodeji, A., Liu, Y.-K., Xia, H.: Knowledge base operator support system for nuclear power plant fault diagnosis. Prog. Nucl. Energy **105**, 42–50 (2018)
13. Silver, D., et al.: Mastering the game of GO with deep neural networks and tree search. Nature **529**(7587), 484–489 (2016)
14. Russakovsky, O., et al.: Imagenet large scale visual recognition challenge. Int. J. Comput. Vis. **115**(3), 211–252 (2015). https://doi.org/10.1007/s11263-015-0816-y
15. Dahl, G.E., et al.: Context-dependent pre-trained deep neural networks for large-vocabulary speech recognition. IEEE Trans. Audio Speech Lang. Process. **20**(1), 30–42 (2011)
16. Baldi, P., Sadowski, P., Whiteson, D.: Searching for exotic particles in high-energy physics with deep learning. Nat. Commun. **5**, 4308 (2014)

Smart Online Monitoring of Industrial Pipeline Defects

Jyoti K. Sinha[1](✉) and Kassandra A. Papadopoulou[2]

[1] Department of Mechanical, Aerospace and Civil Engineering, The University of Manchester, Manchester M13 9PL, UK
Jyoti.Sinha@manchester.ac.uk
[2] Alliance Manchester Business School, The University of Manchester, Manchester M15 6PB, UK
kassandra.papadopoulou@manchester.ac.uk

Abstract. Acoustic Wave Reflection (AWR) approach seems to be the future avenue for long pipeline monitoring, typically for the oil and gas industries. There are several research studies are available on successful detection of defects using this AWR approach in the pipelines based on the laboratory scaled experiments. The method seems to be successfully applied to few industrial scale pipelines as well. The paper is proposing a smart online monitoring system using this AWR approach together with the modern instrumentation and Internet of Things (IoT) features to integrated wireless sensor node, input acoustic wave signal optimisation and then remote collection of the AWR signal to determine the pipe defect location using the piping layout with the geographical positioning system (GPS). The paper presents the proposed smart online monitoring system.

Keywords: Pipeline monitoring · Acoustic Wave Reflection (AWR) · Artificial Intelligence (AI) · Defect detection

1 Introduction

Over the years, pipelines are found to have various forms of failure due to internal and/or external factors. These factors may be unusual working conditions, poor manufacturing processes, environmental affect, etc. The failure of pipelines in oil and gas industries is of critical importance mainly due to the fatalities, injuries and costs [1]. It has also been observed that the design of the pipelines are generally robust and failure generally triggered from relatively smaller defects such crack, hole, localised thinning of pipeline, etc. The reason for these small defects could be poor welding, rubbing, localised defect in pipe material, etc. The European Gas pipeline Incident data Group (EGIG) [2] has also provided the data for gas pipeline failures and describes these failures in relation to type and size of leaks. About 85% of the failures are due to presence of pinhole/crack and hole type leaks [2].

Therefore the identification of this defect spot as quickly as possible in the pipeline is essential for plant reliability, safety, production and reducing maintenance overhead.

R. Karim et al. (Eds.): IAI 2021, LNME, pp. 352–359, 2022.
https://doi.org/10.1007/978-3-030-93639-6_30

Therefore, the industry is always looking for the reliable and robust detection methods to mitigate such failures [3, 4].

Different studies are carried out to locate the pipe defect [5–8]. These are Acoustic Ranger tool (AR), Impact Echo Technique, EMAT technique, Acoustic wave Reflectometry Technique, etc. Acoustic Wave Reflection (AWR) approach seems to be the future avenue for long pipeline monitoring, typically for the oil and gas industries. There are several research studies are available on successful detection of defects using this AWR approach in the pipelines based on the laboratory scaled experiments. The method seems to be successfully applied to few industrial scale pipelines as well [9].

The paper is proposing a smart online monitoring system using this AWR approach together with the modern instrumentation and Internet of Things (IoT) features to integrated wireless sensor node, input acoustic wave signal optimisation and then remote collection of the AWR signal to determine the pipe defect location using the piping layout with the geographical positioning system (GPS). The paper presents the concept of the proposed Artificial Intelligence (AI) integrated Smart Online Monitoring System based on the Acoustics Reflection (AI-SOMAR).

2 Current Approaches

2.1 Intrusive Approach

The most popular device is a robotic instrument called a smart PIGs (Pipeline Inspection Gauges). This is intrusive approach for pipeline inspection [10]. The PIG is a wire-wrapped straw used for cleaning out wax and other contaminants. Depending on the model, smart pigs detect cracks and weld defects through magnetic flux leakage or shear wave ultrasound. It can also measures the roundness of the pipe, pipe wall thickness and metal loss through compression wave ultrasound. Therefore this approach can provide detailed information and conditions of the pipeline. However, it is intrusive and extensive preparation is needed to inspect any pipeline. Therefore it has limited application depending on the pipeline networks e.g., difficult in presence of bend, valves and change in pipe diameter and age of the pipeline network.

2.2 Non-intrusive Approach

AWR is a non-intrusive approach for monitoring but it requires intrusive initial instrumentation. The approach generally requires intrusive installation of loudspeakers and microphones (or hydrophones) depending upon the length of the pipeline as shown in Fig. 1. But it is significantly easier and non-intrusive to carry out pipeline inspections in practice once required instruments are in place. The laboratory experiments show the potential of identifying the defect location(s) quickly in the pipeline [11]. This technique can inspect long pipelines with different configurations up to ~10 km [9]. However, this approach requires either a full phase flow (not applicable on multiphase flows) or a fully empty pipeline.

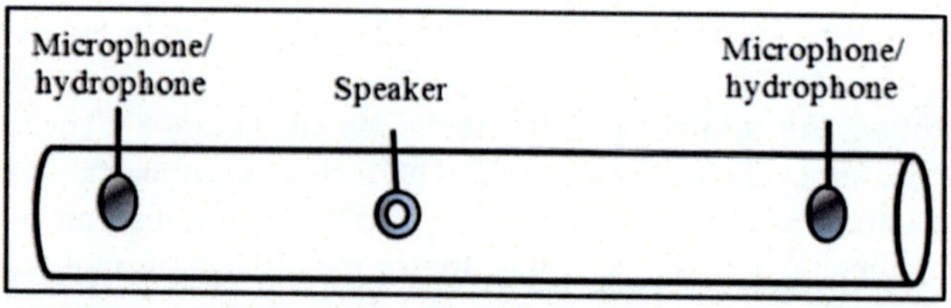

Fig. 1. A pipeline with positioned instruments using AWR.

3 Earlier Studies

This section is presenting the summary of the earlier research studies carried out within The University of Manchester UK [8, 9, 11–14] for the completeness and background of the current proposal of the AI-SOMAR.

3.1 Wave Refection and STFT Analysis [8, 11]

The schematic of the experiments in a pipeline with a simulated leakage hole is shown in Fig. 2 [8, 11]. The measured AWR signals by the microphone for the input signals with different input frequency by the speaker into the open end pipeline are also shown in Fig. 3 [8, 11]. The details of instruments, pipe materials and dimensions, and environmental conditions during the experiments are given in [8, 11]. It is important to note that the location of pulse in the AWR signals are exactly the same for different input signal frequencies as the velocity of the sound travel was the same for all experiments. The difference in their AWR amplitudes is more related with the input frequencies even though the input signal amplitude was equal for all input signals. However, the AWR signal clearly indicates the location of the leakage hole and the open end of the pipe.

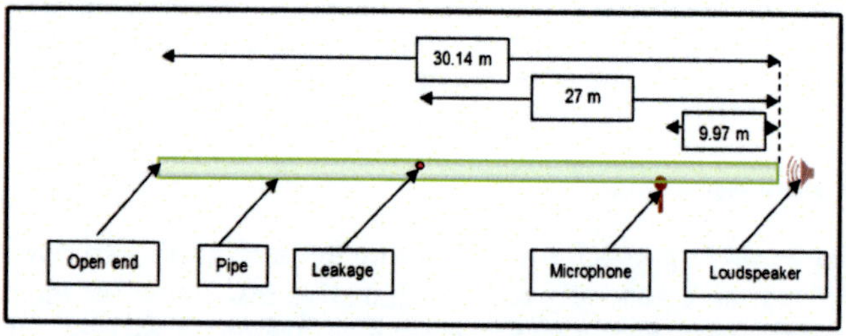

Fig. 2. A pipeline with a simulated leakage hole

The frequency domain analysis using the short time Fourier Transformation (STFT) analysis is also carried out by Papadopoulou *et al.* [11]. A typical spectrogram contour plot of the STFT analysis is shown in Fig. 3 which clearly shows the frequency peak at around 200 Hz at the pulse location. This frequency seems to be constant for most of the input signal frequencies [11]. The peak at around 200 Hz may be indicator of the defect size (Fig. 4).

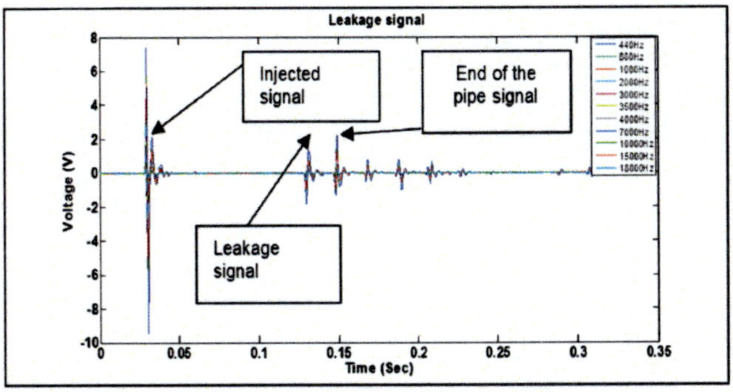

Fig. 3. Measured AWR signals by the microphone

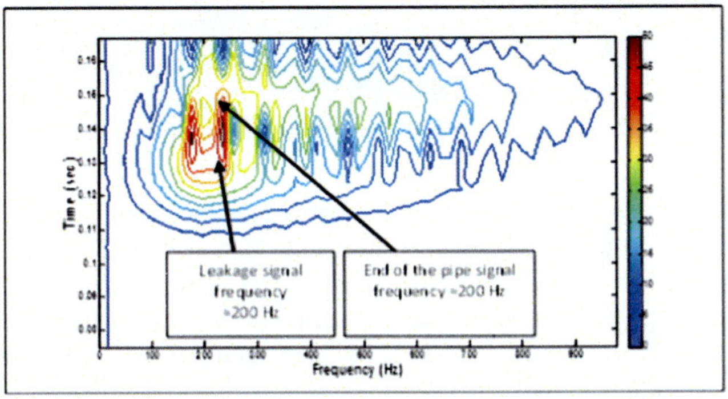

Fig. 4. A typical spectrogram contour plot of the measured AWR signal

3.2 Optimisation of Input Signals [12]

Russell *et al.* [9] have already demonstrated that the AWR approach can work up to approximately 10 km. However, the different amplitudes of the AWR signals in Fig. 3 also highlights the need of appropriate input signal sections typically in terms of the signal frequency. This definitely depends on the pipeline dimensions (radius and length) and material so that the AWR signal can be measured accurately. The minimum size to the defect to be detected is also likely to influence the input signal frequency selection.

Yusoff *et al.* [12–14] have initiated the research in this direction to optimise the frequency of the input signal. The pressure of the sound wave propagation in a pipeline is given by:

$$p(x) = p_i e^{-\alpha x} \tag{1}$$

where p_i is the pressure of initial input acoustic signal propagating along x-axis, The attenuation coefficient can be estimated by Eq. (2).

$$\alpha = \frac{\omega}{cr}\left[\sqrt{\frac{\mu}{2\rho\omega}} + (\gamma - 1)\sqrt{\frac{K}{2\rho\omega C_p}}\right] \tag{2}$$

where ω is the angular frequency, c is the speed of sound of the medium sound wave travelling in, r is the pipe radius, μ is the shear viscosity, ρ is the density, γ is the ratio of specific heat, K is the thermal conductivity and C_p is the heat capacity. The typical trend between the sound wave pressure (with respect to initial pressure) (p/p_i) with the distance (x) travelled is shown in Fig. 5. Therefore Eqs. (1) and (2) can help to determine the appropriate input signal frequency or distance between 2 speakers in a long pipeline.

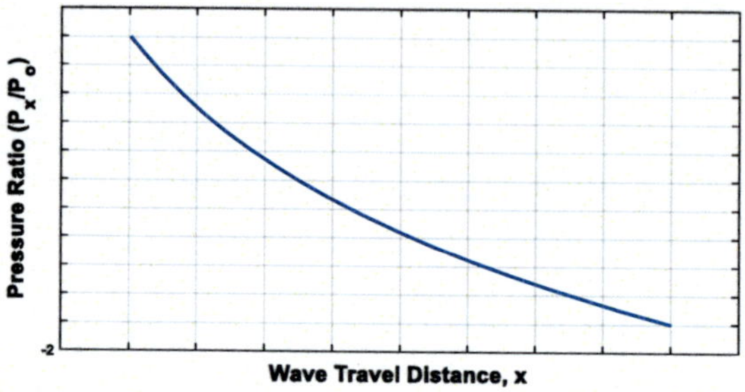

Fig. 5. Typical relation between ARW pressure delay with the wave travel distance

However, Eq. (2) has not considered the additional expected losses during experiments in comparison to the ideal analytical case. Hence Yusoff *et al.* [12–14] have conducted a series for experiments on the pipes with different diameter. The experimentally found decay in the sound wave pressure with distance was slightly more than the predicted by Eq. (1). Yusoff *et al.* [12–14] have then suggested the modification in Eq. (1) by replacing the attenuation coefficient (α) by α_m. The new coefficient is given by Eq. (3).

$$\alpha_m = (\alpha + \beta) \tag{3}$$

where

$$\beta = (0.0395 - 0.0001r)e^{-0.001\frac{2\pi}{\omega}} \tag{4}$$

Equation (4) is based on the observations from the experiments on different pipe diameters and air medium but same material. So the influence of the pipe material and other fluid medium should be examined. This modified equation is likely to determine the input signal frequency.

4 Proposed Approach

4.1 Development of Wireless Sensor Nodes – Smart Sensor

It is proposed to develop a wireless sensor node which consists of commercially available (a) hydrophone, (b) fluid proof speaker, (c) temperature sensor, (d) GPS device and (e) wireless data transmission device. The sensor node should also have fixture to install on pipe permanently with a small motor with power supply unit based on either solar system or vibration-based energy harvesting system. The motor is required to push both sensors to the pipe during experiments and data collection. The proposed simple concept of the wireless sensor node is shown in Fig. 6.

4.2 Smart Monitoring System (SMS)

The schematic of the proposed smart monitoring system (AI-SOMAR) is shown in Fig. 7. The system is creating a network of smart sensors linked together and monitored via a Geographical Information System (GIS) [15]. The Unmanned Air Vehicle (UAV), such as a "Drone" is proposed to use the link between the smart sensor note and the smart monitoring system based in the central monitoring unit (CMU) if the wireless sensing from the CMU is not feasible. The use of the smart monitoring system (including UAV if needed) may lift the problem of inaccessibility of long pipeline. The following are the planned roles of the smart monitoring system.

(i) Trigger the sensors of the pre-installed sensor node to move inside the pipeline.
(ii) Measure the fluid temperature to determine the fluid properties at the time of measurements.
(iii) Provide appropriate input signal to the pipeline and then collect the AWR signals measured by at least 3 hydrophone sensors (one at the speaker location and one each from either side of the speaker location).

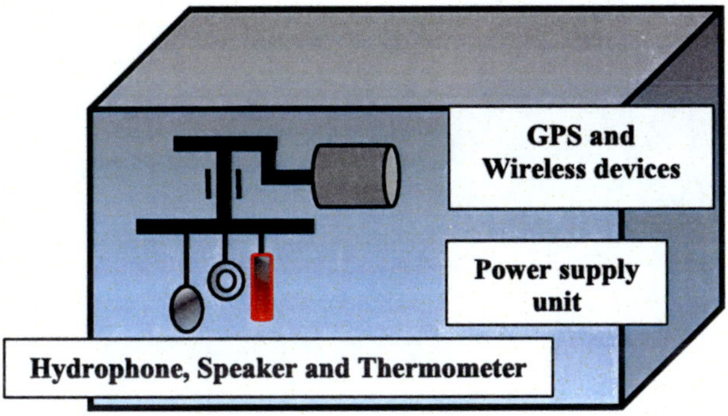

Fig. 6. A typical schematic of the wireless sensor node

The measured data and other information are then transferred wirelessly to the smart monitoring system to analyse the data of the AWR signals from multi-sensors by integrating through the pipeline maps and sensors location from the GIS data. The pipeline network may be likely to complex and measured data from the multi-sensors may make the diagnosis process difficult, therefore it is suggested to use the AI based machine learning approach to do the diagnosis and identify the defect location(s).

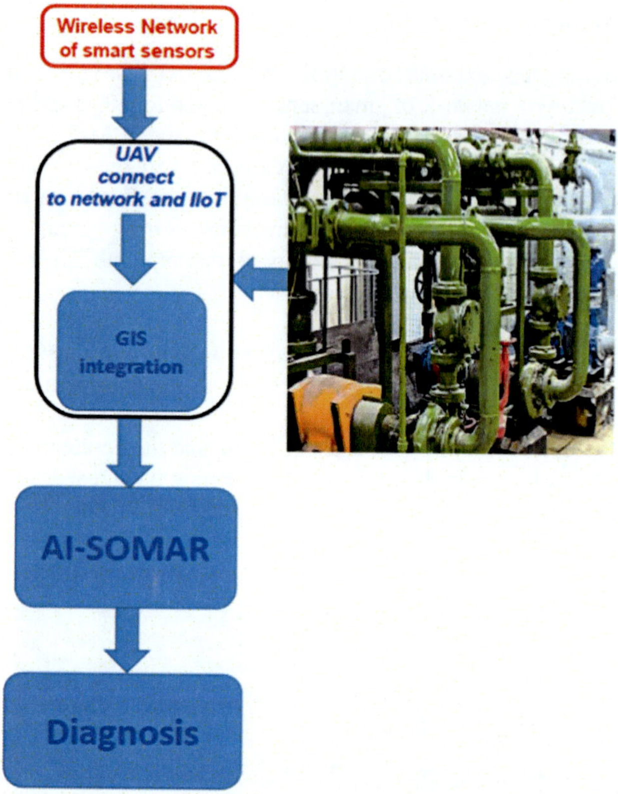

Fig. 7. Proposed smart monitoring system

5 Concluding Remarks

The suggested method of the AI integrated AI-SOMAR is in the direction of Industry 4.0 objectives. The method is using the existing concepts and technologies for possible identification of defect(s) in any pipelines, typically for O&G sectors onshore and offshore and service sectors. The suggested method also lift the limitations of inaccessibility of pipelines and intrusive inspection approaches. The approach can help to prioritise maintenance decisions to safer operations and maintaining pipeline integrity and the automation of some of the inspection aspects.

References

1. Pipeline and Hazardous Material Safety Administration (PHMSA). All Reported Pipeline Incidents by Cause, Washington DC, PHMSA (2017)
2. European Gas Pipeline Incident Group. 10th Report of the European Gas Pipeline Incident Data Group (period 1970 –2016) (2018)

3. Omoya, O., Papadopoulou, K.A., Lou, E.C.W.: Pipeline integrity: the need for an innovative reliability engineering approach. Int. J. Qual. Reliab. Manage. (2019). https://doi.org/10.1108/IJQRM-09-2017-0197
4. Beele, F.V.D., Denis, R.: Numerical modelling and analysis for offshore pipeline design, installation and operation. J. Pipeline Integrity **12**(4), 273–286 (2012)
5. Holroyd, T.: Acoustics Emission and Ultrasonic. Coxmoor publishing company, Oxford (2000)
6. Rajani, B., Kleiner, Y.: Non-destructive inspection techniques to determine structural distress indicators in water mains. Working Paper [NRCC-47068]. National Research Council, Canada, June (2004)
7. Damaschke, J., Beuker, T., Alqahtani, H.: In-line inspection with high resolution (EMAT) technology crack detection and coating disbondmen. Paper presented at 20th International Pipeline Pigging, Integrity Assessment and Repair conference, Houston Marriott Westchase Hotel, Houston, TX, 12–13 February (2008)
8. Papadopoulou, K.A., et al.: An evaluation of acoustic reflectometry for leakage and blockage detection. Proc. Inst. Mech. Eng. C J. Mech. Eng. Sci. **222**(6), 959–966 (2007)
9. Russell, D., Dawson, K., Lingard, L., Lennox, B., Papadopoulou, K.: Acoustic determination of remote sub-sea valve status. [257/79] POMME 2016 - Pipeline Operations and Management - Middle East conference 12–14 April, 2016, Kingdom of Bahrain (2016)
10. Guan, L., Gao, Y., Liu, H., An, W., Noureldin, A.: A review on small-diameter pipeline inspection gauge localization techniques: problems, methods and challenges. In: 2019 International Conference on Communications, Signal Processing, and their Applications (ICCSPA), Sharjah, UAE, pp. 1–6 (2019). https://doi.org/10.1109/ICCSPA.2019.8713703
11. Papadopoulou, K.A., Alzahrani, S.A., Sinha, J.: Pipeline inspection and maintenance via acoustic methods and tools. In: 3rd International Conference on Maintenance Engineering (IncoME-III), 6 and 7 September, Coimbra, Portugal. Conference Proceedings (2018). ISBN 978–989–8200–17–4
12. Yusoff, M.S.A., Sinha, J.K, Papadopoulou, K.A., Mandal, P.: Mono sensor empirical curve-fitting determination of acoustic attenuation coefficient for pipeline monitoring. In: 3rd International Conference on Maintenance Engineering (IncoME-III), 6 and 7 September, Coimbra, Portugal. Conference Proceedings (2018). ISBN 978–989–8200–17–4
13. Yusoff, M.S.A.M., Sinha, J.K., Mandal, P.: Sensitivity analysis of wave travel distance for different acoustic signals in pipelines. In: Proceedings of the 4th International Conference on Integrity, Reliability and Failure (IRF), July 2016, Porto, Portugal (2016)
14. Yusoff, M.S.A.M., Sinha, J.K., Mandal, P.: Sensitivity analysis of peak decay for acoustic input pulse in pipe inspection. In: Proceedings of 1st^{h} International Conference on Maintenance Engineering (IncoME-I), 30–31 August 2016. The University of Manchester, UK (2016)
15. Curkovic, A., Mlinaric, I.: Using GIS in pipeline repair process. GIS Odyssey Geographic Information Systems International Conference and Exhibition Proceedings, 4th–8th September 2017, Trento – Vattaro, Italy (2017)

Validation Framework of Bayesian Networks in Asset Management Decision-Making

Stephen Morey, Gopinath Chattopadhyay(✉), and Jo-ann Larkins

Federation University Australia, Northways Road, Churchill, VIC 3842, Australia
stephenmorey@students.federation.edu.au, {g.chattopadhyay, Jo-ann.larkins}@federation.edu.au

Abstract. Capital-intensive industries are under increasing pressure from capital constraints to extend the life of long-life assets and to defer asset renewals. Assets in most of those industries have complex life-cycle management challenges in aspects of design, manufacture, maintenance and service contracts, the usage environment, and changes in support personnel over the asset life. A significant challenge is the availability and quality of relevant data for informed decision-making in assuring reliability, availability and safety. There is a need for better-informed maintenance decisions and cost-effective interventions in managing the risk and assuring performance of those assets.

Bayesian networks have been considered in asset management applications in recent years for addressing these challenges, by modelling of reliability, maintenance decisions, life extension and prognostics, across a wide range of technological domains of complex assets. However, models of long-life assets are challenging to validate, particularly due to issues with data scarcity and quality. A literature review on Bayesian networks in asset management in this paper shows that there is a need for further work in this area.

This paper discusses the issues and challenges of validation of Bayesian network models in asset management and draws on findings from literature research to propose a preliminary validation framework for Bayesian network models in life-cycle management applications of capital-intensive long-life assets.

Keywords: Bayesian network · Asset management · Life extension · Maintenance · Reliability · Model validation

1 Introduction

Asset management is a field which covers the life cycle management of assets in order to achieve the business objectives. It is a multi-disciplinary field encompassing asset acquisition, maintenance, logistic support, risk management and business management.

Asset management in capital intensive industries is challenging due to continual pressure to manage asset related risks and to ensure the continued economic viability of the business. The scale and complexity of modern assets makes this a challenging task. As an example, consider a government operating a fleet of coastal patrol vessels. These vessels are an integrated system of interacting subsystems, which operate in

R. Karim et al. (Eds.): IAI 2021, LNME, pp. 360–369, 2022.
https://doi.org/10.1007/978-3-030-93639-6_31

hazardous environments, which must sustain and safely carry crew, operate without external support for extended periods, and must be sustained through long service lives. To aid asset managers in their task researchers have developed a range of advanced tools in prognostics and health management, modelling and decision tools for maintenance decision making, and the application of, data analytics, artificial intelligence and internet of things.

This paper focusses on the use Bayesian networks (BN) in asset management decision making. BNs are a type of directed acyclic graph, which model probabilistic relationships between events or variables (called nodes), and the conditional relationships between the nodes (called arcs). Associated with each node is a conditional probability table, which describes the relationship between each node. Figure 1 shows a simple BN and conditional probability table modelling the failure of a piping system, *S*, depending on the function of the relief valve, *R*, and system overpressure, *P*.

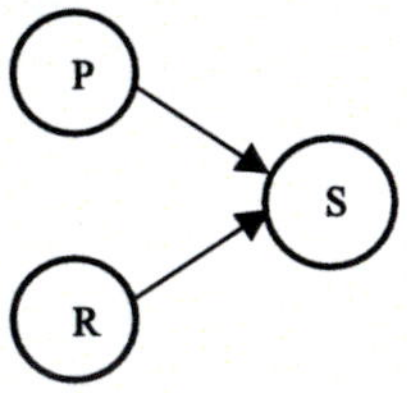

Fig. 1. Bayesian network for piping system

The conditional probability table for node *S* is shown in Table 1.

Table 1. Conditional probability network for piping system

	System overpressure (P)	Yes		No	
	Relief valve failed (R)	Yes	No	Yes	No
System failed (S)	Yes	1.0	0.01	0.0	0.0
	No	0.0	0.99	1.0	1.0

Along with the tables for nodes *P* and *R*, the probability of failure for the system, the marginal distribution, can be calculated as:

$$P(S) = P(P)P(R)P(S|P, R) \tag{1}$$

By the definition of a BN, the arcs do not imply causality [1], but are often interpreted in asset management applications to be causal.

BNs have been applied to a wide range of asset management applications, including reliability modelling, maintenance decision making, asset health prognosis, life extension and risk assessment. A review conducted by Weber et al. [2] noted that the publication of papers on BNs has rapidly increased since the 1990s.

The purpose of modelling in asset management applications is to recreate the system so the effects of interventions can be analysed, for example the effect of changing maintenance on life cycle cost or to identify weak points in system design [3]. Asset management modelling is needed as complex systems cannot be physically tested or experimented on due to their huge cost and scale of a real plant. For example, if one wants to understand the risk associated with operating a chemical plant, one cannot simply build and experiment on a test plant.

A model requires assumptions and simplifications compared to the actual system. This simplification is a necessary aspect of modelling to ensure model tractability. The process that determines whether the model is suitable enough for the intended purpose is model validation. Rosqvist [4], provides a useful definition of validation: "to make the model useful in the sense that the model addresses the right problem, provides accurate information about the system being modelled, and ensures that the model is actually used." Model validation allows the decision maker to understand the context in which their decisions are based on, and gives confidence in making decisions using those models. Despite the importance of validation in modelling, there is little discussion of validation in BN literature.

In the first instance, validation is necessary for establishing the credibility of BN modelling. The models used in traditional reliability prediction, for example block diagrams, have decades of sustained criticism. A quote provided from a report on reliability engineering efforts in the Apollo Program [5, p. 21], shows early criticisms of reliability prediction models: "In particular, many engineers look down on reliability as a kind of space-age astrology in which failure rate tables have been substituted for the zodiac." Other authors have also heavily criticised of traditional reliability modelling [6, 7]. A strong validation methodology could help BN modelling establish credibility and avoid this outcome.

Beyond credibility, the validity of BN models has real-world impacts on asset economics and risks. BN models are used to aid and better-understand interventions that impact on asset integrity and safety, availability, economic and environmental risks. These decisions need to be defensible and justified.

Asset management applications typically, although not always, are ultimately concerned with failure of equipment. This drives maintenance, spares requirements, safety, and risks that asset managers are interested in controlling. However, it is typical of modern assets that there is insufficient and low-quality data. The fact is, most of assets are generally quite reliable, and if they are not, then corrective actions are taken to address failures. Data scarcity and quality are common and well-known issues and discussed frequently in literature to build and validate models [7, 13].

Validation is often interpreted to be a model's fit with data [14], but model validation extends beyond this single aspect. The problem of data scarcity means validation must have a framework that does not solely rely on data fitting.

This paper reviews the validation problem in asset management, discusses the issues and background of validation, and proposes an overall framework for asset management BN validation.

1.1 Background on Validation

As the definitions of categories differs between fields, this list is provided as a guide for any further discussion [14–16].

- Nomological: Establishing the model fits within a wider knowledge domain as established by literature
- Face: Asking appropriate expertise if the model and/or its behaviour appears satisfactory.
- Content, structure or conceptual validity: Establishing the model's structure, theories, assumptions, mathematical and other relationships are appropriate for the model's purpose. Closely related is *model validity*, which demonstrates the conceptual model has been implemented correctly.
- Concurrent: That the model, or subsection of the model, behaves in a manner similar to a different model that is modelling the same type of system.
- Convergent: Comparing the model's structure, discretisation and parameterisation to other models on similar systems.
- Discriminant: Comparing the differences in the models to other models describing similar systems.
- Predictive, behaviour or operational: Demonstrating that the model behaviour concords with the real-life system behaviour. This can include historical data or future system behaviour.
- Data: Establishing the input data is appropriate and justified for the model's use.

Nomological, concurrent, convergent and discriminant validation will be referred to in this paper as *comparative* methods of validation.

2 Literature Review

Key references in the BN field [1, 17], and [18] only discuss validation of BNs in brief detail. For an analyst wishing to validate their asset management BN, there is unfortunately little help in the field.

Pitchforth and Mengersen [11] drew on the field of psychometrics to develop an overall framework for BN validation. Their paper outlines a general, high level framework for validation of BNs. They suggested that nomological, face, concurrent, content, convergent, discriminant and predictive validity should be present in a BN validation framework.

2.1 System Dynamics

System dynamics is a field that aims to model and simulate complex feedback dynamic systems. Conceptually, it has similarities with BNs, as both are aimed at modelling complex dynamic systems (in BNs, this is achieved by implementing dynamic BNs).

System dynamics has a rich field of literature on validation of system dynamics models, which dates back to the 1970's [19]. Researchers have outlined several key validation domains, and various validation tests. Roungas et al. [15] for example, list 64 different validation tests.

Barlas [20] discussed the fundamental validation of system dynamics models and argues that model validation should primarily focus on structure validity, with behaviour validity as a subsequent process once structure validity has been performed.

Sargent's [16] validation framework consists of data validity, operational validity and conceptual model validity which is also defined by Sargent and is similar to Barlas's structure validity.

It is noted that in system dynamics, what is understood to be a rational validation framework is based upon a philosophical argument of what constitutes knowledge and the scientific process. For example Barlas [20], justifies his validation framework by arguing that the philosophy of system dynamics accords to the relativistic and holistic philosophies of science, and therefore concludes that "model validity [is] not absolute and cannot be entirely objective and formal". See also [21] for a full discussion on these fundamentals.

2.2 Risk Analysis

A BN model in asset management is generally deals with risk assessment, and hence it is pertinent to review how the risk analysis field has approached validation of risk models. As with system dynamics, risk assessment literature contains a rich discussion on the validity of risk models [22].

Similarly to system dynamics, what is considered to be a cogent validation framework in risk analysis depends on the philosophical position the analyst takes towards risk. Goerlandt and Montewka [23] outline a spectrum of philosophical positions on what constitutes risk, in the broad categories of realist and constructivist. A risk realist interprets risk as a physical property of nature. A constructivist interprets risk as a shared mental construction. Later, Goerlandt and Montewka [87] reject the position that real risk estimation is possible (therefore taking a constructivist position), but did not explicate a validation framework in place of behaviour validation.

What then constitutes appropriate validation then depends on the philosophical risk interpretation one takes. A constructivist will not aim to validate a risk model with data, as this approach only makes sense if risk is interpreted to be a physical property one can measure, which a constructivist does not recognise.

This is evident in Aven and Heide's work [24] on a conceptual frameworks for the reliability and validity of risk analysis. They defined four validity criteria any risk assessment must meet:

- V1: The degree to which risk numbers are accurate compared to the true underlying risk
- V2: The degree to which assigned subjective probabilities describe the assessor's uncertainties of unknown quantities
- V3 The degree to which the epistemic uncertainty assessments are complete
- V4 The degree to which the analysis addresses the right quantities.

They do not refer to the spectrum of risk positions in [23], but refer instead to the more common frequentist vs. Bayesian positions on probability. They also note that what is considered appropriate validation criteria depend on the position one takes in respect to the meaning of probability.

3 Structure and Data Validation

As noted from our review findings of system dynamics and risk analysis, a correct validation methodology is ultimately determined by the underlying position one takes towards risk.

This paper considers the position that a BN risk model describes a constructivist position with uncertainty towards risk, as described by Goerlandt and Montewka [23]. As a BN is fundamentally based on Bayes' Theorem, the risk position should accord with the position in Bayesian statistics that probability, or risk, is not a property of nature, but a metric to quantify uncertainty. For an argument of this position in Bayesian literature see [25]. Validation then, should not necessarily attempt to compare the model against any 'true risk'.

Behaviour validation is challenging for two reasons. The first is due to data scarcity and quality issues, which means validation against historical data is difficult. The second is that asset management models often describe systems with long term or rare outcomes. For example, Ramirez and Utne [26], use a BN to model the life extension of a seawater pumping system from a design life of 30 years for an additional 20 years of extended life. For similar applications, analysts will not be able to collect sufficient data for behaviour validation.

Drawing on previous work in BN and system dynamics literature [11, 20], this paper proposes that a BN validation framework should primarily consist of structure and data validation.

3.1 Structure Validation

Structure validation establishes the model's structure, relationships, theories and assumptions. In a BN, this concerns aspects such as the node types and node states, model arcs and node relationships and causal assumptions. To meet Aven and Heide's [24] proposed validation criteria, the model must be constructed such that it models epistemic uncertainty.

The simplest form of validation is to check each structural aspect during and at the end of model development. It is important that validation is considered during model development as aspects may be missed if it is only conducted at the end. For example, a check of causal assumptions is much more difficult if the list of assumptions is not kept during model building. To this end, the validation checklist should be constructed prior to model development so that the appropriate information is recorded as the model is constructed.

Beyond a simple check, system dynamics offers a range of specific structural tests which can also be conducted and are more appropriate for large or complex models.

A validation framework should select a range of appropriate tests for the model being developed:

The *structure confirmation analysis* [20], is adapted to BNs by analysing the causal relationships in the model and confirming these causal relationships exist and accurately match the real world.

The *extreme-condition* test [20], inputs extreme values into the model and ensures the model behaves as expected in a real system. For example, extremely low or high failure rates for components are entered and the system behaviour is checked.

The *Turing test* [20], presents the output of the model and a real system and then, an expert is asked if they can distinguish them. In a BN, this could be accomplished by presenting outputs of components failure numbers, expected reliability and expected cost for a system, along with a similar list from the real system and asking if an expert can distinguish them.

Sensitivity analysis [27], systematically adjusts input variables and assess the effect on model behaviour. Any unexpected behaviour may indicate an issue in the model structure.

The *reality check* [20], checks the model output A given input B and ensures the model behaves as expected. In a BN, this could take the form of making a critical component unavailable and checking the system behaviour.

3.2 Data Validation

Data validation ensures the information in the conditional probability tables is correct and appropriate as follows:

- Data is supplied with a quantification of uncertainty, this is a requirement of Aven and Heide's [24] validation criteria for risk models.
- System conditions that produced the data are the same as for the modelled system.

Reliability is a system property. Failure data is dependent on system factors such as environmental conditions, maintenance, operating procedures, manufacturing quality and so forth. Any change in system condition from the time when the dataset was produced to the modelled system might mean the dataset is not appropriate for use.

Data validation is a challenge and very few tests exist to conduct data validation. Therefore, further research is required in this area. Following qualitative checklists are proposed:

- Has epistemic uncertainty in the data been quantified?
- Are the system conditions that produced the data, for example environmental conditions, the same as the conditions for the model?
- Have any changes, for example system configuration or operator procedure, occurred since the data was collected?
- Has the data been taken from the actual system?

An answer 'no' to any of these questions reduces the confidence in the validity of the input data. It is position of this paper that where data does not exist, it is preferable to honestly represent that uncertainty, rather than input data which may be inappropriate and give misleading model results.

3.3 Validation in the Modelling Process

This paper has proposed that the core of the validation framework for a BN model should consist of a selection of structure and data validation tests. These two types of validation assess different facets of the model, and while these are the most suitable to apply to a BN model, other types of validation, such as face and discriminant may also be suitable.

No matter what the specific methods used, it is important that the validation methodology is established early in the modelling process, and it needs to be agreed on by all stakeholders prior to the stage of model interpretation. The recommendation of system dynamics researchers is that validation should occur as a formal stage in the modelling methodology, but only after the model is at a sufficient state of maturity. An iterative cycle of model development and validation then follow until the model is satisfactory for its intended purpose [20].

The methods proposed so far, consisting of the comparative, structural and data validation methods, are all 'subjective' in the sense they ultimately rely on an individual's personal assessment that the validation has been adequately performed. This is contrasted with behaviour validation, which is more objective in the sense that the validation is a comparison between an external data set and the model. Due to the more subjective nature of the validation methods proposed, it is necessary to ensure all stakeholders agree the validation methodology is appropriate and will satisfy the outcome of validation. This consensus agreement is essential due to the constructivist position taken towards risk, in which risk does not exist as an external property of nature but as an assessment of uncertainty. Under this position, 'objectivity' is achieved by as broad a consensus as practically possible.

4 Conclusion

This paper has discussed issues and challenges of validation of BN models in asset management. The system dynamics and risk analysis fields were reviewed in order to understand approaches to validation in those domains. It was found that both fields have a rich field of literature on which to draw from and begin building a validation framework for BN models applicable for asset management.

This paper built on existing literature on BN validation and discussed aspects of structure and data validation of BNs in asset management. It was noted that the validation methodology needs to be clearly established early in the modelling process and should be agreed on by all model stakeholders.

Further research is needed in this area to expand and refine the types of tests, particularly in data validation.

Acknowledgments. Our thanks to the sponsors of IAI2021 Conference for their intellectual and financial support.

Stephen Morey is supported by an Australian Government Research Training Program (RTP) Fee-Offset Scholarship through Federation University Australia.

References

1. Jensen, F., Nielsen, T.: Bayesian Networks and Decision Graphs. Springer Science and Business Media, New York (2007). https://doi.org/10.1007/978-1-4757-3502-4
2. Weber, P., Medina-Oliva, G., Simon, C., Iung, B.: Overview on Bayesian networks applications for dependability, risk analysis and maintenance areas. Eng. Appl. Artif. Intell. **25**(4), 671–682 (2012). https://doi.org/10.1016/j.engappai.2010.06.002
3. Smith, D.: Reliability, Maintainability and Risk. Butterworth-Heinemann, Waltham (2011)
4. Rosqvist, T.: On the validation of risk analysis—a commentary. Reliab. Eng. Syst. Saf. **95**(11), 1261–1265 (2010). https://doi.org/10.1016/j.ress.2010.06.002
5. Childers, F.: History of Reliability and Quality Assurance at Kennedy Space Centre, KHR-20 (2004)
6. O'Connor, P.D.: Statistics in quality and reliability. lessons from the past, and future opportunities. Reliab. Eng. Syst. Saf. **34**, 23–33 (1991)
7. O'Connor, P.D.: Practical Reliability Engineering. Wiley, West Sussex (2002)
8. Kobaccy, K., Murthy, D.: Complex System Maintenance Handbook (Spring Series in Reliability Engineering). Springer-Verlag, London (2008). https://doi.org/10.1007/978-1-84800-011-7
9. Oniśko, A., Druzdzel, M.J., Wasyluk, H.: Learning Bayesian network parameters from small data sets: application of Noisy-OR gates. Int. J. Approximate Reasoning **27**(2), 165–182 (2001). https://doi.org/10.1016/S0888-613X(01)00039-1
10. Musharraf, M., Bradbury-Squires, D., Khan, F., Veitch, B., MacKinnon, S., Imtiaz, S.: A virtual experimental technique for data collection for a Bayesian network approach to human reliability analysis. Reliab. Eng. Syst. Saf. **132**, 1–8 (2014). https://doi.org/10.1016/j.ress.2014.06.016
11. Rebello, S., Yu, H., Ma, L.: An integrated approach for system functional reliability assessment using dynamic Bayesian network and Hidden Markov model. Reliab. Eng. Syst. Saf. **180**, 124–135 (2018). https://doi.org/10.1016/j.ress.2018.07.002
12. Scarf, P.A.: On the application of mathematical models in maintenance. Eur. J. Oper. Res. **99**(3), 493–506 (1997). https://doi.org/10.1016/S0377-2217(96)00316-5
13. Animah, I., Shafiee, M.: Condition assessment, remaining useful life prediction and life extension decision making for offshore oil and gas assets. J. Loss Prev. Process Ind. **53**, 17–28 (2018). https://doi.org/10.1016/j.jlp.2017.04.030
14. Pitchforth, J., Mengersen, K.: A proposed validation framework for expert elicited Bayesian Networks. Expert Syst. Appl. **40**(1), 162–167 (2013). https://doi.org/10.1016/j.eswa.2012.07.026
15. Roungas, B., Meijer, S., Verbraeck, A.: A framework for simulation validation and verification method selection. Presented at the SIMUL 2017: The Ninth International Conference on Advances in System Simulation, Athens, Greece (2017)
16. Sargent, R.G.: Verification and validation of simulation models. In: Proceedings of the 2010 Winter Simulation Conference, 5–8 December 2010, pp. 166–183 (2010). https://doi.org/10.1109/WSC.2010.5679166
17. Fenton, N., Neil, M.: Risk Assessment and Decision Analysis with Bayesian Networks. CRC Press, Boca Raton, Florida (2013)
18. Langseth, H., Portinale, L.: Bayesian networks in reliability. Reliab. Eng. Syst. Saf. **92**(1), 92–108 (2007). https://doi.org/10.1016/j.ress.2005.11.037
19. Sargent, R.G., Balci, O.: History of verification and validation of simulation models. In: Proceedings of the 2017 Winter Simulation Conference. IEEE Press, Las Vegas, Nevada (2017). Article 17

20. Barlas, Y.: Formal aspects of model validity and validation in system dynamics. Syst. Dyn. Rev. **12**(3), 183–210 (1996). https://doi.org/10.1002/(SICI)1099-1727(199623)12:3<183::AID-SDR103>3.0.CO;2-4
21. Barlas, Y., Carpenter, S.: Philosophical roots of model validation: two paradigms. Syst. Dyn. Rev. **6**(2), 148–166 (1990). https://doi.org/10.1002/sdr.4260060203
22. Goerlandt, F., Khakzad, N., Reniers, G.: Validity and validation of safety-related quantitative risk analysis: a review. Saf. Sci. **99**, 127–139 (2017). https://doi.org/10.1016/j.ssci.2016.08.023
23. Goerlandt, F., Montewka, J.: Maritime transportation risk analysis: review and analysis in light of some fundamental issues. Reliab. Eng. Syst. Saf. **138**, 115–134 (2015)
24. Aven, T., Heide, B.: Reliability and validity of risk analysis. Reliab. Eng. Syst. Saf. **94**, 1862–1868 (2009)
25. Jaynes, E.: Probability Theory: The Logic of Science. Cambridge University Press, Cambridge (2003)
26. Ramírez, P.A.P., Utne, I.B.: Use of dynamic Bayesian networks for life extension assessment of ageing systems. Reliab. Eng. Syst. Saf. **133**, 119–136 (2015). https://doi.org/10.1016/j.ress.2014.09.002
27. Balci, O.: Verification, Validation and Testing, in Handbook of Simulation: Principles, Methodology, Advances, Applications and Practice, J, Banks Wiley, New York (1998)

Enterprise Modeling for Dynamic Matching of Tactical Needs and Aircraft Maintenance Capabilities

Ella Olsson[1](✉), Olov Candell[2], Peter Funk[3], and Rickard Sohlberg[3]

[1] Saab AB, Nettovägen 6, 175 41 Järfälla, Sweden
ella.olsson@saabgroup.com
[2] Saab AB, Kungsörsvägen 60, 732 47 Arboga, Sweden
olov.candell@saabgroup.com
[3] Mälardalen University, Högskoleplan 1, 721 23 Västerås, Sweden
{peter.funk,rickard.sohlberg}@mdh.se

Abstract. The increasingly complex context of dynamic, high-tempo military air operations raises new needs for airbase aircraft maintenance and logistic support systems to more rapidly respond to changes in operational needs, with retained support resource efficiency. The future maintenance and support system are thus envisioned with improved net-centric capabilities to facilitate matching of tactical needs with aircraft maintenance capabilities.

Today military Command and Control (C2) of tactical needs against airbase aircraft maintenance capabilities contain many manual activities. This constrains speed of execution as well as drives manning requirements, and there is a need to further develop existing IS and IT support. Thus the studied matching capability address these limitations through an approach based on improved integration of the air vehicle on-board health management system with corresponding ground-based functions, and exploitation of technologies such as big-data analytics, diagnostic-prognostics, Artificial Intelligence (AI), machine learning and reasoning systems in a system-of-system-wide service architecture.

However, understanding the concept of matching of tactical needs and aircraft maintenance capabilities requires insights of complex multi-domain C2 interactions and interrelations between the tactical domain and the aircraft maintenance domain. Whereas each domain is quite well understood, the more detailed interrelations between the domains is less studied. This paper present an approach to this problem by creating useful representations of the underpinning insights, by enterprise modeling of abstract representations and definitions of relevant tactical and maintenance structures, processes, and resources, in the airbase context, to better understand the matching problem, and to address this problem through a holistic perspective.

Keywords: Enterprise modeling · Aircraft maintenance · Artificial Intelligence · System-of-systems

R. Karim et al. (Eds.): IAI 2021, LNME, pp. 370–383, 2022.
https://doi.org/10.1007/978-3-030-93639-6_32

1 Introduction

The capabilities and functionality of air combat forces are to large extent made possible by use of advanced technologies. This comprises of individual technologies, integrated in different aircraft and airbase systems, and together have a direct impact on how well air operations can be conducted. Air units form the core of air combat forces, but they should be considered as part of a larger system where supporting airborne and ground based systems and functions, are required to achieve desired effects. The military conflict in its nature, a unique enterprise in that it is characterized by decision-making that put both individual lives, and ultimately the existence of societies, at risk. Military decision-making - often under extreme pressure in highly dynamic environments - is thus formed not only by incomplete information, but also by the active hostilities and countermeasures of an aggressive opponent. Hence, as successful military operations are dependent on both time, timing and resources, shortened response times enables more flexible and efficient threat handling. From the perspective of the presented study, *tactical needs* are viewed as an air operation commanders expressed intent (through e.g. an order) to perform a number of air combat missions at certain points in time, and at the same time maintain a certain level of readiness for reactive, 'on-demand' missions. *Aircraft Maintenance Capabilities* in turn, are viewed as the ground support means at an airbase, required to perform the maintenance and support activities needed to produce mission ready aircrafts at defined times and places, according to the above order.

Hence, interest, and system needs, for air operations is also widened from the individual aircraft platform towards system-of-systems and interoperability. I.e. requirements on solutions where each system can exchange vast amounts of information and services with other systems, and operate frictionless with other systems from different vendors, with other friendly units and coalition partners. Thus, in this paper we suggest a modeling approach that enables matching of tactical needs and aircraft maintenance capabilities, based on new holistic insights on the potential of a system-of-system-wide service architecture. This by combining enterprise modelling with a net-centric approach to provide new possibilities for a more rapid and efficient matching of tactical needs against limited maintenance capabilities and resources. This matching process will require an improved digital integration of both air vehicle on-board health management systems, as well as corresponding ground-based functions, and further leveraged by technologies such as big-data analytics, diagnostic-prognostics, AI, machine learning and reasoning systems - in a system-of-system-wide service architecture. This to ensure a high operational efficiency without increased labor needs and at the same time improving quality of planning, execution and decision-making. Though the digital perspective requires introduction of partly new and disruptive technologies, such as machine learning and AI, recent years' experience indicates failure rates of big data projects in general and AI projects in particular that are disturbingly high [9]. Hence, the approach must pay due attention to key enterprise aspects when introducing these new technologies. In addition to new intelligent functionality, one must also address basic aspects of process improvement, data quality and data access, to achieve an increased operational efficiency. For this purpose, we propose a holistic development approach that also provides a deeper understanding of customer's core operational needs and unique military domain prerequisites.

Based on a case study of the Saab Gripen fighter aircraft system in a tactical air operation context, this paper describes a practical approach of the problem domain. The approach aim to produce a comprehensive, yet clearly articulated problem statement, by means of enterprise modeling. Enterprise modeling of an aircraft support system in its air operational and airbase context provides an architecture definition that represents the fundamental properties of the matching system in its joint tactical-maintenance environment, embodied in its elements, relationships and design principles. The purpose of the architecture definition process is to enable the generation of a set of system architecture alternatives that frame stakeholder concerns and meet system requirements, all expressed in a set of coherent views. These views and their corresponding elements provides a formal reference able to be used for more elaborate insights, military war gaming, operational analysis, modelling and simulation, etc., for a continued system development [17]. In this paper, we use enterprise modeling from three perspectives: 1. Top-down modeling of static high-level concepts as well as the high-level dynamics of the interactions between tactical air operations and maintenance operations. 2. A middle-out approach that models the maintenance system(s) and organization. 3. A bottom-up approach that focuses on operational use of data-driven and AI-based decision support solutions and services, that facilitate dynamic matching of tactical needs and aircraft maintenance capabilities.

2 Related Work

The top-down approach uses a set of selected operational and capability views from the DoDAF (US Department of Defense Architecture Framework) [3]. It includes the CV-2 high level Capability Taxonomy, the OV-1 High-level Operational Concept Description, followed by an operational node connectivity description (OV-2) and ends in the OV-5; a high-level dynamic view of command, air, and maintenance operations. The middle-out approach comprises of a set of system and organizational views related to the UAF (Unified Architecture Framework) [2] and DoDAF. These views are: the high level maintenance organizational relationship chart (OV-4), loosely based on publicly available air operation doctrines [7, 14], accompanied by the system view SV-1 and its corresponding SV-4, again loosely based on the publicly available air operation doctrines [6, 7, 14]. The bottom-up approach uses a set of proposed aircraft related data-driven and AI-based decision support solutions and Digital Twins [5] able to performs real-time simulations and condition assessment [4] of the aircrafts and their components as well as the maintenance resources.

3 Why Enterprise Modelling?

Enterprise modelling is an abstract description and representation of an organization including processes, recourses, information and models of a specific enterprise. Modeling and visualizing an enterprise enables both new insights, improvements, digitalization and optimizations. An Enterprise in a military context can be seen as delivering a service, providing defense capabilities (e.g. a service able to carry out air operations including maintenance and C2 according to set up requirements).

In order to use Enterprise Models in AI based decision support systems, they need to be formal and have a clear semantic. This is done by adopting an ontological approach and to further model relevant parts of the enterprise model as ontologies. Ontologies are commonly based on first order logic or description logic to which the enterprise model can be translated. Formalization also enables the potential of validating current service levels against requirements and identify divergences in resources, maintenance capabilities and logistic capabilities.

4 Aircraft Maintenance

Aircraft maintenance consist of a large number of maintenance operations scenarios, and their relations and resource requirements. A basic example is to replace a broken tire, where the resources needed is a person able to change or repair the tire, tools, a replacement tire or repair kit, a suitable place and time to carry out the maintenance operation.

Aircraft maintenance is a suitable application for Enterprise modelling since there are clear processes, data models, resource models, as well as extensive and well-documented domain knowledge. Modelling aircraft maintenance in a formal (graphical) notation enables matching the current maintenance capacity against the current tactical needs within given time constraints, expressed in a formal notation. The abstraction in the Enterprise Model also enables strategic planning and improvements and/or reallocations, of capacity to where it can provide the most prioritized effect.

5 Enterprise Modeling of Air and Maintenance Operations - Top-Down Approach

In this chapter, we show how the integration of tactical air operations and aircraft maintenance operations can be modelled with an Enterprise Modeling tool, starting with a top-down approach. (C2). The created top-down representation provides insights about C2 rationales and principles (doctrine, 'what' and 'why' about air ops. in general) that shapes the underlying organizational structure and applied processes. It identifies relations with other operational nodes and interfaces, where the commanders intent ('orders') are communicated and reports received, as well as how more detailed information and data about tactical needs (e.g. Air Tasking Order (ATO)) and maintenance capabilities are exchanged with lower organizational units.

5.1 CV-1: Capability Taxonomy

The Capability Taxonomy models the top-level capabilities of air and maintenance operations, including planning, control and execution of military air operations (Fig. 1).

The semantic in the diagram is that the upper capability aggregates a number of sub-capabilities (the diamond is the aggregation symbol).

Support Operational Air Units: Maintaining military air operations is the main capability of the aircraft maintenance domain. This capability includes the ability to set

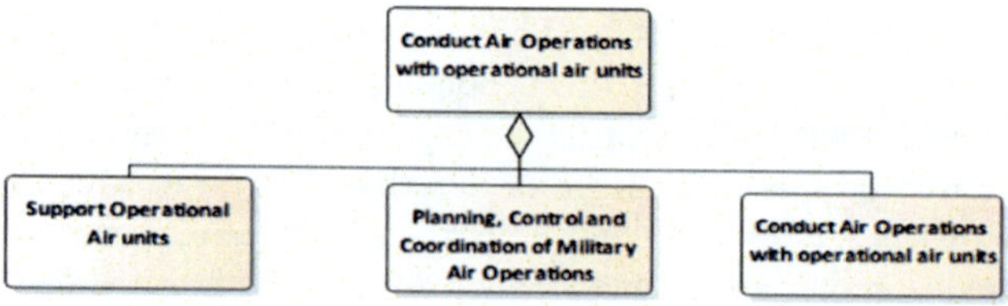

Fig. 1. Top-level capabilities of air and maintenance operations.

up and maintain an air base and serve flight crews. It also includes the ability to plan, execute and follow-up of maintenance operations.

Planning, Control and Coordination of Military Air Operations: The basis for planning and coordination of military air operations is to have the ability to maintain an event-driven readiness and an ability to concentrate resources in time and space in order to uphold territorial integrity of an operational area and to protect related national interests.

Conduct Air Operations with Operational Air Units: This capability includes the ability to plan, execute, evaluate and report military air operations. We will focus on the capability to perform air operations with the JAS Gripen combat air system. The types of missions that the JAS Gripen system is capable of is depicted in the figure below, shortly explained thereafter. All imposing a unique set of type specific requirements and restrictions on the utilization of the related maintenance resources (Fig. 2).

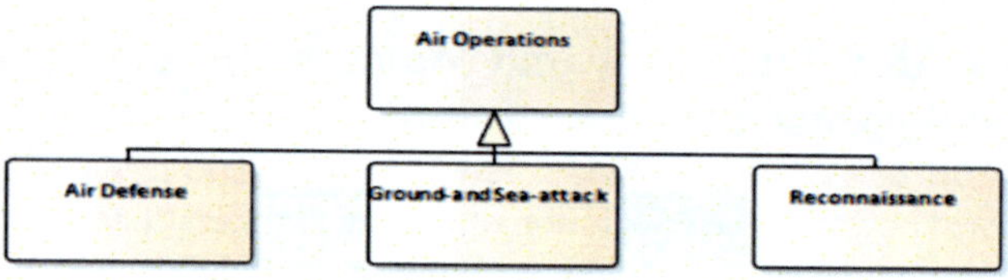

Fig. 2. This JAS Gripen mission taxonomy.

Air Defense – This is the most basic task of the air combat forces. Air defense aims to maintain territorial integrity, ensure the protection of the population; protect important social functions and infrastructure and support other parts of The Armed Forces. It can be divided into defensive and offensive air defense.

Ground- and Sea-attack – Air attack missions strikes targets on the ground and at sea. When carried out as joint efforts with land and naval forces, attack missions usually need to be highly coordinated on both operational and tactical level.

Reconnaissance – Intelligence gathering missions carried out within the air combat forces using aircrafts.

5.2 OV-1: High-Level Operational Concept Description

The OV-1 shows the main operational concepts. It provides a graphical depiction of what the architecture is about and an idea of the actors and operations involved [10]. In this paper OV-1 is a high-level operational concept description of an air and maintenance operations scenario, capturing basic air and maintenance scenario entities and their high-level relations, described in the following sub- sections.

The Air Operations Staff (SE: Flygtaktisk Stab, FTS) and the Commander of Air Operations (Swedish (SE): Flygtaktisk chef, FTCH) command air operations (applying C2 processes) such as Air defense, Attack, and Reconnaissance operations, among other. FTCH's is the level of command above the Air Wings (SE: Flottilj) and fighter squadrons (SE: Stridsflygdiv.).

The Air Wing include units with the ability to set up and maintain one or more air bases to support to Operational Air units. It includes planning and follow-up of tactical level flight-line servicing, maintenance and other base operations. An Air Wing may be co-located, but it can also be divided in smaller units that are geographically dispersed.

The fighter squadron is an operational air unit typically deploying 6–8 aircraft and pilots that carry out various air missions. When deployed, the fighter squadron operates within a mission cycle called a tactical loop (Fig. 5). The tactical loop is initiated by an Air Tasking Order (ATO, emanating from the Air Operations Staff), whereby the individual air mission assignments are planned, prepared and executed, and finally evaluated and reported.

5.2.1 Orders and Reports

Air Tasking Orders (ATO) are issued for air operations and defines individual air missions. The ATO is released as soon as possible in order to enable subordinate units to perform needed planning, coordination, collaboration and manage logistics (Fig. 3).

The Tactical Order (FTO) normally contains FTCH decisions about the operational objective, tactical idea and the disposition of forces, the division and deployment of units, tasks and an update of the operational situation. The FTO regulates both air operations and base operations, and controls resource allocation and prioritizing, use of air bases and logistics. The FTO is issued as necessary. At lower levels of readiness it applies for longer time periods and at higher readiness levels it is published every 24 h or as needed.

Reporting. The cyclic reporting to FTS by subordinate units is the basis for FTCH and FTS's command and follow-up of the operations. A well-functioning reporting is absolutely crucial for FTCH control of ongoing operations, as well as for planning, and continuous development of orders. Reports are submitted to FTS after which internal briefings within FTS are conducted. The reports are then handled and processed by the respective staff functions as required and depending of type of report.

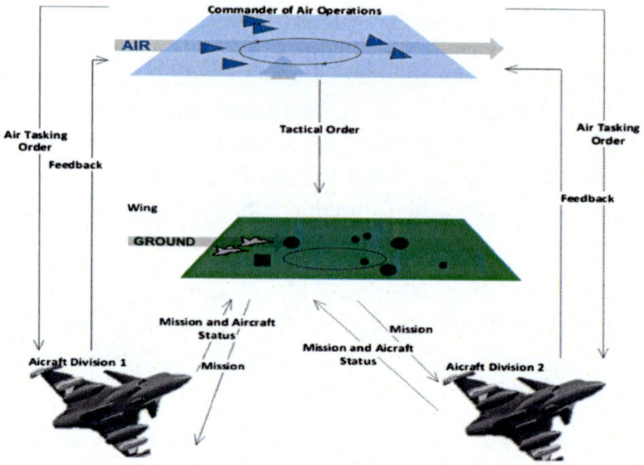

Fig. 3. OV-1 High level flow of command.

5.3 OV-2: Operational Node Connectivity Diagram

The OV-2 depicts operational nodes, which are logical collections of operational activities. Operational nodes produce or consume information and may realize a set of capabilities. The main features of OV-2 are the nodes, the links between them, and the characteristics of the information exchanged [11]. In this work the operational nodes models the main entities of the OV-1, namely the Air Operations Staff (FTS, HQ), the Fighter Squadron and the Air Wing. It also models the main operational activities of each node and the information flow in-between the nodes such as orders, reports and status updates. The OV-2 is depicted in Fig. 4.

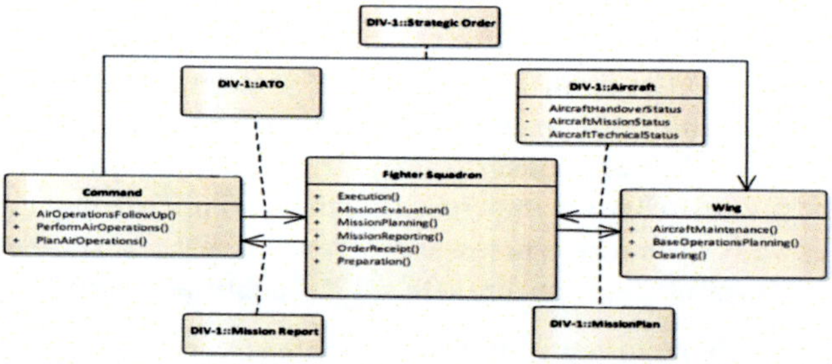

Fig. 4. OV-2, operational nodes (collections of activities).

5.4 OV-5: Operational Activity Model

The NATO definition of the OV-5 (or NOV-5 as in NATO abbreviation) is "The Operational Activity Model describes the operations that are normally conducted in the course of achieving a mission or an operational objective" [11]. In this work, it describes the high-level dynamics of the activities that normally are conducted in the course of achieving one or more tactical air missions. It describes operational activities, as well as the flow of control and dynamics of these high-level activities as depicted in the figure below.

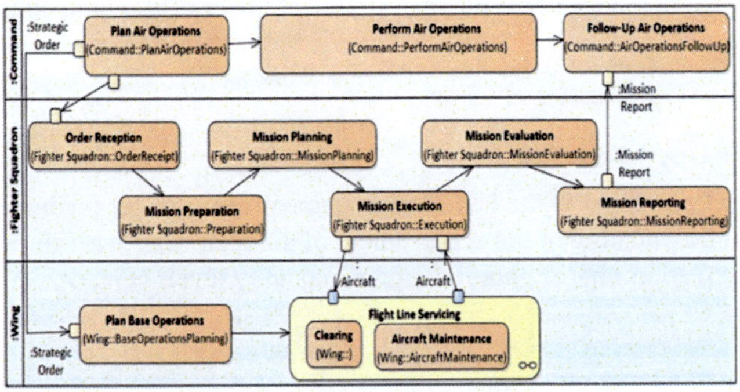

Fig. 5. Flow of control and dynamics of high-level activities.

The swim lanes depicted in Fig. 5 represents the operational nodes described in the previous, but here their respective activities are modelled with the aspect of time and synchronized with the interaction between the nodes. The activities of each operational node are further described in the below sections.

The FTS and FTCH (Tactical Staff and Commander). In order to execute air operations, a number of factors must be considered. To carry out air operations, the armed forces act in a tactical loop that includes operational planning for missions and how missions are conducted and evaluated.

Planning and execution of air operations follow a tactical loop that defines the lifecycle of a mission, from when the mission starts to when it finished. It also includes the data and its analysis [15].

Air Operations Follow-Up. After mission execution, the results of a mission are analyzed, summarized and sent to command for further interpretation. This can in result spawn new measures and the conduct of new air operations can be based on these results. Can also result in specific maintenance measures to ensure availability of all weapons systems for the next mission [15].

6 Middle-Out Approach

The middle-out approach comprises of a set of system and organizational views, namely the OV-4, SV-1 and corresponding SV-4. The middle-out approach enables knowledge-building about the organizational context, structures and processes that transforms order from higher command into actual operations ('what' and 'how' about specific air ops. and missions). It identifies relations with yet other operational nodes and interfaces, and additional insights on how tactical needs (in terms of orders) are transformed to more specific mission production requirements and plans, to finally enable their matching towards actual maintenance capabilities available at lower organizational units.

6.1 OV-4: The Organizational Chart

6.1.1 The Air Operations Staff

The Air Operations Staff is the FTCH management resource for planning, execution, following up and evaluating of the achievement of air operations with air combat forces and, if necessary, additional support units in all standby levels (Fig. 6).

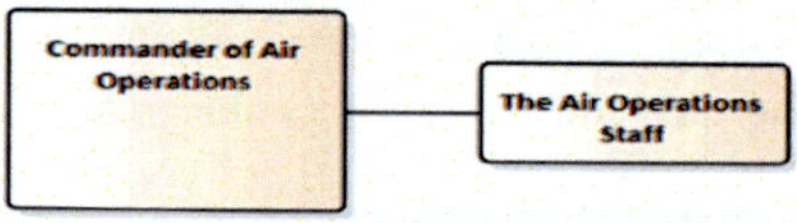

Fig. 6. Air operations Staff organization.

The Commander of Air Operations (FTCH) exerts a command line against its war allies. The commanding is over several timescales from hours to months. It includes of FTCH decisions in large and in tasks. The purpose of FTCH command is to create endurance and freedom of action with the resources needed to carry out the planned air operations.

6.1.2 The Airbase Organization

The wing is responsible for the air bases in an air base group. The wing serves and supports the fighter squadrons as specified by the Commander of Air Operations with regard to orders such as *flight Tactical Order* and *ATO* (Fig. 7).

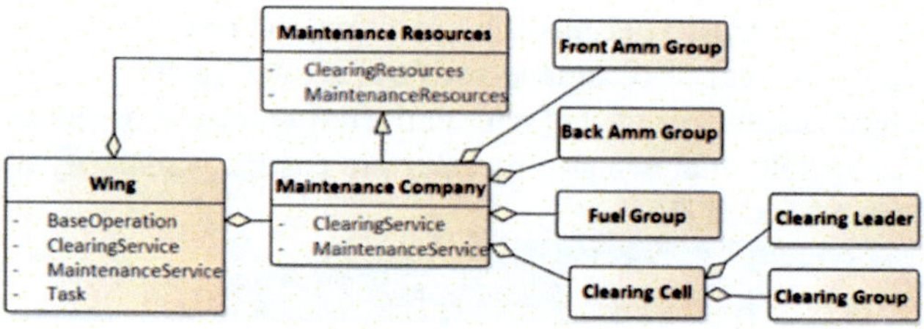

Fig. 7. The air wing organization.

The Aircraft Maintenance Unit. The task of the unit (Company-level) is to operate the fighter aircrafts, which includes Pre-/Post-flight servicing; Aircraft maintenance and other support tasks required to operate at the airbase. Their main function is flight line servicing, and related services for Aircraft, Munitions, Base equipment, Transport etc.

The Fighter Squadron and Organization. A fighter squadron consist of Squadron Command; Fighter Unit(s) and Mission Support Element(s) (MSE). A squadron may as needed be divided to operate from different locations (Fig. 8).

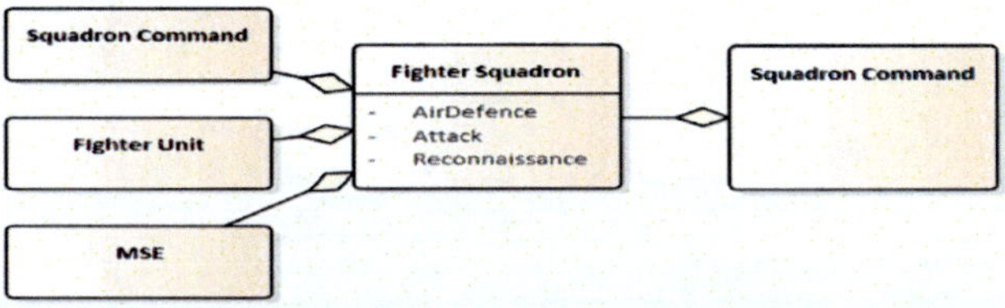

Fig. 8. Example of task organizations

6.2 SV-1: The Air Base System

The purpose of the air base system is to support the fighter squadron in the task of performing air operations. This includes providing landing and take-off capability for fighter aircrafts and protecting the air bases from various types of incidents, threats or attacks. The air base system includes Fighter squadrons; Air base groups; Resources such as aircraft ammunition, aircraft spare parts, fuel, ammunition etc.; Methods and tactics used for the use of the system. Airbase types (Main bases, Side bases, Reserve bases) are differentiated by infrastructure, number of runways, protection and logistics supplies. E.g. a Main base can be used 24/7 and in all weather conditions by the types of aircraft operating at the air base (Fig. 9).

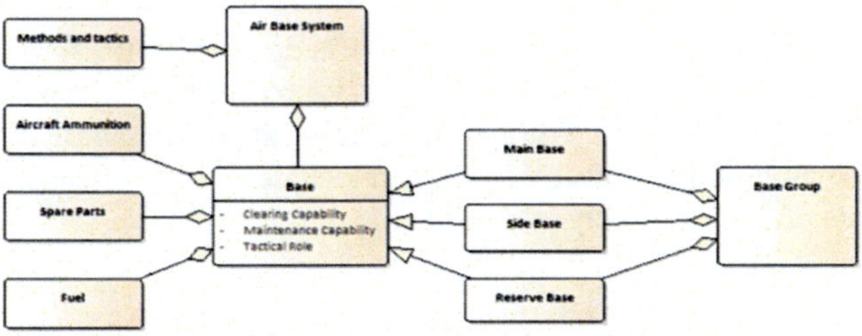

Fig. 9. SV-1: the air base system.

A side base normally has the same capacity as the main base but has none or one a smaller proportion of air base units are grouped on the base. Reserve bases normally have lower capacity than main and side bases. The airbase also usually needs preparatory work to function, e.g. preparation for using the roads as runways, and an alternative for increasing the dispersion of fighter squadrons (Fig. 10).

6.3 SV-4: Prepare Flight Line Servicing

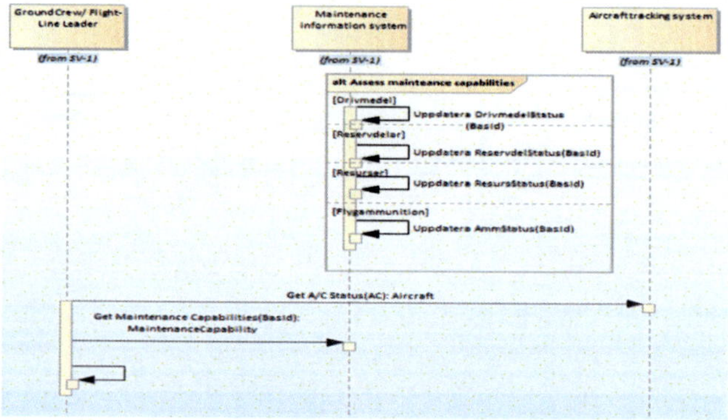

Fig. 10. Example of Flight Line services.

This section describes the concept of execution among the software components. SV-4 provides opportunities to model use of actual resource types at an airbase.

The Ground Crew/Flight-Line Leader. The Ground Crew/Flight-Line Leader functional node performs maintenance, e.g. Pre-/Post-flight Servicing (Sv: Taktisk Operativ flyguh.tjänst,).

The Maintenance Information System. Is a database that keeps track of the airbase maintenance capability status by means of continuously updating its database when change occurs on the airbase. This can be addition/removal of resources (workers and consumables), spare parts, ammunition, fuel etc.

The Aircraft Tracking System. Is a database that keeps track of the squadron aircrafts, including the aircraft model, its mission status (mission type and weapon config.) and its technical status (fuel for now). The aircraft tracking system database is continuously updated with the ambition to reflect mission and technical status of each aircraft in the squadron as accurate and as timely as possible.

7 Bottom-Up Approach

The bottom-up approach identifies the interdependencies (matching) between the definitions for specific mission production requirements and plans, and specific maintenance capabilities at a certain time and location ('how'). I.e. between the tactical need (expressed as a mission ATO) that have to be fulfilled (through sorties by defined, individual aircraft configurations), and the status information about actual resources at an airbase (servicing and maintenance units and personnel, maintenance equipment, facilities, spares, e t c.) defined as maintenance capabilities. We abstract these entities through a set of data-driven services as listed below (Fig. 11).

Fig. 11. Matching management with reality.

Logistic situational awareness (LSA) maintaining representations of the state and related spatial aspects of ingoing architecture entities. This function has the capability to increase combat effectiveness and to extend the operating time of the fighter aircrafts by the following capabilities [16]: Provide asset status; Provide base resource status such as spare parts, fuel, resources and organizational resources etc.; Provide maintenance schedule.

Tracking of relevant architecture entities such as aircrafts, maintenance resources etc. Among others, it provides the following asset tracking functions [16]: Asset technical and operational status; Asset configuration status; Asset logs and history; Etc.

Movement commanding of spatial transitions of architecture entities related to maintenance.

Planning functionality related to maintenance operations within the context of air operations. The planning functionality has the capability to increase mission ready time and it provides, among others the following functions [16]: Provide maintenance schedule functions such as update maintenance schedule; Include manufacturer scheduling in maintenance scheduling; Include assets, base and resource status in the scheduling algorithm; Optimize scheduling based on the above features.

7.1 Graphical Modelling and Matching

A graphic modelling language is a data modelling language used to capture the semantics, i.e. the meaning of the data without having the focus on the data itself.

Information modeling is the next step of detailing parts of the enterprise model. An information model describes, often in formal way concepts, relations between concepts, rules, operations and constraints for the domain. Data structures offer a way of organizing and managing data including relations between data, storage, retrieval and modification. In many data structures relations between data is hidden in programming code and it is system programmers' responsibilities to keep track of relationships in program code. Using e.g. conceptual modeling enables the separation of relations between data and program code. This enables automated transformation and formal connection between models and implementations in a net-centric approach [18] (Fig. 12).

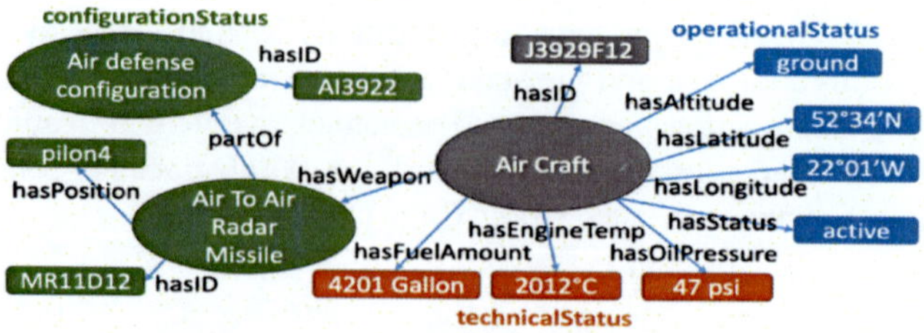

Fig. 12. Example of an instantiated ontology.

Ontologies are efficient graph structures, developed after the realization that knowledge representation is a key in building intelligent AI systems. They are able to capture essential domain knowledge in a formal way in contrast to programming languages to often hiding the semantic and domain knowledge in manually produced programming code. Ontologies formally describe specific domain, categories, properties and relations.

An ontology can be instantiated with objects and values in a graph database and opens up the potential of matching the physical world represented in the data structure to validate which military air operations can be carried out. The instantiated ontology also makes up of a considerable part of the knowledge about the current status of particular aircrafts in the Digital Twins concept.

8 Summary and Conclusions

We have outlined the different levels from Management to Reality and how they relate and how they are linked using a set of architecture framework views modeled from a top-down and a middle out approach. With these results, we have been able to identify key interfaces between the maintenance and the tactical air domains. These key interfaces has further aided us in providing requirements for continued research and development of a data foundry platform along with a set of services able to form a basis for decision support for operational maintenance activities and offer the capability to validate, if current maintenance capabilities meet the requirements of air operations. Further, this foundry has the ability to be used for simulation purposes to offer predictive properties.

Acknowledgments. Our thanks to VINNOVA NFFP7 (2017-04880) and SAAB for their funding of this research project.

References

1. Wagenhals, L.W., Levis, A.H.: Service Oriented Architectures, the DoD Architecture Framework 1.5, and Executable Architectures. Wiley InterScience (2008)
2. UAF Architecture Framework. https://www.omg.org/spec/UAF
3. DoDAF architecture framework. https://dodcio.defense.gov/library/dod-architecture-framework/
4. Holmes, G., Sartor, P., Reed, S., et al.: Prediction of landing gear loads using machine learning techniques. In: 2016 Structural Health Monitoring, pp. 568–582 (2016). ISSN 1475-9217

5. Castaño, M., Karim, R.: Digital Twins in Maintenance. Luleå University of Technology (2018)
6. Fbassystem, R.: Reglemente Flygbassystem. Försvarsmakten, M7739–353130 (2018)
7. Reglemente Taktik för Luftoperationer TR LuftOp 2017, Försvarsmakten M7739-353126 (2017)
8. Roslund, A.: Using domain knowledge functions to account for heterogeneous context for tasks in decision support systems for planning. B.Sc. thesis, Mälardalen University (2018)
9. Thomas, C.: Do Your Data Scientists Know the 'Why' Behind Their Work? Redman. Harvard Business Review, 16 May 2019. https://hbr.org/2019/05/do-your-data-scientists-know-the-why-behind-their-work. Accessed 12 Jan 2020
10. https://dodcio.defense.gov/Library/DoD-Architecture-Framework/dodaf20_ov1/
11. http://trak-community.org/index.php/wiki/NAF:NOV-2_Operational_Node_Connectivity_Description_Subview
12. Gruninger, M.: Enterprise Modelling. Chapter 16, Handbook of Enterprise Architecture, December 2007. http://stl.mie.utoronto.ca/publications/modelling.pdf
13. Chmielewski, M.: Ontology applications for achieving situation awareness in military decision support systems. In: Nguyen, N.T., Kowalczyk, R., Chen, S.-M. (eds.) ICCCI 2009. LNCS (LNAI), vol. 5796, pp. 528–539. Springer, Heidelberg (2009). https://doi.org/10.1007/978-3-642-04441-0_46
14. Joint Air Operations, Joint Publication 3–30, 25 July 2019. https://www.jcs.mil/Portals/36/Documents/Doctrine/pubs/jp3_30.pdf
15. https://saabaircraftindustry.com/en/roads-to-new-capability/value-for-customers/tactical-loop-operational-capability/
16. IBM Technical Workshop, Gain an Information Advantage with Architecture Using IBM System Architect and DoDAF 2, Lab Exercises, Control Number: 09-15-2014-Version
17. INCOSE Systems Engineering Handbook, 4th edn. (2015)
18. Walch, M., Karagiannis, D.: How to connect design thinking and cyber-physical systems: the s* IoT conceptual modelling approach. In: Proceedings of the 52nd Hawaii International Conference on System Science, January 2019

Design and Economic Analyses of Wind Farm Using Meta-heuristic Techniques

Suchetan Sasis(✉), Sachin Kumar, and R. K. Saket

Department of EE, IIT (BHU), Varanasi, UP, India
{suchetans.eee16,sachinkumar.rs.eee18,rksaket.eee}@itbhu.ac.in

Abstract. The application of the conventional non-linear methods to obtain an optimized power output for a wind farm using small scale wind turbines may lead to production of local maxima or local minima. Thus, an optimized solution has a restricted space as far as applied constraints are concerned. There can be a wide variation in the applied constraints based on site-specific parameters variations. In this paper, the Meta-Heuristic approaches (MHA) are implemented with suitable constraints and accordingly derived equations are compared. In these approaches, it is possible to generate a random sample space, which is not affected by local maxima or minima as in non-linear methods. The three methods of MHA, namely (a) Crow search, (b) Harmony search, and (c) Gravitational search have been compared in this paper. The comparison shows that an optimized power output and cost are obtained. Further, it is concluded that GSA provides and optimal and feasible solution for design of small scale wind turbine being used in wind farm.

Keywords: Wind turbine · Non-linear mathematical model · Optimization · Meta-heuristic · HSA · CSA · GSA

1 Introduction

According to the United Nations projections, the number of people on the earth is likely to increase by 2 billion from the current 7.7 billion to 9.7 billion by 2050. An effective means to provide sustainable energy would be to improve the performance of renewable energy assets in order to derive maximum benefits from their applications [1]. An overview of the literature provides many such efforts to optimize wind farm's performance [2, 11–14]. A discussion on the development costs, cost of energy, the value of energy, type of credits and break-even costs is described elsewhere [3], which has been briefly discussed in this paper too. It is to be noted that many efforts have been put in layout optimization of a wind farm. Nevertheless, optimization of the layout is important. Here, we have identified important parameters dominantly affecting the performance of a wind farm.

A small scale turbine, i.e. a turbine which generates less than 10 KW of power, unlike conventional wind turbines in which layout as well as parameter optimization needs to be applied. Small scale turbine are more compact in nature and there is an observably significant difference between the magnitudes of D and h.

R. Karim et al. (Eds.): IAI 2021, LNME, pp. 384–392, 2022.
https://doi.org/10.1007/978-3-030-93639-6_33

In view of enhancing power output of such turbines, two parameters are singled out and then the meta-heuristic approach can be gainfully applied. In this paper based on the literature review and the application of engineering judgment, two parameters, namely Hub-height (h) and Diameter (D) of turbines were singled out for the application of meta-heuristic techniques here in this paper. The meta-heuristic approach has been conceived after analysis of a naturally occurring and repetitive process with the use of computational methods like programming on MATLAB 2016A software to optimize performance in case of wind farms as has been briefly discussed here. Out of many available meta-heuristic approaches, the Harmonic search and crow search methods have been described. Application steps for each have been provided in this paper after defining the applicable equations. This paper, moreover, while concluding highlights conclusions of CSA and GSA are comparable with more confidence in optimizing power output in comparison to HSA.

The rest of the paper is as follows. The basics of wind farm analysis including design and economic are discussed briefly in Sect. 2. Section 3 explains the three meta-heuristic methods implemented in this paper. Section 4 shows a comparative analysis and provides an insight on design and cost analyses of wind farm using a small scale wind turbine. At last, conclusion with future scopes is described in Sect. 5.

2 Wind Farm Analyses Using a Small Scale Turbine

2.1 Design Analysis

A moving object with mass has:

$$\text{Kinetic Energy} = (1/2) * \text{Mass} * (\text{Velocity})^2. \tag{1}$$

The wind power is mathematically expressed as described elsewhere given by [4], with the assumption that the there is a theoretical conversion of the complete wind energy to kinetic energy:

$$\text{Power} = (1/2) * \text{Swept Area} * \text{Air Density} * (\text{Velocity})^3 \tag{2}$$

where power is given in watts, the Air density in kilograms per cubic meter, the velocity in meters per second and the swept area in square meters.

$$\text{Swept Area} = (\text{Radius})^2 * \pi \tag{3}$$

The swept area is the area covered by the turbine rotor, i.e. the area swept by the blades while rotating. For a wind turbine, an exponential variation in wind speed with hub height may be defined relative to wind speed measured at a reference height. The reference height is usually chosen to be 10 m. Thus, the equation will be:

$$Vw(h) = V10(h/H(10))^{\mathbf{a}} \tag{4}$$

$V_w(h)$ = velocity of the wind in [m/s] at an arbitrary height of h. V10 = velocity of the wind at height of H(10) = 10 m, $\boldsymbol{a}$ is the Hellman exponent. The exponent, $\boldsymbol{a}$, is an empirically derived coefficient that varies with the stability of the atmosphere. For neutral stability conditions, ***a*** is approximately 1/7, or 0.143. Based on (1), (2), (3), and (4). The wind density is assumed to be 1.23 kg/m^3 at mean sea level. The final mathematical model, which is the objective function for this study is reproduced hereby [4].

$$\text{Max } P(D, h) = 11.22 * D^2 h^{(0.429)} \tag{5}$$

The above Eq. (5) has constraint equations which are as follows

$$\begin{aligned} &0.06 * D^{2.96} + 2.65 * D^{2.35} + 0.48 D^{2.65} + 103.05 * D + 303.96 * D^{1.06} \\ &+ 0.01 * D^{2.88} + 3.17 * D^2 h < 360089.4 \end{aligned} \tag{6}$$

As understandable from Eq. (6) the constraints are governed primarily by rotor diameter (D) along with the last term inclusion of hub height. It is also noted that the Fig. 360089.4 represents the total cost of installation of a mid-sized wind turbine in US Dollars [4]. As Eq. (6) is not sufficient as far as constraints are considered. It would be worthy if Eq. (6) is complemented with the following equation in order to prevent the rotor blade from striking the ground.

$$h > (D/2) + 3 \tag{7}$$

2.2 Economic Analysis

The rotor consists of four main components as detailed elsewhere [5], Blades, Hub, Pitch mechanisms and bearings Spinner, nose cone. The blade cost consists of blade material costs along with labor costs, which both are related to the rotor radius estimated by the following relationship:

$$\begin{aligned} &\text{Blade cost} = \text{blade material cost} + \text{labor cost} \\ &= [(0.4019 * R^3 - 21051) + (2.7445 * R^{2.5025})]/(1 - 0.28) \end{aligned} \tag{8}$$

The estimated relation assumes 28% of overhead. The hub cost is related to the hub mass, which is calculated from the following approximate relation:

$$\text{Hub-mass} = 0.945 * (\text{blade-mass}/2.61) + 5680.3 \tag{9}$$

Where the blade mass is estimated as:

$$\text{Blade-mass} = 0.4948 * R^{2.53} \text{ per blade} \tag{10}$$

The overall cost of pitch bearing for a three-blade turbine is calculated as a function of rotor diameter

$$\begin{aligned} &\text{Cost of the complete pitch system for three blades is} \\ &= 2.28 \times \left(0.2106 \times \text{rotor diameter}^{2.6578}\right) \end{aligned} \tag{11}$$

The cost of a Nose cone depends on nose cone mass

$$\text{Nose cone mass} = 18.5 * \text{rotor diameter} - 520.5 \quad (12)$$

$$\text{Nose cone cost} = \text{nose cone mass} * 5.57 \quad (13)$$

The calculation of Total costs related to the installed components, the sum of costs Eqs. (8–13) total cost of the rotor as in Eq. (6) or the constraint equation can be developed.

3 Meta-heuristic Methods

The word "heuristic" is a Greek word and it means "to know", "to find", "to discover" or "to guide an investigation [6]. Specifically, "Heuristics are techniques which seek good (near-optimal) solutions at a reasonable computational cost without being able to guarantee either feasibility or even optimality in many cases to state how close to optimality a particular is the feasible solution" [7]. Mathematical optimization (MO) or mathematical programming is the selection of the best element (with regard to some criterion) from a set of available alternatives. Optimization problems of sorts arise in all quantitative disciplines [13].

In the simplest case, optimization imponderables consist of maximizing or minimizing a problem-defining real function in a case-by-case basis by systematically choosing input values from within an allowed set and computing the value of the function.

The generalization of optimization theory and techniques to other formulations constitutes a large area of applied mathematics. More generally, optimization includes finding "best possible" values of the respective objective function, given a defined domain (or input-range). Thus, generalization includes a variety of assorted types of objective functions and different types of domains. Among numerous types of search methods, some notables are: Dragonfly Algorithm (DA), Biogeography-based optimization (BBO), Brain Storm Optimization Algorithm (BSO), Elephant Herd Algorithm (EHO), Evolutionary Strategy (ES), Earthworm Optimization Algorithm (EWA), Flower Pollination Algorithm (FPA), Krill Herd algorithm (KH), Lightning Search Algorithm (LSA), Monarch Butterfly Optimization (MBO), Moth-Flame Optimization algorithm (MFO), Population-Based Incremental Learning (PBIL), Sine Cosine Algorithm (SCA), Whale Optimization Algorithm (WOA), The three search methods are used in this paper as described in this section.

3.1 Harmonic Search

Harmonic search (HS) first developed by [8] in 2001. Harmony search is a music-based or sound-based meta-heuristic optimization method. The aim of music, in general, is to search for a perfect or near-to-perfect state of harmony which serves as an inspiration for (HS). The efforts to feel harmony in music is analogous to figuring out the optimality in an optimization process. In other words, a jazz musician's improvisation, more or better than the scripted path can compared to the search process in an optimization algorithm. Music harmony is a combination of sounds considered pleasing from an aesthetic point of view.

Harmony in nature is a special relationship between numerous sound waves that have different non-identical frequencies. Dating back to the Greek philosopher and mathematician Pythagoras (582 BC-497 BC), many people have researched this phenomenon. The French composer and musicologist Jean-Philippe Rameau (1683–1764) established the classical harmony theory; the musicologist Tirro has documented a thorough history of American jazz. Musical performances seek to attain a most effective state (fantastic harmony) determined by aesthetic estimation, as the optimization algorithms seek an effective state (global optimum-minimum cost or maximum benefit or efficiency) determined by objective function evaluation. Aesthetic estimation is determined by the set of the sounds played by joined instruments (bandwidth and Pitch Adjustment Rate). Just as objective function evaluation is determined by the set of the values produced by component variables; the sounds for better aesthetic estimation can be improved through repeated practice, just as the values for better objective function evaluation can be improved by repeated iteration and the steps in the procedure of HS are as follows:

Step 1. Initialize a Harmony Memory (HM).
The sample space of variables D and h based on Eq. (6) and Eq. (7) was generated.
Step 2. Improvise a new harmony from HM.
With the additional inputs regarding bandwidth and pitch adjustment parameters the sample space generated in step 1 was passed through Eq. (5) to generate a solution space for the next step.
Step 3. If the new harmony is better than minimum harmony in HM, include the new harmony in HM, and exclude the minimum harmony from HM.
Step 4. If stopping criteria are not satisfied, go to Step 2.

3.2 Crow Search

Crow search (CS) is a user-friendly, simple concept and is easy to implement this metaheuristic technique by which we have obtained promising results while solving the power output optimization problem. It is based on the cunning and rather intelligent behaviour of crows. Crows can memorize faces, use tools, communicate in sophisticated ways. They hide and retrieve food across seasons [9]. The following steps are followed:

Step 1: Initialize the problem and adjustable parameter.
The objective function Eq. (3) along with the constraint Eqs. (4), (5), along with constraint that the minimum height has to be more than 10 m and less than 100 m. The rotor diameter has to be more than 40 m and less than 90 m.
Step 2: Initialize position and memory of crows.
Based on the above constraints a random set of memory locations is created.
Step 3: Evaluate fitness (objective) function.
The random memory locations are placed in the objective function (here the fitness function).
Step 4: Generate a new position.
The values of the objective function are compared.
Step 5: Check the feasibility of a new position.
Step 6: Evaluate fitness function of new positions.

Step 7: Update memory.
Step 8: Check termination criterion.

3.3 Gravity Search Algorithm

Gravity Search (GS) algorithm is an optimization algorithm based on the law of gravity and mass interactions. This algorithm is based on the Newtonian gravity theory: "Every particle in the universe attracts every other particle with a force that is directly proportional to the product of their masses and inversely proportional to the square of the distance between them" [8]. The different steps of the proposed algorithm are the followings:

Step 1: Search space identification.
The upper and lower bounds applied to the two variables under consideration.
Step 2: Randomized initialization.
Initialization of the sample space (agents) using Eq. (5).
Step 3: Fitness evaluation of agents.
Step 4: Update G(i), Best(i), Worst (i) and M(i) for i = 1, 2,…, N. G is defined as the gravitational mass which stores the value (either maximum or minimum) till the ith iteration. M represents the updated value at the ith iteration.
Step 5: Calculation of the total force in different directions.
Step 6 Calculation of acceleration and velocity.
Step 7 Updating agents' position.
Step 8 Repeat steps 4 to 8 until the termination criteria are achieved.

4 Results and Discussion

The Table 1 is for a small scale turbine, i.e. a turbine which generates less than 10KW of power, unlike conventional wind turbines in which layout as well as parameter optimization needs to be applied. Small scale turbine are more compact in nature and there is an observably significant difference between the magnitudes of D and h as can be seen in the table corroborated in the literature too [16]. The results can be compared with those obtained elsewhere [17] and included in the last row of the table in which the output power per turbine is around 1.315 KW, where as the power obtained here in this paper is 1.52 KW. The hub height of the turbine considered is 50 m for [17].

As per Table 1, a comparison of the final results of all three techniques used in this optimization initiative is shown with the following notables:

- **Crow Search:** it can be observed that the values of D and h vary in a manner such that their power output vs iteration plot (first 50 for illustration purpose here) varies between 1 KW to 1.40 KW initially while rising towards the final value as shown in Table 1 third row. The final value is 1.47 KW per turbine, The awareness probability **(AP)** causes the variation in the movement of the solution towards the final value. The variations observed here more than that of Gravitational search but less than that of Harmony search. Here, the awareness probability (AP) represents how likely a crow is to remember the place where the food that was hidden earlier. The value of AP generally is between 0 and 1 The value of AP does not change with subsequent iterations. The similarity with crow search was considered in the context of optimization of wind power output, the search of the next memory location for the next successive search of 'D' and 'h' in place of food by the crow. The value of AP was kept at 0.8. For lower values of AP, it was observed that the optimal solution would change by a significant amount for each time the program was re-executed.

- **Gravitational Search:** Gravitational search, it can be observed that the values of D and h do not vary as much as that of Crow and harmony search. The same can be observed in the power vs iteration curve as well. The results of Gravitational search are the best in terms of cost and power output as shown in Table 1. This is because the gravity constant depends on boundary conditions of the objective function i.e. Best(i) and worst(i) (where i represents the iteration) [8]. In the context of this program (i.e. GS), the mass element M(i), as shown in 5, represented the next memory location for the successive search of'D' and'h' passed through the inequalities (5) and (7) which decide the boundary values of the solution. M(i) is analogous to mass as in the law of gravitational force. On the other hand, pitch adjustment rate in Harmony search causes a step drop in the curve initially as seen in Fig. 1. Fortuitously, it did not extend further in progressive stage changes, instead it rose to the final value gradually after that and then steadily. The final value of HS though varying, still remained within the tolerable limits of the optimal solution and did not result in step-changes, rather a gradual increase towards the final value was observed
- A sudden variation in **Hamonic Search** as shown in Fig. 1 this variation can be attributed to the sudden pitch correction by Pitch Adjustment Parameter (PAR), which in turn, affects the next element to be selected from the Harmonic memory location. As the value of PAR varies be-tween maximum and minimum values specified in the beginning, it does not correct the next value in further iterations as promptly as Gravity Search. The simple reason as mentioned above is because G depends on the best and the worst values at every respective iteration.
- The cost observed in all three methods was calculated to be less than the original cost per turbine of taken value [expression (6)], equal to360089.4 in U.S dollars as observed from Table 1. Thus, in the way, reducing the cost of electricity generation by wind was achieved, and optimal 'D' and 'h' values resulting in optimal power output was found out.

Table 1. A comparative analysis of three metaheuristic methods

Method	h (m)	D (m)	OP (MW)	Cost ($)
HS	29.42	55.62	1.5	40657.26 – 2873794
CS	29.056	53.776	1.38	166588.2
GSA	30.045	53.616	1.39	343060
GA in [16]	50	40	1.31	Not available

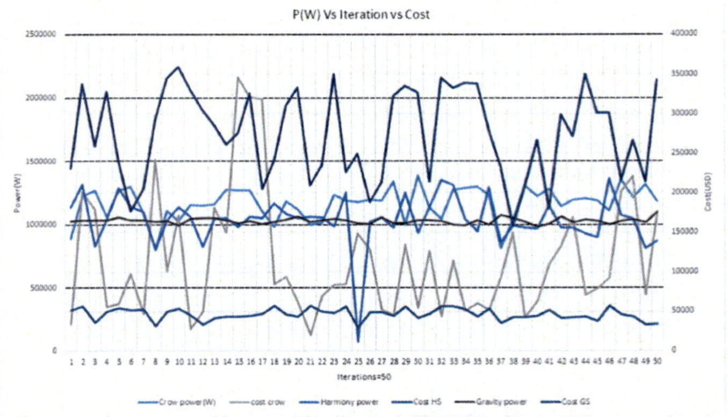

Fig. 1. Power vs iteration vs cost.

The authors are of the view that there are many research areas in this respect still outstanding, especially applying meta-heuristic techniques to more case study applications related to electricity generation by wind. But, by of developed customised programming as done and presented in this paper, the cost reduction and performance enhancement of wind energy may be simply done. There is potential to be scaled-up technically and commercially.

5 Conclusion

The application of meta-heuristic techniques to a small-scale wind generation unit has indicated that the while keeping the diameter constant, the variation in power does not significantly vary (increase or decrease in proportion to hub-height) while the same cannot be said for a constant height and variable diameter, in this case the power output can be varies significantly in proportion to the diameter. The best meta-heuristic method is Gravitational Search as it gives the most optimum values of power along with reduced cost. These methods can be further scaled up for a other applications in renewable energy along with applications to large scale wind turbines as well.

The meta-heuristic approaches as applied in this paper is able to suggest a methodology for optimizing the power output a smallscale wind-turbine in the way of enhancing its performance.

References

1. Blanchard, B.S., Modi, Y.D., Patel, J., Nagababu, G., Jani, H.K.: Wind farm layout optimization using teaching-learning based optimization technique considering power and cost (2020)
2. Kusiak, A., Song, Z.: Design of wind farm layout for maximum wind energy. Renew. Energy **35**(3), 685–694 (2010)
3. Quraeshi, S.: Costs and economics of wind turbine generators for electrical power production. In: Intersol Eighty-Five. Elsevier (1986)

4. Zaman, H., Shakouri, H.: A simple nonlinear mathematical model for wind turbine power maximization with cost constraints. In: 1st International Conference on Energy, Power and Control (EPC-IQ), pp. 255–258. IEEE (2010)
5. Fingersh, L., Hand, M., Laxson, A.: Wind turbine design cost and scaling model. National Renewable Energy Lab. (NREL), Golden, CO (United States), Technical report (2006)
6. Lazar, A.: Heuristic knowledge discovery for archaeological data using genetic algorithms and rough sets. In: Heuristic and Optimization for Knowledge Discovery, pp. 263–278. IGI Global (2002)
7. Russell, S., Norvig, P.: Introduction to Artificial Intelligence (1995)
8. Nikolić, V., Sajjadi, S., Petković, D., Shamshirband, S., Ćojbašić, Ž, Por, L.Y.: Design and state of art of innovative wind turbine systems. Renew. Sustain. Energy Rev. **61**, 258–265 (2016)
9. Trejos-Grisales, L., Guarnizo-Lemus, C., Serna, S.: Overall description of wind power systems. Ingeniería Ciencia **10**(19), 99–126 (2014)
10. Topaloğlu, F., Pehlivan, H.: Analysis of wind data, calculation of energy yield potential, and micro siting application with WASP. Adv. Meteorol. **2018** (2018)
11. Chen, Y., Li, H., Jin, K., Song, Q.: Wind farm layout optimization using genetic algorithm with different hub height wind turbines. Energy Convers. Manage. **70**, 56–65 (2013)
12. Shetty, R.P., Sathyabhama, A., Pai, P.S.: Wind power optimization: a comparison of meta-heuristic algorithms. In: IOP Conference Series: Materials Science and Engineering, vol. 376, no. 1, p. 012021. IOP Publishing, June 2018
13. Khanali, M., Ahmadzadegan, S., Omid, M., Nasab, F.K., Chau, K.W.: Optimizing layout of wind farm turbines using genetic algorithms in Tehran province, Iran. Int. J. Energy Environ. Eng. **9**(4), 399–411 (2018)
14. Wilson, W.D., et al.: Evolutionary computation for wind farm layout optimization. Renew. Energy **126**, 681–691 (2018)
15. Giladi, C., Sintov, A.: Manifold learning for efficient gravitational search algorithm. Inf. Sci. **517**, 18–36 (2020)
16. Zaman, H., Shakouri, H.: A simple nonlinear mathematical model for wind turbine power maximization with cost constraints. In: 2010 1st International Conference on Energy, Power and control (EPC-IQ), pp. 255–258. IEEE (2010)
17. Chen, Y., Li, H., Jin, K., Song, Q.: Wind farm layout optimization using genetic algorithm with different hub height wind turbines. Energy Convers. Manag. **70**, 56–65 (2013)

Estimation of User Base and Revenue Streams for Novel Open Data Based Electric Vehicle Service and Maintenance Ecosystem Driven Platform Solution

Lasse Metso(✉), Ari Happonen, and Matti Rissanen

LUT University, Yliopistonkatu 34, 53850 Lappeenranta, Finland
{lasse.metso,ari.happonen,matti.rissanen}@lut.fi

Abstract. This paper presents a novel maintenance and service supporting concept for electric car owners as customers. At the same time, the platform will support independent car service companies in providing their services. In data side, this particular concept is based on ideologies of open data and models of sharing economies. These models with open data policies would be used to utilize data in cloud service platforms, and the system would link the commercial partners together, using the platform and its services. This link will then connect to the customer needs and e.g. support of the generation of pre-emptive maintenance services and help users to get the best possible user experience with their purchase, as it offers lifelong value and good ownership experience for the end users. The service utilizes artificial intelligence enhanced data-analysis layers, which link everything together and raise the user experience of end users, the service provides and the 3rd party partners, completely in a new level. This paper considers analysis of the initial cost and revenue streams for the platform concept, by estimating the user base through the cumulative number of electric cars in use and the new manufacturers and service companies needed to upkeep the raising demand.

Keywords: Electric car · Electric vehicle BEV · Digitalization · Maintenance · Service · Cost structure · Cost estimation · Artificial intelligence · Big data-analysis · Open data · Industry 4.0 · Climate change · Climate warming · CO_2 emissions · Future prediction · Ecosystem modelling · Business model

1 Introduction

The global warming effect has resulted well-known climate changes, which have raised average temperatures in global scale. The reasons for the change and questions of actions we take to medicate, stop the effects or even turn the temperatures back to previous levels has become a very hot topic of many heated discussion, both in science and in

R. Karim et al. (Eds.): IAI 2021, LNME, pp. 393–404, 2022.
https://doi.org/10.1007/978-3-030-93639-6_34

non-academic circles and politics too. These are definitely not "just a coffee table discussions" as discussions are already been some time in global politics tables, affecting many new legislations, carbon neutrality agreements, energy source considerations and consumables manufacturing and maintenance activities [1–4]. Lot of new business models, related to the monitoring of emissions and automatically has been generated [5] and more and more digitalization related research [6–9] is generated to try to steer away from previous generations material wasting living habits like changing music, games and movie business sectors more and more digitalized nowadays. Additionally, digitalization is also driven into design process to generate unique products more efficiently and automatically [10] and to decarbonize current operational models [11]. Then related to the political and regulatory settings, many cities around the world, e.g. Berlin, Hamburg, Copenhagen, Rome, Athens, Paris, Madrid and Mexico City, have set up limitations for personal vehicle owners how, when and for what purposes they have access and right to drive within inner city regions, with their petrol, diesel or natural gas powered vehicles. Several countries, Norway, Denmark, India, Israel, Netherlands, France, UK, Taiwan, China, Germany and some US states have plans, regulation preparations and open public statements already given, that they will be starting to make restrictions or even complete hard on limitations for people to drive with combustion engine powered cars within their inner city regions between the years 2025 to 2050 [12, 13].

The restrictions and limitations on combustion engine powered cars makes people reconsider their need for owning such a car. Some will look for other means of transport (i.e. public transport) and those who want or need a car, will consider changing theirs to electric cars. The current car service and maintenance infrastructure serves combustion engine cars very well but does not optimally support the increasing number of electric cars. Therefore, there is a need for a platform to support the emerging electric car service and maintenance ecosystem. To model the feasibility of the platform in high level, in this publication the authors are mainly focusing into the calculation of the ecosystem level costs and their source elements costs distributions. It is the hypothesis of the authors, that this sort of case specific cost related understanding [10] will be needed, to give an overview and clear ballpark for the costs and to model the income streams for the concept for feasibility analysis. In this paper EV is used for electric vehicles when not specified what kind of electric vehicle is in question. To be more precise, BEV is battery electric vehicle, PHEV is plug-in hybrid electric vehicle and HEV is hybrid electric vehicle, which can only be charged during drive.

2 Governmental Subventions to Boost Transition Towards EVs

Currently, the change towards higher percentage of sold EV in the world, is mostly driven by certain nations Governmental subvention programs. Of course there is a sizable group of people, who want to invest on EVs on any country in the world, just for the environmental reasons, but for the large crowds of people, vehicle is an house hold investments, which has big impact into the daily lives and as such families will be strongly concerned, what is their available budget and what are the assumed vehicle investments cost structure components (vehicle itself, insurance and usage taxes and maintenance and service fees). Because of this, the governmental subvention programs [14, 15] are highly important tools to enhance the EV car numbers on streets, as they are the tool which allows to change the effectual total cost model the end user faces with EV investment, but they do not do everything. There are many other aspects in play in the big picture, and one of them is the easiness of living with EV compared to the internal combustion engine vehicle, plus the differences in services and maintenance support (if any exists).

Within the scope of subvention programs, for example Germany alone is investing around 3,5 billion euros for their nationwide EV charging infrastructure before 2030 [16]. With the last few years of time, the EV quality, functionality and desirability increasing actions has results a rapid rise in the numbers of electric cars purchased by the generic public. For example, in 2018, Norway achieved record braking level of 31, 2 percent of EVs for all their sold cars within that year [17]. Considering the current and predicted EV sales, following Table 2 and Table 3 present the current sales trend and year 2020 and 2030 predictions [1].

Table 1. Predicted market share in UK of the sold electric cars per year [1].

Technology	2020	2030
Full hybrid (HEV)	5–20%	20–50%
Plug-in hybrid (PHEV)	1–5%	15–30%
Electric vehicles (BEV)	2–7%	10–40%

Another forecast is based on data from The International Energy Agency (IEA), more specifically from their Technology report—May 2019 [15]. For additional details, see the Table 2.

Table 2. IEA forecast of electric cars market share for different market sectors in 2030 [15].

Location	New policies scenario	EV30@30 scenario
China	28%	42%
Europe	26%	50%
Japan	21%	37%
North America	29%	30%

Table 3. The sold cars in EU in the 2020 [18].

Fuel type	Market share
BEV + PHEV	10,5%
HEV	11,9%
Petrol	47,5%
Diesel	28,0%
Alternative fuels	2,1%

In Table 1 data, the forecast of electric cars market share was predicted to grow 2–7% in 2020. But in fact, this prediction has actually been already surpassed in 2020 (Table 3). The table shows, that in 2020, the real numbers of BEV and PHEV together is 10,5% which is at the top end of the prediction. Considering the current numbers and how they surpass the previous predictions, plus the political atmosphere within the BEV related topics and sizable investments on BEV development and development in technologies related and needed by BEVs, the given 2030 prediction, mentioned in the Table 1, looks to be in correct ballpark. And speaking about the forecasts, e.g. in Europe, the market share of electric cars are forecasted to be anywhere from 26% to up to the level 50% in 2030. And considering the previously mentioned facts and current situation, this is well in line with the forecast shown in the Table 1.

A little bit uncertainty to forecast is the hydrogen as fuel to electric car. Now the market share of cars using hydrogen is very low because hydrogen has not been available [19]. On other hand, the hydrogen ecosystem is having the challenge on how much energy is wasted when hydrogen is produced and stored in large scale operations [20]. But in big picture, hydrogen could be one option to lower emissions somewhere in our future.

3 Forecasting the Future of the Finnish Electric Car Markets

In year 2017 at Finnish vehicle markets, mere 776 electric cars were sold in the whole year. That is 0,64 percent from the all sold cars in the given year. But it seems that this is just a start, and Finnish markets are slowly starting to catch up the more developed BEV sales areas. The market for BEVs are definitely growing up in Finland, the sales of electric cars are increasing, and actually, BEVs sales already totalled to be 4245 units in 2020, based on national statistics information [21]. Calculating from statistic the annual grow percentage of sold BEV has been more than 100% in last years. Several car manufacturers for example Ford in Europe, Jaguar Land Rover, General Motors have informed to focus to manufacture only or mainly to BEVs. They told they wont produce combustion engine cars by 2030 or 2035. [22] Volvo have more ambitious plan they want to quit gasoline-only cars manufacturing by 2025 [23].

For a short history view of vehicle sales markets in Finland, the average number of sold passenger cars during years 2000 to 2018 has been 122 461 cars per a year. The biggest amount of sold cars was in year 2005, in which the sales did peak to 148 161 cars per year. And the least sold cars per year was experienced in 2009 which was the year of global recession [24]. In that specific year, only 90 574 cars were sold [25].

In Table 4, the forecast of sold BEVs in Finland for years 2020 to 2030 are presented. In Finland rapid changes in car market are possible. An example of rapid market share transition is the invasion of Japanese cars into Finland at 1960's. Japanese cars increased their market share rapidly [26]. Another case in Finland car market is car brands from South Korea. They raised market share in 2000's from zero to 7% in less than 10 years [25]. Same kind of changes are expected to be happening with electric cars too.

In Finland cars in everyday use are quite old, in 2018, 12,2 years old, compared to international averages, in EU average is 10,8 and newest cars on average are in UK, 8,0 years old [27]. The average age of vehicle EOL (End of Life) is 18 years in Finland. New cars are almost 100% in use in first seven years when the decrease begins [28].

Table 4. Finnish car importers collected forecasts of EV, BEV and PHEV cars in use in Finland 2030.

Forecast 2030	BEV	PHEV
Finnish energy and climate strategy, target	700 000	
Climate panel	850 000	
SITRA	700 000	100 000
ILMO-working group	474 000	190 000
Gaselli	300 000	230 000
Finnish car importers association, base scenario	136 000	228 000
Finnish car importers association, road map	229 000	349 000
Average	484 143	219 400

The authors have evaluated the development of electric cars in Finland and created an evaluation of sold BEVs in Finland in near future. The total amount of sold cars in Finland is selected 125 000 cars in year. Based on car manufacturers plans to produce only electric cars in 2030 the forecast of electric car will be very high, almost every sold car in Finland will be electric cars. In estimation the market share in 2020 was about 1% and it will raise to 95% by 2030. The number of BEVs in use in Finland is almost 700 000 in 2030. See Table 5. On other forecasts same kind of number exist, for example Finnish car importers have collected forecasts of electric car number in Finland in 2030. Because of accidents some cars are wrote off, but the wastage is assumed to be replaced with used cars import from abroad.

The number of electric cars in the authors calculation is more than average of the numbers of BEV in forecats, the difference is 206 107. The difference between forecasts is actually quite high. The authors think the optimistic forecasts (in Table 4) will probably come true because car manufacturers will not make combustion engines at the end of 2030 and only a few wants to buy those because limitations to use those.

Table 5. The estimation of BEVs market share 2021 to 2030 in Finland.

Year	BEVs market share	Sold BEVs	BEVs in use
2021	10%	12 500	21 500
2022	20%	25 000	46 500
2023	30%	37 500	84 000
2024	40%	50 000	134 000
2025	50%	62 500	196 500
2026	60%	75 000	271 500
2027	70%	87 500	359 000
2028	80%	100 000	459 000
2029	90%	112 500	571 500
2030	95%	118 750	690 250

4 Data Sharing Electric Car Service Platform

In this maintenance and electric vehicle easiness of serviceability focusing concept, the data in the system is collected from multiple data sources. The vehicles themselves are the main raw data source, feeding the system and AI algorithms with the big data-analysis engine for base line evaluations. The AI is important part in the whole ecosystem, as it provides higher level data-analysis capabilities to the ecosystem, compared to what human experts could do alone. First of all, large scale data-analysis give ecosystem partners possibilities to enhance their business efficiency, at the same time when they boost their industrial scale operations ecological aspects [29]. Secondly, using AI in long term, with the data collected from vehicle related maintenance, longevity and material selection factors will give car manufacturers and 3rd party partners new combined data source to design cars and 3rd party accessories which are more environmentally friendly and more circular in nature [30].

For the system initialization phase, the manufacturers of the vehicles will set the base parameters for the system and can then later on start to update data analysis efficiency with additional algorithms for calculating data derived information units (when the vehicle production matures during vehicle lifetime in markets) and so on. On vehicle owner side, the end user or the vehicle owner will have an access into the data and they can then feed addition specifics for the system, e.g. related to vehicle self-maintenance activities, in use time felt uncertainty on some usage relate aspects, wonders on "strange" vehicle related behaviour events (e.g. classify note to the system "battery drainage higher than nominal for given driving patterns and local weather conditions?"). The vehicle original manufacturers approved service and spare parts suppliers can build new level of service model, on top of the given new additional data, to enhance the service quality for all participating collaborators in this maintenance and service ecosystem. If the vehicle owner is willing to approve the data usage for verified 3rd party service providers, the owner could receive offers for special discounts in yearly MOT inspection rounds, winter / summer tires enhance actions, insurance plus car exchange offers, plus of course and normal maintenance operations etc.

The platform can offer electric car owners a wide view of new status information about their vehicle(s) (real-time "on-line" facts, short term activities summaries and possible to-do list for e.g. two next weeks, and more comprehensive data logs for long term things and maintenance activities and their predicted items for future times and so on). The ecosystem in the platform is able to provide the users with the details of the availability of services, the information of service prices, offers for items in to-do list and so on.

For more DIY (Do-It-Yourself) savvy users, the platform could give access for the users electric car service manuals (or offer manuals for purchase, if they are commercially produced materials, not open source option), training materials (e.g. videos made by specialists for vehicle specific instructions how to driving that vehicle ecologically, guidance for efficient packing of the car, introductions to features and functions and so on). Additionally, users could get access to spare part lists from service platform, to make it as easy as possible for users to order specific parts or at least to study the parts locations and do some problem mapping in weekends to get faster and more productive support from the car maintenance shops, within their opening hours. On the other hand, car manufacturers could offer after-market services, upgrades for the vehicles in mass purchases with discount processes and use the platform to build service networks faster than ever before (for new vehicles). Servicing the vehicle is especially important for vehicles, which have not been previously sold on markets and also for manufacturers who are currently aiming to new markets with new vehicle style and/or configuration.

This should make it easier for new electric car owners, who are using the ecosystem of electric car service platform, to find services for their cars and on other hand it should enable new players (BEV manufacturers) to enter new market areas too. And finally, it should be easier to join in to the maintenance network, if you are building new company for BEV maintenance, as the platform offers faster access to inform people, about new possibilities to maintain good operational condition of their vehicles.

5 Revenue to Platform

The platform needs to define which invoice models to use, for example charge users, based on advertisements. Who are charged? Are all users charged or only specific groups of customers only? And which pricing structures are used? Are charged only when joining or time-based periods or are transactions charged? [31].

It is impossible to make perfectly precis future predictions for the calculations [32] of the revenues on the platform, because e.g. the quantity of sold electric cars in itself is impossible to forecast with 100% accuracy. Another challenge for perfect predictions, is the difficulty to estimate the possible number of users in this sort of ecosystem related platform [32]. But still, we can do quite good estimates, be reasoning with the possible minimal and maximal numbers and take a conservative number in between. For this, in revenue calculations, the market share of electric cars is estimated to be 40% in 2030 and the number of BEVs in use, was estimated to be 286 750 vehicles. The potential users of platform are estimated to be around 10% of BEV owners. The minimum estimate for potential users is used for estimates, as most of the new car owners might want to use only the brand related service, within their vehicles guarantee period. After the guarantee period the independent services and service-related platforms are expected to become more popular.

For revenue streams, it is estimated that the platform charges electric car owners, service providers and manufacturers with different ways and fee structures. This estimate is based on typical revenue stream models, use in nowadays platforms like YouTube, Airbnb, Uber, Trivago and so on, plus on related research studies [33, 34]. In our example, the car owners are charged by monthly payment and by number of transactions. The fees for the car owner are 10€ in month and 5€ per action. Car services are charged only by transactions. Car manufacturers are charged based on the sold cars, that are connected into the platform, e.g. 400€ per car, when the user receives 2 years free access to the platform, with the car. Also, any maintenance etc. service, which is reserved from the platform, for the car, produces 2€ fee for every starting 100€ estimated service cost. See Table 6.

Table 6. The charging principles of charging the platform users.

Payer	Monthly payment	Paid by action	Fee to join
Car owner	10€	5€	
Car service		2€/100€	
Car manufacturer			400€

In Table 6 income from car owners is based on three transaction in month and monthly payments. The income from car services is calculated three transactions to each user of the platform. This is based on 2€ per 100€ worth of service and estimation in table is 10€/car owner using platform. The income from car manufacturers is calculated by 400€ / car from manufacturer in platform, if manufacturers want to join into the platform. In 2025, the estimation of manufacturers cars joining to platform is 500 cars and in 2030 2000 cars. Some of new players like Aiways want to become to European market without own service network [35]. See Table 6.

The revenues to the platform are difficult to calculate because the potential of the users in platform in not possible to calculate exactly. However, the potential users are mainly front ha customers of new players. The traditional car manufacturers have created their own service network and won't give up customers for those. The market share of new players can raise quickly and that makes impossible to forecast the revenue for platform. Any way there are potential for this kind of cloud service because of the raise of BEVs sold until 2030.

6 Costs of Platform

The costs from platform are difficult to evaluate, and the EV car users in Finland are not all potential users of platform. The platform can be developed to be scalable and be customised to other countries too.

The critical for the success of the platform is to create business model attract participants [36]. Participation in platform ecosystem increased sales of participations [37]. In literature costs and cost components are not widely defined to service platforms. However, some topics affect to costs of platform are presented. The scalability has influence to costs [36]. Another topic to increase the costs is the openness of platform. The number of actions at same time, the use of third-party software, numbers of open APIs, technology and the age of service can affect to costs of the platform [38, 39].

The costs are composed of different forms, for example search and information costs, social relations. The behavior of users incur uncertainty to the platform owner as well as technological and market issues [40].

7 Discussion and Conclusion

With the research, authors produced a novel new concept for electric car service. The concept is based on data openness and data sharing with efficient data analysis, digitalization and AI connections. Both electric car owners and different vehicle related service providers, car manufacturers and car importers are all able to join to the platform. The idea of the platform is to collect wide range of data from the electric vehicles. Data like the service history, service manuals and service instructions into one place, where the information can easily be accessed, when needed. This enhances the experience delivery, related to the vehicle, for all platform users and especially improves the experience of owning the vehicle. The owner can e.g. check the status of their vehicle assets and get discount offers for forthcoming service(s). User can compare prices of services and can select and book service most fitting for their needs. User can also get immediate notifications if service provider needs to discuss about issues/reschedule/changes in the service. One example could be a delay on maintenance parts delivery to the service centre. The data will also allow the service providers to enhance their operation efficiency and service offering. Independent car services have access to new business opportunities to start to offer electric vehicle maintenance operations. And also, new manufacturers / vehicle brand owners, which do not have their own service network, can establish themselves in new markets, by partnering with the independent car service providers. The service companies could download service manuals and instructions to maintain new electric cars and owners would get access to the basic maintenance information to understand where to look, in case there is a usage challenge with the vehicle and service is not easily immediately available.

For the platform growth rate estimates, a forecast of yearly sold electric cars was built for the study. The estimate is based on knowledge about the average yearly nationally sold cars, from the 10 years' time period. The average is 122 461 sold cars in a year, which is considered conservative as the history data includes few years that have been impacted by the great recession. The authors still did keep the conservative line and based their estimate on 125 000 yearly sold cars. The growth of electric cars sales is expected to be quite linear into 2030, which is also considered to be a conservative estimate. In this study the market share of sold electric cars in 2030 is forecasted to be 95%. The market share of electric cars could easily be seen to be higher, if governmental pressure for cleaner traffic will be strong from 2020 to 2030. However, both limitations and the availability of new electric cars can together increase the market share of electric cars to 95% in 2030. Compared to e.g. to Norway, the market share of electric cars is already very high, because of the government costs compensation advantages for the end users. Same kind of increase can happen in Finland too, if encouragement for highly clean traffic will be the key interest of national and EU political decision makers.

Future research is suggested to be focused on developing new application and services for the platform concept. For example, to boost transition to BEVs some parts of the platform could be opened for combustion engine- based cars too, to promote the full benefits that become available when people do transition to BEVs. Another new service is to collect and share information of rechargers and availability of those. platform can offer services to combustion engine cars, at least service offerings and information of available car services.

References

1. Kay, B., Hill, N., Newman, D.: Powering ahead the future of low-carbon cars and fuels (2013)
2. Li, M., Boriboonsomsin, K., Wu, G., Zhang, W.B., Barth, M.: Traffic energy and emission reductions at signalized intersections: a study of the benefits of advanced driver information. Int. J. Intell. Transp. Syst. Res. **7**(1), 49–58 (2009)
3. Assaad, Z., Bil, C., Eberhard, A., Moore, M.: Reducing fuel burn through air traffic flow optimisation incorporating wind models. Procedia Eng. **99**, 1637–1641 (2015)
4. Stepanian, C.M., Ong, F.N., Mendez, L.M., Khurana, S.: Assessing Traffic and Air Quality in Central Copenhagen (2015)
5. Eskelinen, T., Räsänen, T., Santti, U., Happonen, A., Kajanus, M.: Designing a business model for environmental monitoring services using fast MCDS innovation support tools. Technol. Innov. Manage. Rev. **7**(11), 36–44 (2017)
6. Minashkina, D., Happonen, A.: Operations automatization and digitalization – a research and innovation collaboration in physical warehousing. AS/RS and 3PL logistics context. LUT Res. Rep. Ser. Rep. **86**, 66 (2018). ISBN 978–952–335–293–3, ISSN 2243–3376
7. Metso, L., Happonen, A., Ojanen, V., Rissanen, M., Kärri, T.: Business model design elements for electric car service based on digital data enabled sharing platform. In: Cambridge International Manufacturing Symposium, Cambridge, UK, pp. 26–27 (2019)
8. Happonen, A., Minashkina, D., Nolte, A., Angarita, M.A.M.: Hackathons as a Company – University Collaboration Tool to Boost Circularity Innovations and Digitalization Enhanced Sustainability. In: 13th International Engineering Research Conference, 27 November 2019, Subang Jaya, Malaysia, p. 11 (2019)
9. Kortelainen, H., Happonen, A., Hanski, J.: From asset provider to knowledge company—transformation in the digital Era. In: Mathew, J., Lim, C.W., Ma, L., Sands, D., Cholette, M.E., Borghesani, P. (eds.) Asset Intelligence through Integration and Interoperability and Contemporary Vibration Engineering Technologies. LNME, pp. 333–341. Springer, Cham (2019). https://doi.org/10.1007/978-3-319-95711-1_33

10. Piili, H., Happonen, A., Väistö, T., Venkataramanana, V., Partanen, J., Salminen, A.: Cost estimation of laser additive manufacturing of stainless steel. Phys. Procedia **78**, 388–396 (2015). https://doi.org/10.1016/j.phpro.2015.11.053
11. Minashkina, D., Happonen, A.: Decarbonizing warehousing activities through digitalization and automatization with WMS integration for sustainability supporting operations. In: 7th International Conference on Environment Pollution and Prevention (ICEPP 2019), 18–20 December 2019, Melbourne, Australia, pp. 1–7 (2020). https://doi.org/10.1051/e3sconf/202015803002
12. Coren, M.: Nine countries say they'll ban internal combustion engines. So far, it's just words, Quartz, 7 August 2018
13. Mohareb, E.A., Kennedy, C.A.: Scenarios of technology adoption towards low-carbon cities. Energy Policy **66**, 685–693 (2014)
14. Trommer, S., Jarass, J., Kolarova, V.: Early adopters of electric vehicles in Germany unveiled. World Electr. Veh. J. **7**(4), 722–732 (2015)
15. IEA. Global EV Outlook 2019—Scaling-Up the Transition to Electric Mobility. IEA, London, UK (2019)
16. Nurminen, T.: Saksa nostaa sähköautotukea jopa 50 prosenttia ja investoi 3,5 miljardia latauspisteisiin. Kauppalehti (2019)
17. Knudesen, C., Doyle, A.: Norway's electric cars zip to new record: almost a third of all sales, Reuters, 2 January 2019
18. ACEA, 2021, 4th February 2021, Press release, Fuel types of new cars: electric 10.5%, hybrid 11.9%, petrol 47.5% market share full-year (2020). https://www.acea.be/press-releases/article/fuel-types-of-new-cars-electric-10.5-hybrid-11.9-petrol-47.5-market-share-f
19. Aalto, T., Lallo, M., Hatakka, J., Laurila, T.: Atmospheric hydrogen variations and traffic emissions at an urban site in Finland. Atmos. Chem. Phys. **9**(19), 7387–7396 (2009)
20. Ball, M., Wietschel, M.: The future of hydrogen–opportunities and challenges. Int. J. Hydrogen Energy **34**(2), 615–627 (2009)
21. Statistics Finland's free-of-charge statistical databases, 2020, 11ck -- First registrations of passenger cars by driving power, 1990–2020 (2020). https://pxnet2.stat.fi/PXWeb/pxweb/fi/StatFin/StatFin__lii__merek/statfin_merek_pxt_11ck.px/table/tableViewLayout1/
22. Boudette, N., Ewing, J.E.: Ford Says it will Phase out Gasoline-Powered Vehicles in Europe, The New York Times, 17 February 2021
23. Molnar, C.: Volvo says it wants to be through selling gasoline-only cars by 2025, The company is bringing EV motor development completely in-house to hit its goals. Driving, 4 November 2020
24. Gore, C.: The global recession of 2009 in a long-term development perspective. J. Int. Dev. **22**(6), 714–738 (2010)
25. Autoalan tiedotuskeskus, 2019, Henkilöautojen ensirekisteröintikehitys (2019). http://www.aut.fi/tilastot/ensirekisteroinnit/henkiloautojen_ensirekisterointimaaran_kehitys
26. Sahi, J.: Verkostot kaukaiseen itään: Suomen kauppasuhteet Japaniin 1919–1974 (2016)
27. Autoalan tiedotuskeskus, 2020, Average age of passenger cars in EU is 10,7 years (2020). http://www.aut.fi/en/statistics/international_statistics/average_age_of_passenger_cars_in_some_european_countries
28. Pöllänen, M., Kallberg, H., Kalenoja, H., Mäntynen, J.: Autokannan tulevaisuustutkimus Tulevaisuuden autokantaan vaikuttavat tekijät ja skenaarioita vuoteen 2030. Ajoneuvohallintokeskus Tutkimuksia ja selvityksiä, Nro 4/2006 (2006)
29. Ghoreishi, M., Happonen, A.: Key enablers for deploying artificial intelligence for circular economy embracing sustainable product design: three case studies. AIP Conf. Proc. **2233**(1), 1–19 (2020). https://doi.org/10.1063/5.0001339

30. Ghoreishi, M., Happonen, A.: New promises ai brings into circular economy accelerated product design: review on supporting literature. E3S Web Conf. **158**, 1–10 (2020). https://doi.org/10.1051/e3sconf/202015806002
31. Agustí, A.: Predicting the future from the past (2017)
32. Costanza, R., et al.: Twenty years of ecosystem services: how far have we come and how far do we still need to go? Ecosyst. Serv. **28**, 1–16 (2017)
33. Wan, X., Cenamor, J., Parker, G., Van Alstyne, M.: Unraveling platform strategies: a review from an organizational ambidexterity perspective. Sustainability **9**(5), 734 (2017)
34. Kohtamäki, M., Parida, V., Oghazi, P., Gebauer, H., Baines, T.: Digital servitization business models in ecosystems: a theory of the firm. J. Bus. Res. **104**, 380–392 (2019)
35. Bremner, R.: Aiways U5 2019 review: First taster from Chinese firm Aiways provides an interesting insight into its forthcoming affordable electric SUV, Autocar (2019). https://www.autocar.co.uk/car-review/aiways/u5/first-drives/aiways-u5-2019-review
36. Chauhan, S., Goyal, S.: Platform ecosystems: the expert opinion. J. Inf. Technol. Case Appl. Res. **20**(2), 86–89 (2018)
37. Ceccagnoli, M., Forman, C., Huang, P., Wu, D.J.: Cocreation of value in a platform ecosystem! The case of enterprise software. MIS Quart. 263–290 (2012)
38. Nambisan, S., Baron, R.A.: On the costs of digital entrepreneurship: role conflict, stress, and venture performance in digital platform-based ecosystems. J. Bus. Res. (2019)
39. Hein, A., et al.: Digital platform ecosystems. Electron. Mark. **30**(1), 87–98 (2019). https://doi.org/10.1007/s12525-019-00377-4
40. Rindfleisch, A., Heide, J.B.: Transaction cost analysis: past, present, and future applications. J. Mark. **61**, 30–54 (1997)

Multiclass Bearing Fault Classification Using Features Learned by a Deep Neural Network

Biswajit Sahoo and A. R. Mohanty(✉)

Department of Mechanical Engineering,
Indian Institute of Technology, Kharagpur, Kharagpur 721302, India
biswajitsahoo1111@iitkgp.ac.in, amohanty@mech.iitkgp.ac.in

Abstract. Accurate classification of faults is important for condition based maintenance (CBM) applications. There are mainly three approaches commonly used for fault classification, viz., model-based, data-driven, and hybrid models. Data-driven approaches are becoming increasingly popular in applications as these methods can be easily automated and achieve higher accuracy at different tasks. Data-driven approaches can be based on shallow learning or deep learning. In shallow learning, useful features are first calculated from raw time domain data. The features may pertain to time domain, or frequency domain, or time-frequency domain. These features are then fed into a machine learning algorithm that does fault classification. In contrast, deep learning models don't require any hand-crafted features. Representations are learned automatically from data. Thus, deep learning models take raw time domain data as input and produce classification results as output in an end-to-end manner. This makes interpretation of deep learning models difficult. In this paper, we show that the classification ability of deep neural network is derived from hidden representations. Those hidden representations can be used as features in classical machine learning algorithms for fault classification. This helps in explaining the classification ability of different layers of representations of deep networks. This technique has been applied to a real-world bearing dataset producing promising results.

Keywords: Fault classification · Data-driven methods · Deep learning · Support vector machines

1 Introduction

Condition based maintenance (CBM) is the process of doing maintenance only when it is needed [1]. Adoption of this maintenance strategy leads to significant monetary gains and reduction in unplanned downtime. Ease of implementation and ability to accurately identify faults are some of the important requirements for wide adoption of CBM in industry. Data-driven methods for fault classification are suitable for these applications as these methods can easily be automated and at the same time achieve near perfect accuracy at classifying faults. Data-driven methods are general purpose in the sense that the same technique can be applied for classification of faults in different

R. Karim et al. (Eds.): IAI 2021, LNME, pp. 405–414, 2022.
https://doi.org/10.1007/978-3-030-93639-6_35

types of machines with minor modifications. This provides significant advantages over model-based methods that require different formulations for different fault types and are sensitive to external factors that might inevitably affect the operation of the machine. Therefore, data-driven methods are increasingly being applied to real applications.

Data-driven approaches can broadly be divided into two groups: shallow learning approaches and deep learning approaches. In shallow learning approach, relevant features are first calculated from raw time domain signal and these features are then fed into a classification algorithm to classify different faults. Accuracy of this approach depends heavily on the chosen features. Thus, it requires significant domain knowledge. Deep learning based approaches, on the other hand, take raw time domain data as input and produce the desired classification result in an end-to-end manner. However, working of deep learning models is difficult to interpret.

Several attempts have been made at interpreting the results of DNN [2–4]. Almost all methods are applicable to image based data where there is a spatial structure in the data and activations of different layers can be visualized to see the abstract representations learned by the network. Those methods mainly focus on backpropagating gradient information for explaining the network outputs.

Deep learning based methods have also been used in mechanical applications. Janssens and others [5] used CNN on frequency spectrum image for fault classification. Wavelet packet energy maps have also been used as input to convolution layers for multiclass classification problems [6]. The above two methods have employed preprocessing on input time domain before using it for classification by deep learning model. It is also possible to directly use time domain data for fault classification. By stacking one-dimensional time series signals as rows of a matrix, researchers have obtained two dimensional image signal [7]. Wen et al. [8] have used time domain signal to come up with a 2D image and then have applied CNN for multiclass classification. Deep learning techniques without using any preprocessing have achieved state of the art results in condition monitoring.

To the best of our knowledge, no prior work has tried to explain the results of deep neural networks for the case of vibration signal. In this paper, we show that output of layers of deep neural network (DNN) are similar to features that are used in shallow learning approaches. While in shallow learning, one has to explicitly design the features from data before using it with a classification algorithm, in deep learning the features are learned from data. We also show that the features learned by DNN, when used with support vector machine (SVM), achieve near perfect classification accuracy on a real-world IMS bearing dataset [9]. This accuracy is even higher than the accuracy of the DNN model itself.

2 Theory

2.1 Deep Neural Networks

Deep neural networks are extensions of neural networks with many hidden layers and novel architectures. The architectures vary from each other depending on how weights are shared in a layer and how it is connected to subsequent layers. Number of parameters in a densely connected feed-forward network can increase exponentially with additional layers thus making the training process difficult. However, sharing weights within a layer and reducing the size of layers gradually significantly reduce the number of parameters and make it convenient to train a network. For different applications, different architectures have been proposed. Some commonly used architectures are convolutional neural networks (CNN), recurrent neural networks (RNN), long short term memory (LSTM) network, deep belief network (DBN), etc. For our application, involving two-dimensional matrix like data, convolutional neural networks work best among other architectures.

2.2 Convolutional Neural Network

Convolutional neural network (CNN) is particularly useful for grid-like data. It is a special class of DNN in which weights are shared in a layer. While feedforward neural networks take linear combination of outputs from previous layers to calculate the input for present layer, CNNs use a kernel that is convolved with a region of input to produce an output. This convolution operation reduces the size of the output layer depending on the kernel size. As the same kernel is convolved with different regions of an image by sliding it, total number of trainable parameters are equal to the number of parameters in the kernel. Output of this convolution is called a feature map or a filter. The convolution operation can be expressed as [10]

$$S(i,j) = \sum_{m}\sum_{n} I(i+m, j+n)K(m,n)$$

Where I is the image and K is the kernel of size $(m \times n)$. If it is required to produce outputs of same size as input, padding can be applied to the input. Most commonly used padding method is zero-padding. In zero padding, additional rows and columns of zeros are added to the input so that input size increases. The amount of padding added depends on the size of the kernel used. Convolution is then applied to the padded input and after shrinkage due to kernel, output of same size as that of input is obtained. Outputs of convolution are then passed through an activation function. Most commonly used activation function is rectified linear unit (ReLU) that returns every positive activation as it is but makes all negative activations zero. Another layer commonly used in CNN is pooling layer. Pooling is the operation of returning a single value from a rectangular region. If the maximum value of the rectangular region is returned, pooling is called max pooling. On the other hand, if the average value of the region is returned, it is called average pooling. Pooling helps in making the representation invariant to small translations of the input. For classification applications, entries of the last convolution layer are flattened and passed through several feedforward layers before giving actual classification result. These last feedforward layers greatly influence the parameter count of the network.

2.3 Support Vector Machine

Support vector machine (SVM) is a classification algorithm that is suitable for multiclass classification. SVM is the generalization of support vector classifier that itself is the generalization of maximal margin classifier [11]. In a binary classification scenario, if the two classes are linearly separable, maximal margin classifier finds a hyperplane such that the orthogonal distance from the hyperplane to nearest points in each class is maximized.

Given a set of n labeled data points $\left\{\left(\boldsymbol{x}^{(i)}, y^{(i)}\right)\right\}_{i=1}^{n}$, where each $\boldsymbol{x}^{(i)}$ is a $(p \times 1)$ vector. p is the number of features and $y^{(i)}$ is the respective label that can take, in binary classification, values from $\{+1, -1\}$. Maximal margin classifier finds the parameters $\boldsymbol{\beta} = [\beta_1, \ldots, \beta_p]^T$, such that $\boldsymbol{\beta}^T \boldsymbol{x}^{(i)} + \beta_0$ belongs to the correct class for each i and exceeds certain margin (M).

When two classes are not strictly separable, maximal margin classifier can still be used to produce a separating hyperplane allowing some misclassification. This is the basis of support vector classifier. Each misclassification will incur some penalty. So support vector classifier finds an optimal solution that allows some misclassification yet remaining within the penalty. The modified quadratic programming problem for support vector classifier becomes

$$\min_{\beta_0, \boldsymbol{\beta}} \frac{1}{2} \|\boldsymbol{\beta}\|^2 + C \sum_{i=1}^{n} \xi_i$$

$$\textit{subject to}, \; y^{(i)}\left(\boldsymbol{\beta}^T \boldsymbol{x}^{(i)} + \beta_0\right) \geq 1 - \xi_i, \; i = 1, 2, \ldots, n$$

$$\xi_i \geq 0$$

where C is a parameter that controls the maximum number of allowed misclassifications and ξ_i are slack variables. This optimization problem can be solved by applying convex optimization techniques. When the data are not linearly separable, the principle of support vector classifier can be generalized to obtain nonlinear classification boundary. This can be achieved by using kernels. Using kernels, such as radial basis functions (RBF), the feature space is transformed to a higher dimensional space, sometimes infinite dimensional, assuming that in the higher dimensional space the data is linearly separable. If a solution is found in the higher dimensional space, it is projected back to the original feature space taking the shape of nonlinear classification boundary. Computation with kernels in higher dimensions is manageable because of the fact that kernels require only the inner products between data points. Those inner products are computed in original feature space instead of the transformed high dimensional space. Multiclass classification is handled by SVM by using either one versus one or one versus all approaches.

3 Description of Data

The IMS data set [9] provides three test to failure tests of four Rexnord ZA-2115 bearings mounted on a horizontal shaft. Rotational speed was kept constant at 2000 RPM. Sampling frequency is reported to be set at 20 kHz. Some authors have verified the sampling frequency to be 20.48 kHz [12] Accelerometers were attached to bearing casing and data were collected till failure occurred in one of the bearings. For test one, inner race fault occurred in bearing 3 and roller defect occurred in bearing 4. In test three, outer race defect occurred in bearing 3. For this study, normal, inner race fault and rolling element fault data were taken from test 1 and outer race fault data were taken from test 3. Data taken from channel 1 of test 1 from 12:06:24 on 23/10/2003 to 13:05:58 on 09/11/2003 were considered normal. For inner race fault and rolling element fault, data were taken from 08:22:30 on 18/11/2003 to 23:57:32 on 24/11/2003 from channel 5 and channel 7 respectively. Outer race fault data were taken from channel 3 of test 3 from 14:51:57 on 12/4/2004 to 02:42:55 on 18/4/2004. There are a total of 750 files in each category. From each file data for respective column are read and segmented into lengths of 1024. Remainder of data from each file is dropped. Subsequently, the segment of length 1024 is resized to size (32 × 32) that is fed into the CNN model. For each category, out of 750 files, first 650 files are used for training and validation and rest 100 files are used for testing. From 650 training files, 100 validations files are chosen randomly. As the data length in each column is fixed, number of training, validation and test data are summarized in Table 1.

Table 1. Details of dataset used

	Number of training data of size (32 × 32)	Number of validation data of size (32 × 32)	Number of test data of size (32 × 32)
Normal	11000	2000	2000
Inner race fault	11000	2000	2000
Outer race fault	11000	2000	2000
Rolling element fault	11000	2000	2000

4 Methodology

CNN is first trained on training data. Once trained, it is then used for inference. Inference involves passing input through the network and computing the resulting output. Inference through the network is fast as weights of the model have already been learnt. When outputs are computed at a hidden layer, the results are called activations. Activations for test data are computed at the penultimate layer of CNN. The test activations are then used in a support vector machine (SVM) for classification of test data. Deep learning library TensorFlow [13] was used to train the deep learning model. Our code is publicly available and can be found at: https://github.com/biswajitsahoo1111/data_driven_features_ims. The CNN architecture used in this study is given in Table 2.

Table 2. Deep neural network architecture

Layer No	Layer characteristics
1	Convolution (32 filters, kernel size = 5)
2	MaxPooling (Pool size = (2,2))
3	Convolution (16 filters, kernel size = 5)
4	MaxPooling (Pool size = (2,2))
5	Flatten layer
6	Dense layer (Units = 120)
7	Dense layer (Units = 84)
8	Dense layer (Units = 16)
9	Output layer (Units = 4)

Activation function in each layer is rectified linear unit (ReLU) except for the output layer where activation function is softmax. Total number of trainable parameters in the model are 73360. During training the CNN, sparse categorical cross-entropy objective is minimized using Adam optimizer with a learning rate of 0.001. The model is trained for 12 epochs.

After training, test data is passed through the model and outputs at the penultimate layers are calculated. As the penultimate layer has 16 units, output data size from this layer is (8000 × 16). This now acts as a feature matrix. SVM is applied to this feature matrix and results of SVM is compared to the results of the CNN model. For training the SVM, a small fraction of input data of size (4000 × 16) are used with equal representation for each fault type. Parameters of SVM include radial basis function as kernel, cost value of 1 and gamma of 0.01.

5 Results and Discussions

The DNN model trained on training data achieved a training accuracy of 99.86% and validation accuracy of 99.87%. On test data, the model achieved an accuracy of 99.84%. The trained model is used to compute activations for input data. Pretrained model takes as input data of size (32 × 32 × 1) where the last dimension corresponds to the number of filters in data. For grey images, this number is 1 and for color images, it is 3. As the penultimate layer of the original model has 16 neurons in it, every input produces an output of size 16. These activations are used in SVM for fault classification. The SVM is trained on a fraction of training activations. Out of 650 training files, 50 files are chosen at random for each fault type and data in these files are used to find activations of training data. SVM is trained using this data and tested on the activations of test data. Test data is completely new both for the CNN model and for SVM model. Overall test accuracy of 99.94% is achieved using SVM on test activations. Resulting confusion matrix is shown in Fig. 1.

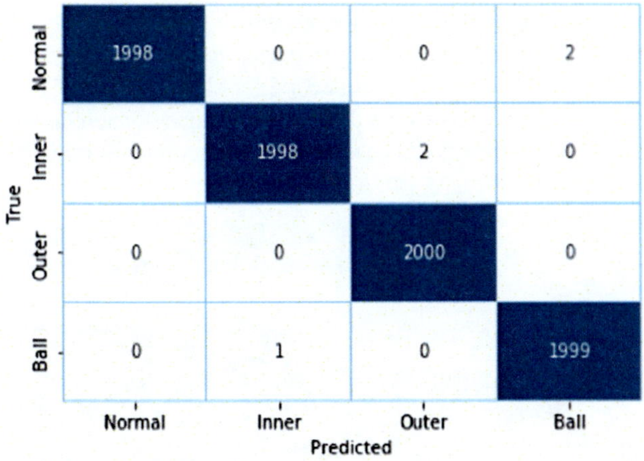

Fig. 1. Confusion matrix of results of SVM on activations of test data

This shows that activations of a CNN are good candidates to be used as features in a classification algorithm. There is no need to design relevant features for a classification algorithm. Instead, the features can be learned in a data-driven way. Overall accuracy of the SVM model on test activations is even higher than the accuracy of CNN model on same test data of different size. But this comes at the cost of more computational expenses. An SVM model needs to be trained and fine-tuned for this. The extra computational expenses increase the accuracy by 0.1%. Therefore, this method is best suited for explaining the activations of a network as opposed to using it for the classification task.

The data-driven features for each category also form distinct clusters when projected onto a low dimensional surface. t-SNE [14] projection of test activations is shown in Fig. 2.

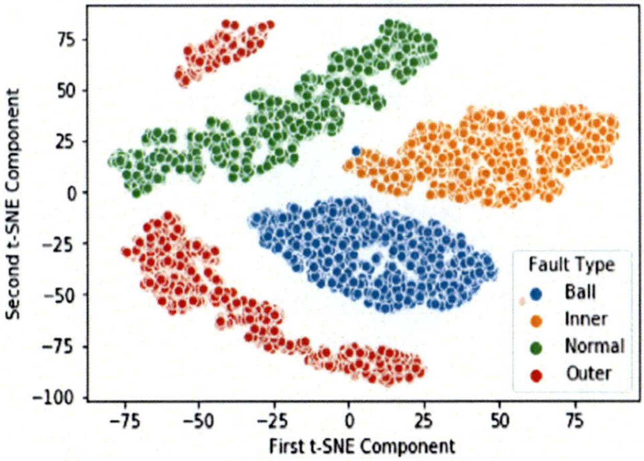

Fig. 2. t-SNE projection of activations of test data

Activations of test data can be further analysed to learn more about the working of different neurons of the network. If activations from a neuron are always zero irrespective of data, it means that for the neuron is not needed for classification of fault in CNN thus making it redundant. Active neurons are those that produce nonzero activation for given data. In this way, we can separate active neurons from redundant ones.

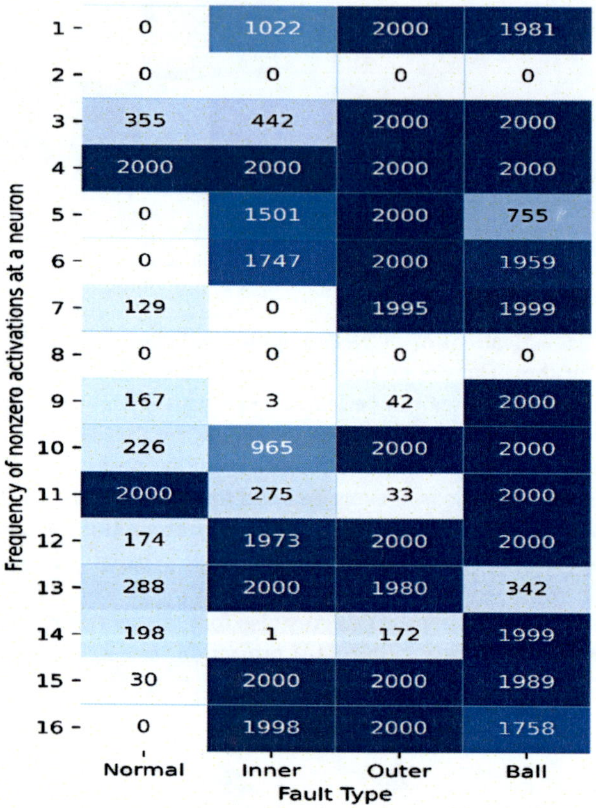

Fig. 3. Frequency count of nonzero neuron activations of test data

Figure 3 shows the frequency of activation of neurons of the penultimate layer for different fault types. For normal case, neuron 4 and 11 get activated every time whereas neurons 1, 2, 5, 6, 8, and 16 never get activated. Activation frequencies of other neurons are also comparatively small. Frequency count for other fault types can be read similarly. It should be noted that neuron 2 and 8 never get activated irrespective of fault type. This means that these neurons are redundant neurons for test data considered. The redundant neurons can be excluded from the network without affecting its accuracy.

6 Conclusion

Interpreting the working of a neural network is a difficult task. Though successive layers of a deep network learn abstract representations of input, visualizing those is possible for image inputs that have some spatial structure in it. For temporal signals like bearing fault signals, such a visualization is infeasible. Therefore, in this paper, we proposed activations of penultimate layer as a tool for explaining the result of the neural network. These activations can be used as features for further analysis. Computing these activations is very fast as these are obtained from a previously trained model. The features are learnt in a data-driven way by the network thus precluding any feature design step. Accuracy of classification algorithm using these features is also at par with the accuracy of deep neural network. Activations can also be used as a tool to find out redundant neurons in a network. These characteristics make data-driven features an excellent tool for interpreting the results of the neural network.

References

1. Mohanty, A.R.: Machinery Condition Monitoring: Principles and Practices. CRC Press, Boco Raton (2018)
2. Erhan, D., Bengio, Y., Courville, A., Vincent, P.: Visualizing higher-layer features of a deep network, University of Montreal, p. 1341 (2009)
3. Zeiler, M.D., Fergus, R.: Visualizing and understanding convolutional networks. In: Fleet, D., Pajdla, T., Schiele, B., Tuytelaars, T. (eds.) ECCV 2014. LNCS, vol. 8689, pp. 818–833. Springer, Cham (2014). https://doi.org/10.1007/978-3-319-10590-1_53
4. Sundararajan, M., Taly, A., Yan, Q.: Axiomatic attribution for deep networks. arXiv:1703.01365 [Cs] (2017). Accessed 10 Feb 2020
5. Janssens, O., et al.: Convolutional neural network based fault detection for rotating machinery. J. Sound Vib. **377**, 331–345 (2016). https://doi.org/10.1016/j.jsv.2016.05.027
6. Sun, W., et al.: An intelligent gear fault diagnosis methodology using a complex wavelet enhanced convolutional neural network. Materials **10**, 790 (2017). https://doi.org/10.3390/ma10070790
7. Liu, R., Meng, G., Yang, B., Sun, C., Chen, X.: Dislocated time series convolutional neural architecture: an intelligent fault diagnosis approach for electric machine. IEEE Trans. Industr. Inf. **13**, 1310–1320 (2017). https://doi.org/10.1109/TII.2016.2645238
8. Wen, L., Li, X., Gao, L., Zhang, Y.: A new convolutional neural network-based data-driven fault diagnosis method. IEEE Trans. Industr. Electron. **65**, 5990–5998 (2018). https://doi.org/10.1109/TIE.2017.2774777
9. These data come from National Aeronautics and Space Administration Website. (n.d.). https://ti.arc.nasa.gov/tech/dash/groups/pcoe/prognostic-data-repository/#bearing. Accessed 25 Nov 2018
10. Goodfellow, I., Bengio, Y., Courville, A.: Deep Learning. The MIT Press, Cambridge (2016)
11. Hastie, T., Tibshirani, R., Friedman, J.H.: The Elements of Statistical Learning: Data Mining, Inference, and Prediction, 2nd edn. Springer, New York (2009). https://doi.org/10.1007/978-0-387-84858-7
12. Gousseau, W., Antoni, J., Girardin, F., Griffaton, J.: Analysis of the rolling element bearing data set of the center for intelligent maintenance systems of the University of Cincinnati. In: CM2016 (2016)

13. Abadi, M., et al.: TensorFlow: a system for large-scale machine learning. arXiv:1605.08695 [Cs] (2016). Accessed 17 Jan 2021
14. van der Maaten, L., Hinton, G.: Visualizing data using t-SNE. J. Mach. Learn. Res. **9**, 2579–2605 (2008)

SVM Based Vibration Analysis for Effective Classification of Machine Conditions

D. Ganga[1(✉)] and V. Ramachandran[2]

[1] Department of EEE, NIT Nagaland, Dimapur, India
ganga@nitnagaland.ac.in
[2] Department of CSE, DMI College of Engineering, Chennai, India

Abstract. In condition monitoring, alarms have been the significant references for assessment of critical conditions. Generation of false alarms or suppression of defectiveness have often occurred while maintenance of machines due to incorrect setting of alarms appropriate to machine conditions. The evolution of data analytics and related technologies for real-time data acquisition as per industrial 4.0 standards will resolve such deceptive problems perceived till date. This paper attempts to elucidate a framework for identification and classification of vibration thresholds as per machine conditions using Internet of Things (IoT) based data acquisition and vibration analytics using support vector machines. The classification algorithm for the detection of higher vibrating range of machines has been tested with different sets of vibration features extracted. Incorporation of IoT devices facilitate unbounded and flexible data acquisition for carrying out detailed analytics in cloud environment. The kernel based support vector machine classifies the features extracted from signal processing into higher and lower vibrating levels of machine and fixes the higher vibration levels as thresholds for condition monitoring. The proposed alarm fixation model automatically classifies the maximum vibrating ranges of machines under different operating conditions. This automated condition monitoring model illustrates the efficacy of cloud environment for execution of exhaustive machine learning algorithms for condition monitoring. The performance analysis of the proposed model on the extracted vibration features validates the competence of machine learning algorithms for precise fixation of vibration thresholds and classification of machine conditions for effective maintenance.

Keywords: Vibration thresholds · Signal processing · Machine learning · IoT · Support Vector Machine

1 Introduction

Industry 4.0 transforms physical systems using smart technologies in multiple industrial sectors for efficient performance and maintenance. The electrical machines are undoubtedly the predominant systems in various plants and hence precise monitoring of their behavioral aspects has always been endeavored with utmost importance using cyber-physical techniques such as data mining and machine learning. The evolving Industrial IoT technology facilitates appreciable digital connectivity of machines to cloud and

R. Karim et al. (Eds.): IAI 2021, LNME, pp. 415–423, 2022.
https://doi.org/10.1007/978-3-030-93639-6_36

collection of abundant machine data from sensors mounted on them. Though machine learning technique such as Support Vector Machine had been a strategy in the research of machine maintenance as cited by Afrooz Purarjomandlangrudi et al. [1] to perform anomaly detection and condition assessment, incorporation of classification technique in the present environment of IoT and cloud will render better data handling, computation capacity and execution time. Due to accumulation of variety of machine data in IIoT, the efficiency of the algorithm in learning the insights of data patterns also becomes higher. The outcome of machine learning is highly dependent on the data features used for learning. This paper focuses on the classification of machine condition with non-stationary vibration signal encountered during dynamic machine conditions using machine learning algorithm. The challenges in the classification of non-stationary vibration signal has been addressed through combination of non-stationary signal processing, machine learning and IoT. Machine learning is a sector of data analytics which mines the data to automatically identify the embedded data patterns for making predictions and appropriate decisions. The algorithms either learn from the labeled data set as in supervised learning or infers the data patterns hidden in the unlabeled data set as in unsupervised learning. Some algorithms categorized as reinforcement learning interacts with the users, obtain and learns from the feedback to generate context specific solutions. The cloud based implementation of machine learning algorithm on the machine's vibration data acquired through IoT gateway shows the potential of IoT in predictive maintenance of machines. The integration of machine learning algorithm in the cloud substantiates the adoption of IoT for the transformation of conventional machine maintenance into IoT based maintenance. This is observed to surpass the conventional methods of machinery diagnostics in the aspects scalability, enhanced analytics, integrated monitoring and precise decision making on anticipated threats at the incipient stage itself. Due to ease and competence of data analytics, maintenance of machines under appropriate operational limits is guaranteed with IoT and cloud based machine maintenance. The industries presently explore and devise various analysis techniques for condition monitoring of machines subjected to dynamic conditions. The invent of IoT technologies forces transition in the current condition monitoring strategies and such development of new techniques is becoming inevitable.

In machinery diagnostics, vibration signals have always been potential indicators of machine conditions. Although vibration signals of rotating machines are highly non-stationary, several processing methods have been adopted to extract the signal features. The machines often undergoing different operating modes possess dynamic vibration behavior which calls for optimum threshold fixation, to evade the implications of false alarms and ensure safe operations. It is predominantly observed that the same type of electrical machines vibrates differently when placed in changing environment and operational modes. Hence it is understood that instead of maintaining fixed alarms specified either by manufacturer or in historical data, flexible computation of adaptive thresholds is of profound importance and will suffice the core objective of predictive maintenance. Hence, for accurate and faster condition monitoring of electrical machines more focus is required on the application of efficient techniques for precise information extraction and detection of adaptive thresholds as per individual machine's working characteristics.

2 State of the Art

Data driven analysis and classification methodologies have been employed in variety of systems with different objectives. Youichi Nonaka and Sudhanshu Gaur [2] have stated that industries carry out better management of production performances, optimize processes and load distribution using machine learning techniques. Convolutional Neural Network is used by Haedong Jeong et al. [3] for autonomous pattern recognition of orbit images and thereby to classify fault modes of rotating machines. As discussed by Gang Zhao et al. [4] classification rules and data mining techniques have been used in machinery fault diagnostics for quite a long period. In their work, the standard values of vibration frequencies at different fault conditions were used for classification of faults using Decision Tree Classifier. In line with that, Deokwoo Jung et al. [5] had described about the calculation of RUL of the pumps using the vibration frequency spectrum. Based on the labelled features, pumps were classified and decisions were made for further maintenance. In general, spectral features of the vibration become distinct only on occurrence of faults. Hence the analysis and classification carried out are observed to be suitable for post fault analysis. Moreover, non-stationary spectral features occurring during dynamic conditions become a challenge while implementing the classification methods. Achmad Widodo et al. [6] have reviewed elaborately about the application of SVM in machinery fault diagnosis highlighting the problem that has been addressed. They had pointed on the drawback of having huge features extracted from the signal on the classification accuracy. In this aspect, application of Principal Component Analysis for extraction of optimal features and reduction of dimensionality had been emphasized. The impacts of selection of data features on speed of classification and accuracy have been discussed. It is also explained that uncertain data inputs need combination of intelligent methods and Support Vector Machine (SVM) for achieving better results. As per Bain's report [7], predictive maintenance said to be the most awaited application of IoT in 2016 has lost its impetus in the due to course of time because of integration issues. Yet it had predicted Industrial Internet of Things (IIOT) market to double by 2021. The report says that augmented reality, energy management solutions and software deficiencies are the issues need to be addressed for the wide acceptance of IIOT. The concept of connecting motor data to cloud using Programmable Logic Controller (PLC) in the IIOT environment is illustrated by Ameeth Kanawaday et al. [8] The potentials available with IoT for prognostics and health management of assets in various industrial sectors are discussed by Daeil Kwon et al. [9] The significant progress made by various industries in the fields of manufacturing, automobiles, smart grids, robotics and industrial asset management were also highlighted. The impending challenges in IoT based health management in various aspects such as analytics, security, platforms etc., have been well discussed. Ganga D and Ramachandran V [10] have proposed an IoT based vibration analytics for electrical machines in which the non-stationary vibration signals were processed by statistical classification in order to extract the number of oscillations happening between different amplitude levels of vibration. The proposed algorithm was tested with the vibration data acquired by two different devices namely myRIO and IoT2040 at different sampling rates which assures the effectiveness of the algorithm in extracting the signal features. As an extension, this paper discusses about the application of classification technique on the features extracted through which the denser vibration regions of machine for different operating conditions are classifies as well as the thresholds were determined.

3 Experimental Set up and Implementation

The framework of the experimental set up used is shown in Fig. 1 which has an IoT device for the acquisition of vibration data from three phase induction motor and transfer to cloud environment for inherent classification of motor conditions.

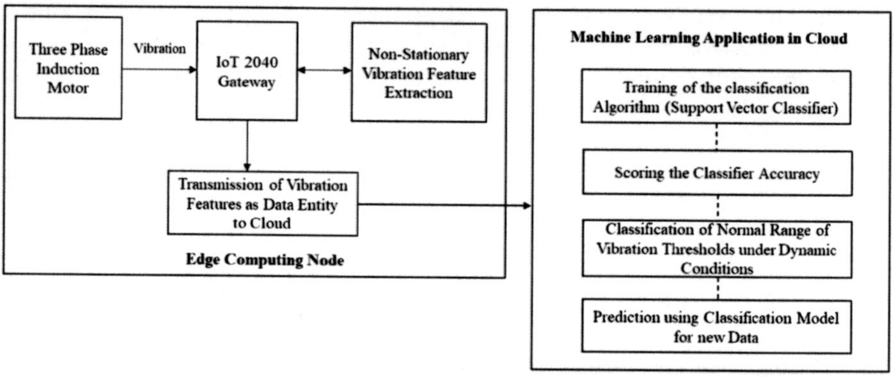

Fig. 1. Schematic flow of IoT based classification of machine condition

The IoT 2040 gateway used as an edge computing node collects the raw non-stationary shaft vibration data from three phase induction motor. The DIAdem and LabVIEW computational engine in the edge computing node processes the acquired raw vibration data using statistical classification algorithm and computes the oscillation percentage at each class. The python application running in IoT2040 passes the vibration data to LabVIEW framework to process the vibration data into features namely 'oscillation percentage' and 'class mean' for various operating conditions of the motor. It also reads the processed features from the LabVIEW framework and streams to the cloud platform intermittently for classification of vibrating regions of the motor and thereby its condition.

4 Supervised Classification Model for Condition Assessment and Threshold Determination

A supervised classification model is developed for the non-stationary machine vibration data for classification of higher vibration levels during different operating conditions so that a range of normal vibrating levels are classified as alarm levels of normal conditions instead of distinct values. Exploring the pattern of data features, Support Vector Machine has been chosen as the algorithm for classification. The algorithm identifies the decision boundaries as a hyperplane in an 'n' dimensional space or a line in the 2-D space to distinctly classify the data features considered for analysis. The dot product of support vectors and the input data that forms the kernel optimally adjusts the orientation of the hyperplane for better data classification. This kernel based classifier is applied in this work to classify the vibration feature space according to the classes labeled in the training dataset. The SVM is implemented in Python using 'pandas', 'numpy', sklearn, and matplotlib python packages. By plotting the data in scatter plot and observing the nature of the data pattern, the kernel for classification chosen is 'linear'. While training the machine learning model, the SVM model parameters, regularization and gamma are tuned in line with the data nature so that the accuracy and performance of the classification model is improved. This kernel based support vector classifier with tuned parameters trains the vibration data features to segregate into classes of "higher oscillation percentage – Class 1" and "lower oscillation percentage – Class 0". This supervised binary classification algorithm implemented in cloud classifies the vibration data features streamed by IoT2040 from the edge computing node which enables to identify the range of vibration threshold and classify the machine condition. The strategy of automatic classification of machine vibrations perceived during dynamic conditions as normal or abnormal using machine learning algorithm aids in reliable condition classification. This feasibility of data analysis is facilitated with the availability of adequate resources for data collection, advanced computation and storage platforms.

The vibration data of a healthy 1 HP three phase induction motor are acquired by IoT2040 at different machine operating and environmental conditions. The length of the vibration data acquired at different conditions is 2,00,000 samples. Instead of transmitting 2,00,000 samples of vibration data of non-stationary nature to cloud and increasing the complexity of learning the data pattern, the data features are processed at the edge node and data features of reduced length are pushed to cloud by IoT 2040. The machine learning application is executed in cloud on this compressed data to classify the vibration levels that hold higher oscillation percentage for different machine operating conditions of starting and loading by choosing the linear kernel classifier. The amplitude levels for which the oscillation density are higher than 10 percent have been labeled as "higher oscillation percentage – Class 1" for each of the operating condition. The classification contours for the chosen operating conditions and the denser vibration levels are illustrated Fig. 2 and Table 1.

4.1 Validation of Classification with New Data Features

Different set of vibration features extracted from the same motor have been tested with support vector machine model for classification of its operating condition. In spite of the minor deviations in the extracted values of new vibration features namely 'class mean' and 'oscillation density', the classification algorithm has performed well in the classification of higher vibration regions of the machine. The results of classification contours are shown in Fig. 3 and the range of denser vibration levels at different operating conditions are given in Table 1.

Table 1. Range of denser vibration levels

Case	Operating condition	Classified levels of denser vibration (LSB)	Alarm limit
1	Starting to no load condition	14 to 232	−4 to 232
	Starting to no load under external disturbance condition	−2 to 213	
	Loaded condition	−4 to 206	
2	Starting to no load condition	18 to 225	−1 to 225
	Starting to no load under external disturbance condition	−1 to 216	
	Loaded condition	1 to 218	

It is observed from the results that the segregation of denser vibration levels using SVM as classification algorithm has facilitated easy identification of safe operating region of motor under different conditions. The SVM classification analysis carried out on two sets of data features pertaining to three operating conditions show the denser vibration regions to range between −4 to 232 LSB (Least Significant Byte) in case 1 and −1 to 225 LSB in case 2 as given in the classification results tabulated in Table 1.

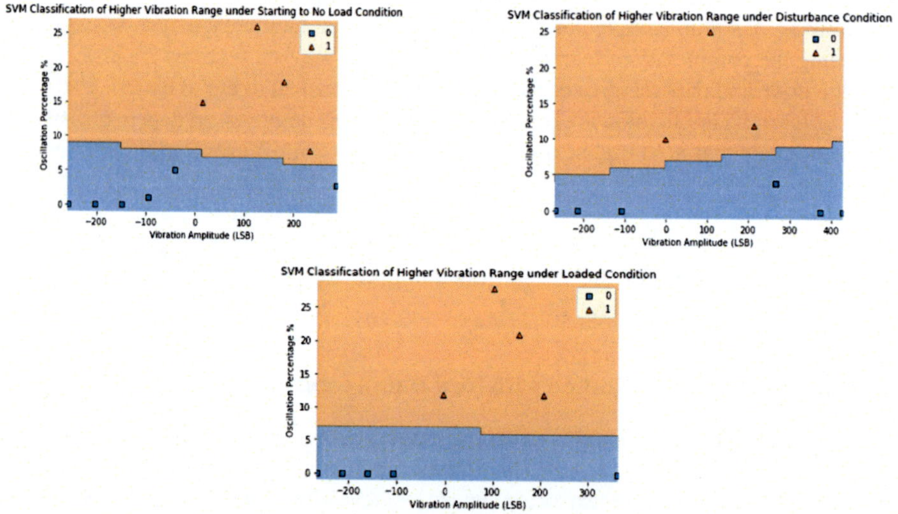

Fig. 2. Classification of denser vibration amplitude ranges – Test Case 1

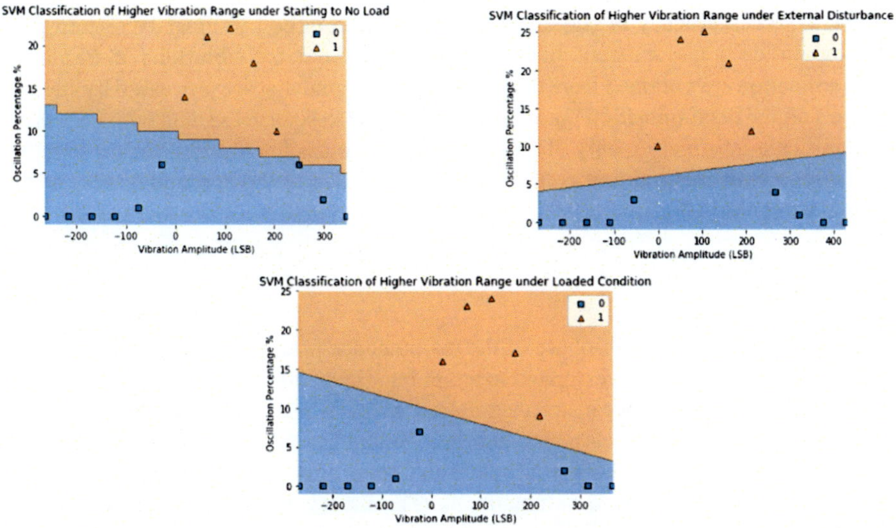

Fig. 3. Classification of denser vibration amplitude ranges – Test Case 2

Due to borderline variations observed between the two alarm limits identified in Table 1, better margin of −4 to 232 LSB has been chosen as the safer vibrating region of the machine under normal operation. In both the cases, the support vector classifier is found to fit accurately. It is conceived that the application of classification technique on vibration features leads to identification of reliable thresholds which enables effective determination of machine condition as normal or anomalous. The performance of the support vector classifier is measured in terms of model accuracy, precision, recall and F1 score. The results given in Table 2 substantiate the precise fitness of the classification model on the vibration features of the electrical machine.

Table 2. Performance of support vector classifier

Class	Precision	Recall	F1-score	Support
0	1	1	1	1
1	1	1	1	2
Accuracy				
Macro Avg.	1	1	1	3
Weighted Avg.	1	1	1	3

The observed efficacy in classification is perceived to be significantly related with data preparation in major extent. In the analysis carried out, the vibration features taken for classification are obtained from the raw vibration signal that are processed by statistical based signal decomposition algorithm which aided to squash the data with necessary and significant information only. It is realized that the classifier has resulted an effective performance with less training complexity (Table 2) due to the appropriateness of the training data chosen.

5 Conclusion

Decision making is the major phase of the maintenance practice towards which variety of analytic techniques are investigated to attain intelligent and reliable decisions, higher uptime, scalability and reduce cost consumption. In this work, binary classification using support vector machine as classifier has been implemented to classify the vibration amplitude levels of the machine under different conditions so that optimal monitoring of machine conditions happen automatically. The illustrated IoT framework incorporating machine learning technique proves to be effective in providing these expected outcomes in machinery prognostics. Cognitive data reduction due to edge computing is also observed as one of major factors for classification accuracy obtained in this IoT framework. In the experimental study made with physical machines, IoT and cloud network in real time, the vibration data with length of 3,00,000 samples are reduced cognitively to the size of data set of size 70×2 and classification analysis has been carried out using the reduced data. It is well known that data reduction reduces the transmission overheads such as network delay, bandwidth, data storage, data handling and computational time which are often said to be the major challenges in IoT and cloud based systems. Moreover, the accuracy of classification is also seen to be result of precise data reduction. This shows the potential of edge computing node in IoT based data analysis for decision making. Hence the framework devised in this research work with the developed non-stationary data analysis and IoT-cloud integration techniques will facilitate the end users or organizations to accomplish universal, easy and trustable predictive maintenance of electrical machines. In contrast to the widely adopted classification problems where the classifiers are used to determine the class to which the data belongs to, this paper proposes classification approach for identification of thresholds to be set as alarms.

References

1. Purarjomandlangrudi, A., Ghapanchi, A.H., Esmalifalak, M.: A data mining approach for fault diagnosis: an application of anomaly detection algorithm. Measurement **55**, 343–352 (2014)
2. Nonaka, Y., Gaur, S.: Factories of the future (2017). https://www.hitachinext.com/en-us/pdfd/presentation/factories-of-future.pdf. Accessed June 2021
3. Jeong, H., Park, S., Woo, S., Lee, S.: Rotating machinery diagnostics using deep learning on orbit plot images. Procedia Manuf. **5**, 1107–1118 (2016)
4. Zhao, G., Jiang, D., Li, K., Diao, J.: Data mining for fault diagnosis and machine learning for rotating machinery. Key Eng. Mater. **293–294**, 175–182 (2005)

5. Jung, D., Zhang, Z., Winslett, M.: Vibration analysis for IoT enabled predictive maintenance. In: Proceedings of IEEE 33rd International Conference on Data Engineering, pp. 1271–1282 (2017)
6. Widodo, A., Yang, B.-S.: Support vector machine in machine condition monitoring and fault diagnosis. Mech. Syst. Sig. Process. **21**, 2560–2574 (2007)
7. Paul, F.: Techwatch, Why IoT-enabled predictive maintenance hasn't taken off (2019). https://www.networkworld.com/article/3340132/why-predictive-maintenance-hasn-t-taken-off-as-expected.html. Accessed June 2021
8. Kanawaday, A., Sane, A.: Machine learning for predictive maintenance of industrial machines using IoT sensor data. In: Proceedings of 8th IEEE International Conference on Software Engineering and Service Science, pp. 2327–2594 (2018)
9. Kwon, D., Hodkiewicz, M.R., Fan, J., Shibutani, T., Pecht, M.G.: IoT-based prognostics and systems health management for industrial applications. IEEE Access **4**, 3659–3670 (2016)
10. Ganga, D., Ramachandran, V.: IoT-based vibration analytics of electrical machines. IEEE Internet Things J. **5**(6), 4538–4549 (2018)

An Experimental Study on Railway Ballast Degradation Under Cyclic Loading

Elahe Talebiahooie[1(✉)], Florian Thiery[1], Hans Mattsson[2], and Matti Rantatalo[1]

[1] Division of Operation and Maintenance Engineering, Luleå University of Technology, 971 87 Luleå, Sweden
{Elahe.Talebiahooiei,florian.thiery,Matti.Rantatalo}@ltu.se
[2] Division of Mining and Geotechnical Engineering, Luleå University of Technology, 971 87 Luleå, Sweden
Hans.Mattsson@ltu.se

Abstract. To investigate the degradation of railway track ballast layer, triaxial tests, full-scale railways test facilities or models of a section can be used. In this study, a half-section scaled model of a ballasted track consisting of a steel sleeper and a metal box filled with aggregates was developed. The impact of cyclic loading on the ballast degradation was investigated using the experimental model considering ballast rearrangement and breakage. The frequency of the cyclic loading was set to 8 Hz, which mimics a train passage with an axle wheel spacing of 2.6 m and a speed of 70 km/h. MakAdam aggregates from Skutskär quarry in Sweden were used in the tests. The size of the ballast aggregates was scaled down by the parallel-gradation scaling technique. The results demonstrate that vertical ballast deformation exhibited an exponential trend at the beginning of the cyclic loading followed by a linear trend. The ballast aggregates mainly broke at their corners rather than in the middle after 100,000 loading cycles.

Keywords: Ballasted track · Cyclic loading · Degradation · Breakage

1 Introduction

The substructure of ballasted tracks consists of ballast, subballast and subgrade, and the superstructure consists of the rail, sleeper, and fastening system. Ballast plays a major role in the transmission of loads from the train passage to the subgrade. It is important to understand the behaviour of the ballast layer under loading, such as wear of the aggregates and settlement of the ballast.

To understand the degradation of the ballast layer, different experimental approaches have been employed. For instance, triaxial and shear-box tests have been employed to measure the shear strength and friction angle of the ballast.

Railway test facilities are also used to investigate ballast degradation. These test facilities can be used to investigate the degradation of ballast under loading or determine the correlation between the frequency response of the track and the condition of the trackbed. Bold and Paul [4] studied the railway trackbed frequency response of clean, spent, and mixed ballast. They achieved a 94% correlation between the fouling index

R. Karim et al. (Eds.): IAI 2021, LNME, pp. 424–433, 2022.
https://doi.org/10.1007/978-3-030-93639-6_37

and the stiffness of the ballast. Tutumluer et al. [13] assessed the degradation of four ballast materials at the Transportation Technology Center, Pueblo, Colorado. This study verified the results from discrete element model (DEM) of the ballasted track.

Railway test facilities are expensive, and the tests are time consuming. Consequently, some studies have considered just a small section of the track to investigate. Al-Saoudi and Khawla [1] reported that this type of model can give sufficient insight into the ballast degradation behaviour even though there are some limitations. The selected section of the track can be a half-section of the track, consisting of two sleepers and a rail embedded in a box of ballast aggregates, similar to the test apparatus used in the study by Liu et al. [9], or even half of one sleeper instead of two sleepers, used in the study by Fathali et al. [5].

In most experimental studies on ballast degradation, the real size of the ballast aggregates is used. Due to difficulties arising from the size of the apparatuses, Sevi [10] investigated the scaling of the ballast aggregates in triaxial tests, and concluded that the parallel-gradation scaling technique is a useful approach for scaling down the ballast aggregates.

The amount of degradation of ballast aggregates is affected by several properties of the aggregates, the surface texture and roughness. The distribution of contact stress also affects the breakage rate of the aggregates, where the angularity and shape are key factors. Although angular aggregates are easily crushed compared to round aggregates, they promote the interlocking phenomena. Besides individual properties of ballast aggregates, the bulk properties, such as the grain-size distribution and coefficient of uniformity, affect the degradation. The bulk properties affect the contact stress and coordination number (number of contacts for each particle).

In this study, the parallel-gradation scaling technique was employed to scale down the ballast aggregates for three sets of tests. The model consists of a ballast layer and a half-section of the sleeper with rail. The data gathered from the tests were further used for (DEM) simulations of the ballast degradation.

2 Experimental Investigation

Figure 1 shows a schematic of the model and section of the track considered for the tests. The model dimensions are half the dimensions of the section of a conventional ballasted track.

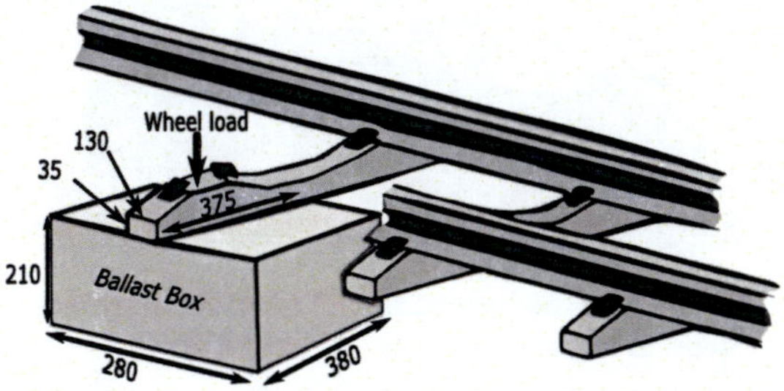

Fig. 1. Section of track selected for tests and its boundary condition. Dimensions are in mm.

Fig. 2. Oven used to dry the washed aggregates.

Fig. 3. Metal box and the sleeper experimental set-up. Interior dimensions of the metal box and the dimensions of the sleeper are in mm.

The ballast aggregates were washed and dried in an oven for 12 h (Fig. 2). Thereafter, they were sieved and then mixed based on the parallel-gradation scaling technique so that the grading of the mixture was parallel to the grain-size distribution category F from the Swedish standard of aggregates for railway ballast [11].

The metal box was filled with the prepared ballast aggregates to a height of 170 mm, and the sleeper was placed on top of the ballast. The metal box and sleeper are shown in Fig. 3.

The apparatus used to load the sleeper is shown in Fig. 4. On both sides of the metal box, two metal handles were welded to fix the box in place during the loading by the apparatus. The loading capacity of the apparatus was 600 kN, which is equivalent to a maximum stress of 12308 kPa during the sleeper/ballast contact.

Instead of vibrations, a preloading stage was used to pack the ballast aggregates before the cyclic loading. After the preloading stage and 100,000 loading cycles, the aggregates from each test were sieved and the amount of particle breakage was determined.

Fig. 4. Apparatus which was used for preloading and cyclic loading on the sleeper.

2.1 Aggregates Description

Ballast aggregates for the tests were collected from the Skutskär quarry in Sweden. Based on the petrography documents provided by the supplier, the aggregates are made of diorite. They are mainly composed of plagioclase feldspar, amphibole, and biotite. The quartz content is lower than 5% and the sulphide content is about 1%.

The effective size D_{10} is the size of aggregates of which 10% of the particles are smaller than that, and other sizes, such as D_{30} and D_{60}, are defined in a similar way [3]. The C_U and coefficient of curvature (C_Z) are used to describe the general slope and shape of the grain-size distribution curve as follow:

$$C_U = \frac{D_{60}}{D_{10}} \tag{1}$$

$$C_Z = \frac{{D_{30}}^2}{D_{60}D_{10}} \tag{2}$$

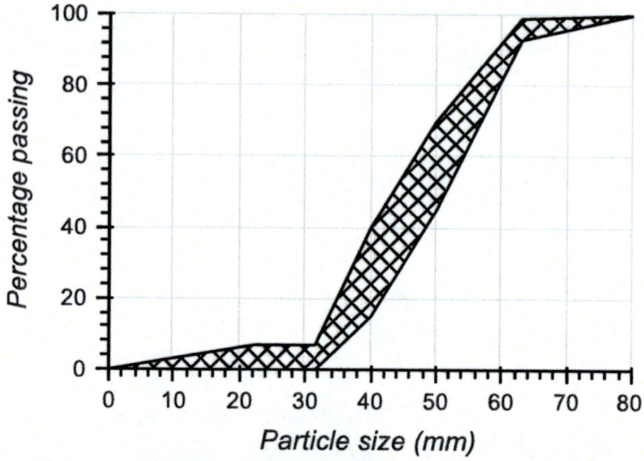

Fig. 5. Grain-size distribution category F from Swedish standard of aggregates for railway ballast [11].

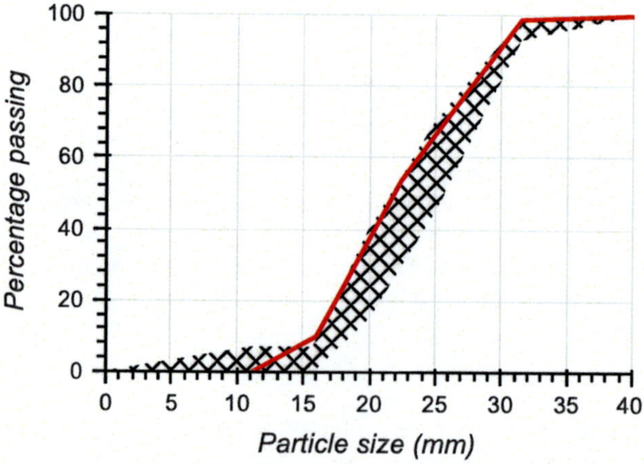

Fig. 6. Grain-size distribution category F scaled down by parallel-gradation scaling technique.

The grain-size distribution from category F is shown in Fig. 5, which is hatched with diagonal crossing. Based on the standard, the size of more than 85% of the particles in category F should range between 31.5 and 63 mm. The size of aggregates used in the experiments is from category F, it was but scaled down by 0.5 (Fig. 6), since the size of the track section used to perform the tests was scaled down by 0.5. The red line in Fig. 6 indicates the initial grain-size distribution for the tests. The C_U and C_Z of the initial ballast aggregates were 1.49 and 0.95, respectively.

2.2 Loading

Assuming an axle wheel spacing of 2.6 m and a train speed of 75–94 km/h, the typical loading frequency of traffic load is about 8–10 Hz for a normal train [2]. The loading frequency can be calculated by dividing the train speed by the axle wheel spacing.

Aursudkij et al. [2] reported that in laboratory tests, frequency does not affect the results since defects in the track, such as wheel flat or welded joints, do not exist in the laboratory. However Thakur et al. [12] performed cyclic triaxial tests with different frequencies on samples. Increasing the frequency from 10 to 40 Hz during cyclic loading increased the axial strain. Kim and Tutumluer [8] also showed that increasing the frequency of loading from 2 to 10 Hz increases the vertical permanent strain of the specimen.

Based on the capacity of the apparatus, a loading frequency of 8 Hz was used in this study. Since the settlement in the first cycles of loading is larger, the ability of hydraulic oil flow of the apparatus to produce the required deformation rate needs to be verified. Loading with the selected frequency was appropriate for the apparatus.

In the preloading stage, the sleeper was moved slowly toward the surface of the ballast and the load gradually increased from 1 kN to 21 kN with a rate of 1 kN/s. Then the loading decreased to 12 kN at a rate of −1 kN/s (Fig. 7). The midpoint of the cyclic loading on the sleeper was 11 kN. At this point, the cyclic loading had started. 100,000 loading cycles with a maximum of 18 kN and a minimum of 4 kN was performed on the sleeper as seen in Fig. 8.

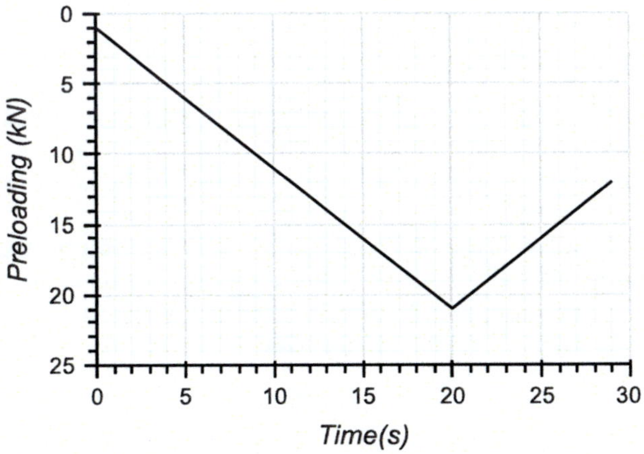

Fig. 7. Preloading on the sleeper.

The length and the width of the sleeper are 375 mm and 130 mm, respectively. The stress on the sleeper-ballast contact cycles was between 82 and 369 kPa, which is consistent with the contact stress in the experimental study by Indraratna et al. [6].

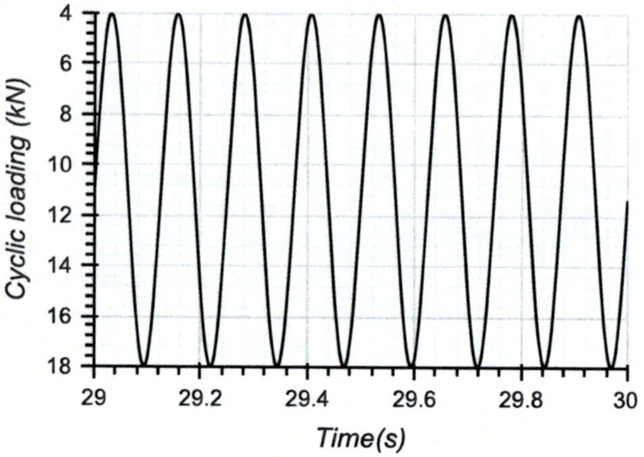

Fig. 8. Cyclic loading on the sleeper.

3 Results

The preloading on the sleeper is shown in Fig. 9. During the first 20 s of preloading, a sudden increase in the exported preloading data was observed. This sudden increase is attributed to either the breakage of the aggregates or the sudden relocation of ballast aggregates resulting from the release of interlocks between particles. In the second part of the preloading, where the stress on the sleeper decreased, there was no sudden increase as shown in Fig. 9.

The settlement of the sleeper in the preloading stage is shown in Fig. 10. The sudden increase in preloading (Fig. 9) correspond to the abrupt change in the settlement rate (Fig. 10). The settlements of the sleeper for the three samples, during the cyclic loading are shown in Fig. 11. The settlement in the loading stage exhibited an exponential behavior at the beginning of loading and a linear behavior after 10,000 cycles. About 85% of the settlement in the three tests occurred in the first 10% of the cyclic loading.

The settlement of the sleeper during the cyclic loading involves elastic and plastic deformation of the ballast. Plastic deformation of the ballast is important to the tamping and predictive maintenance, and the elastic deformation affects dynamic track train interactions. The elastic deformation in the tests was approximately 0.1 mm based on the unloading response in the cycles.

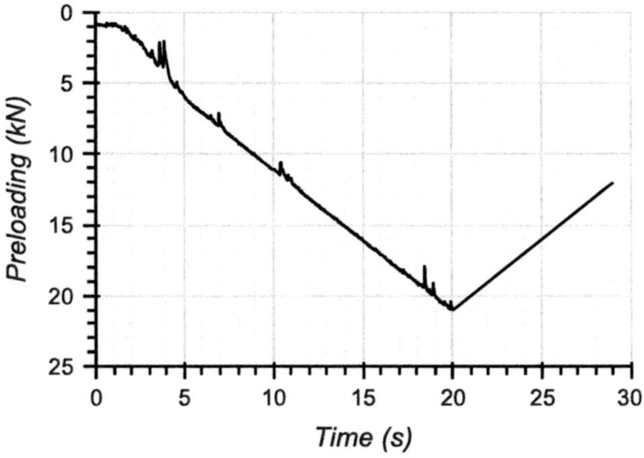

Fig. 9. Preloading data exported from apparatus.

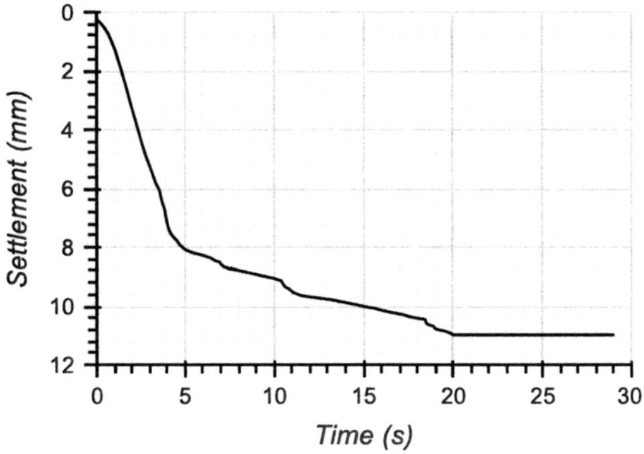

Fig. 10. Settlement of sleeper in preloading stage.

After 100,000 loading cycles for each specimen, the specimens were sieved and the particle-size distribution were obtained (Table 1). The initial aggregates used in the tests contained no grains smaller than 11.2 mm. During loading, particles smaller than 11.2 mm were produced in the samples.

Based on the sieve analysis, the percentage of aggregates retained on the sieve with the size of 31.5 and 22.4 mm decreased between 0.6% and 4.4%, and it slightly increased for smaller sieve sizes. The particles produced during loading were much smaller than half of the largest particles in the initial mixture. Due to the sieve analysis, the ballast aggregates mostly broke at corners.

Table 1. The sieve size analysis of 3 samples at the end of loading compared with the initial condition

Sieve size	Initial	Sample-1	Sample-2	Sample-3
mm	% retained	% retained	% retained	% retained
Pan	0	0.27	0.23	0.27
4	0	0.05	0.04	0.05
5.6	0	0.09	0.07	0.10
8	0	0.4	0.22	0.24
11.2	10.17	12.32	9.89	11.65
16	43.60	46.11	44.75	45.96
22.4	45.05	40.58	44.23	41.51
31.5	1.16	0.16	0.55	0.18
40	0	0	0	0

Indraratna et al. [7] studied the effect of the frequency of loading on the breakage of ballast aggregates. They found that lower frequency loading results in the breakage of the corners of the ballast aggregates, and higher frequency loading results in the splitting of the particles. A frequency of 8 Hz was considered low, which is consistent with our results.

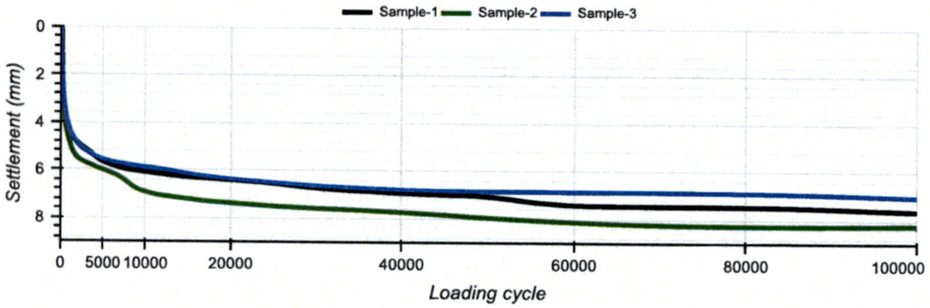

Fig. 11. Settlement of the sleeper in the loading stage.

4 Conclusion and Future Work

In this study, a half section of the track under a sleeper and rail was selected to investigate the sleeper settlement and the breakage and wear of the ballast aggregates under cyclic loading. The dimensions of the model were scaled down to half of the conventional ballasted track. Hence, the ballast aggregates were also scaled down using the parallel-gradation scaling technique.

The ballast aggregates were washed and dried before sieving. The initial grain-size distribution of the ballast used in the tests was parallel to that of the category F railway ballast based on the Swedish standard [11].

To pack the aggregates before cyclic loading, preloading was performed on the sleeper. After preloading, 100,000 loading cycles were performed on the sleeper at a frequency of 8 Hz. The ballast aggregates were sieved, and based on the grain-size distribution it can be concluded that most of the breakage occurred at the corners of the ballast. For the aggregates smaller than 16 mm, 11.2, 8 and 4 mm sieves were used.

The vertical settlement of the sleeper during cyclic loading showed an exponential trend at the beginning, followed by a linear trend. During the tests, about 85% of the settlement occurred in the first 10% of cycles.

The data obtained in this study will be used to simulate the degradation and wear of the ballast using DEM. The type of aggregates breakage herein will determine the model to be used for the simulations to mimic the ballast aggregates. The magnitude of elastic deformation of the sleeper, beside the plastic deformation, also affects the choice of models for the simulation.

References

1. Al-Saoudi, N.K., Hassan, K.H.: Behaviour of track ballast under repeated loading. Geotech. Geol. Eng. **32**(1), 167–178 (2014). https://doi.org/10.1007/s10706-013-9701-z
2. Aursudkij, B., McDowell, G.R., Collop, A.C.: Cyclic loading of railway ballast under triaxial conditions and in a railway test facility. Granular Matter **11**, 391–401 (2009). https://doi.org/10.1007/s10035-009-0144-4
3. Craig, R.F.: Craig's Soil Mechanics. CRC Press, Boca Raton (2004)
4. De Bold, R.P.: Non-destructive evaluation of railway trackbed ballast. The University of Edinburgh (2011)
5. Fathali, M., Nejad, F.M., Esmaeili, M.: Influence of tire-derived aggregates on the properties of railway ballast material. J. Mater. Civ. Eng. **29**(1), 04016177 (2017)
6. Indraratna, B., Hussaini, S.K.K., Vinod, J.: The lateral displacement response of geogrid-reinforced ballast under cyclic loading. Geotext. Geomembr. **39**, 20–29 (2013)
7. Indraratna, B., Thakur, P.K., Vinod, J.S.: Experimental and numerical study of railway ballast behavior under cyclic loading. Int. J. Geomech. **10**(4), 136–144 (2010)
8. Kim, I.T., Tutumluer, E.: Field validation of airport pavement granular layer rutting predictions. Transp. Res. Rec. **1952**(1), 48–57 (2006)
9. Liu, S., Huang, H., Qiu, T., Gao, L.: Comparison of laboratory testing using smartrock and discrete element modeling of ballast particle movement. J. Mater. Civ. Eng. **29**(3), D6016001 (2017)
10. Sevi, A.F.: Physical modeling of railroad ballast using the parallel gradation scaling technique within the cyclical triaxial framework. Ph.D. thesis, Missouri University of Science and Technology (2008)
11. Swedish Standards Institute: Svensk Standard Makadamballast för järnväg Aggregates for Railway Ballast (2003)
12. Thakur, P.K., Vinod, J.S., Indraratna, B.: Effect of confining pressure and frequency on the deformation of ballast. Géotechnique **63**(9), 786–790 (2013)
13. Tutumluer, E., Qian, Y., Hashash, Y., Ghaboussi, J., Davis, D.D.: Field validated discrete element model for railroad ballast. In: Proceedings of the Annual Conference of the American Railway Engineering and Maintenance-of-Way Association, pp. 18–21 (2011)

Research on Visual Detection Method of Cantilever Beam Cracks Based on Vibration Modal Shapes

Rongfeng Deng[1], Yubin Lin[1(✉)], Baoshan Huang[1], Hui Zhang[2], Fengshou Gu[3(✉)], and Andrew D. Ball[3]

[1] School of Industrial Automation, Beijing Institute of Technology, No. 6, Jinfeng Road, Zhuhai, Guangdong, China
{Rongfeng.Deng2,Yubin.Lin}@hud.ac.uk

[2] Programme of Computer Science and Technology, BNU-HKBU United International College, No. 2000, Jintong Road, Zhuhai, Guangdong, China
amyzhang@uic.edu.cn

[3] Centre for Efficiency and Performance Engineering, University of Huddersfield, Huddersfield H1 3DH, UK
{F.Gu,A.Ball}@hud.ac.uk

Abstract. A visual detection method is proposed in this paper to identify cracks in a cantilever beam crack. The method takes full advantages of high spatial resolution of image sensing, relying only on cost-effective ordinary frame rate camera to record the process of free vibration of the cantilever beam, and combines with singular value decomposition method to obtain the vibration mode shapes of the cantilever beam. Then modal shape differences from baseline are taken as the features for detection and diagnosis The effectiveness of the first-order vibration mode shape difference in cantilever beam crack size and location detection is verified by both simulation and experiment.

Keywords: Vibration mode shape · Crack detection · Cantilever beam · Singular value decomposition

1 Introduction

As the most common construction in building structures and mechanical structures, beams may develop cracks during service due to environmental effects or human factors, which in turn reduce the stiffness and strength and affect the durability and safety of the structure. In the past decades, many studies focusing on beam crack detection have been conducted. According to the type of signal acquisition, the methods of beam crack detection can be divided into two categories: contact detection and non-contact detection, where the contact detection mainly relies on vibration sensors, acoustic emission sensors [1, 2], optical fibers [3] and Piezoelectric sensors [4], while non-contact mainly relies on laser doppler vibrometer [5, 6], millimetre-wave radar [7] and image sensors [8]. Compared to contact detection methods, non-contact detection methods do not require

R. Karim et al. (Eds.): IAI 2021, LNME, pp. 434–444, 2022.
https://doi.org/10.1007/978-3-030-93639-6_38

the installation of sensors on the surface of the object under test, which not only shortens the detection cycle, but also reduces the detection cost.

The methods for crack detection of structures mainly based on structural parameters such as damping factors, nature frequencies and mode shapes. Many researches have been conducted to detect and localize damage based on the change of nature frequencies [9, 10], These methods take advantage of the fact that natural frequencies are easy to measure, but there were at least two reasons that limit the application of such methods. First, the damage of a structure may cause very small changes in natural frequencies, especially for larger structures like wind turbine blades or wind turbine towers. These changes may not be discovered due to measurement accuracy limitations. Second, variations in the mass distribution or even ambient temperature may cause uncertainties of the frequency change, which will increase the difficulty of data processing.

In order to overcome the above-mentioned difficulties, many researchers [11, 12] have turned their attentions to variation of vibration mode shapes. Related studies [13] have confirmed that the mode shapes are more sensitive to structural damage than the change in natural frequencies. However, these studies have used acceleration sensors or laser vibrometers for data acquisition. In order to obtain a high-precision mode shape, it is necessary to arrange a large number of sensors or increase laser vibration measurement points. obviously, such methods will be severely limited in practical applications.

This study will propose a visual-based vibration mode shape acquisition method, which gives full play to the advantages of high spatial resolution of image sensors, allowing the acquisition of high precision vibration mode shapes by relying only on cameras with ordinary frame rates, and based on which the detection of cracks in cantilever beams can be achieved.

2 Theoretical Background

2.1 Free Vibration of a Cantilever Beam

According to the beam bending theory in mechanics of materials, the free vibration of a cantilever beam with equal cross section satisfies the Euler-Bernoulli equation (Eq. 1) without considering damping.

$$\frac{\partial^4 y}{\partial x^4} + \frac{1}{a^2}\frac{\partial^2 y}{\partial t^2} = 0 \tag{1}$$

where $a = \sqrt{EI/\rho A}$, E is the modulus of elasticity of beam material, I is the moment of inertia of the beam cross-section, ρ is the material density, A is the cross-sectional area of the cantilever beam. Under the boundary condition of the cantilever beam, the natural frequency of the cantilever beam can be obtained as

$$\omega_i = \beta_i^2 a \quad i = 1, 2 \ldots \tag{2}$$

and the mode shape function satisfies the equation below

$$Y_i(x) = -\frac{\cos(\beta_i l) + ch(\beta_i l)}{\sin(\beta_i l) + \mathrm{sh}(\beta_i l)}[\mathrm{sh}(\beta_i x) - \sin(\beta_i x)] + \mathrm{ch}(\beta_i x) - \cos(\beta_i x) \tag{3}$$

where l is the length of the cantilever beam, $\beta_i l$ is determined by $cos(\beta_i l)ch(\beta_i l) = -1$. Once the cantilever beam has a fault, its natural frequency will change, which will lead to the corresponding change of vibration mode shape.

2.2 Mode Shape Extraction Based on SVD

Free vibration process of a cantilever beam recorded by a camera can be regarded as a local time-varying motion $vp(y,t)$ from the change of image intensity $I(y + vp(y,t))$, where y is the pixel coordinate. Meanwhile the vibration motion could also be expressed as a linear combination of all the modal responses by modal superposition

$$vp(y, t) = \mathbf{q}(t)\phi(y) = \sum_{i=1}^{k} q_i(t)\phi_i(y) \tag{4}$$

where $\mathbf{q} \in \mathbf{R}^{n\times k}$ is the modal response matrix, $\boldsymbol{\phi} \in \mathbf{R}^{k\times m}$ is the mode shape matrix, n represents the number of frames, m represents the number of pixels, k is the number of vibration modes. $q_i(t)$ represents the i-th modal coordinate, and $\phi_i(y)$ represents the i-th mode shape. Obviously, $\boldsymbol{\phi}$ has a high requirement for spatial resolution, and high-resolution images just meet this requirement.

In order to reconstruct the mode shapes from the free vibration process of the cantilever beam, the SVD method was selected in this study because of its excellent performance in orthogonal decomposition. According to SVD, a matrix $A \in R^{m\times n}$ could be decomposed into the following form:

$$A = USV^T \tag{5}$$

where U is an $m\times m$ orthogonal matrix, V is an $n\times n$ orthogonal matrix, and S is an $m\times n$ diagonal matrix with non-negative principal diagonal elements arranged from the largest to the smallest, and the non-principal diagonal elements are all 0. It has been proved that for an undamped or very lightly damped structure with uniform mass distribution, V obtained by SVD is the mode shape vector matrix [14].

3 Simulation Study

In order to verify the influence of cracks on the vibration mode shape of the cantilever beam, the simulation analysis was carried out on the Matlab platform according to reference [15]. The total length of the cantilever beam with uniform mass distribution l = 550 mm, the width b = 25 mm, and the thickness h = 1 mm. The elasticity modulus of the material was 206 GPa, and the mass density was 7800 kg/m^3. Divide the cantilever beam into 550 units along its length, the first four mode shapes were obtained together with their corresponding frequencies as shown in Fig. 1.

The section width at 110 mm from the fixed end of the cantilever beam was reduced by 10% to simulate a cracked failure at the corresponding location. The same excitation was applied to the free end of the healthy cantilever beam and the cracked cantilever beam, and the vibration processes were recorded at a sampling rate of 20 Hz for 20 s. The

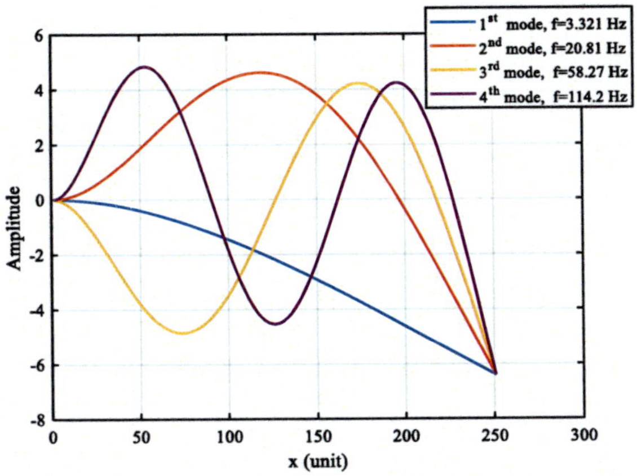

Fig. 1. The first four mode shapes of the cantilever beam.

first four vibration mode shapes were obtained by the SVD method described in Sect. 2.2. The differences of the first four mode shapes from the baseline and the cracked cantilever beam are shown in Fig. 2(a). Although the first four mode shapes of the baseline and the cracked cantilever beam are in good agreement, the differences between them still exist, as shown in Fig. 2(b), and it can be seen that there is a clear abrupt change in the first four mode shapes at the crack point. Such abrupt changes can be located by calculating the second order derivative of the difference of each mode shape, as shown in Fig. 2(c). It can be seen that the crack can be accurately located at each order of the difference of the vibrational mode shapes, and the higher order of the difference of the vibrational mode shape is more sensitive to the crack.

By reducing the width, the crack faults at the corresponding position of the cantilever beam were simulated. Taking $5\%l$ as the step, a total of 19 cracks were set from the position $5\%l$ away from the fixed end to the position $5\%l$ away from the free end. There were 3 types of cracks at each place, and the proportions of crack depth to the width of cantilever beam were 20%, 40% and 60% respectively.

The first four mode shapes of the cantilever beam in each state were calculated, and the difference between the mode shape of the cantilever beam with crack and the corresponding mode shape of the baseline at the cracked unit is shown in Fig. 3. As can be seen that only the difference of the first mode shape between the cantilever beam with cracks and the baseline has no zero crossing, which means that the difference of the first mode shape can theoretically reflect the crack fault of any position of the cantilever beam.

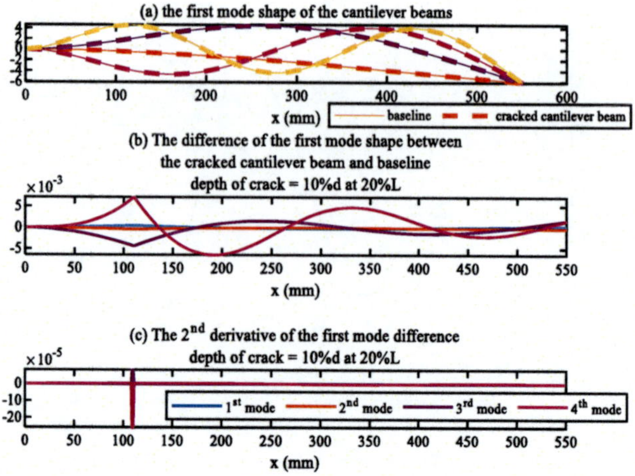

Fig. 2. Crack detection based on the difference of modes.

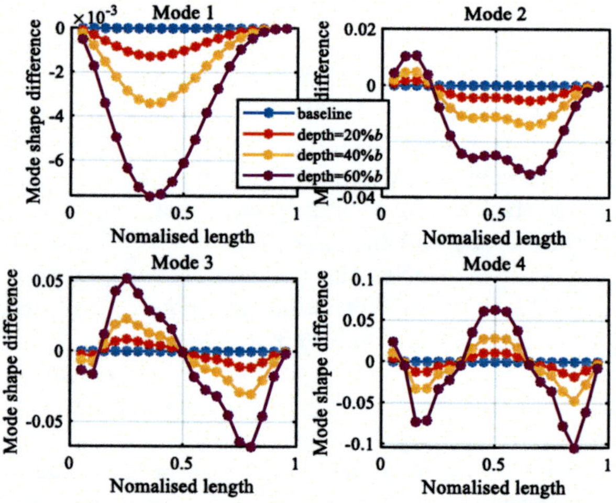

Fig. 3. The difference between the mode shape of the cantilever beam with crack and the corresponding mode shape of the baseline at the cracked unit.

4 Experimental Verification

4.1 Experimental Setup

Several common steel rulers with a length of 635 mm, a width of 25 mm and a thickness of 1.5 mm were used as cantilever beams for experimental verification. To ensure that each cantilever beam has the same length, a groove with a length of 85 mm, a width of 25 mm and a depth of 1 mm was set on the flat plate for placing the steel rules, as shown in Fig. 4(a) and (b), and another flat plate was used to cover the steel rules, as shown in

Fig. 4(c). These two flat plates not only ensure the stability of the rulers' position, but also provide conditions for applying pressure with a vise.

(a) (b) (c)

Fig. 4. Equipment for fixing the steel ruler. (a) A flat plate with a groove; (b) Place the steel ruler into the groove to form a cantilever beam with a length of 550 mm; (c) A flat plate used for covering.

Through the pair of plates mentioned above, each ruler forms a cantilever beam of 550 mm in length. The free ends of the cantilever beams were excited with a rubber hammer and the vibration process of the cantilever beams were recorded with a rolling shutter camera, which has a frame rate of 20, and a resolution of 2048 × 3072. To mitigate the effects of motion blur, the exposure time for each line of the image was set to 1 ms. The experimental scene is shown in Fig. 5.

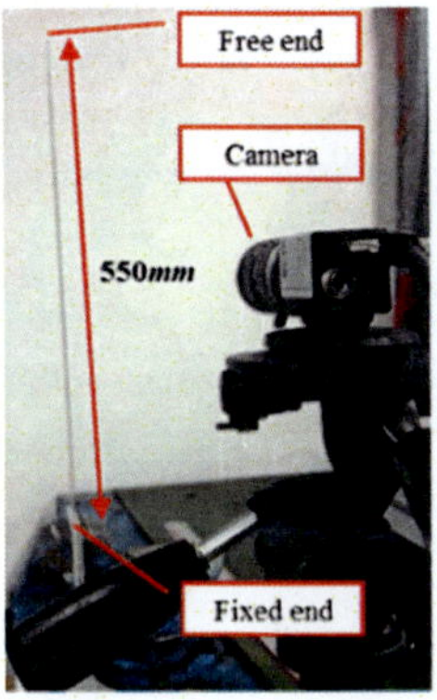

Fig. 5. Scene of experiment.

Three cracks with depths of 5 mm, 10 mm and 15 mm were created at three different positions of the cantilever beams, as shown in Fig. 6. Since all cantilever beams are the same length of 550 mm and the 0 scale is at the free end, 250 mm, 350 mm and 450 mm represent 54%, 36% and 18% of the total length (L) from the fixed end, respectively.

4.2 Data Processing and Analysis

4.2.1 Vibration Process Extraction Based on Image Edge Detection

Since the cantilever beam vibrates against a white background, the vibration information of the cantilever beam can be obtained by extracting the edge position of each frame

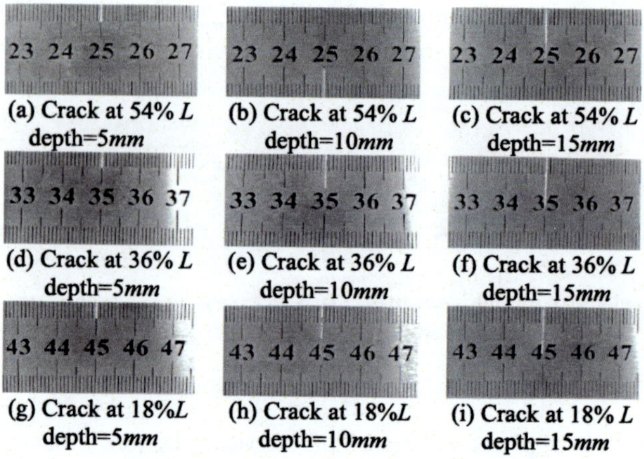

Fig. 6. Set crack faults on the cantilever beams.

in the video. Figure 7 shows frames 51–54 of the video shot (the left and right sides of the image are removed, and the width of remaining image is 600 px), showing the movement of the cantilever beam from left to right.

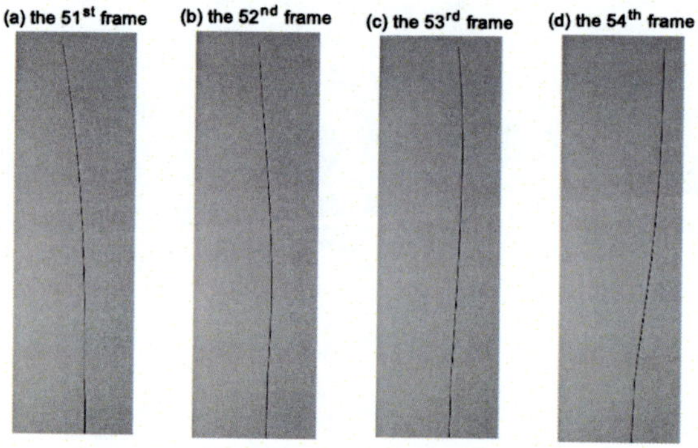

Fig. 7. Four consecutive frames in the video.

The one-dimensional gradient operator [−1,0,1] was used to convolve each image in the video to get the edge magnitude matrix, in which the edge in the vertical direction was highlighted. The edge magnitude of line 1000 in frames 51–54 are shown in Fig. 8. It can be clearly seen from the figure that the cantilever beam moves from left to right at the position of line 1000 during the short period, and the left and right edges of the cantilever generate two strong edge responses respectively, corresponding to the positive and negative peaks in the figure.

In order to improve the positioning accuracy of the edge, a certain number of points are selected on both sides of the positive and negative peak points to estimate the edge

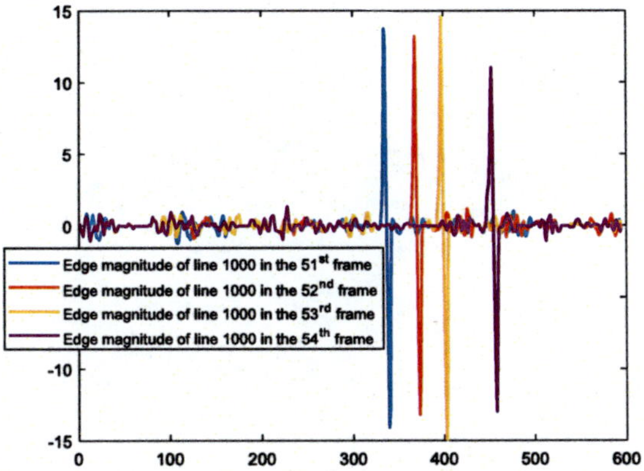

Fig. 8. Edge detection result of the line 1000 in frames 51–54.

position. Assuming that the edge magnitude on both sides of the edge is normally distributed, the edge position within the observation range can be estimated by the following formula.

$$\mathrm{EL}(i) = \sum_{j=peak(i)-2}^{peak(i)+2} x_j\,\mathrm{EM}(i,j) / \sum_{j=peak-2}^{peak+2} x_j \tag{6}$$

where x_j represents the position of the point (in pixels), $EM(i,j)$ represents the edge magnitude of the point (i,j), $peak(i)$ represents the location of the positive or the negative peak point of the i-th line in edge magnitude matrix (in pixel), $EL(i)$ represents the exact position of the left or right edge of the i-th line. The actual position of the cantilever beam at line i of the image can be calculated by the following formula

$$CBlocation(i) = (EL_{left}(i) + EL_{right}(i))/2 \tag{7}$$

where $CBlocation(i)$ represents the position of cantilever beam at row i, $EL_{left}(i)$ and $EL_{right}(i)$ represent the left and right edge of the cantilever beam at row i, respectively. Obviously, the vibration process of the cantilever beam will be expressed with sub-pixel accuracy.

According to the above method, the displacement of cantilever beam can be obtained from each frame in the video. The vibration process of baseline is shown in Fig. 9.

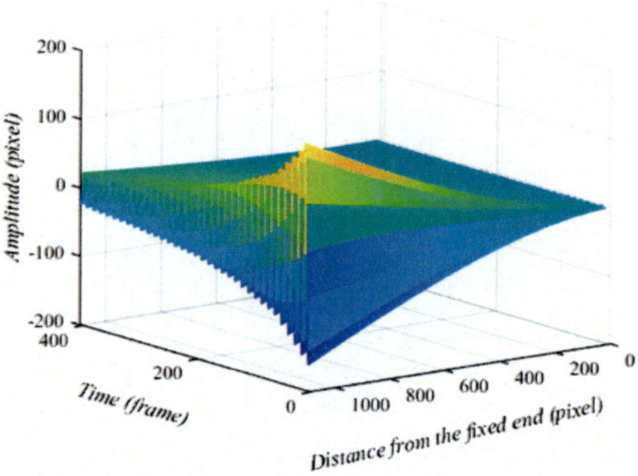

Fig. 9. The vibration process of baseline.

4.2.2 Crack Detection Based on SVD

Three repeat tests were carried out for the cantilever beam under a certain state, and the first-order mode shapes were extracted from the vibration process by SVD. Average the three first-order mode shapes as the first-order mode shape of the cantilever beam under a certain state. In order to highlight the effect of crack fault on the first-order mode

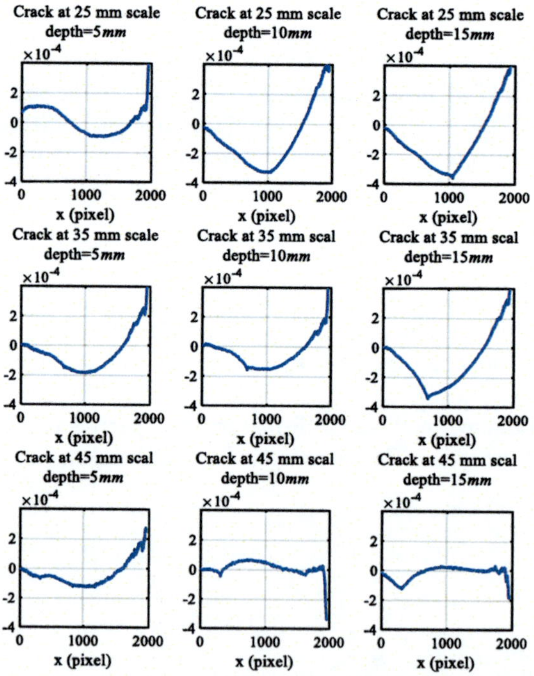

Fig. 10. Difference of the 1st mode shapes between cracked cantilever beams and the baseline.

shape, the first-order mode shape of baseline was subtracted from the first-order mode shape of cantilever beam with crack. The results are shown in Fig. 10.

As can be seen from Fig. 10, although there is little difference between the first-order mode of the cantilever beam with cracks and that of the baseline, the location and degree of the abrupt change in the difference still clearly give information about the location and depth of cracks. At least two conclusions can be summarized based on Fig. 10: (1) The deeper the crack depth at the same location, the greater the difference of the first-order mode shape; (2) For a crack with the same depth, the closer it is to the fixed end, the smaller the difference of the first-order mode shape is. At the same time, it can also be seen from the experimental results that the difference of the first-order mode shape near the free end is less reliable to reflect the crack fault. This is because the projection length in the vertical direction of the cantilever beam decreases after it leaves the initial position, and the mode shape near the free end deviates greatly from the actual mode shape. Fortunately, cracks are more likely to occur near the fixed end [16], which does not affect the practical application of the approach presented in this paper.

5 Conclusions

In this paper, a visual detection method of cantilever beam cracks based on vibration mode shapes is proposed, using a cost-effective common frame rate camera to record the vibration process of the cantilever beam and SVD to extract the model shapes. The feasibility of using the difference of first-order vibration mode shapes to detect cracks in cantilever beams is verified by both simulation and experiment. The main conclusions of this paper are as follows: (1) The vision-based vibration analysis technique can give full play to the high spatial resolution of images to obtain highly accurate mode shapes; (2) Although its sensitivity to cracks at non-node locations is not as good as that of higher-order vibration shapes, the first-order vibration shapes are more suitable for crack detection because they do not have nodes. (3) The difference of the first-order vibration mode shape can reflect not only the location of the crack, but also the degree of the crack.

Acknowledgments. Supports from the National Natural Science Foundation of China (62076029), Scientific Research Platforms and Projects Funding of Guangdong Province (2021KTSCX186) and innovating major training projects of Beijing Institute of Technology, Zhuhai (XKCQ-2019–06) are gratefully acknowledged.

References

1. García, D., Tcherniak, D.: An experimental study on the data-driven structural health monitoring of large wind turbine blades using a single accelerometer and actuator. Mech. Syst. Sig. Process. **127**, 102–119 (2019). https://doi.org/10.1016/j.ymssp.2019.02.062
2. Joshuva, A., Sugumaran, V.: A lazy learning approach for condition monitoring of wind turbine blade using vibration signals and histogram features. Meas. J. Int. Meas. Confed. **152**, 107295 (2020). https://doi.org/10.1016/j.measurement.2019.107295
3. Wang, H., Xiang, P., Jiang, L.: Strain transfer theory of industrialized optical fiber-based sensors in civil engineering: a review on measurement accuracy, design and calibration. Sens. Actuators, A Phys. **285**, 414–426 (2019). https://doi.org/10.1016/j.sna.2018.11.019

4. Aulakh, D.S., Bhalla, S.: 3D torsional experimental strain modal analysis for structural health monitoring using piezoelectric sensors. Measurement **180**, 109476 (2021). https://doi.org/10.1016/j.measurement.2021.109476
5. Doliński, Ł., Krawczuk, M., Zak, A.: Detection of delamination in laminate wind turbine blades using one-dimensional wavelet analysis of modal responses. Shock Vib. **2018** (2018). https://doi.org/10.1155/2018/4507879
6. Dilek, A.U., Oguz, A.D., Satis, F., Gokdel, Y.D., Ozbek, M.: Condition monitoring of wind turbine blades and tower via an automated laser scanning system. Eng. Struct. **189**, 25–34 (2019). https://doi.org/10.1016/j.engstruct.2019.03.065
7. Zhang, L., Wei, J.: Measurement and control method of clearance between wind turbine tower and blade-tip based on millimeter-wave radar sensor. Mech. Syst. Signal Process. **149**, 107319 (2021). https://doi.org/10.1016/j.ymssp.2020.107319
8. Li, M., Feng, G., Gu, F., Ball, A.: Investigating into minimum detectable displacement signal in image-based vibration measurement. In: Zhen, D., Wang, D., Wang, T., Wang, H., Huang, B., Sinha, J.K., Ball, A.D. (eds.) IncoME-V 2020. MMS, vol. 105, pp. 882–894. Springer, Cham (2021). https://doi.org/10.1007/978-3-030-75793-9_82
9. Cawley, P., Adams, R.D.: The location of defects in structures from measurements of natural frequencies. J. Strain Anal. Eng. Des. **14**(2), 49–57 (1979). https://doi.org/10.1243/03093247V142049
10. Rubio, L., Fernández-Sáez, J., Morassi, A.: Identification of an open crack in a beam with variable profile by two resonant frequencies. JVC/Journal Vib. Control **24**(5), 839–859 (2018). https://doi.org/10.1177/1077546316671483
11. Solís, M., Algaba, M., Galvín, P.: Continuous wavelet analysis of mode shapes differences for damage detection. Mech. Syst. Sig. Process. **40**(2), 645–666 (2013). https://doi.org/10.1016/j.ymssp.2013.06.006
12. Dahak, M., Touat, N., Benkedjouh, T.: crack detection through the change in the normalized frequency shape. Vibration **1**(1), 56–68 (2018). https://doi.org/10.3390/vibration1010005
13. Kim, J.T., Ryu, Y.S., Cho, H.M., Stubbs, N.: Damage identification in beam-type structures: frequency-based method vs mode-shape-based method. Eng. Struct. **25**(1), 57–67 (2003). https://doi.org/10.1016/S0141-0296(02)00118-9
14. Ravindra, B., Lappagantu, R.: On the physical interpretation of proper orthogonal modes in vibrations. J. Sound Vib. **219**(1), 189–192 (1999). https://doi.org/10.1006/jsvi.1998.1895
15. Joseph, S., Deepak, B.S.: Virtual experimental modal analysis of a cantilever beam. Int. J. Eng. Res. Technol. **6**(06), 921–928 (2017)
16. Shohag, M.A.S., Hammel, E.C., Olawale, D.O., Okoli, O.I.: Damage mitigation techniques in wind turbine blades: a review. Wind Eng. **41**(3), 185–210 (2017). https://doi.org/10.1177/0309524X17706862

Author Index

R. Karim et al. (Eds.): IAI 2021, LNME, pp. 445–446, 2022.
https://doi.org/10.1007/978-3-030-93639-6